# Neuroendocrine-Immune Axis

Rahul Shukla • Saba Naqvi
Monika S. Deore • Ravinder K. Kaundal
Editors

# Neuroendocrine-Immune Axis

## Towards Understanding the Pathogenesis of Neurodegenerative Disorders

Springer

*Editors*
Rahul Shukla
Department of Pharmaceutics
National Institute of Pharmaceutical Education and Research (NIPER)-Raebareli
Lucknow, UP, India

Monika S. Deore
Department of Pharmacology and Toxicology
National Institute of Pharmaceutical Education and Research (NIPER)-Raebareli
Lucknow, UP, India

Saba Naqvi
Department of Pharmacology and Toxicology
National Institute of Pharmaceutical Education and Research (NIPER)-Raebareli
Lucknow, UP, India

Ravinder K. Kaundal
Department of Pharmacology and Toxicology
National Institute of Pharmaceutical Education and Research (NIPER)-Raebareli
Lucknow, UP, India

ISBN 978-981-95-5087-6 ISBN 978-981-95-5088-3 (eBook)
https://doi.org/10.1007/978-981-95-5088-3



This Springer imprint is published by the registered company Springer Nature Singapore Pte Ltd.
The registered company address is: 152 Beach Road, #21-01/04 Gateway East, Singapore 189721, Singapore

# Preface

In recent days, various cross talks including neuroimmune, the intestinal neuro-immune axis, central sensory motor, neuroendocrine-immune, etc. are going on in the field of neurological research. Neurodegeneration is a very broad terminology covering various neurological diseases and disorders, which lead to progressive loss of neurons. Such progressive loss of neurons causing both structural and functional alterations might be due to various altered pathophysiological process, which occurs during the lifetime of an individual who comes across various toxic insults. Even if such toxic insults do not directly arm the neurons but their effects on various other systems including the immune, endocrine, gut, etc. consequently cause harm to the nervous system. While significant strides have been made in understanding these complex processes, the multifaceted nature of neurodegenerative diseases demands a deeper exploration of the neuroimmune and neuroendocrine crosstalk.

The volume brings together contributions from leading experts worldwide, delving into topics such as the integration of peripheral and central immune alterations, cytokine modulation using nanotechnology, and the transformative role of nano-theranostics. Additionally, chapters on gut health, mitochondrial function, and advanced pharmacotherapies highlight the interdisciplinary approach needed to address the challenges in neurodegeneration research. By bridging basic science with cutting-edge clinical innovations, this book aspires to serve as a foundational resource for researchers, clinicians, and students alike.

It is our hope that the discussions herein not only foster a better understanding of the neuroendocrine-immune axis but also inspire innovative research that bridges the gap between bench and bedside. The challenges posed by neurodegenerative diseases are immense, but so too are the opportunities for transformative solutions when the power of interdisciplinary collaboration is harnessed.

Lucknow, Uttar Pradesh, India — Rahul Shukla
Lucknow, Uttar Pradesh, India — Saba Naqvi
Lucknow, Uttar Pradesh, India — Monika S. Deore
Lucknow, Uttar Pradesh, India — Ravinder K. Kaundal

# Contents

# Editors and Contributors

## About the Editors

**Rahul Shukla** is working as an Assistant Professor at the Department of Pharmaceutics, NIPER Raebareli, India. Currently, he is the SERB International Research fellow for the year, 2022–2023 at School of Pharmacy and Biomedical Sciences, University of Central Lancashire, United Kingdom. He did PhD in Pharmaceutical Sciences at CSIR CDRI and J.N.U New Delhi, India, MPharm at IIT BHU, Varanasi, India, and BPharm at Jamia Hamdard, New Delhi. He has the experience of as Research Scientist at Dr Reddy's Laboratories, India, and D.S Kothari Post-Doctoral Fellow at Panjab University, India. He has more than ten years of research and academic experience. He has more than 250 publications including book chapters to his credit. He has filed twelve patents and one copyright. His current research interests include targeted drug delivery systems, particle engineering, controlled delivery for neurodegenerative diseases, dendrimer mediated drug delivery, solubilization and bioavailability enhancements. He is a Life member of many societies and organizations importantly, Indian Red Cross Society, Association of Pharmaceutical Teachers of India (APTI), and Indian Pharmacy Graduates' Association (IPGA). He received Gyanodaya Award (Excellence in Service) from NIPER Raebareli. He is Editorial board of Journal Scientific Reports. Listed in Stanford University Top 2% Scientist List released on 2023, 2024 and 2025 in Pharmaceutical Sciences Worldwide.

**Saba Naqvi** is working as an Assistant Professor in the Department of Pharmacology & Toxicology/Regulatory Toxicology. She obtained her master's from Jamia Hamdard, New Delhi, and her PhD in 2013 from Jamia Hamdard in collaboration with AIIMS, New Delhi. She has done her post-doc at Jamia Millia Islamia and IIT Roorkee and published more than 37 papers and 14 book chapters in reputed international journals like Elsevier, Springer, Wiley, and Frontiers. She has two Indian patents. She is working in the field of nanomedicine, nanotoxicology, and neuropharmacology, and has synthesized ceramic calcium phosphate nanoparticles for "Suicide Gene Therapy" in cancer therapeutics. She has been honored with the prestigious Women Scientist Award from DST, AMR DXC Summer School Winner Award sponsored by DBT, India, and University of Edinburgh, UK.

**Monika S. Deore** is currently pursuing a PhD in pharmacology and toxicology from NIPER Raebareli, India. She did her MS (Pharm) in regulatory toxicology at NIPER Mohali, India, and BPharm at L.N.J.D.P College of Pharmacy, Manur. She has past research experience in C.R.O.-Syngene International Ltd. as a senior research associate. Her current research includes neurodegenerative diseases, neuroimmunology, safety assessments, nanotoxicology and nanopharmacology, Gut dysbiosis, reproductive toxicology, etc.

**Ravinder K. Kaundal** is working as an Assistant Professor at NIPER Raebareli. He did his MS (Pharm) and PhD in Pharmacology and Toxicology at NIPER-SAS Nagar, India. He received his postdoctoral training at Yale School of Medicine, New Haven, and Icahn School of Medicine at Mount Sinai, New York. He has made an outstanding contribution in the field of pharmacological research and has elucidated the involvement of various pathogenic pathways like oxidative stress, PARP overactivation, inflammation, and apoptosis in cerebral ischemic-reperfusion injury and diabetic neuropathy. Gene regulation is another of his interests, and he has developed various tools/methods to study the mechanisms of epigenetic memory. His work is published in journals of high repute and is highly cited. He is a reviewer and editorial board member of several international journals.

His research interests are Neuropharmacology, Drug repurposing, Molecular biology, Ischemic-reperfusion Injuries, and Gene Regulation.

## Contributors

**Kaifi Ali** Department of Pharmacology and Toxicology, National Institute of Pharmaceutical Education and Research (NIPER-R), Lucknow, Uttar Pradesh, India

**Harapriya Baral** Department of Pharmacology and Toxicology, National Institute of Pharmaceutical Education and Research, Raebareli (NIPER-R), Lucknow, UP, India

**Bhaskar Jyoyi Dutta** Department of Pharmacology and Toxicology, National Institute of Pharmaceutical Education and Research, Hajipur, Bihar, India

**Debapriya Garabadu** Department of Pharmacology, Central University of Punjab, Bathinda, India

**K. M. Gijith Mohan** Drug Delivery and Nanomedicine Research Laboratory, Department of Pharmaceutics, National Institute of Pharmaceutical Education and Research (NIPER)-Raebareli, Lucknow, India

**Sakshi Goswami** Department of Toxicology, Chaudhary Charan Singh University, Meerut, India

**Krishna Jadhav** Pharmaceutical Nanotechnology lab, Institute of Nano Science and Technology (INST), Mohali, Punjab, India

**Keerti Jain** Drug Delivery and Nanomedicine Research Laboratory, Department of Pharmaceutics, National Institute of Pharmaceutical Education and Research (NIPER)-Raebareli, Raebareli, Lucknow, India

**Manav Jain** Department of Pharmacology, Postgraduate Institute of Medical Education and Research (PGIMER), Chandigarh, India

**Smita Jain** Department of Pharmacy, Banasthali Vidyapith, Banasthali, Rajasthan, India

Department of Pharmacy, SVKM's NMIMS, Shirpur Campus, Rajasthan, India

**Vineet K. Jain** Department of Pharmaceutics, Delhi Pharmaceutical Sciences and Research University (DPSRU), New Delhi, India

**Ravinder K. Kaundal** Department of Pharmacology and Toxicology, National Institute of Pharmaceutical Education and Research (NIPER)-Raebareli, Lucknow, UP, India

**Jasleen Kaur** Department of Pharmacology and Toxicology, National Institute of Pharmaceutical Education and Research, Raebareli, Lucknow, India

**Anoop Kumar** Department of Pharmacology, Delhi Pharmaceutical Sciences and Research University (DPSRU), New Delhi, India

**Ashutosh Kumar** Department of Pharmacology and Toxicology, National Institute of Pharmaceutical Education and Research (NIPER) SAS Nagar, Mohali, Punjab, India

**Jayant Kumar** Department of Pharmaceutical Sciences Hemwati Nandan Bahuguna Garhwal University (A Central University), Srinagar, Uttarakhand, India

**Krushna Ch. Maharana** Department of Pharmacology and Toxicology, National Institute of Pharmaceutical Education and Research, Hajipur, Bihar, India

**Ranika Maurya** Department of Pharmacology and Toxicology, National Institute of Pharmaceutical Education and Research (NIPER-R), Lucknow, Uttar Pradesh, India

**Aishwarya Mhatre** Department of Pharmacology and Toxicology, National Institute of Pharmaceutical Education and Research (NIPER-R), Lucknow, Uttar Pradesh, India

**Abdul Naeem** Laboratory for Stem Cell & Restorative Neurology, Department of Biotechnology, Era's Lucknow Medical College & Hospital, Era University, Lucknow, India

**Saba Naqvi** Department of Pharmacology and Toxicology, National Institute of Pharmaceutical Education and Research (NIPER-R), Lucknow, Uttar Pradesh, India

Department of Regulatory and Toxicology, National Institute of Pharmaceutical Education and Research, Raebareli, Lucknow, India

**Umashanker Navik** Department of Pharmacology, Central University of Punjab, Bathinda, India

**Neha** Department of Medical Elementology and Toxicology, School of Chemical and Life Sciences, New Delhi, India

**Bhaskaranand Pancholi** Department of Pharmacology, Central University of Punjab, Bathinda, India

**Suhel Parvez** Department of Medical Elementology and Toxicology, School of Chemical and Life Sciences, New Delhi, India

**Kisan Pawar** Department of Medicinal Chemistry, National Institute of Pharmaceutical Education and Research, Raebareli, India

**Pinky** Department of Medical Elementology and Toxicology, School of Chemical and Life Sciences, New Delhi, India

**Harvinder Popli** Department of Pharmaceutics, Delhi Pharmaceutical Sciences and Research University (DPSRU), New Delhi, India

**S. V. S. Rana** Department of Toxicology, Chaudhary Charan Singh University, Meerut, India

**Syed Shadab Raza** Laboratory for Stem Cell & Restorative Neurology, Department of Biotechnology, Era's Lucknow Medical College & Hospital, Era University, Lucknow, India

**Abha Sharma** Department of Medicinal Chemistry, National Institute of Pharmaceutical Education and Research, Raebareli, India

**Anjali Sharma** School of Pharmaceutical Sciences, MVN University, Palwal, Haryana, India

**Ashwani Sharma** School of Pharmaceutical Sciences, MVN University, Palwal, Haryana, India

**Nisha Sharma** Department of Pharmacology, Postgraduate Institute of Medical Education and Research (PGIMER), Chandigarh, India

**Swapnil Sharma** Department of Pharmacy, Banasthali Vidyapith, Banasthali, Rajasthan, India

**Raghuraj Singh** Pharmaceutical Nanotechnology lab, Institute of Nano Science and Technology (INST), Mohali, Punjab, India

**Ritu Singh** Department of Pharmacy, Banasthali Vidyapith, Banasthali, Rajasthan, India

**Sanjiv Singh** Department of Pharmacology and Toxicology, National Institute of Pharmaceutical Education and Research, Hajipur, Bihar, India

**Hemant Soni** Department of Pharmacology and Toxicology, National Institute of Pharmaceutical Education and Research (NIPER-R, Lucknow, Uttar Pradesh, India

**Somidi Srinu** Department of Medicinal Chemistry, National Institute of Pharmaceutical Education and Research, Raebareli, India

**Keshari Kumar Sriwastawa** Department of Pharmacology and Toxicology, National Institute of Pharmaceutical Education and Research (NIPER) SAS Nagar, Mohali, Punjab, India

**Abhishek Thakran** Department of Regulatory and Toxicology, National Institute of Pharmaceutical Education and Research, Raebareli, Lucknow, India

**Abhishek Tiwari** Department of Pharmacology and Toxicology, National Institute of Pharmaceutical Education and Research, Hajipur, Bihar, India

**Kanika Verma** Department of Pharmacy, Banasthali Vidyapith, Banasthali, Rajasthan, India

**Rahul Kumar Verma** Pharmaceutical Nanotechnology lab, Institute of Nano Science and Technology (INST), Mohali, Punjab, India

**Yeshvandra Verma** Department of Toxicology, Chaudhary Charan Singh University, Meerut, India

**Zitin Wali** Department of Medical Elementology and Toxicology, School of Chemical and Life Sciences, New Delhi, India

# Neuroimmune Crosstalk and Gut Dysbiosis in Neurodegenerative Diseases

Jasleen Kaur, Abhishek Thakran, and Saba Naqvi

**Abstract**

Loss of neurons and decline in motor and cognitive functions are the major hallmarks of the neurodegenerative disorders. Over a long time span, focus has been given to the intricate associations of neurons and immune system which is a bidirectional approach. Neuroimmune system which is also highly associated with the gut anatomy has a major role to play in the therapeutic intervention of neurodegenerative disorders. Herein, we have attempted to review the key aspects of gut and neuroimmune system by understanding their anatomy, signaling cascades, innate and adaptive immune responses, and molecular basis of neuroimmune interactions. Moreover, focus has been laid on the dualistic role of microglia which are the resident immune cells and possess neuroprotective as well as neurotoxic roles to play. Furthermore, the emerging immunotherapies and their clinical significance in the decline of neurodegenerative disorders has been highlighted. Altogether, representing the dynamic and complex interplay between immune and nervous system having gut as the major player. This has relevance in developing novel and effective therapies.

**Keywords**

Neuroimmune system · Gut · Adaptive immunity · Innate immunity · Neurodegeneration · Immunotherapies

J. Kaur · S. Naqvi (✉)
Department of Pharmacology and Toxicology, National Institute of Pharmaceutical Education and Research (NIPER)-Raebareli, Lucknow, Uttar Pradesh, India

A. Thakran
Department of Regulatory and Toxicology, National Institute of Pharmaceutical Education and Research, Raebareli, Lucknow, India

 
R. Shukla et al. (eds.), *Neuroendocrine-Immune Axis*,
https://doi.org/10.1007/978-981-95-5088-3_1

# 1 Introduction

Major worldwide burden is posed by neurodegenerative disorders such as Alzheimer's, amyotrophic sclerosis, and Parkinson's disease. The steady deterioration of cognitive and motor functions and neuronal dysfunction are the major hallmarks of the neurodegenerative disorders. Neuro-immune crosstalk and the gut-brain axis are the 2 intriguing and interlinked aspects having considerable promise for comprehending and treating these diseases, even if the underlying causes are multivariate and intricate. Microglia, the resident cells of the CNS, as well as the peripheral immune cells are among the major players of the neuroimmune system. The dynamic interplay between the microglia and neural cells is responsible for the progression of neurodegenerative cascades via enhancing neuroinflammation. Moreover, the involvement of peripheral immune cells, like T cells and monocytes, has been highlighted here. Infiltration of these cells in the brain parenchyma and their interactions with the neurons and resident immune cells significantly impact the disease outcome.

Moreover, the vagus nerve, the autonomic nervous system, and immunological signaling molecules are the frontline mechanisms that gut-brain axis utilizes to communicate with the brain and gastrointestinal tract. The gut-brain axis is significantly shaped by the gut bacteria, which has an impact on cognitive abilities, behavior, and brain development. The noteworthy impact of gut dysbiosis on neurodegenerative illnesses has recently come to light in newer research arenas. It is now clear that changes in the composition of the gut microbiota can influence neuroinflammation, synaptic dysfunction, and the buildup of aberrant protein aggregates in the brain, all of which are indicators of neurodegeneration. In addition to this, gut microbiota influences the activation state of microglia, affecting the neurodegenerative insults [1].

This chapter precisely focuses on the devilling mechanisms of neuroimmune crosstalk and its influence on neuronal health and progression of neuroinflammation-derived neurodegeneration. Additionally, immunotherapeutic implications and their clinical significance to develop potential neurotherapeutic approaches have been discussed which might hold the promise to mitigate the challenges in neurological conditions.

# 2 Neuroanatomy of the Gut and Interaction with Adaptive and Innate Immunity

The enteric nervous system (ENS) is often called our "second brain." Situated within the walls of the gastrointestinal (GI) tract, the ENS forms an extensive network of neurons, surpassing even the total neuron count found in the spinal cord and peripheral nervous system. We will delve into the unique autonomy of the ENS, which allows it to operate independently from the central nervous system (CNS), governing various digestive processes without direct input from the brain. This capability enables the ENS to initiate local reflexes, such as triggering muscle contractions to move food along the digestive tract when its walls are stretched. The gastrointestinal tract consists of various anatomical layers, including the mesentery, serosa, muscularis, submucosa, lamina propria, and epithelium [2]. This complex structure is innervated by several peripheral neurons that work adjacently with local immune cells to coordinate and regulate responses within the GI tract [3].

Enteric neurons located within the gastrointestinal (GI) tract are organized into ganglionated networks that encircle the intestine. These neurons are categorized into two layers known as the myenteric plexus, positioned between the circular and longitudinal muscle layers, and the submucosal plexus, located in the submucosa. These two plexuses are intricately interconnected by interneurons, motor neurons, and enteric glial cells. Working in unison, they form reflex circuits responsible for controlling peristalsis (intestinal contractions) and secretory functions in the gut. Furthermore, these enteric neurons also receive and integrate signals from external sensory, parasympathetic, and sympathetic neurons to regulate overall gut physiology (Fig. 1).

Neurotransmitters within the enteric nervous system are released by sensory, parasympathetic, sympathetic neurons and their significance in transmitting signals between neurons has an essential role in stimulating and inhibiting GI motility [4]. Key neurotransmitters like serotonin, acetylcholine, norepinephrine, dopamine, and other neuropeptides (substance P, CGRP) secreted by the neuron contribute to regulating digestion and overall gut health [5].

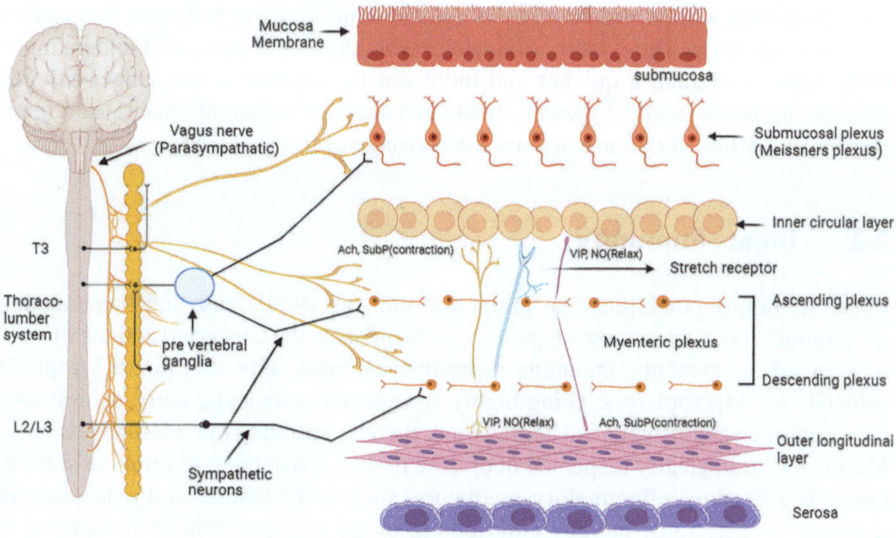


**Fig. 1** Neuroanatomy of gut

Parasympathetic neurons and sympathetic neurons arise from the brainstem and spinal cord, respectively. They serve as mediators for signal transduction from the brain to the gastrointestinal tract. Sympathetic neurons are involved in initiating the body's stress response; they also play a role in inhibitory intestinal functions, causing a decrease in intestinal motility and secretion. These sympathetic neurons transmit their signals through catecholamines like dopamine, epinephrine, and norepinephrine, which bind to α- or β-adrenergic receptors in the gut. On the other hand, parasympathetic neurons have opposing effects, promoting digestive activities and stimulating intestinal movements. Parasympathetic neurons arise from the brain stem showing effects on GIT such as stimulation of GUT motility, enhance secretion, and digestion process [6].

## 2.1 Interaction with Innate and Adaptive Immunity

The immune system is a complex network of defense mechanisms that protect the body from harmful pathogens. Innate immunity serves as the first line of defense, acting swiftly and non-specifically to combat invaders. For example, when a person gets a cut on their skin, the innate immune response quickly kicks in to prevent infection. Neutrophils and macrophages, types of phagocytes, rush to the site of injury and engulf any invading bacteria, preventing them from spreading further. Additionally, the complement system is activated, marking the pathogens for destruction, and recruiting other immune cells to assist in the defense. This immediate and generalized response buys time for the adaptive immune system to develop a more targeted and specific defense. Once the adaptive immunity is activated, specific antibodies will be produced by B cells to neutralize the pathogen that caused the initial infection. These antibodies, along with memory T cells, will remain in the body, ready to launch a quicker and more potent response if the same pathogen attempts to invade again. Together, innate and adaptive immunity ensure a comprehensive and efficient defense against a wide range of threats [7, 8].

## 2.2 Innate Immunity

Innate immunity constitutes the initial and non-specific defense mechanisms that act promptly upon encountering pathogens. It involves the concerted actions of various crucial components, including macrophages, mast cells, and innate lymphoid cells (ILCs). Macrophages, being highly specialized phagocytic cells, exhibit efficient pathogen engulfment and degradation, thereby initiating the immune response. Mast cells, strategically dispersed in diverse tissues, participate in immune surveillance by releasing inflammatory mediators, such as histamine and cytokines, in response to infections, thereby contributing to the orchestration of inflammation and the recruitment of other immune effectors. ILCs, localized within mucosal tissues, are key players in maintaining mucosal immune homeostasis, mounting immune responses against infections, and regulating the intricate immune milieu in the mucosal microenvironment. Together, these innate immune components synergistically form a rapid and robust defense network, efficiently safeguarding the organism against potential threats [8, 9].

## 2.3 Neuronal Interaction with Macrophages

In the gastrointestinal tract, the close interaction between neurons and macrophages is essential for maintaining tissue balance and defending against infections. Specialized macrophages in different intestinal layers, particularly the muscularis externa, work alongside nerves to protect the tissue and regulate intestinal motility. Neuronal signals play a vital role in influencing macrophage function, and reciprocally, macrophages also support neuronal health. For instance, muscularis macrophages promote intestinal motility through the release of bone morphogenic protein 2 (BMP2), while enteric neurons help sustain macrophages through the production of the macrophage survival factor CSF1. Furthermore, this neuron-macrophage interplay protects enteric neurons during inflammation and infection-induced damage. Self-maintaining macrophages distributed in various layers play a critical role in tissue support, including neuronal health, adhesion, and angiogenesis. Furthermore, implication towards potential implications in gastrointestinal disorders like IBD or IBS. Overall, the intricate communication between neurons and macrophages plays a crucial role in preserving gastrointestinal tissue homeostasis and responding effectively to challenges [8, 9].

## 2.4 Neuronal Interaction with Mast Cell

In the gastrointestinal (GI) tract, intricate and bidirectional interactions exist between mast cells and neurons, playing pivotal roles in various physiological and pathological processes. Mast cells, strategically located in the mucosal and submucosal layers, establish proximity to sensory and autonomic nerve endings. Upon activation, mast cells swiftly release inflammatory mediators, including histamine, serotonin, and tryptase, sensitizing sensory neurons and contributing to visceral pain. In response, sensory neurons produce neuropeptides like substance P, calcitonin gene-related peptide (CGRP), vasoactive intestinal peptide (VIP), and corticotropin-releasing hormone, further fostering mast cell activation in a positive feedback loop. This neuro-immune axis has significant implications in conditions such as food allergy and irritable bowel syndrome (IBS) [10].

In the context of food allergy, vagus nerve stimulation has been observed to attenuate allergic responses, potentially by inhibiting mast cell expansion. Conversely, vagotomy does not appear to affect the development of allergic responses, suggesting involvement of alternative neural pathways. Sensory neurons' production of CGRP has been proposed as a regulatory mechanism for mast cell degranulation and expansion. Notably, in IBS, nerve-mast cell interactions have garnered considerable attention, as mast cell expansion near mucosal nerves in IBS patients correlates with the severity and frequency of abdominal pain. Both mast cells and enteric neurons contribute to IBS progression through the production of VIP, which promotes mast cell degranulation and exacerbates the condition. IBS patients exhibit elevated systemic VIP levels, increased VIPR2-expressing mast cells in the colon, and heightened bacterial translocation into the tissue. Inhibition of VIPR2 in biopsy samples curtails mast cell degranulation and prevents bacterial translocation, underscoring the significance of VIP-mast cell signaling in maintaining mucosal barrier integrity during IBS [11].

To unravel the underlying mechanisms governing this complex nerve-mast cell communication, rigorous investigation through targeted genetic or pharmacological studies in animal models is indispensable. The heterogeneity of both neurons and mast cells adds an additional layer of complexity, necessitating a meticulous dissection of specific neuronal populations and mast cell subsets involved in this intricate interplay. Moreover, exploring the involvement of this axis in other intestinal inflammatory diseases will furnish valuable insights into the broader implications and therapeutic potential of targeting this network for managing GI disorders effectively [10, 11].

## 2.5 Neuronal Interaction with ILC

Gastrointestinal (GI) neurons and innate lymphoid cells (ILCs) are critical components of the gut's immune-neural axis. ILCs, a subset of immune cells, are strategically distributed throughout the GI tract and play pivotal roles in maintaining mucosal homeostasis, orchestrating immune responses, and providing defense against infections. Emerging evidence highlights the dynamic interactions between ILCs and enteric neurons, revealing a complex interplay that influences GI functions and immune regulation [12].

ILCs are equipped with a repertoire of receptors that enable them to sense and respond to signals from the local microenvironment, including neurotransmitters and neuropeptides released by enteric neurons. The communication between neurons and ILCs is bidirectional, with neuronal signals modulating ILC activation, cytokine production, and migration, thereby shaping the immune milieu in the gut. In turn, ILC-derived cytokines can impact enteric neurons, affecting their activity and function, which highlights the reciprocal nature of this intercellular communication. A prominent subset of ILCs in the gut, ILC3s, assumes a pivotal role in maintaining gut barrier integrity and combating pathogenic infections. Notably, ILC3s receive neuro-immune inputs from enteric neurons through neurotransmitters such as acetylcholine and vasoactive intestinal peptide (VIP). The activation of ILC3s by these neuronal signals modulates their cytokine repertoire, including the production of IL-22 and IL-17, which are crucial for preserving the mucosal barrier and providing defense against microbial threats [13, 14].

Conversely, the enteric nervous system (ENS) displays responsiveness to ILC-derived cytokines. Notably, IL-17 produced by ILC3s can influence the excitability of enteric neurons, thereby impacting gut motility and peristalsis.

The intricate interplay between GI neurons and ILCs assumes significance in the maintenance of gut immune homeostasis and host defense. Dysregulation of this cross-talk has been implicated in various gut-related disorders, including inflammatory bowel diseases (IBD) and gastrointestinal infections. Understanding the mechanistic underpinnings of the immune-neural interactions in the gut holds substantial promise for advancing therapeutic interventions. Targeting this complex network may offer novel strategies to manage GI disorders effectively and bolster gut health [15].

## 2.6 Adaptive Immunity

Adaptive immunity is an intricate and antigen-specific defense mechanism of the immune system, arising following exposure to pathogens or foreign substances. It encompasses two distinct arms: humoral immunity, orchestrated by B cells capable of producing antigen-specific antibodies for targeted antigen neutralization, and cell-mediated immunity, mediated by T cells that directly engage and eliminate infected or aberrant cells. The hallmark of adaptive immunity lies in its capacity to recognize and respond to antigens, facilitating a focused and long-lasting immune response. The establishment of immune memory is a pivotal attribute, enabling the immune system to mount a swifter and more potent defense upon re-encounter with the same pathogen [16].

## 2.7 Neuronal Interaction with T Cells

The intricate interplay between the peripheral nervous system and the immune system, particularly involving T cells, bears significant implications for maintaining gut homeostasis and effective host defense. A notable example is the cholinergic anti-inflammatory reflex, wherein vagal efferent nerves communicate with the spleen, prompting T cells to produce acetylcholine (ACh). This ACh release subsequently engages nicotinic ACh receptors on macrophages, leading to the inhibition of pro-inflammatory cytokine production, such as TNF. This reflex is operational in the intestinal context as well, with ChAT+ T cells recruited to the colon during infections. The absence of T cell-derived ACh renders mice more vulnerable to pathogens, emphasizing the unique role of ACh in host defense [16, 17].

Furthermore, ACh directly influences T cell function via muscarinic and nicotinic receptors. The M3 muscarinic ACh receptor (M3R) plays a crucial role in host defense against intestinal infections, as evident from M3R-deficient mice displaying compromised T cell-mediated immune responses and memory essential for protection against specific pathogens. ACh and M3R agonists facilitate T cell production of IL-13 and IFNγ, but the precise intracellular signaling mechanisms remain to be fully elucidated. Viral infections also induce neuro-immune signaling that affects enteric neurons, resulting in acute intestinal pathology, bowel dilation, and reduced motility. CD8+ T cells have been identified as key players in causing enteric neuron death during neurotropic flavivirus infections. Abnormal CD8+ T cell responses may also contribute to gut dysmotility in inflammatory bowel disease (IBD), where CD8+ T cell recruitment to the gut correlates with extensive enteric neuron death and delayed intestinal transit [16, 17].

The involvement of gut microbes in modulating T cell function through neuronal pathways has also come to light. Different bacterial species elicit distinct effects on enteric neuron gene expression, influencing the frequency of RORγt+ Treg cells, which inversely correlates with Tac1 expression, encoding neuropeptides that influence Tregs. These findings underscore the significant role of the enteric nervous system in regulating T cell differentiation [18].

Further investigations are necessary to comprehend the precise neuronal signaling mechanisms governing the regulation of various T cell subsets, including Th1 cells, Th2 cells, Tregs, and Th17 cells. Unraveling the direct communication between neurons and T cells and its impact on gut motility, visceral pain, and neuroimmune signaling represents a crucial avenue for understanding and potentially treating inflammatory gastrointestinal disorders such as IBD and irritable bowel syndrome (IBS) [19].

## 2.8 Neuronal Interaction with B Cells

In the gastrointestinal tract (GIT), the intricate crosstalk between neurons and B cells represents a compelling area of scientific investigation, elucidating the dynamic interactions between the nervous and immune systems within this essential organ. Although our comprehension of this communication is still unfolding, emerging evidence suggests that neurons can exert significant influence on B cell function and immune responses in the GIT through diverse mechanisms [20].

A key facet of this neuron-B cell interaction in the GIT involves the autonomic nervous system's sympathetic and parasympathetic branches. Sympathetic nerves secrete norepinephrine, which can modulate B cell activity by engaging adrenergic receptors on B cells. This interaction potentially impacts critical aspects of B cell behavior in the GIT, including development, antibody production, and immune regulatory functions. Similarly, the parasympathetic nervous system releases acetylcholine, which also exerts effects on B cell responses through muscarinic acetylcholine receptors. Sensory neurons in the GIT are pivotal players in influencing B cell function. Neuropeptides released by sensory neurons, such as substance P and calcitonin gene-related peptide (CGRP), can interact with specific receptors on B cells, thereby regulating their proliferation and activation within the GIT. Additionally, sensory neurons have the capacity to shape the local microenvironment in the GIT, influencing the recruitment and functional behavior of immune cells, including B cells, within the gut-associated lymphoid tissues [20, 21].

The reciprocity of this neuron-B cell crosstalk is evident, as B cells in the GIT express receptors for neurotransmitters and neuropeptides, enabling them to actively respond to neural signaling. This intriguing interplay suggests that B cells in the gut have the capacity to perceive and react to changes in neural activity, potentially contributing to the fine-tuning of immune responses based on neuronal inputs. The dynamic interplay between neurons and B cells in the gastrointestinal tract provides valuable insights into the sophisticated regulatory mechanisms governing immune responses in this pivotal organ. Ongoing research in this domain will further unravel the precise impact of neural signals on B cell function and immune regulation in the GIT, potentially paving the way for novel therapeutic approaches targeting gastrointestinal disorders and immune-related conditions that profoundly influence this vital system [20].

## 3 Molecular Basis for Neuroimmune Interactions

Intricate interactions between immune system and brain impact the onset of neurodegenerative disorders. Various growing evidences suggest that the adaptive and immune responses play a crucial part in these disorders. Uncontrolled immune responses can aggravate the immunopathology of neurodegenerative disorders. The term 'immunology' denotes a series of changes that identifies 'self', eliminates 'nonself', and maintains stability. Throughout the entire body immune system has its branches rejecting and damaging the tumor cells, pathogens and their metabolites and maintains the stability in the homeostasis [22]. Central nervous system is an immune-privileged place and can endure the antigens without inducing any inflammatory immune response. Previously, immune cells were only investigated in pathological conditions, such as acute infections or acute brain inflammation. However, in present investigations, it has been reported that immune cells derived from bone marrow play instrumental role in neuroprotection, impart relief in brain injury, and maintain healthy synaptic plasticity [23]. Brain and immune system have a 2-way of connection. Kipins et al have demonstrated that in the absence of mature T cells, mice have shown impaired cognitive abilities, but this effect can be reversed upon passive T cell transfer in mice which establishes an association between peripheral adaptive immunity and brain maintenance to reverse neurodegenerative psychiatric stress which may depend on the CD4+ T cells [24]. Recent evidences suggest that neuronal stress is ameliorated by healthy synaptic plasticity, CNS cognition, and neurogenesis which are dependent on the immune responses in the brain. Additionally, neuronal cells such as astrocytes, neurons, and oligodendrocytes play crucial role in the brain immune functions [25]. Most crucial signaling molecules in immune system are cytokines which neuronal cells recognize and have major impact on brain functions such as memory and cognitive impairments, and in response neuropeptides and neurotransmitters can influence the brain nervous system [23]. The neuronal expression of cytokine receptors, pattern recognition receptors (PRRs) such as NLRs and TLRs, initiates the production of molecules such as pathogen-associated molecular patterns (PAMPs) which have regulatory role in both immune and neuronal systems [26].

Pattern recognition receptors (PRRs) are mainly responsible for initiating the initial inflammatory response as activated by neurodegenerative processes. The most common characteristic of innate immune system is the recognition of PRRs.

The family of PRRs comprises of retinoic acid-inducible gene I (RIG-I)-like receptors (RLRs), TLRs, C-type lectin receptors (CLRs), and nucleotide-binding oligomerization domain NLRs, which are associated with various neurodegenerative disorders [27]. The expression of TLR is mainly related to the signaling regulators such as MDD. The main player in the neuroinflammation-induced cognitive dysfunction is the TLR4/NF-κB signaling cascade [28]. Another regulator is the NLRs which are significant for the regulation of anxiety, hypothalamic-pituitary-adrenal (HPA) axis activation. Furthermore, complement cascade is an imperious element for adaptive and innate response which is associated with both acquired and innate defenses [29]. Various biologically active molecules are released which have significant contribution in tissue homeostasis and immune surveillance. Brain is vulnerable for complement-mediated damages which has a crucial role in microglia-mediated synaptic function. An investigation has been conducted which supports the fact that complement C3 is elevated in Alzheimer's disease responsible for the cognitive decline and synaptic loss [30]. Another critical intercellular transmitter are cytokines which regulate the inflammatory response and immune system. Generally, 4 categories of cytokines are present which activate the signaling pathways such as JAK-STAT signaling, NF-κB and MAPK class for IL-1 and TNFα family, ERK pathway, and TGF-β pathway [31].

Thus, regulation of these factors is highly dependent on brain-immune responses underlying the neurodegenerative disorders. These interactions are important to study as they can be a potential treatment strategy.

## 4 Dualistic Role of Glia Cell in Neuroimmune Cross Talk

Glia cells, found in both the enteric nervous system (ENS) and the central nervous system (CNS), are essential players in maintaining the proper functioning and health of these distinct components of the nervous system. In the ENS, enteric glial cells play a vital role in supporting the intricate neural network within the gastrointestinal (GI) tract. These glial cells provide structural support to enteric neurons, modulate synaptic transmission, and help regulate the balance of neurotransmitters in the gut. Through these functions, enteric glial cells influence gut motility, secretion, and immune responses, thus contributing to overall gut health. Dysfunction of enteric glial cells has been implicated in gastrointestinal disorders such as irritable bowel syndrome (IBS) and inflammatory bowel disease (IBD), underscoring their importance in GI health [32].

In the CNS, various types of glial cells, including astrocytes, oligodendrocytes, and microglia, contribute to the overall well-being of neurons and neural circuits. Astrocytes, the most abundant glial cells in the CNS, provide crucial support to neurons by maintaining the extracellular environment, controlling ion concentrations, and participating in synaptic transmission and plasticity. Oligodendrocytes, on the other hand, play a role in efficient neural communication by producing myelin, a lipid-rich sheath that wraps around axons in the CNS, allowing for rapid conduction of nerve impulses. Meanwhile, microglia serve as the resident immune cells of the CNS, constantly surveilling the brain environment. They act as the brain's primary defense against pathogens, and upon detecting damage or infection, microglia become activated and release pro-inflammatory or anti-inflammatory molecules, influencing neural homeostasis and the brain's immune response (Fig. 2).

The presence of glial cells in both the ENS and CNS highlights their significance in supporting and regulating neural functions in different parts of the nervous system. From ensuring proper gut motility to maintaining efficient neural communication and immune response in the brain, glial cells' roles are diverse and critical for overall nervous system health. Understanding the complex interactions between glial cells and neurons in both systems holds promise for advancing our knowledge of neurological disorders and developing potential therapeutic interventions [32, 33].

Glia cells exhibit a dualistic role in the process of neurodegeneration, encompassing both detrimental and protective mechanisms that impact the progression of neurodegenerative diseases. In the central nervous system (CNS), microglia, a type of glial cell, can contribute to neurodegeneration by releasing pro-inflammatory cytokines and reactive oxygen species in response to neuronal stress or damage, leading to chronic neuroinflammation, which exacerbates neuronal loss in

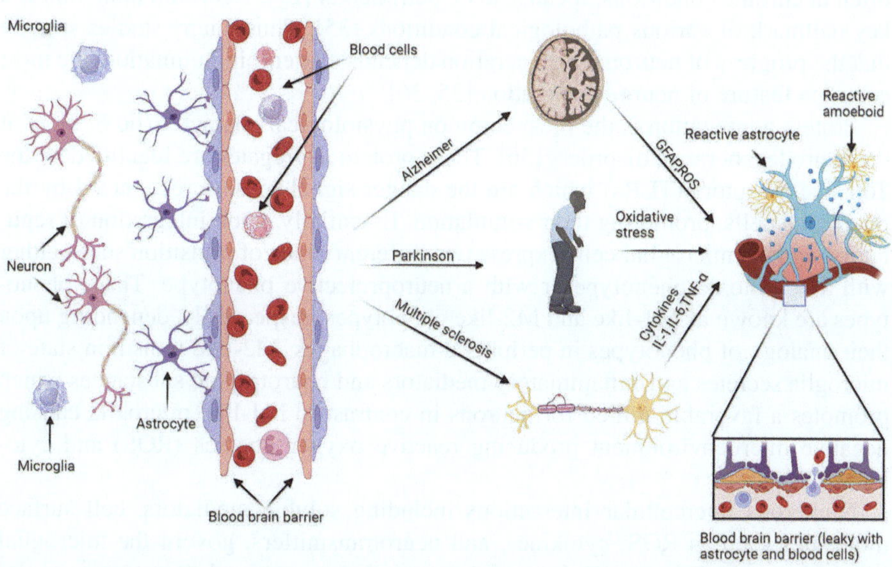


**Fig. 2** Dualistic role of glial cells (astrocyte and microglia cells reduce ROS and free radicals with the help of the antioxidant system. Astrocytes are abundant in the brain. They maintain the integrity of the blood-brain barrier but in disease conditions, glial cell transforms into reactive form and start producing cytokines, NOS, ROS, and other factors which help in the progression of neurodegenerative diseases)

conditions such as Alzheimer's, Parkinson's, and amyotrophic lateral sclerosis (ALS) [34]. Conversely, other glial cells, such as astrocytes and oligodendrocytes, exert neuroprotective functions. Astrocytes support neuronal survival by regulating neurotransmitter clearance and extracellular ion concentrations, thereby safeguarding neurons against excitotoxicity. Oligodendrocytes are essential for maintaining myelin integrity around axons, ensuring proper nerve signal conduction [32].

These neuroprotective roles are not limited to the central nervous system, as glial cells in the enteric nervous system (ENS) also demonstrate analogous functions. Enteric glial cells contribute to neurotransmission modulation and provide support to enteric neurons in the gastrointestinal tract. Additionally, they play a role in gut motility regulation and immune responses. Notably, disturbances in ENS glial cell functions have been observed in neurodegenerative diseases such as Parkinson's, implying their involvement in the pathologies of both the central and enteric nervous systems during neurodegeneration [33].

## 5 Role of Neuroimmune System in Regulating Neurodegeneration

There is a fundamental role of immune system in regulating inflammation, which may be protective against tumors and pathogens and repairs the tissue damage but often in chronic conditions, it can also be detrimental [27]. Neuroinflammation is a key hallmark of various pathological conditions [35]. Thus, many studies suggest that the progress of neuronal degeneration depends on neuroinflammation, the most common feature of neurodegeneration [35, 36].

Protein aggregation is the most common physiological characteristic of most of the neurodegenerative disorders [36]. These protein aggregates are identified by the Toll-like receptors (TLRs) which are the danger signal sensors recognized by the microglial cells, promoting their stimulation. Essentially, upon integration of regulatory signals, microglial cells express two different kinds of transition states either with a neurotoxic phenotype or with a neuroprotective phenotype. These phenotypes are known as M1-like and M2-like phenotypes, respectively, depending upon their analogy of phenotypes in peripheral macrophages. M2-like transition state of microglia secretes anti-inflammatory mediators and neurotrophic substances which promotes a favorable milieu for neurons in contrast to M1-like microglia causing negative microenvironment producing reactive oxygen species (ROS) and cytokines [37].

Numerous intercellular interactions including soluble mediators, cell surface molecules such as ROS, cytokines, and neurotransmitters, govern the microglial interactions which is a crucial step for neuroinflammation and promotes neurodegeneration. Cross-talk between the neurons and microglia with the blood-brain-barrier (BBB), with astrocytes and T lymphocytes infiltrating the CNS parenchyma, are few examples of intercellular interactions governing the microglial activation. Various investigations have shown that T cells have crucial role in controlling the neuroinflammation associated with the neurodegenerative diseases and subsequently in affecting the neuronal density [38].

Recent investigations have shown that microglia reducing therapies and transcriptomic analysis in mice receiving intravenous injection of BMCs (bone marrow cells) have laid new concept that macrophages originating in the periphery infiltrate in brain and maintain distinct microglial identity, thereby demonstrating a newer category of cell type having potential of replacing microglia [39]. Using a coculture system of mouse and human microglia and traumatic spinal cord injury model of mice, another study has revealed that microglia and monocyte-derived macrophages upon establishing communication, suppress the microglial activation and thus reduce the phagocytic activity and expression of inflammatory molecules. Preventing such interaction, functional recovery is impaired and microglial activation is enhanced [40]. These findings are important for the strategy to develop macrophage engrafts for CNS, which might help to target microglia-mediated interactions as a novel strategy.

Thus, inflammatory processes show dual role in exerting beneficial properties like regulating neuronal survival, removing deleterious factors such as myelin debris, protein misfolded aggregates which destroy the microenvironment. Ongoing research is more profound for molecular, cellular, and genetic elements of immune response regulatory mechanisms which might be relevant to circumvent injury to neuronal cells and activate the recovery processes to ameliorate the neuroinflammation [41].

## 6 Various Immunotherapies in Neurodegenerative Disorders

Neurodegenerative disorders are the progressive chronic condition that have certain features of steady neuronal and synaptic loss in certain regions of the spinal cord and brain. These include Huntington's, Parkinson's, Alzheimer's, and amyotrophic lateral sclerosis. While the exact mechanisms may remain unknown but neuroinflammation, oxidative stress, abnormal protein accumulation, and genetic disorders are some major factors for progression of these diseases. And due to involvement of active immune responses and exploration of different immunotherapies has evidently found to be very crucial in improving patient outcomes. Immunotherapies have a promising approach to improve the neurodegenerative events. They deal with the various immune responses to target specific protein misfolding and aggregation, reducing neuroinflammation, and potentially ameliorating the neurodegenerative disorders. Several immunotherapies have gained clinical importance and few of the major immunotherapies include anti-amyloid beta (Aβ) immunotherapies, anti-tau immunotherapies, and alpha-synuclein immunotherapies [42].

Alzheimer's disease displays distinct feature of accumulation of anti-amyloid beta (Aβ) plaques in the brain. Immunotherapies concerning this include monoclonal antibodies which facilitate the clearance of detrimental proteins with the aid of microglia, the prominent immune cells of the brain. This potentially decreases the plaque burden and restores cognitive function. Some potentially relevant therapies include Aducanumab, Solanezumab, and Crenezumab. Aducanumab is a targeted monoclonal antibody which specifically targets the Aβ in its aggregated form. This therapy has potentially reduced the amyloid plaques and has been seen in clinical trials. Accelerated approval for this therapy has been granted by the Food and Drug Administration (FDA) to treat Alzheimer's disease. On the other hand, Solanezumab and Crenezumab have also been in the clinical trials but not much beneficial endpoints have been obtained. Further research is going on to explore its effects in specific patient populations [43].

Other than this anti-tau immunotherapies have been explored in various neurodegenerative disorders, such as certain forms of frontotemporal dementia and Alzheimer's disease. Abnormal forms of tau proteins generate tangles in the neurons, resulting in neuronal cell death and cellular dysfunction. This kind of therapy targets and clears the pathological neurofibrillary tangles. Some of the clinical trials going on this kind of therapy include Semorinemab and Gosuranemab. Both are monoclonal antibodies which selectively remove pathological tau and slow down the progression of cognitive dysfunctions [44].

Similarly, in other neurodegenerative disorders such as Parkinson's disease and multiple system atrophy, alpha-synuclein immunotherapy has shown promise. Accumulation of pathological aggregates of alpha-synuclein is the major pathological hallmark of Parkinson's disease and these immunotherapies target to induce a strong immune response against the pathological aggregates to induce the clearance of these proteins. Vaccine developed addressing this issue is AFFITOPE PD01A which is responsive to induce the immune response against alpha-synuclein and is currently in the early stages of clinical trials against Parkinson's disease [45].

Other developing immunotherapeutic approaches to combat outcomes of neurodegenerative diseases include SOD1 protein therapy and immune modulation. Mutations of SOD1 gene occur in amyotrophic lateral sclerosis (ALS) which results in the accumulation of SOD1 protein aggregates. Immunotherapies such as Tofersen (BIIB067) have been employed for antisense oligonucleotides which reduces the expression of SOD1, potentially improving the motor neuron survival. Similarly, modulating immune response by utilizing treatments such as Ravulizumab (ULTOMIRIS) has been beneficial for C9orf72 mutations in ALS [46, 47]. More detailed immunotherapies are described in Table 1.

Although immunotherapies show promise for neurodegenerative disorders but delivery of vaccines and antibodies to the brain is the main challenge due to the intricate makeup of blood-brain barrier which restricts the entry of larger molecules. More research is required to make therapy more effective.

**Table 1** Types of immunotherapies in neurodegenerative disorders

| S.No | Neurodegenerative disorder | Immunotherapy/kind of treatment | Protein targeted | Clinical trial phase | Clinical trial outcomes | Limitations | Reference |
|---|---|---|---|---|---|---|---|
| 1. | Alzheimer's disease | Aducanumab/ `Monoclonal antibody | Beta-amyloid | Phase III | Decline in formation of amyloid plaques<br>Cognitive benefits | High cost<br>Side effects<br>Conflicting trial results | [48] |
| | | Solanezumab/ Monoclonal antibody | Beta-amyloid | Phase III | No significant cognitive improvement<br>Slight improvements in mild AD | Lack of therapeutic window | [49] |
| | | Ganterenumab/ Monoclonal antibody | Beta-amyloid | Phase III | Decline in amyloid plaques | Side effects<br>High efficacy dose | [50] |
| | | BAN2401/ Monoclonal antibody | Beta-amyloid | Phase II | Dose-dependent reduction in amyloid plaque formation<br>Cognitive benefits | Side effects<br>More evidence required | [49] |
| | | AAB-003 (Ponezumab)/ Monoclonal antibody | Beta-amyloid | Phase II | Primary endpoint not achieved | Side effects<br>No significant cognitive benefits | [51] |
| | | Crenezumab/ Monoclonal antibody | Beta-amyloid | Phase II | Reduction in amyloid plaque<br>Primary endpoint not achieved | Dose optimization not achieved<br>Lack of robust evidence | [52] |
| | | Masitinib/Tyrosine kinase inhibitor | Microglia | Phase IIb/III | Promising clinical evidences | Further validation required | [53] |
| | | Aduhelm (Aducanumab) | Beta-amyloid | Phase III | Reduction in amyloid plaque<br>Cognitive benefits in fewer population | Conflicting trial results<br>High cost<br>Side effects | [54] |
| | | AL001 (donanemab)/ Monoclonal antibody | N3pG-tau | Phase II/III | Cognitive improvement<br>Reduced N3pG-tau levels | Potential side effects | [32] |
| | | Anavex 2–73/Sigma-1 receptor agonist | Sigma-1 receptor | Phase II | Cognitive functions improvement | Potential off-target effects | [55] |
| | | GRF6019/i.v. immunoglobulin | Tau protein | Phase II | | More evidences are required to confirm the safety and efficacy | [56] |

(continued)

**Table 1** (continued)

| S.No | Neurodegenerative disorder | Immunotherapy/kind of treatment | Protein targeted | Clinical trial phase | Clinical trial outcomes | Limitations | Reference |
|---|---|---|---|---|---|---|---|
| 2. | Progressive supranuclear palsy | ABBV-8E12/ Monoclonal antibody | Tau protein | Phase II | | More clinical evidences are required | [57] |
| | | BIIB092 (gosuranemab)/ Monoclonal antibody | Tau protein | Phase II | Tau aggregate reduction in CSF | Limited clinical efficacy<br>More clinical evidences are required | [58] |
| 3. | Parkinson's disease | PRX002 (PASADENA)/ Monoclonal antibody | Alpha-synuclein | Phase II | Primary endpoint not achieved | More clinical optimization required | [59] |
| | | ABBV-0805/ Monoclonal antibody | Alpha-synuclein | Phase I | Early-stage trial | Long-term parameters are not well established | [60] |
| | | NPT088/Monoclonal antibody | Alpha-synuclein | Phase I | Early-stage trial | More evidences required to confirm the efficacy | [61] |

## References


1. Padhi, P., et al. (2022). Mechanistic insights into gut microbiome dysbiosis-mediated neuro-immune dysregulation and protein misfolding and clearance in the pathogenesis of chronic neurodegenerative disorders. *Frontiers in Neuroscience, 16*, 836605.
2. Barajon, I., et al. (2009). Toll-like receptors 3, 4, and 7 are expressed in the enteric nervous system and dorsal root ganglia. *The Journal of Histochemistry and Cytochemistry, 57*(11), 1013–1023.
3. Drokhlyansky, E., et al. (2020). The human and mouse enteric nervous system at single-cell resolution. *Cell, 182*(6), 1606–1622 e23.
4. De Vadder, F., et al. (2018). Gut microbiota regulates maturation of the adult enteric nervous system via enteric serotonin networks. *Proceedings of the National Academy of Sciences of the United States of America, 115*(25), 6458–6463.
5. Baral, P., Udit, S., & Chiu, I. M. (2019). Pain and immunity: Implications for host defence. *Nature Reviews. Immunology, 19*(7), 433–447.
6. Cervi, A. L., Lukewich, M. K., & Lomax, A. E. (2014). Neural regulation of gastrointestinal inflammation: Role of the sympathetic nervous system. *Autonomic Neuroscience, 182*, 83–88.
7. Green, D. P., et al. (2019). A mast-cell-specific receptor mediates neurogenic inflammation and pain. *Neuron, 101*(3), 412–420 e3.
8. Stakenborg, N., Viola, M. F., & Boeckxstaens, G. E. (2020). Intestinal neuro-immune interactions: Focus on macrophages, mast cells and innate lymphoid cells. *Current Opinion in Neurobiology, 62*, 68–75.
9. Viola, M. F., & Boeckxstaens, G. (2020). Intestinal resident macrophages: Multitaskers of the gut. *Neurogastroenterology and Motility, 32*(8), e13843.
10. Drabsch, Y., et al. (2013). Transforming growth factor-beta signalling controls human breast cancer metastasis in a zebrafish xenograft model. *Breast Cancer Research, 15*(6), R106.
11. Forsythe, P. (2019). Mast cells in neuroimmune interactions. *Trends in Neurosciences, 42*(1), 43–55.
12. Cardoso, V., et al. (2017). Neuronal regulation of type 2 innate lymphoid cells via neuromedin U. *Nature, 549*(7671), 277–281.
13. Airaksinen, M. S., & Saarma, M. (2002). The GDNF family: Signalling, biological functions and therapeutic value. *Nature Reviews. Neuroscience, 3*(5), 383–394.
14. Ibiza, S., et al. (2016). Glial-cell-derived neuroregulators control type 3 innate lymphoid cells and gut defence. *Nature, 535*(7612), 440–443.
15. Xu, H., et al. (2019). Transcriptional atlas of intestinal immune cells reveals that neuropeptide alpha-CGRP modulates group 2 innate lymphoid cell responses. *Immunity, 51*(4), 696–708 e9.
16. Ramirez, V. T., et al. (2019). T-cell derived acetylcholine aids host defenses during enteric bacterial infection with Citrobacter rodentium. *PLoS Pathogens, 15*(4), e1007719.
17. Darby, M., et al. (2015). The M3 muscarinic receptor is required for optimal adaptive immunity to helminth and bacterial infection. *PLoS Pathogens, 11*(1), e1004636.
18. Yissachar, N., et al. (2017). An intestinal organ culture system uncovers a role for the nervous system in microbe-immune crosstalk. *Cell, 168*(6), 1135–1148 e12.
19. Sanchez-Ruiz, M., et al. (2019). Enteric murine ganglionitis induced by autoimmune CD8 T cells mimics human gastrointestinal dysmotility. *The American Journal of Pathology, 189*(3), 540–551.
20. Reboldi, A., et al. (2016). IgA production requires B cell interaction with subepithelial dendritic cells in Peyer's patches. *Science, 352*(6287), aaf4822.
21. Bost, K. L., & Pascual, D. W. (1992). Substance P: A late-acting B lymphocyte differentiation cofactor. *The American Journal of Physiology, 262*(3 Pt 1), C537–C545.
22. Chaplin, D. D. (2010). Overview of the immune response. *The Journal of Allergy and Clinical Immunology, 125*(2 Suppl 2), S3–S23.
23. Croese, T., Castellani, G., & Schwartz, M. (2021). Immune cell compartmentalization for brain surveillance and protection. *Nature Immunology, 22*(9), 1083–1092.

24. Kipnis, J., et al. (2004). T cell deficiency leads to cognitive dysfunction: Implications for therapeutic vaccination for schizophrenia and other psychiatric conditions. *Proceedings of the National Academy of Sciences of the United States of America, 101*(21), 8180–8185.
25. Peferoen, L., et al. (2014). Oligodendrocyte-microglia cross-talk in the central nervous system. *Immunology, 141*(3), 302–313.
26. Xu, Z. Z., et al. (2015). Inhibition of mechanical allodynia in neuropathic pain by TLR5-mediated A-fiber blockade. *Nature Medicine, 21*(11), 1326–1331.
27. Takeuchi, O., & Akira, S. (2010). Pattern recognition receptors and inflammation. *Cell, 140*(6), 805–820.
28. Zhang, J., et al. (2018). Porphyromonas gingivalis lipopolysaccharide induces cognitive dysfunction, mediated by neuronal inflammation via activation of the TLR4 signaling pathway in C57BL/6 mice. *Journal of Neuroinflammation, 15*(1), 37.
29. Defendi, F., et al. (2020). The immunopathology of complement proteins and innate immunity in autoimmune disease. *Clinical Reviews in Allergy and Immunology, 58*(2), 229–251.
30. Shi, Q., et al. (2017). Complement C3 deficiency protects against neurodegeneration in aged plaque-rich APP/PS1 mice. *Science Translational Medicine, 9*(392), eaaf6295.
31. Zhao, F., et al. (2022). Brain-immune interaction mechanisms: Implications for cognitive dysfunction in psychiatric disorders. *Cell Proliferation, 55*(10), e13295.
32. Lee, K. H., Cha, M., & Lee, B. H. (2021). Crosstalk between neuron and glial cells in oxidative injury and neuroprotection. *International Journal of Molecular Sciences, 22*(24), 13315.
33. Tremblay, M. E., Cookson, M. R., & Civiero, L. (2019). Glial phagocytic clearance in Parkinson's disease. *Molecular Neurodegeneration, 14*(1), 16.
34. Geloso, M. C., et al. (2017). The dual role of microglia in ALS: Mechanisms and therapeutic approaches. *Frontiers in Aging Neuroscience, 9*, 242.
35. Glass, C. K., et al. (2010). Mechanisms underlying inflammation in neurodegeneration. *Cell, 140*(6), 918–934.
36. Lucin, K. M., & Wyss-Coray, T. (2009). Immune activation in brain aging and neurodegeneration: Too much or too little? *Neuron, 64*(1), 110–122.
37. Bhaskar, K., et al. (2010). Regulation of tau pathology by the microglial fractalkine receptor. *Neuron, 68*(1), 19–31.
38. Gonzalez, H., et al. (2014). Neuroimmune regulation of microglial activity involved in neuroinflammation and neurodegenerative diseases. *Journal of Neuroimmunology, 274*(1–2), 1–13.
39. Cronk, J. C., et al. (2018). Peripherally derived macrophages can engraft the brain independent of irradiation and maintain an identity distinct from microglia. *The Journal of Experimental Medicine, 215*(6), 1627–1647.
40. Greenhalgh, A. D., et al. (2018). Peripherally derived macrophages modulate microglial function to reduce inflammation after CNS injury. *PLoS Biology, 16*(10), e2005264.
41. Scheiblich, H., et al. (2020). Neuroimmune connections in aging and neurodegenerative diseases. *Trends in Immunology, 41*(4), 300–312.
42. Kwon, S., et al. (2020). Immunotherapies for aging-related neurodegenerative diseases-emerging perspectives and new targets. *Neurotherapeutics, 17*(3), 935–954.
43. Valiukas, Z., et al. (2022). Immunotherapies for Alzheimer's disease-A review. *Vaccines (Basel), 10*(9), 1527.
44. Imbimbo, B. P., et al. (2023). Initial failures of anti-tau antibodies in Alzheimer's disease are reminiscent of the amyloid-beta story. *Neural Regeneration Research, 18*(1), 117–118.
45. Meissner, W. G., et al. (2020). A phase 1 randomized trial of specific active alpha-synuclein immunotherapies PD01A and PD03A in multiple system atrophy. *Movement Disorders, 35*(11), 1957–1965.
46. Blair, H. A. (2023). Tofersen: First approval. *Drugs, 83*(11), 1039–1043.
47. Chen, J. J. (2020). Overview of current and emerging therapies for amytrophic lateral sclerosis. *The American Journal of Managed Care, 26*(9 Suppl), S191–S197.
48. Budd Haeberlein, S., et al. (2022). Two randomized phase 3 studies of aducanumab in early Alzheimer's disease. *The Journal of Prevention of Alzheimer's Disease, 9*(2), 197–210.

49. Jeremic, D., Navarro-Lopez, J. D., & Jimenez-Diaz, L. (2023). Efficacy and safety of anti-amyloid-beta monoclonal antibodies in current Alzheimer's disease phase III clinical trials: A systematic review and interactive web app-based meta-analysis. *Ageing Research Reviews, 90*, 102012.
50. Wang, G., et al. (2022). Evaluation of dose-dependent treatment effects after mid-trial dose escalation in biomarker, clinical, and cognitive outcomes for gantenerumab or solanezumab in dominantly inherited Alzheimer's disease. *Alzheimers Dement (Amsterdam), 14*(1), e12367.
51. Delnomdedieu, M., et al. (2016). First-In-Human safety and long-term exposure data for AAB-003 (PF-05236812) and biomarkers after intravenous infusions of escalating doses in patients with mild to moderate Alzheimer's disease. *Alzheimer's Research & Therapy, 8*(1), 12.
52. Teng, E., et al. (2023). The use of episodic memory tests for screening in clinical trials for early Alzheimer's disease: A comparison of the Free and Cued Selective Reminding Test (FCSRT) and the Repeatable Battery for the Assessment of Neuropsychological Status (RBANS). *The Journal of Prevention of Alzheimer's Disease, 10*(1), 41–49.
53. Dubois, B., et al. (2023). Masitinib for mild-to-moderate Alzheimer's disease: Results from a randomized, placebo-controlled, phase 3, clinical trial. *Alzheimer's Research & Therapy, 15*(1), 39.
54. Yadollahikhales, G., & Rojas, J. C. (2023). Anti-amyloid immunotherapies for Alzheimer's disease: A 2023 clinical update. *Neurotherapeutics, 20*, 914.
55. Collina, S., et al. (2013). Sigma receptor modulators: a patent review. *Expert Opinion on Therapeutic Patents, 23*(5), 597–613.
56. Hannestad, J., et al. (2021). Safety and tolerability of GRF6019 infusions in severe Alzheimer's disease: A phase II double-blind placebo-controlled trial. *Journal of Alzheimer's Disease, 81*(4), 1649–1662.
57. Florian, H., et al. (2023). Tilavonemab in early Alzheimer's disease: Results from a phase 2, randomized, double-blind study. *Brain, 146*(6), 2275–2284.
58. Liu, K. Y., et al. (2023). Key questions for the evaluation of anti-amyloid immunotherapies for Alzheimer's disease. *Brain Communications, 5*(3), fcad175.
59. Schenk, D. B., et al. (2017). First-in-human assessment of PRX002, an anti-alpha-synuclein monoclonal antibody, in healthy volunteers. *Movement Disorders, 32*(2), 211–218.
60. Nordstrom, E., et al. (2021). ABBV-0805, a novel antibody selective for soluble aggregated alpha-synuclein, prolongs lifespan and prevents buildup of alpha-synuclein pathology in mouse models of Parkinson's disease. *Neurobiology of Disease, 161*, 105543.
61. Straccia, G., et al. (2022). Precision medicine in Parkinson's disease: From genetic risk signals to personalized therapy. *Brain Sciences, 12*(10), 1308.

# Harnessing Nanotechnology in Modulating Neuroimmune Axis

Abhishek Tiwari, Krushna Ch. Maharana, Bhaskar Jyoyi Dutta, and Sanjiv Singh

**Abstract**

Neuroimmune system is considered as one of the vital systems that is mainly responsible for inducing various disorders to the body. The proper ameliorative effect on the neuroimmune axis is still a mystery among the researchers because it involves the penetration to the BBB and other parameters. In this context the nanotechnology has proven a beneficial tool. In the current scenario, the nanotechnology has influenced in treating the various neurologic as well as immunologic disorders. It helps to create innovative nanostructures with molecular, cellular, or atomic functions to regulate and repair damaged brain circuits. Materials are manipulated on a nearly atomic scale using nanotechnology. The formulation of various nano medicines with the help of nanotechnology serves by increasing the bioavailability of the drug and treating the diseases. The ability to tune molecular interactions is made possible by the integration of nanotechnology with physiology and cell biology. The current article's goals include approaches to effectively interact with neuronal and immune cells, modulate molecular-based diagnostic methods, biomaterials, and hybridized compounds used for specific delivery of medications and tiny molecules through the barriers of the brain, as well as neuronal regeneration and neuroprotection.

**Keywords**

Nanotechnology · Nanoparticles · Neuroimmune Axis · Neuroscience

A. Tiwari · K. C. Maharana · B. J. Dutta · S. Singh (✉)
Department of Pharmacology and Toxicology, National Institute of Pharmaceutical Education and Research, Hajipur, Bihar, India
e-mail: sanjivsingh@niperhajipur.ac.in



R. Shukla et al. (eds.), *Neuroendocrine-Immune Axis*,
https://doi.org/10.1007/978-981-95-5088-3_2

## 1 Introduction

Extensive research has long explored the relationship linking the neurological and immune systems, significantly improving our comprehension of neurogenic inflammation and pain causes. A key development in neuroimmunology has been the identification of long-range sensory-autonomic reflexes, where sympathetic and parasympathetic neurons are central in regulating immune responses. A major milestone in this field was achieved by Kevin Tracey in the early 2000s with the introduction of the "inflammatory reflex" concept, which has greatly influenced our insight of the interplay between the neuroimmune [1]. Within this framework, the vagus nerve is recognized as a critical component, modulating TNF-α a splenic production during peripheral inflammatory events [2]. Sensory signals from the vagus nerve detect inflammatory states and relay information to the brainstem and autonomic centers, resulting in feedback mechanisms that inhibit TNF-α generation. In particular, the spleen's production of norepinephrine and epinephrine causes ChAT-positive T cells to produce acetylcholine, which inhibits the manufacture of TNF-α by splenic macrophages [3]. In recent years, scientific efforts have increasingly focused on decoding the molecular pathways underlying communication between immune cells and neurons in both directions, TNF-α and IL-1 are examples of lipid-derived mediators and cytokine receptors expressed by neurons, allowing immune signals to alter neuronal activity, including the enhancement of pain sensitivity. Furthermore, neurons themselves produce pattern-recognition receptors, typically associated with immune cells, allowing both systems to monitor microbial threats simultaneously [4]. This bidirectional signaling is further supported by the presence of neurotransmitter, neuropeptide, and neural mediator receptors on immune and epithelial cells. Neuroimmunology, with roots in the discovery of systemic and localized reflexes, continues to evolve as we deepen our insight of the underlying molecular processes that coordinate nervous and immune system interactions.

## 2 Neuronal Integration with Immune System

The innate immune and the neurological network are both necessary for the quick identification and reaction to threat signals. While neurons use ion channels to detect ATP and other chemicals linked to damage, immune cells use receptors like Toll-like receptors to recognize infections [5]. These mechanisms share a common goal: the swift identification and reaction to potential threats. Interactions between innate immune cells and neurons likely enable a coordinated detection of environmental signals, promoting quicker tissue repair and recovery following infections. The cooperation between these two systems thus strengthens the body's capacity to recognize harmful stimuli and trigger protective and restorative responses.

In recent years, researchers have increasingly focused on the interactions between three specific types of innate immune cells within the gut—mast cells, innate lymphoid cells (ILCs), and macrophages—and their communication with both the CNS

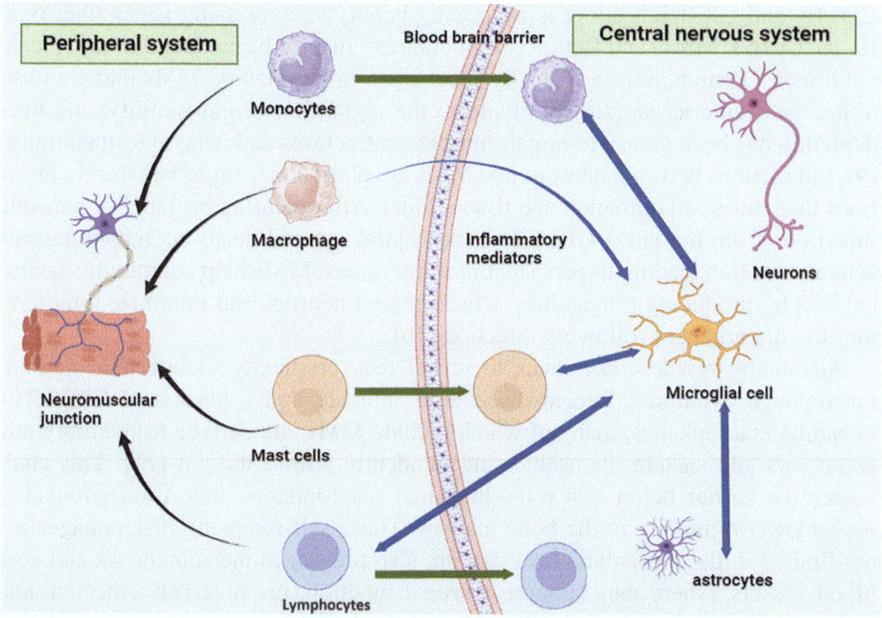


**Fig. 1** A schematic diagram of immune crosstalk between PNS and CNS

and PNS, as depicted in Fig. 1. Although neuronal communication with various innate immune cells has been previously documented, contemporary studies have placed particular emphasis on understanding how these three cell types interact with neurons in the intestinal environment. By investigating these neuroimmune dynamics in the gut, scientists aim to unravel the intricate signaling networks that govern the relationship between neurons and innate immune cells within this specialized tissue.

## 3 Neuron–Macrophage Interactions

Macrophages within the GI tract play a vital role in preserving tissue stability and defending the body against invading pathogens. These immune cells display a wide range of functions and phenotypes, shaped by their specific location within various intestinal layers. Notably, nerves and macrophages are often found in close association throughout the gut, both under healthy and disease conditions. Recent findings highlight how neural signaling significantly influences macrophage behavior, while, conversely, macrophages also help sustain neuronal integrity in both physiological and pathological states [6].

The earliest insights into two-way communication between neurons and macrophages in the gut came from studies on a specialized subset known as muscularis macrophages (MMs)—a layer that encompasses the enteric nervous system and smooth muscle tissue [7]. These MMs, characterized by markers such as MHCII,

CD11b, and CX3CR1, adopt a protective phenotype, expressing genes like Arg1, IL-10, Cd163, Mrc1, and Retnla [8]. In contrast, macrophages found in the intestinal lumen's lamina propria typically show a pro-inflammatory. MMs that are positioned near enteric ganglia contribute to the regulation of gut motility, and their depletion has been shown to impair muscle contractions and delay intestinal transit [9]. Interactions between neurons and MMs are particularly important during infections that cause inflammation and tissue injury. After clearing bacterial or parasitic infections from the gut, VGLUT2+ is susceptible to cell death via inflammasome activation, often leading to persistent motility issues. MMs help counter this neuronal loss by producing polyamines, which protect neurons and minimize long-term motility disturbances following infection [10].

Advancing research continues to reveal the complexity of intestinal neuron-macrophage dynamics. Recent discoveries indicate that a portion of CX3CR1+ intestinal macrophages, many of which include MMs, are derived from embryonic precursors and sustain themselves independently within the gut [11]. This challenges the earlier belief that most intestinal macrophages arise from circulating monocytes originating in the bone marrow. These self-renewing macrophages are not limited to the muscularis layer but are also present in the submucosa and near blood vessels, where they support neuronal function, promote cell adhesion, and facilitate blood vessel formation. Interestingly, this macrophage population expresses a distinct set of genes (e.g., Nova1, Chrm2, and Efr3b), which do not fully align with previously characterized MM gene profiles [12].

Loss of these self-maintaining macrophages has been linked to caspase-3-dependent apoptosis in enteric neurons, disrupting neuronal function such as anion secretion and causing reduced gut motility [11]. Further investigation is needed to clarify how different neuron types—including sensory neurons involved in pain perception, enteric neurons, and autonomic neurons (both parasympathetic and sympathetic)—interact with macrophages under both healthy and diseased conditions. In particular, exploring how dysfunction in these neuroimmune interactions may contribute to GI disorders, including IBD and other gut-associated disorders, continues to be of significant concern.

## 4 Neuron–Mast Cell Interactions

The GI tract's mucosal and submucosal layers include mast cells at crucial locations, closely associated with sensory and autonomic nerve endings. This proximity enables dynamic crosstalk between nerves and mast cells. Findings indicating an elevated density of mast cells close to mucosal nerves in the colons of IBS patients support the idea that these interactions play a major role in the development of IBS [13]. Notably, the closer mast cells are to nerves, the more severe and frequent the abdominal pain experienced by IBS patients. VIP, a peptide, can be secreted in enteric neurons as well as mast cells., a molecule linked to the pathophysiology of

IBS that stimulates mast cell degranulation [14, 15]. Recent studies have reported elevated systemic levels of VIP, and an increased number of VIP receptor 2-expressing mast cells shows elevated bacterial movement rates in IBS patient tissues [1]. VIPR2 inhibitor treatment of biopsy specimens has effectively decreased bacterial translocation and mast cell degranulation, indicating that VIP-mast cell signaling is essential for preserving the integrity of the mucosal barrier during IBS [1]. Although these findings highlight how crucial the nerve-mast cell connection is in GI disorders, further findings are required to clarify the exact pathways, especially through genetic and pharmacological investigations in animal models. Additionally, given the diversity of both mucosal mast cell populations and neurons, future research will need to search specific circuits associated with this complex interaction.

## 5 Neuron–ILC Interactions

Innate lymphoid cells (ILCs) are specialized lymphocytes residing in the barrier, where they contribute to immune defense and tissue repair following damage. Recent advances have emphasized the importance of neuron-ILC interactions in several barrier organs, including the gut. Neurons have been shown to communicate actively with both gr-2 ILCs and gr-3 ILCs to maintain intestinal homeostasis and fight infections. Particularly important in preventing intestinal helminth infections are cholinergic enteric neurons that generate the neuropeptide known as neuromedin U (NMU). High amounts of the NMU receptor 1, which binds NMU specifically, are synthesized by ILC2s. In vitro, NMU exposure ILC2 proliferation in a way that is reliant on Nmur1. Additionally, NMU treatment in vivo strengthens immune responses that protect against Nippostrongylus brasiliensis oral infections. The physical proximity of ILC2s to NMU-secreting neurons suggests that these neurons may direct ILC2-mediated immune responses. Additionally, enteric neuron organoid cultures have been shown to produce NMU in response to stimulation with N. brasiliensis or alarmin IL-33 products via a MyD88-dependent pathway. These findings indicate that a specific subset of enteric neurons is capable of sensing danger signals and promoting type 2 immunity by signaling to ILC2s [16].

## 6 Neuronal Communication and Acquired Immunity

Immune system within the GI tract is essential for preserving immune tolerance, supporting the integrity of the gut barrier, and establishing long-term immunity against pathogens and antigens. Despite these critical roles, our understanding of how neurons interact with components of adaptive immunity remains limited. This section highlights current insights into the crosstalk between neurons and two key adaptive immune players: T cells and B cells.

## 7 Neuronal Interactions with T Cells

Peripheral nervous system engagement has been demonstrated for CD4+ and CD8+ T cells, essential for acquired immune activity [17]. Moreover, it is still unclear what these neuron–T cell connections in the gut mean functionally. The cholinergic anti-inflammatory reflex is a prominent example, where sympathetic adrenergic fibers carry signals from the vagus nerve to the spleen via the celiac ganglia. Choline acetyltransferase T cells are stimulated to form acetylcholine by this cascade, which activates their β2-adrenergic receptors [18]. ACh then inhibits the synthesis of TNF via acting on nicotinic receptors on macrophages. Present research has confirmed that similar mechanisms operate in the intestine: during *Citrobacter rodentium* infection, ChAT+ T cells migrate to the colon, and mice lacking ChAT in T cells demonstrate increased vulnerability to the infection. This underscores the critical, non-redundant role of T cell-derived ACh in mucosal defense [19].

It has been demonstrated that Ach directly affects T cell activities via interacting with nicotinic and muscarinic receptors. Specifically, the M3 receptor (M3R) is essential for protecting against infections like *Nippostrongylus brasiliensis* and *Salmonella Typhimurium*. Mice deficient in M3R exhibit weakened immune memory and impaired T cell responses, reducing their resistance to these infections. Studies have demonstrated that ACh and M3R agonists can enhance the secretion of IL-13 and IFN-γ by T cells, although the exact intracellular signaling mechanisms behind these effects remain to be fully elucidated [20].

## 8 Neural Interaction with B Cells

B cells contribute to adaptive immunity by transforming from naïve cells that produce IgM to IgA-secreting plasma cells in response to activation [18]. Recent discoveries show that sympathetic nerve fibers in the spleen can influence B cell development. These neurons emit norepinephrine, which activates ChAT+ T cells. These cells eventually generate ACh, which helps B cells differentiate into plasma cells that produce antibodies [21]. In the colon, B cells are positioned near mucosal neurons and Peyer's patches, suggesting potential neural involvement in regulating their activity [22]. Laboratory studies support this hypothesis, showing that vasoactive intestinal peptide (VIP) can induce IgA synthesis by human B cells. Additionally, it has been demonstrated that substance P encourages the expansion of immunoglobulin-secreting B cells in vitro. Furthermore, adrenergic stimulation via β2-adrenergic receptors on B cells enhances IL-4-dependent immunoglobulin production. When noradrenaline is depleted, Th2-mediated antibody responses are impaired, reinforcing the importance of this pathway [23].

Beyond their immunological functions, B cells may also influence enteric neuronal health. Animals exhibit decreased intestinal motility and alterations in the expression of glial fibrillary acidic protein throughout the enteric nervous system in EAE, and a mouse model of MS. These effects are believed to result from the action of immunoglobulins produced by B cells, as serum from EAE mice or multiple sclerosis patients shows heightened reactivity against enteric neurons and glial cells. Interestingly, mice with EAE and B cell-deficient do not display these motility disturbances, indicating a role for B cells in regulating neural function in the gut. However, the exact mechanisms and whether enteric neurons directly influence B cell function remain unclear. Future research is needed to clarify whether enteric neurons shape B cell differentiation, antigen presentation, or antibody production. Given the abundance of B cells in the Peyer's patches and lamina propria, these locations may serve as critical sites for neuron–B cell communication [24].

## 9 Neuroimmune Axis-Related Disorders

The term "neuroimmune diseases" refers to a broad category of disorders that impact the CNS and PNS. These include conditions that provide serious global health issues, such as myasthenia gravis (MG), MS, and Guillain-Barré syndrome. Numerous immune cell types, including dendritic cells, circulating monocytes, natural killer (NK) cells, microglia/macrophages, and T and B lymphocytes, are activated and interact throughout the development of these illnesses [25].

These conditions often arise from a combination of contributing factors, such as neuroinfectious diseases, autoimmunity, paraneoplastic syndromes, neurodegenerative conditions, and neuropsychiatric disorders. A common feature among them is inflammation within the CNS. The underlying mechanisms may include cellular immunity, antibody-mediated responses, or pathogen-driven processes, each with distinct biological pathways depending on the specific disease [26]. Disorders like AD, PD, ALS, HD, and MS are all associated with neuroinflammation. T and B cells, together with immunosuppressive glial cells such as oligodendrocytes, astrocytes, and microglia, are key players in the synthesis of pro-inflammatory cytokines in these disorders. Immune system activation can weaken the BBB and blood-nerve barrier, allowing immune cells to enter the nervous system and aid in the development of illness [27].

## 10 Multiple Sclerosis (MS)

Multiple sclerosis is a long-term autoimmune abnormality impacting the CNS, marked by neuroinflammation and progressive neurodegeneration. According to the 2020 MS Census, a new diagnosis was made every five minutes worldwide, contributing to the estimated 2.8 million people living with MS. Notably, global prevalence rose by 30% after 2013 [28]. MS is known to progress through an intricate relationship between genetic predisposition, environmental exposures, and behavioral factors, which begin affecting individuals during a prodromal phase. This phase may involve symptoms such as depression, anxiety, gastrointestinal disturbances, and increased levels of serum sNfL, a marker of neuroaxonal injury [29].

Although MS clearly has an inflammatory component early in its course, it is debated whether it is a classic autoimmune disease. Insights from viral models suggest that MS could represent a chronic viral or pathogen-associated illness. The disease process appears to involve diverse immune cell populations, including cells of the innate immune system, B lymphocytes, and resident CNS immune cells, all of which may contribute to triggering and sustaining inflammation [30].

The landscape of disease-modifying therapies for MS has expanded greatly since the approval of interferon-β1b in 1993. More than 15 DMTs are now available, many of which function by targeting the peripheral immune system. These therapies work through different mechanisms, broadly categorized as immunomodulatory, immune cell-depleting, or inhibiting immune cell trafficking [26].

## 11 Alzheimer's Disease (AD)

AD is a progressive neurodegenerative abnormality defined by the accumulation of neurofibrillary tangles and Aβ plaques, as well as cellular debris and inflammatory molecules. These Aβ plaques are classified into diffuse, neuritic, and burnt-out forms based on their reactivity to anti-Aβ antibodies and staining with silver, thioflavin-S, and Congo red [31]. In addition, AD is linked to increased levels of inflammatory markers such as soluble CD8 (sCD8) antigens, interleukin-1 (IL-1), IL-2, and IL-6, autoantibodies that target the brain's thymic cells and myelin basic protein (MBP) [28].

M1-type microglia contribute to AD pathology by producing pro-inflammatory agents like NO, TNF, IL-1, IL-6, and superoxide, while expressing increased levels of receptors like C3R, CD14, and CD36, all of which intensify neuronal damage. Mutations in TREM2 and the APOE4 allele impair microglial phagocytosis, leading to further accumulation of Aβ plaques. Although phagocytosis of Aβ may initially provide protection, it can also exacerbate oxidative stress, worsening neural injury.

Activated mast cells release inflammatory mediators through rapid degranulation via the CD47/integrin complex, contributing to enhanced neuroinflammation. In parallel, heightened activity of CD4+ and CD8+ T cells correlates with elevated Aβ buildup, indicating an active but dysregulated immune surveillance mechanism. Neutrophils also amplify AD-associated inflammation by secreting mediators like myeloperoxidase, IL-2, and macrophage migration inhibitory factor (MIF).

Astrocyte activation by Aβ peptides through NF-κB signaling leads to complement component C3 release, which in turn influences microglial polarization. Interactions between astrocytes and microglia can significantly aggravate both plaque pathology and neuroinflammation, further accelerating disease progression [28].

## 12 Inflammatory Bowel Disease

The term "IBD" represents a collection of long-term inflammatory diseases, mainly ulcerative colitis and Crohn's disease. Crohn's disease is linked to a Th1- and Th17-mediated immune response, which is typified by increased levels of cytokines such as IL-12, IFN-γ, and TGF-β. It frequently causes inflammation in the colon and distal small intestine. Colon-specific ulcerative colitis is usually linked to a type 2 immune response, which is typified by increased cytokine levels, including IL-5 and IL-13. The enteric nervous system (ENS) may be directly impacted by these long-term inflammatory diseases. Interestingly, it has been demonstrated that IL-1 and IL-6 directly depolarize enteric neurons in guinea pigs, resulting in the production of action potentials. An antagonist of IL-1's receptor inhibits the activity that IL-1 induces. Remarkably, both excitatory and inhibitory synaptic transmission seem to be suppressed by IL-1 and IL-6. Recent research has shown that glial cells in mice can release soluble macrophage colony-stimulating factor (M-CSF) through connexin-43 channels. This causes intestinal inflammation by activating muscularis macrophages and polarizing them toward a pro-inflammatory. During chronic colitis caused on by dextran sodium sulfate (DSS), elevated levels of pro-inflammatory IL-1 were found to enhance M-CSF secretion, which was associated with a decreased pain threshold [32].

In the pathophysiology of IBD, neurological alterations, such as changes in the expression of ENS surface markers, are implicated. A higher percentage of neurons in the human myenteric plexus have been shown to express substance P (SP) and NOS, but not VIP. Conversely, in the submucosal plexus, only VIP-expressing neurons are elevated. Functional alterations in enteric neurons are also evident during IBD. For instance, in guinea pig models of TNBS-induced colitis, intrinsic sensory neurons (ISNs) exhibit hyperexcitability, while other neuronal populations display enhanced excitatory synaptic plasticity. Moreover, enteric glial cells play a pivotal role in regulating intestinal inflammation. Experimental ablation or dysfunction of these glial cells in mice leads to jejuno-ileitis, highlighting their protective role. This is consistent with clinical findings showing reduced glial cell density in patients with Crohn's disease. Importantly, ex vivo treatment of mucosal biopsies from these patients with glial-derived S-nitrosoglutathione has been shown to restore mucosal barrier integrity [10].

## 13 Cardiovascular Disorder

The primary pathways facilitating interaction between the brain and the circulatory system include the HPA axis, the ANS, and the brain's circumventricular organs. The peripheral neural system tightly innervates immune organs. Through its innervation of the circulatory system, the nervous system may directly cause cardiovascular disease, or it may indirectly cause it by modulating the immunological reactions involved in cardiovascular pathology. For the treatment of cardiovascular diseases, novel methods based on bioelectronic or ultrasound-driven control of immunological organs are emerging [31].

## 14 Nanotechnology

Nanotechnology refers to the research, development, design, synthesis, description, and use of devices and materials with functional structures that are organized at the nanoscale—typically on the order of one billionth of a meter in at least one dimension. At this scale, manipulation of the fundamental molecular architecture becomes possible, enabling precise control over the physical and chemical characteristics at the macroscopic level of materials. Consequently, the behavior of individual molecules and their interactions gain critical importance in determining the overall bulk properties of the resulting material or device. Numerous uses of nanotechnology have been made in medicine [33].

Nanotechnology is a rapidly advancing field that offers innovative and efficient tools for investigating brain function in both health and disease. Notably, biosensors, multi-electrode arrays, memristive devices, and brain-machine interfaces play critical roles in expanding our understanding and capabilities in neuroscience research. NPs are particles with a diameter of 1 to 1500 nm. They may be created from practically any substance, including non-biodegradable substances like Au, Ag, C, Fe, and silica. They can also be formed from polysaccharides, polyamino acids, and poly-hydroxy acids. This article will concentrate on the use of NPs in immunomodulation-related treatments, which have been transformed by the capacity to synthesize them from biocompatible and biodegradable polymers like polylactide-co-glycolide (PLGA). NPs can also be engineered to reach a diverse range of medicinal agents, like immunosuppressants and chemotherapeutics, proteins such as antibodies and hormones, peptides for vaccine development, DNA for gene therapy applications, miRNAs, and molecular tools to target CRISPR systems [34].

These materials and devices can be engineered having a high level of functional specificity, which enables them to engage in molecular (subcellular) interactions with cells for use in physiology and medicine. This facilitates a level of integration between technology and biology that was once thought to be unattainable. To develop these groundbreaking innovations, expertise from traditional disciplines such as physics, chemistry, biology, and materials science has converged, giving rise to the interdisciplinary field of nanotechnology, which must be comprehended in this context. Nanotechnology holds forth a variety of possibilities, including completely new tools and capabilities in addition to enhancing existing ones. The basic characteristics and bioactivity of medicines and other materials can be changed via nanoscale manipulation. These technologies can provide control over a variety of medication or agent properties, including changes in blood volume residence time and solubility, as well as controlled release over short- or long-term periods regulated release that is environmentally induced or delivery that is extremely site-specific [33].

## 15 Size and Shape

The cellular absorption mechanism of NP treatment has a tremendous impact on the subsequent immunological result. Immune-modulating properties of NPs that enter cells through cytosolic routes differ from those acquired through phagocytosis. The ideal size and shape for particle uptake depends on the cell type, according to studies using a wide range of standard cell lines (HeLa, CHO, Caco-2, and MCF-7) and NPs (derived from gold, polystyrene, polymer, silicon, titanium, and iron oxide). Particles' makeover influences their biodistribution and the process by which they are taken up. Clathrin-mediated/Caveola processes are the most effective ways to take up NP with a diameter of less than 100 nm in nonphagocytic cell types. Spherical nanoparticles (NPs) exhibit superior absorption kinetics compared to rod-shaped NPs, regardless of their size, when interacting with APCs such as macrophages and DCs. However, size has a lesser impact on this process. Any therapeutic approach aiming to modulate the immune response should focus on directing NPs to phagocytes, a topic explored further below.

NP toxicity is also impacted by size. The diameter of nanoparticles is less than 100 nm often interact with cellular structures such as the mitochondria and nucleus, potentially leading to respiratory and gene toxicity. The risk of these interactions diminishes as the nanoparticle size increases, likely because larger NPs are more prone to being engulfed by phagocytes, which helps isolate the particles from the sensitive cytoplasm [34].

## 16 Charge

The surface charge of NPs is a dynamic physicochemical property that may be changed by the particle microenvironment, including the protein corona. In general, the total charge of the NP surface may be changed by adding more carboxyl (negatively charged) or amine (positively charged) groups. It has been demonstrated through studies utilizing various NP charges that this factor affects both absorption pathways and subsequent immunological effects. For instance, cationic NPs seem to be absorbed more easily through clathrin-mediated mechanisms than anionic particles. Additionally, positively charged antigen-loaded NPs considerably enhance Th1 responses following mucosal (pulmonary) or intradermal injection, but anionic particles only marginally enhance T and B cell responses under comparable circumstances. Cationic particles' potential to control positive co-stimulatory molecules and their preferential absorption by DC have both been linked to their capacity to promote Th1 responses. Cationic NPs can interfere with the functioning of mitochondria and the endoplasmic reticulum, resulting in the release of pro-inflammatory cytokines, ROS, and cell death. These effects may contribute to the adjuvant properties of cationic NPs; however, any clinical application of these effects must consider potential off-target toxicities. Anionic NPs, on the other hand, are often linked to negligible or no toxicity. Additionally, it has been demonstrated that under-30 mV charged NPs show anti-inflammatory qualities, and can aid in the establishment of immunological tolerance to an antigen when paired with it [34, 35].

## 17 Nanotechnology Modulating Neuroimmune Axis

In the present scenario, the involvement of the nanoscience-based drug delivery evolves as a major breakthrough by which we are able to deliver various drugs by focusing on specific tissue and enhancing their bioavailability, solubility, and altering the pharmacokinetic property [36]. In the present context, the two major challenges, i.e., the BBB and the blood cerebrospinal fluid barrier (BCSFB), have been regarded as the major physiological barriers in targeting the neuroimmune axis and limit the therapeutic agents in reaching the proper site of the brain. In this context, nanotechnology presents an important strategy for improving drug delivery to the brain and improving bioavailability along the neuroimmune axis [37]. Owing to their nanoscale dimensions, nanoparticles are an effective carrier for therapeutic agents, including drugs, immunomodulators, and antigens, as they are readily internalized by cells. Moreover, nanoparticles offer versatile administration routes beyond systemic delivery such as intranasal administration—which enables direct nose-to-brain transport. This approach enhances drug solubility, enables targeted delivery, minimizes adverse consequences linked to large systemic dosages, and allows for controlled release, making nanosystems highly advantageous for neurological applications.

The interplay between the neuroimmune system and nanoparticle-based technologies is an area of active investigation. Compared to conventional drug administration, nanoparticles employed as drug delivery systems significantly enhance pharmacokinetics and bioavailability, thereby improving therapeutic outcomes. Additionally, nanoparticles can act as vehicles for immunomodulation specific to antigens. Numerous advantages of antigen-loaded nanoparticles include prolonged antigen release, adjuvant and antigen shared delivery, the development of confined antigen depots at the injection site, efficient presentation of B-cell epitopes, and enhanced uptake and activation of cell-mediated immune responses against non-cellular antigens. These antigens and their epitopes, which are particular to the autoimmune response in neuroimmune axis, are attached to the nanoparticles. These nanoparticles can trigger the development of Treg cells and dendritic cells, control T-cell activity, and restore immunological tolerance [38].

Remyelination is a further area of intense scientific inquiry into neuroimmune axis, which is also associated with the discovery of novel therapies. Remyelination has emerged as a key therapeutic target in the modulation of the neuroimmune axis. Recent studies find activated neural stem and progenitor cells in the vicinity of lesion sites contribute significantly to the remyelination process. Promoting remyelination not only aids in restoring neural function but also provides an additional protective mechanism against axonal injury associated with myelin degradation

[39]. A growing number of therapeutic agents are being identified that target CNS inflammation and influence various stages of the remyelination process, such as promoting oligodendrocyte precursor cells' (OPCs') differentiation. Interestingly, certain nanoparticles have also demonstrated intrinsic remyelinating properties, further expanding their potential as multifunctional therapeutic platforms [40]. A gold nanocrystal solution with a clean surface and facets was discovered to have a strong remyelinating reaction to demyelinating drugs as a potential nanocatalytic therapeutic option in neuroimmune modulation. In both open-field and kinematic gait investigations, the oral administration of gold nanocrystals enhanced the motor abilities of cuprizone-treated mice.

## 18 Use of Nanomedicine and Its Implication in Neuroimmune Axis

In the modulation of the neuroimmune axis, the use of nanomedicine is very crucial. This generally uses various approaches to target the tissue. Among these it includes polymer-based nanosystem and lipid-based nanosystem as shown in Fig. 2.

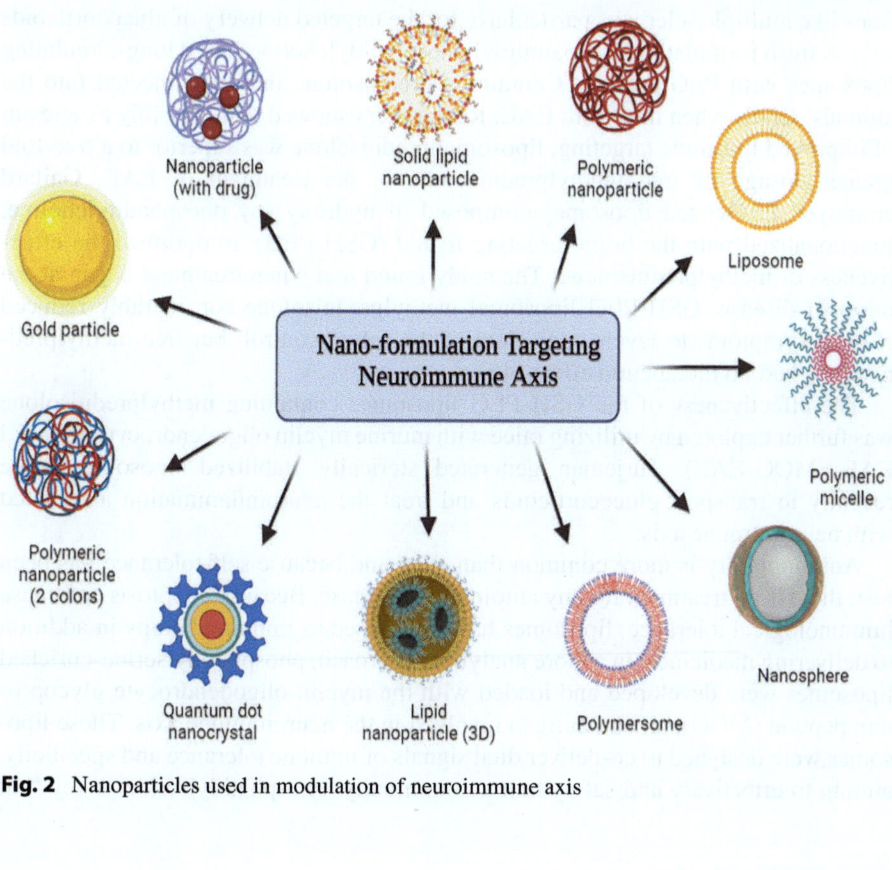


**Fig. 2** Nanoparticles used in modulation of neuroimmune axis

## 19 Lipid-Based Nanosystem

Lipid-based nanosystems, with liposomes, SLNs, NLCs, and nanoemulsions, represent some of the most effective nanocarriers for delivering therapeutic agents in neuroimmune disorders while improving the delivery of drugs to the brain. These carriers possess the capability to traverse the BBB by spontaneously interacting with and penetrating brain capillary endothelial cells, thereby improving central nervous system delivery while minimizing peripheral side effects [37]. Table 1 explains about the various nanoformulation modulating neuroimmune axis.

## 20 Liposome

Liposomes are among the most extensively researched drug transport nanosystems, offering key benefits such as biodegradability and biocompatibility. Their versatility is enhanced by the ability to modify their surface with functional biomaterials, enabling features such as reduced clearance in circulation and enhanced targeting to diseased tissues. Alongside lipid nanoparticles, liposomes—especially those PEGylated—have been identified as promising drug delivery vehicles for conditions like multiple sclerosis, particularly for the targeted delivery of glucocorticoids [55]. A fresh formulation of dipalmitoyl phosphatidylcholine-based long-circulating liposomes with PEG coating. Containing prednisolone that was injected into the animals' CNSs when they with EAE. Researchers showed that, possibly as a result of improved liposome targeting, liposomal prednisolone was superior to a five-fold greater dosage of free methylprednisolone in the treatment of EAE. Gailard employed PEGylated liposomes composed of hydroxy soy phosphatidylcholine, functionalized with the brain-targeting ligand (GSH-PEG), to optimize the effectiveness of methylprednisolone. The study found that when treatment began at the onset of disease, GSH-PEG liposomal methylprednisolone considerably reduced clinical symptoms to levels equivalent to the saline control, but free methylprednisolone had no therapeutic effect [45].

The effectiveness of the GSH-PEG liposomes containing methylprednisolone was further explored by utilizing mice with murine myelin oligodendrocyte-induced EAE (MOG-EAE). Turjeman generated sterically stabilized liposomes more recently to transport glucocorticoids and treat the neuroinflammation associated with neuroimmune axis.

Autoimmunity is more common than ever, and because self-tolerance has been lost, there is no treatment for any autoimmune disease. Because apoptosis can cause immunological tolerance, liposomes have been used as immunotherapy in addition to delivering medicines. In a more analytical approach, phosphatidylserine-enriched liposomes were developed and loaded with the myelin-oligodendrocyte glycoprotein peptide (MOG), an autoantigen involved in the neuroimmune axis. These liposomes were designed to co-deliver dual signals of immune tolerance and specificity, aiming to effectively and safely halt autoimmunity. Phosphatidylserine, a key "eat

**Table 1** Classification of various nanoformulation used in neuroimmune modulation

| Nanoformulation | Classification | Examples | Reference |
|---|---|---|---|
| Lipid-based nanosystems | Nanolipid carriers (NLCs) and solid lipid nanoparticles (SLNs) | Intranasal delivery of nanolipid carriers encapsulating TFM, a dihydroorotate dehydrogenase inhibitor | [41] |
| | | DMF and tocopherol acetate are delivered using nanolipid carriers offer improved brain targeting and bioavailability | [42] |
| | | Solid lipid nanoparticles (SLNs) encapsulating methylthioadenosine have been formulated for oral administration | [43] |
| | | SLNs with PEG have been used to improve the administration of the glucocorticoid methylprednisolone. | [44] |
| | Liposomes | PEGylated liposomes functionalized with the glutathione (GSH-PEG) which targets the brain | [45] |
| | | Glutathione-conjugated PEGylated liposomes encapsulating methylprednisolone offer targeted delivery to the brain | [46] |
| | Others | HDL-mimicking peptide–phospholipid scaffold (HPPS) | [47] |
| | | Nanosystems have been engineered for the containment and regulated release of miR-219a-5p, facilitating remyelination by enhancing oligodendrocyte functionality and differentiation | [48] |
| Polymer-based nanosystem | Poly (lactic-co-glycolic acid) (PLGA) polymeric nanoparticles | PLGA nanoparticles encapsulating the immunodominant myelin oligodendrocyte glycoprotein | [49] |
| | | PLGA nanoparticles carrying the myelin proteolipid protein epitope PLP139–151 are utilized to modulate autoimmune responses in central nervous system. | [50] |
| | | Dual-sized polymeric microparticles fabricated from PLGA (dMPs) are designed to facilitate controlled immune modulation. | [51] |
| | | PLG-based nanoparticles delivering disease-specific antigens, also referred to as silver nanoparticles (ag-NPs) | [43] |
| | | PLG microparticles presenting encephalitogenic peptides are engineered to suppress pathological T cell activity | [52] |
| | Others | Polymeric nanoparticles modified with specific peptides are designed to act on blood clots and ECM components | [53] |
| | | PCL-based nanoparticles encapsulating rhMBP | [54] |

me" and "tolerate me" signal found on the membrane of apoptotic cells, facilitates phagocytosis by APC such as dendritic cells, thereby promoting immune tolerance.

## 21 Nanolipid Carriers (NLCs) and Solid Lipid Nanoparticles (SLNs)

The solid matrix lipid-based nanoparticulate system was created from an O/W emulsion, in which solid lipids are substituted for the liquid lipid phase to keep the system solid at body temperature. While SLNs are the first generation of lipid nanoparticles, made of solid lipids, NLCs are second-generation lipid nanoparticles made of a combination of liquid and solid lipids together with surfactants. By controlling the quantity of liquid lipid they contain, NLCs are able to retain their solid state at both body temperature and environmental temperature. In an O/W emulsion, SLNs—colloid particles with enhanced physical stability—are created by substituting solid lipids for liquid lipids. The lipid core of SLNs is stabilized by surfactants and usually consists of waxes, steroids, and fatty acids. The lipid composition and organization in the matrix differ between SLNs and NLCs, leading to distinct morphological structures. NLC formulations frequently contain liquid lipids such as olive oil, sesame oil, almond oil, peanut oil, soybean oil, oleic acid, maize oil, soy lecithin, phosphatidylcholine, and vitamin E, and solid lipids such as stearic acid, stearyl alcohol, glycerol monostearate, and mono-stearin.

Recently, methylthioadenosine has been encapsulated in SLNs for oral delivery targeting brain in the treatment of multiple sclerosis. To prepare the SLNs, a combination of stearic acid, phospholipid 90G, and Tween-80 is used and a well-established microencapsulation method is employed. Additionally, intranasal nanolipid carriers loaded with TFM, a dihydroorotate dehydrogenase inhibitor with anti-inflammatory properties, have been developed. These TFM-loaded NLCs were fabricated utilizing the ultrasonication method for melt emulsification, with the formulation optimized through the Box-Behnken statistical design.

## 22 Other Lipid-Based Nanocarriers

Pomegranate seed oil, also referred to as nano-PSO, was formulated and administered in an autoimmune EAE model as an O/W nanoemulsion (Fig. 2). PSO is rich in punicic acid, a polyunsaturated fatty acid known for its potent antioxidant properties. Pathological evaluations showed that nano-PSO treatment notably diminishes demyelination and lipid peroxidation in the brains of affected animals, even in the presence of immune cell infiltration within the CNS. Nano-PSO may offer a promising therapeutic approach for beginning stages of neuroimmune disorders [56]. Because monocytes are essential to the development of neuroimmune diseases, an immunomodulatory carrier that targets monocytes with an HPPS was developed. For individuals in the early stages of neuroimmune disorders, nano-PSO could present a viable therapeutic option. Using optical imaging and fluorescent dye-loaded nanoparticles, the in vivo distribution of the particles within monocytes was confirmed. The administration of curcumin via HPPS nanoparticles effectively prevented inflammatory monocytes from crossing the BBB in EAE mice. This was achieved by downregulating the expression of NF-κB, ICAM-1, and MAC-1 in the monocytes [47].

Similarly, various nanosystems have been developed for the encapsulation, release, and promotion of remyelination by miR-219a-5p. These include biologically designed extracellular vesicles that overexpress miR-219a-5p, PLGA polymeric nanoparticles, and mRNA-based liposomes made of DSPC lipid. These cutting-edge delivery methods seek to improve miR-219a-5p's therapeutic potential in remyelination [48].

## 23 Polymer-Based Nanosystem

Various polymers and their corresponding nanoparticles have been widely studied for their ability to cross the BBB, deliver drugs to the CNS, and treat neurodegenerative diseases such as neuroimmune disorders. By applying various preparation techniques, a variety of natural and synthetic monomers and polymers may be used to create polymer nanoparticles. Additionally, these nanoparticles can be surface-functionalized to improve their ability to target the brain. Drug absorption and release processes in the brain include buildup in the brain capillaries and endocytosis or transcytosis via endothelial cells. This results in transfer into the brain parenchyma, which is fueled by a gradient of high concentration. Additionally, lipid solubilization brought on by surfactants causes membrane fluidization, which aids in opening tight junctions and makes medication transport to the central nervous system easier [57].

## 24 Poly (Lactic-co-Glycolic Acid) (PLGA) Polymeric Nanoparticles

PLGA nanoparticles are promising drug delivery vehicles due to their biocompatibility, biodegradability, and approval by the U.S. FDA. A key advantage of using PLGA nanoparticles is their ability to encapsulate a broad range of therapeutic agents efficiently.

Antigen-specific tolerogenic immunization procedures are referred to as "inverse vaccination" and have a significant therapeutic potential for neuroimmune disorders. In neuroimmune disorders, several myelin-associated proteins—including PLP, myelin basic protein, and myelin oligodendrocyte glycoprotein (MOG)—have been identified as key targets of autoreactive T lymphocytes. Antigen-specific therapies are highly desirable for treating autoimmune diseases, as they offer targeted immunomodulation without inducing broad immunosuppression [58]. In this context, Cappellano and colleagues developed PLGA nanoparticles encapsulating either recombinant interleukin-10 (IL-10), functioning as an inverse adjuvant, or the immunodominant MOG peptide (MOG35–55). These formulations were evaluated in C57BL/6 mice for both prophylactic and therapeutic applications in a chronic, progressive model of neuroimmune disease [49]. In order to induce long-lasting and antigen-specific immune tolerance, even in the presence of strong Toll-like receptor agonists, and to control both cellular and humoral immune responses, biodegradable PLGA nanoparticles loaded with the immunodominant 139–151 epitope of myelin proteolipid protein (PLP139–151) and the tolerogenic immunomodulator rapamycin are used as an inverse adjuvant. dMPs are polymeric microparticles made of PLGA that are loaded with a variety of immunomodulatory substances, including particular antigens and tolerizing substances that are directed at both intracellular and extracellular tolerogenic receptors. The internalization and polarization of the immune cells that become attached to the particles, as well as the future course of the disease, may be impacted by different nanoparticle formulations and the degree of antigen loading at the carrier [59].

## 25 Other Polymeric Nanoparticles

In neuroimmune disorders, achieving targeted drug delivery while minimizing off-target effects is critical. One promising strategy is use of peptide-modified polymeric nanoparticles designed to recognize blood clots and ECM components such as nidogen. In particular, fibrin-targeting peptide-functionalized PEG-b-PCL nanoparticles were created. Animals with experimental autoimmune encephalomyelitis (EAE), a condition marked by increased fibrin and nidogen/entactin-1 levels, were given these nanoparticles. In both ex vivo and in vivo tests, the functionalized nanoparticles showed noticeably better binding to their molecular targets and to lesion sites than the non-functionalized nanoparticles [53].

Recombinant rhMBP, purified from transgenic cow's milk with a vacuum-assisted cation exchange process, was encapsulated within poly(caprolactone) nanoparticles to achieve sustained release kinetics. This formulation was developed as a therapeutic vaccine aimed at mitigating EAE in mice. Additionally, poly (butylcyanoacrylate) (PBCA) nanoparticles were employed for the delivery of the thymic peptide thymulin, enhancing its systemic circulation time in mice with relapsing-remitting EAE (rEAE).

## References

1. Jacobson, A., Yang, D., Vella, M., & Chiu, I. M. (2021). The intestinal neuro-immune axis: Crosstalk between neurons, immune cells, and microbes. *Mucosal Immunology, 14*(3), 555–565.
2. Chu, C., Artis, D., & Chiu, I. M. (2020). Neuro-immune interactions in the tissues. *Immunity, 52*(3), 464–474.
3. Lu, J., & Wu, W. (2021). Cholinergic modulation of the immune system–A novel therapeutic target for myocardial inflammation. *International Immunopharmacology, 93*, 107391.

4. Constant, D. A., Nice, T. J., & Rauch, I. J. C. (2021). Innate immune sensing by epithelial barriers. *Current Opinion in Immunology, 73*, 1–8.
5. Donnelly, C. R., Chen, O., & Ji, R.-R. (2020). How do sensory neurons sense danger signals? *Trends in Neurosciences, 43*(10), 822–838.
6. Buckley, M. W., & DB, M. G. (2022). Immune dynamics in the CNS and its barriers during homeostasis and disease. *Immunological Reviews, 306*(1), 58–75.
7. Partney, H., & Yissachar, N. (2022). *Regulation of host immunity by the gut microbiota. Evolution, biodiversity and a reassessment of the hygiene hypothesis* (pp. 105–140). Springer.
8. Matheis, F., Muller, P. A., Graves, C. L., Gabanyi, I., Kerner, Z. J., Costa-Borges, D., et al. (2020). Adrenergic signaling in muscularis macrophages limits infection-induced neuronal loss. *Cell, 180*(1), 64–78. e16.
9. Ma, T., Xue, X., Tian, H., Zhou, X., Wang, J., Zhao, Z., et al. (2023). Effect of the gut microbiota and their metabolites on postoperative intestinal motility and its underlying mechanisms. *Journal of Translational Medicine, 21*(1), 349.
10. Wang, H., Foong, J. P., Harris, N. L., & Bornstein, J. C. (2022). Enteric neuroimmune interactions coordinate intestinal responses in health and disease. *Mucosal Immunology, 15*(1), 27–39.
11. De Schepper, S., Verheijden, S., Aguilera-Lizarraga, J., Viola, M. F., Boesmans, W., Stakenborg, N., et al. (2019). Self-maintaining gut macrophages are essential for intestinal homeostasis. *Cell, 176*(3), 676.
12. Gabanyi, I., Muller, P. A., Feighery, L., Oliveira, T. Y., Costa-Pinto, F. A., & Mucida, D. (2016). Neuro-immune interactions drive tissue programming in intestinal macrophages. *Cell, 164*(3), 378–391.
13. Di Nardo, G., Cremon, C., Staiano, A., Stanghellini, V., Borrelli, O., Strisciuglio, C., et al. (2023). Role of inflammation in pediatric irritable bowel syndrome. *Neurogastroenterology and Motility, 35*(3), e14365.
14. Hasler, W. L., Grabauskas, G., Singh, P., & Owyang, C. J. N. (2022). Mast cell mediation of visceral sensation and permeability in irritable bowel syndrome. *Neurogastroenterology and Motility, 34*(7), e14339.
15. Singh, S., & Pawar, A. (2019). Effect of radio frequency electromagnetic field from mobile phones' base station antennas on maturation of rat erythrocytes. *Comparative Clinical Pathology, 28*(5), 1395–1401.
16. Singh, S., Aggarwal, P., & Ravichandiran, V. (2021). Immunological response of the respiratory tract in the SARS-CoV-2 infection. *Coronaviruses., 2*(9), 8–17.
17. Maalmi, H., Wouters, K., Savelberg, H. H., van der Velde, J. H., Reulen, J. P., Mess, W., et al. (2021). Associations of cells from both innate and adaptive immunity with lower nerve conduction velocity: the Maastricht Study. *BMJ Open Diabetes Research & Care, 9*(1), e001698.
18. Reboldi, A., Arnon, T. I., Rodda, L. B., Atakilit, A., Sheppard, D., & Cyster, J. G. (2016). IgA production requires B cell interaction with subepithelial dendritic cells in Peyer's patches. *Science, 352*(6287), aaf4822.
19. Ramirez, V. T., Godinez, D. R., Brust-Mascher, I., Nonnecke, E. B., Castillo, P. A., Gardner, M. B., et al. (2019). T-cell derived acetylcholine aids host defenses during enteric bacterial infection with Citrobacter rodentium. *PLoS Pathogens, 15*(4), e1007719.
20. Darby, M., Roberts, L. B., Mackowiak, C., Chetty, A., Tinelli, S., Schnoeller, C., et al. (2021). ILC3-derived acetylcholine promotes protease-driven allergic lung pathology. *The Journal of Allergy and Clinical Immunology, 147*(4), 1513–6.e4.
21. Zhang, X., Lei, B., Yuan, Y., Zhang, L., Hu, L., Jin, S., et al. (2020). Brain control of humoral immune responses amenable to behavioural modulation. *Nature, 581*(7807), 204–208.
22. Ma, B., von Wasielewski, R., Lindenmaier, W., & Dittmar, K. E. (2007). Immmunohistochemical study of the blood and lymphatic vasculature and the innervation of mouse gut and gut-associated lymphoid tissue. *Anatomia, Histologia, Embryologia, 36*(1), 62–74.
23. Kohm, A. P., & Sanders, V. M. (1999). Suppression of antigen-specific Th2 cell-dependent IgM and IgG1 production following norepinephrine depletion in vivo. *Journal of Immunology, 162*(9), 5299–5308.

24. Spear, E. T., Holt, E. A., Joyce, E. J., Haag, M. M., Mawe, S. M., Hennig, G. W., et al. (2018). Altered gastrointestinal motility involving autoantibodies in the experimental autoimmune encephalomyelitis model of multiple sclerosis. *Neurogastroenterology and Motility, 30*(9), e13349.
25. Li, H., Liu, S., Han, J., Li, S., Gao, X., Wang, M., et al. (2021). Role of toll-like receptors in neuroimmune diseases: therapeutic targets and problems. *Frontiers in Immunology, 12*, 777606.
26. Charabati, M., Wheeler, M. A., Weiner, H. L., & Quintana, F. J. (2023). Multiple sclerosis: Neuroimmune crosstalk and therapeutic targeting. *Cell, 186*(7), 1309–1327.
27. Wijeweera, G., Wijekoon, N., Gonawala, L., Imran, Y., Mohan, C., De Silva, K. R. D., et al. (2023). Therapeutic implications of some natural products for neuroimmune diseases: A narrative of clinical studies review. *Evidence-based Complementary and Alternative Medicine, 2023.*
28. Singh, V. K. (1994). Studies of neuroimmune markers in Alzheimer's disease. *Molecular Neurobiology, 9*(1–3), 73–81.
29. Meghana, A., Obulapathi, U., & Singh, S. (2021). Indian cow urine as a therapeutic alternative in treatment of human diseases: A review. *AYUHOM., 8*(2), 57.
30. Pachner, A. R. (2021). The neuroimmunology of multiple sclerosis: Fictions and facts. *Frontiers in Neurology, 12*, 796378.
31. Carnevale, D. (2022). Neuroimmune axis of cardiovascular control: mechanisms and therapeutic implications. *Nature Reviews Cardiology, 19*(6), 379–394.
32. Handa, M., Aparnasai, R. G., Panicker, N., Singh, S., & Ruwali, M. (2022). Recent trends of extracellular vesicles for therapeutic intervention of brain-related diseases. In *Nanomedical drug delivery for neurodegenerative diseases* (pp. 119–128). Elsevier.
33. Saini, R., Saini, S., & Sharma, S. J. (2010). Nanotechnology: the future medicine. *Journal of Cutaneous and Aesthetic Surgery, 3*(1), 32.
34. Getts, D. R., Shea, L. D., Miller, S. D., & King, N. J. (2015). Harnessing nanoparticles for immune modulation. *Trends in Immunology, 36*(7), 419–427.
35. Liu, Y., Hardie, J., Zhang, X., Rotello, V. M., & editors. (2017). Effects of engineered nanoparticles on the innate immune system. *Seminars in Immunology, 34*, 25–32.
36. Demetzos, C., & Pippa, N. J. D. (2014). Advanced drug delivery nanosystems (aDDnSs): A mini-review. *Drug Delivery, 21*(4), 250–257.
37. Ortiz, G. G., Pacheco-Moisés, F. P., Macías-Islas, M. Á., Flores-Alvarado, L. J., Mireles-Ramírez, M. A., González-Renovato, E. D., et al. (2014). Role of the blood–brain barrier in multiple sclerosis. *Archives of Medical Research, 45*(8), 687–697.
38. Shakya, A. K., & Nandakumar, K. S. (2018). Antigen-specific tolerization and targeted delivery as therapeutic strategies for autoimmune diseases. *Trends in Biotechnology, 36*(7), 686–699.
39. Deshmukh, V. A., Tardif, V., Lyssiotis, C. A., Green, C. C., Kerman, B., Kim, H. J., et al. (2013). A regenerative approach to the treatment of multiple sclerosis. *Nature, 502*(7471), 327–332.
40. Jolanda Münzel, E., & Williams, A. (2013). Promoting remyelination in multiple sclerosis—Recent advances. *Drugs, 73*, 2017–2029.
41. Gadhave, D. G., & Kokare, C. R. (2019). Nanostructured lipid carriers engineered for intranasal delivery of teriflunomide in multiple sclerosis: Optimization and in vivo studies. *Drug Development and Industrial Pharmacy, 45*(5), 839–851.
42. Ghasemian, E., Vatanara, A., Navidi, N., & Rouini, M. R. (2017). Brain delivery of baclofen as a hydrophilic drug by nanolipid carriers: Characteristics and pharmacokinetics evaluation. *Journal of Drug Delivery Science and Technology, 37*, 67–73.
43. Kumar, P., Sharma, G., Gupta, V., Kaur, R., Thakur, K., Malik, R., et al. (2019). Oral delivery of methylthioadenosine to the brain employing solid lipid nanoparticles: Pharmacokinetic, behavioral, and histopathological evidences. *AAPS PharmSciTech, 20*, 1–7.
44. Gandomi, N., Varshochian, R., Atyabi, F., Ghahremani, M. H., Sharifzadeh, M., Amini, M., et al. (2017). Solid lipid nanoparticles surface modified with anti-Contactin-2 or anti-Neurofascin for brain-targeted delivery of medicines. *Pharmaceutical Development and Technology, 22*(3), 426–435.

45. Gaillard, P. J., Appeldoorn, C. C., Rip, J., Dorland, R., van der Pol, S. M., Kooij, G., et al. (2012). Enhanced brain delivery of liposomal methylprednisolone improved therapeutic efficacy in a model of neuroinflammation. *Journal of Controlled Release, 164*(3), 364–369.
46. Lee, D.-H., Rötger, C., Appeldoorn, C. C., Reijerkerk, A., Gladdines, W., Gaillard, P. J., et al. (2014). Glutathione PEGylated liposomal methylprednisolone (2B3–201) attenuates CNS inflammation and degeneration in murine myelin oligodendrocyte glycoprotein induced experimental autoimmune encephalomyelitis. *Journal of Neuroimmunology, 274*(1–2), 96–101.
47. Lu, L., Qi, S., Chen, Y., Luo, H., Huang, S., Yu, X., et al. (2020). Targeted immunomodulation of inflammatory monocytes across the blood-brain barrier by curcumin-loaded nanoparticles delays the progression of experimental autoimmune encephalomyelitis. *Biomaterials, 245*, 119987.
48. Osorio-Querejeta, I., Carregal-Romero, S., Ayerdi-Izquierdo, A., Mäger, I., Nash, L. A., Wood, M., et al. (2020). MiR-219a-5p enriched extracellular vesicles induce OPC differentiation and EAE improvement more efficiently than liposomes and polymeric nanoparticles. *Pharmaceutics, 12*(2), 186.
49. Cappellano, G., Woldetsadik, A. D., Orilieri, E., Shivakumar, Y., Rizzi, M., Carniato, F., et al. (2014). Subcutaneous inverse vaccination with PLGA particles loaded with a MOG peptide and IL-10 decreases the severity of experimental autoimmune encephalomyelitis. *Vaccine, 32*(43), 5681–5689.
50. Maldonado, R. A., LaMothe, R. A., Ferrari, J. D., Zhang, A.-H., Rossi, R. J., Kolte, P. N., et al. (2015). Polymeric synthetic nanoparticles for the induction of antigen-specific immunological tolerance. *Proceedings of the National Academy of Sciences of the United States of America, 112*(2), E156–EE65.
51. Cho, J. J., Stewart, J. M., Drashansky, T. T., Brusko, M. A., Zuniga, A. N., Lorentsen, K. J., et al. (2017). An antigen-specific semi-therapeutic treatment with local delivery of tolerogenic factors through a dual-sized microparticle system blocks experimental autoimmune encephalomyelitis. *Biomaterials, 143*, 79–92.
52. Getts, D. R., Martin, A. J., McCarthy, D. P., Terry, R. L., Hunter, Z. N., Yap, W. T., et al. (2012). Microparticles bearing encephalitogenic peptides induce T-cell tolerance and ameliorate experimental autoimmune encephalomyelitis. *Nature Biotechnology, 30*(12), 1217–1224.
53. Führmann, T., Ghosh, M., Otero, A., Goss, B., Dargaville, T. R., Pearse, D. D., et al. (2015). Peptide-functionalized polymeric nanoparticles for active targeting of damaged tissue in animals with experimental autoimmune encephalomyelitis. *Neuroscience Letters, 602*, 126–132.
54. Al-Ghobashy, M. A., ElMeshad, A. N., Abdelsalam, R. M., Nooh, M. M., Al-Shorbagy, M., & Laible, G. (2017). Development and pre-clinical evaluation of recombinant human myelin basic protein nano therapeutic vaccine in experimental autoimmune encephalomyelitis mice animal model. *Scientific Reports, 7*(1), 1–16.
55. Schmidt, J., Metselaar, J. M., Wauben, M. H., Toyka, K. V., Storm, G., & Gold, R. J. B. (2003). Drug targeting by long-circulating liposomal glucocorticosteroids increases therapeutic efficacy in a model of multiple sclerosis. *Brain, 126*(8), 1895–1904.
56. Binyamin, O., Larush, L., Frid, K., Keller, G., Friedman-Levi, Y., Ovadia, H., et al. (2015). Treatment of a multiple sclerosis animal model by a novel nanodrop formulation of a natural antioxidant. *International Journal of Nanomedicine, 10*, 7165.
57. Khalid, M., & El-Sawy, H. S. (2017). Polymeric nanoparticles: Promising platform for drug delivery. *International Journal of Pharmaceutics, 528*(1–2), 675–691.
58. Johnson, D., Hafler, D. A., Fallis, R. J., Lees, M. B., Brady, R. O., Quarles, R. H., et al. (1986). Cell-mediated immunity to myelin-associated glycoprotein, proteolipid protein, and myelin basic protein in multiple sclerosis. *Journal of Neuroimmunology, 13*(1), 99–108.
59. Saito, E., Kuo, R., Kramer, K. R., Gohel, N., Giles, D. A., Moore, B. B., et al. (2019). Design of biodegradable nanoparticles to modulate phenotypes of antigen-presenting cells for antigen-specific treatment of autoimmune disease. *Biomaterials, 222*, 119432.



# Decoding Neurodegeneration: Unveiling Cytokine Dynamics, Neuroinflammation, and the Role of Nanotherapeutic Approaches

Harapriya Baral and Ravinder K. Kaundal

**Abstract**

Neuroinflammation is a cellular process that mediates neuronal damage within the confine central nervous system (CNS), commonly observed as a significant hallmark in various neurodegenerative diseases. The excessive production of proinflammatory cytokines such as interleukins (IL), chemokines, tumor necrosis factor (TNF-α), transforming growth factor (TGF-ß), interferons (IFN), and colony-stimulating factor (CSF), by both neuroglial cells and peripheral immune cells, initiates neuronal cell death cascade and expedites the process of neurodegeneration. Despite the pivotal role of inflammation in the initiation and progression of neurodegenerative diseases, pharmaceutical interventions with anti-inflammatory drugs yield limited therapeutic efficacy, primarily due to their inability to traverse the blood-brain barrier (BBB), poor bioavailability, and various side effects. In recent years, nanoparticles have garnered interest as a strategy to mitigate the immunotoxicity associated with traditional nanomedicine, presenting a promising avenue for safer and more effective BBB penetration. In this book chapter, we provide a comprehensive exploration of diverse cytokine receptor families, elucidating their physiological functions and examining their involvement in neurodegenerative diseases. Furthermore, we delve into nanotechnology interventions targeting inflammation-related neurodegenerative diseases, offering insights into both their applications and limitations.

**Keywords**

Cytokines · Neuroinflammation · Neurodegenerative diseases · Nanoparticles

H. Baral · R. K. Kaundal (✉)
Department of Pharmacology and Toxicology, National Institute of Pharmaceutical Education and Research (NIPER)-Raebareli, Lucknow, UP, India
e-mail: ravinder.kaundal@niperraebareli.edu.in

 
R. Shukla et al. (eds.), *Neuroendocrine-Immune Axis*,
https://doi.org/10.1007/978-981-95-5088-3_3

## 1 Introduction

Cytokines, a diverse group of small signaling proteins with molecular weights ranging from 6 to 70 kDa, are indispensable mediators in the intricate landscape of cell communication and immune response. Within the cerebral milieu, resident glial cells, comprising astrocytes and microglia, collaboratively interact with neurons to precisely orchestrate the release of cytokines [1]. Additionally, elements of the neurovascular unit, such as vascular endothelial cells, infiltrating leukocytes, and pericytes contribute to the release of these signaling proteins. The diversity in the cellular origins of cytokines underscores the multifaceted dynamics governing immune responses. These molecules play a crucial role in weaving a complex network of interactions essential for overall homeostasis, regulating immune function, and maintaining physiological balance within the central nervous system (CNS) [2]. Beyond their immunomodulatory role, cytokines wield a significant influence on neurotransmitter metabolism, impacting the synthesis, reuptake, and release of neurotransmitters. This influence contributes to the precise regulation of neuronal signaling [3]. Cytokines also play crucial roles in regulating neurogenesis and astrocyte differentiation [4]. Additionally, they are involved in maintaining synapse homeostasis and modulating synaptic plasticity, thereby regulating cognitive function, learning, and memory processes [5]. Cytokines are also reported to be involved in maintaining synapse homeostasis by facilitating the process of microglial synapse engulfment [6]. Furthermore, cytokines actively regulate the permeability of the blood-brain barrier (BBB), exerting control over its structural components [7]. This regulatory function underscores their intricate role in modulating substance exchange between the bloodstream and the brain, impacting overall neuronal homeostasis. Cytokines are involved in neuronal proliferation, differentiation, and cell death, which play an essential role in neuronal development and maintenance [8]. This chapter presents a comprehensive exploration of diverse cytokine receptor families, elucidating their physiological functions, and examining their involvement in neuroinflammation across various CNS disorders. Additionally, we delve into nanotechnology interventions targeting inflammation-related neurological diseases, offering insights into both their applications and limitations. Through this meticulous exploration, we aim to enhance our understanding of the nuanced impact of cytokines on the intricate balance of the CNS.

## 2 Cytokine Receptor Family

Cytokines showcase a diverse classification based on both their cellular source and structural characteristics. From a cellular perspective, they are categorized into two primary types: Type-I and Type-II cytokines. Additionally, cytokines undergo classification based on their structural attributes or specific functions. In terms of structure, they fall into categories such as chemokines, interferons (IFN), interleukins (IL), transforming growth factor (TGF-ß), colony-stimulating factor (CSF), and tumor necrosis factor (TNF-α). Each of these structural classifications corresponds to distinct roles in immune regulation, antiviral responses, and cell growth. Functionally, cytokines can be classified as proinflammatory or anti-inflammatory. Proinflammatory cytokines, including IL-1β, IL-6, IL-8, IL-12, and TNF-α, contribute to inflammatory reactions. On the other hand, anti-inflammatory cytokines like IL-4, IL-10, IL-13, and IL-1R antagonists work to inhibit inflammation.

Cytokines execute their biological functions by binding to specific cell surface receptors on target cells. The cytokine receptor family encompasses a wide range of receptors, each specific to different cytokines. These receptors are classified into several types based on their structural and functional characteristics. The main types include (1) hematopoietin and interferon family (well known as type-I and type-II cytokine receptor family respectively), (2) IL-1R and associated Toll-like receptors (TLR), (3) receptor tyrosine kinase (IL-17 receptor), (4) receptor serine kinase (TGF-β receptor), (5) TNF-α receptor family, and (6) chemokines receptor family [9]. Table 1 presents an in-depth depiction of various cytokine receptor families and their corresponding physiological functions. This detailed representation offers a comprehensive overview of diverse receptors, outlining their roles and contributions to physiological processes.

## 3 Cytokine Signaling

Figure 1 illustrates the classification of cytokines into major families based on their structural interactions with receptors. The primary cytokine families include Type-I/II cytokines, IL-1 cytokines, IL-17 cytokines, TGF-β cytokines, and chemokines. In the Type-I/II cytokines family, janus kinase (JAKs) and signal transducer and activator of transcription (STATs) are activated, leading to receptor phosphorylation. Phosphorylated STATs form dimers, translocating to the nucleus and initiating transcription. The IL-1 cytokines family involves IL-1R1 and IL-1R3 activation, forming myeloid differentiation primary response 88 (MyD88) oligomers that recruit interleukin-1 receptor-associated kinase 4 (IRAK-4) and IRAK-1. Myddosome-induced IRAK-1 oligomers trigger auto-hyperphosphorylation, leading to polyubiquitination and tumor necrosis factor receptor-associated factor 6 (TRAF6) recruitment. Ubiquitinated TGF-beta activated kinase 1 binding protein 2 (TAB2) and TAB3 activate MAP kinase kinase (MKKs), regulating gene expression. In the IL-17 cytokines family, IL-17R with IL-17RA and IL-17RC subunits binds to ligands, and similar expression to fibroblast growth factor genes (SEFIR) domains interact with adaptor protein 1 (Act1). Act1, with a TRAF-binding site associated with TRAF6, activates NFκβ and MAPK: AP1 pathways for transcriptional induction of target genes. The TGFβ cytokine family involves TGFβ binding, activating TGFβR and phosphorylating mothers against decapentaplegic homolog 2 (SMAD2) and TRAF6. This process regulates transcriptional gene modifications in proliferation, differentiation, and apoptosis. In the chemokine receptor family, receptors activate Gαi, inhibiting adenylate cyclase and reducing cyclic adenosine monophosphate (cAMP). Gαi also activates Src for extracellular signal-regulated kinase (ERK1/2) phosphorylation. Gβγ signaling, crucial for chemotaxis, directly

**Table 1** Cytokine receptor families and their physiological function

| SL.No. | Cytokine receptor family | Cytokines | Function | Ref. |
|---|---|---|---|---|
| 1 | Hematopoietin family (Type-I) | IL-6 | Anti-inflammatory effect by classical signaling, e.g., organ development, acute phase response, metabolic regulation, and immune response | [10] |
| | | | Proinflammatory effect by trans signaling, e.g., leukocyte infiltration, T cell apoptosis inhibition, and inhibits the action of regulatory T cell | |
| | | IL-11 | Stimulates hematopoiesis and platelet production | [11] |
| | | IL-27 | Proinflammatory action by CD4$^{+}$ T cell proliferation, Th1 polarization, NK cell production, and macrophage activation | [12] |
| | | | Anti-inflammatory action by inducing programmed cell death | |
| | | IL-3 | Development of basophils, mast cells, and eosinophils | [13] |
| | | IL-5 | Mediates activation, proliferation, and survival of eosinophils | [14] |
| | | Granulocyte macrophage (GM)-CSF | Proliferation of granulocytes, monocyte-promoting, dendritic cell recruitment, and maturation | [15] |
| | | IL-2 | T cell proliferation, early stage of effector differentiation, and mediates activation-induced cell death | [16] |
| | | IL-4 | Differentiation of T cells into Th2 and also mediates a positive feedback loop in Th2 | [17] |
| | | IL-7 | Homeostatic T cell proliferation, differentiation, and survival | [18] |
| | | IL-9 | Promotes proliferation and accumulation of mast cells and leukocytes | [19] |
| | | IL-15 | Differentiation and proliferation of T cells, NK cells, maturation of dendritic cells and macrophages | [20] |
| | | IL-21 | Promotes production of antibodies from B cell, supports cytotoxicity of CD8$^{+}$ T cell | [21] |
| | | IL-12 | INF-γ production, Th1 differentiation | [22] |
| | | IL-23 | Recruits neutrophils, development and maintenance of Th17 cells | [23] |
| | | IL-35 | Differentiation and proliferation of Treg cells that inhibit activation of Th17 cells | [24] |
| | | IL-13 | Regulate response of lymphocyte, myeloid cell, and macrophage activation | [25] |
| | | IL-31 | Regulate cell proliferation and hematopoiesis, promote secretion of chemokines and MMPs | [26] |

| | | | | |
|---|---|---|---|---|
| 2 | IFN family (Type-II) | IFN-α/β | Induces chemokinesis of T cells and its migration, antiviral activity against HCV and HEV | [27] |
| | | IFN-γ | Antiviral activity against the dengue virus increases MHC class II expression | |
| | | IL-10 | Suppresses the positive role of macrophages in innate immunity<br>Suppresses release of proinflammatory mediators from TNF-α, IL-1β, and GM-CSF | [28] |
| | | IL-19, 20, 22,24,26 | Proliferation and differentiation of keratinocytes, activation of epithelial cells, production of inflammatory cytokines | [29] |
| | | IL-28, 29,30 | Antiviral activity, i.e., inhibition of viral replication | [30] |
| 3 | (IL-1 and TLR family | IL-1α/β | Increase cytokines production and phagocytosis, neutrophils survival | [31] |
| | | IL-18 | Mast cell maturation and survival<br>Increased cytokines and histamine production | |
| | | IL-33 | Production of IFN-γ and Th2-type cytokines, eosinophils survival | |
| 4 | Receptor tyrosine kinase (IL-17 receptor) | IL-17A | Wound healing, epithelial proliferation, inflammation to combat infection, and ECM remodeling | [32] |
| | | IL-17B, C, D | | |
| | | IL-17E | | |
| | | IL-17F | | |
| 5 | Receptor serine kinase (TGF-β receptor) | TGF-β1, β2, β3 | Regulation of T cell proliferation, survival, and homing | [33] |
| | | IL-32 | Induce degradation of κβ and phosphorylation of mitogen-activated protein p38 | [34] |
| | | IL-16 | T cell growth facilitates antigen recognition by T cell migration of Treg cells | [35] |
| 6 | Chemokine receptor family | CXC (CXCL-1 to CXCL-17) | Neutrophil, NK and CD8 trafficking, Th1 response, NKT, and ILC migration and survival | [36] |
| | | XCL-1 & XCL-2 | Cross-presentation by CD8$^+$ DC | |
| | | CX3CL-1 | T cell and monocyte migration | |
| | | CC (CCL-1 to CCL-28) | Th2, Treg and monocyte trafficking, migration of Th2, eosinophils, basophils and Treg, T cell/DC interaction | |

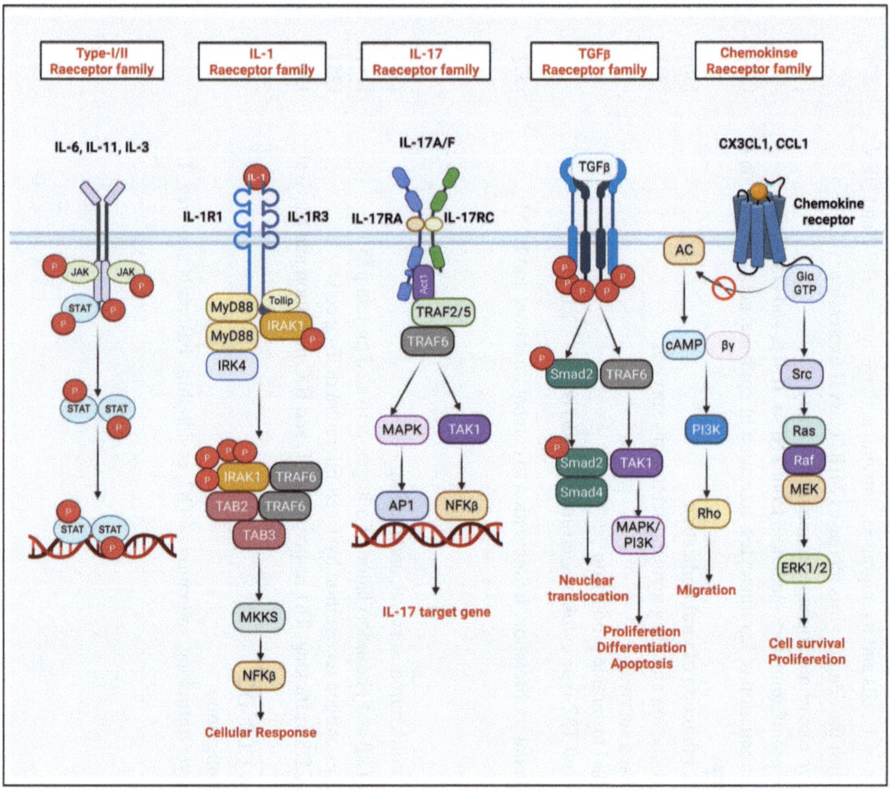


**Fig. 1** Cytokine receptor signaling

activates phosphatidylinositol 3-kinase (PI3K), triggering Rho activation involved in cell migration and proliferation.

## 4 Role of Cytokines in Neuroinflammation

Neuroinflammation is the brain's defense mechanism, coordinating microglia, astrocytes, and neurons to form a protective network, that preserves the functionality of CNS in response to potential injuries [37]. However, the severity and persistent neuroinflammation also play a crucial role in the emergence and progression of chronic neurological disorders by modulating various cellular processes and signaling pathways [38].

Neuroinflammation occurs when immune cells enter the brain due to a damaged BBB that activates microglia and astrocytes then produce proinflammatory cytokines (IL-1α/β, IL-6, GM-CSF, and TNF-α), contributing to brain tissue damage [39]. In addition to activated neuron, glial cells also play a crucial role in influencing the integrity of BBB [40]. Increasing evidence suggests that the breakdown of BBB is a prevalent characteristic of neurodegeneration mediated by neuroinflammation. Patients with neurodegenerative conditions like Alzheimer's disease (AD) [41], multiple sclerosis (MS) [42], Parkinson's disease (PD) [43], and amyotrophic lateral sclerosis (ALS) [44] exhibit observable disruption in the BBB that mediates neuroinflammation. In addition to this, injury, infection, or psychological stress also conciliate proinflammatory cytokines production and turn-out changes in brain function and behavior through the inflammatory response [45]. Research on animals suggested that chronic unpredictable mild stress (CUMS) increases levels of IL-1β, IL-6, and TNF-α in the bloodstream, and repeated social defeat stress leads to elevated circulatory IL-6 levels [46]. This chapter extensively delves into the multifaceted roles that cytokines play in significant neurodegenerative conditions. It illuminates the intricate interaction between neuroinflammation and cytokines, offering crucial insights into neurodegeneration mechanisms for targeted treatment strategies.

## 4.1 Alzheimer's Disease (AD)

AD is the leading cause of dementia, featuring a complex pathophysiological mechanism characterized by the accumulation of amyloid-β and hyperphosphorylation of tau protein. Notably, inflammation emerges as the third most prevalent mechanism contributing to AD development, intricately associated with the amyloidogenesis process. The inflammatory aspects involve activated microglia, astrocytes, and peripheral blood cells that release inflammatory mediators [47]. Microglia get activated upon immune response mediated by amyloid-β plaque and fail to start the repair mechanism (i.e., amyloid-β clearance) by overproduction of proinflammatory cytokines. Whereas astrocytes, the neuro-supportive cell, fail to modulate neurotransmitter recycling and metabolic support upon prolonged exposure of amyloid-β in AD brain [48]. Proinflammatory cytokines, IL-1α/β, IL-6, and GM-CSF, levels have been reported to be increased in PSI/APPsw mouse brains that accumulate more amyloid-β, which may suggest neuroinflammation can be driven by amyloid-β deposition [49]. Shirong et al.( [50]) have reported that TLR2 enhances amyloid-β-directed inflammation in HEK293 cells in-vitro and presented TLR-2 as the primary receptor for amyloid-β aggregates. Peripheral blood cells have gained attention in AD pathogenesis as inflammatory cytokines IFN-γ, IL-1β, vascular cell adhesion molecule (VCAM), and TNF-α are reported to be raised in the blood of AD patients [51]. The hyperphosphorylation of tau protein is regulated by receptors (CX3CR1 and TREM2) on the microglia cell surface by holding post-transcriptional modification and propagation of tau protein [52]. Proinflammatory cytokines TNF-α and IL-1β mediate synaptic deficit in response to amyloid-β through the p38 mitogen-activated protein kinase (MAPK) cascade, leading to cognitive and memory impairment, key symptoms of AD [53].

## 4.2 Parkinson's Disease (PD)

PD is typically characterized by disrupted dopaminergic neurotransmission, presence of Lewy bodies containing α-synuclein protein, and neuroinflammation [54]. Abnormal immune regulation in the periphery and brain intercedes inflammatory cascade, and IL-1β, IFN-γ, IL-6, CCL-2, CXCL-8, and TNF-α cytokine levels are found to be increased in both serum and brain that results in neurotoxicity correlated with PD [55]. Loss of dopaminergic neurons in substantia nigra pars compacta (*SNpc*) is arbitrated by proinflammatory cytokines IL-1α, IL-1β, IFN-γ, IL-6, and TNF-α, and inhibition of inflammatory signal can save neuronal loss [56].

## 4.3 Ischemic Brain Injury

Ischemic brain injury is a major cause of death and disability globally. It is characterized by cerebral hypoperfusion, which causes a lack of nutrients and oxygen, leading to neuronal cell death [57]. Ischemic brain injury engages natural and adaptive immunity with diverse immune cells such as microglia, neutrophils, lymphoid cells, and dendritic cells, contributing to neuroinflammation [58]. Activated microglia increase reactive oxygen species (ROS) and inflammatory cytokines like IL-1β, TNF-α, and IFN-γ, amplifying cytotoxic responses [59]. On the other hand, neutrophils, key ROS sources, can harm the neurovascular unit and BBB, leading to edema and hemorrhagic transformation. Neutrophil-released proteases like matrix metalloproteinase (MMP) and elastase worsen BBB impairment by degrading the basement membrane and extracellular matrix [60]. In the brain parenchyma, infiltrating T cells induce ischemic damage by releasing proinflammatory agents, including ROS, IL-21, IL-17, and IL-1β, that contribute to neurovascular unit cell destruction and direct neuronal injury [61].

## 4.4 Multiple Sclerosis (MS)

MS is a persistent inflammatory condition affecting the CNS, marked by inflammation and neurodegeneration. This can result in clinical impairments, including motor, sensory, and cognitive symptoms [62]. Continuous production of cytokines is a persistent characteristic observed throughout the progression of MS. Numerous investigations indicated an elevation of proinflammatory cytokines in serum and cerebrospinal fluid. For instance, elevated levels of TNFα have been reported in active lesions within the CNS and detected in the serum and CSF of individuals with MS, which is associated with the severity of the disease [63]. Experimental studies demonstrating the effectiveness of blocking TNFα in alleviating MS pathology lent strong support to the pathogenic role of proinflammatory cytokines [64]. Among patients with MS, there is an increase in GM-CSF production by $CD4^+$ T cells, and application of immunotherapy reduces the frequencies of T cells that produce GM-CSF [65]. This approach suggests a way to manage the immune response in

MS. However, IFN-β acts as a protective cytokine, regulating inflammation. Research in experimental autoimmune encephalomyelitis (EAE) has demonstrated that IFN-β has the ability to inhibit immune-mediated demyelination. Notably, IFN-β1b was the initial sanctioned medication for individuals with MS, diminishing the intensity and frequency of relapses and slowing disease progression [66]. This underscores the substantial therapeutic effectiveness of IFN-β in managing MS. Both preclinical and clinical data emphasize the significant involvement of cytokines in MS. Yet, due to functional redundancy often seen in cytokines, development of innovative therapies is challenging.

## 5 Nanomedicine Intervention in Neuroinflammation

The initiation of inflammatory cascades, coupled with the activation of glial cells, leading to the release of proinflammatory cytokines, contributes to the accumulation of misfolded proteins observed in neurodegenerative disease. Mitigating inflammation has the potential to ameliorate these pathological effects, suggesting that immunomodulation could be a viable therapeutic intervention for the treatment of neurodegenerative disorders [67]. While specific medications and biological agents can decrease inflammation, their effectiveness is hindered by challenges such as traversing the BBB and blood-cerebrospinal-fluid barrier.

However, utilizing nanomedicine is an efficient strategy for facilitating drug delivery across the BBB that enhances the distribution and pharmacokinetics of drugs when administered orally, intravenously, or intranasally. Therefore, significant research has been conducted on nanomedicine for treating neurodegenerative disorders stemming from inflammation [68]. Nanotechnology provides diverse methods and technologies for therapeutic administration and diagnostic imaging in immunomodulation [69]. Nanotechnology facilitates drug delivery to the brain in neurological disorders by utilizing biodegradable nanocarriers capable of crossing BBB, and their surface modification enhances compatibility with loaded drugs [70]. Nanotechnology methods hold potential in detecting, tracking, and treating inflammation-derived pathogenesis of neurodegenerative disease. The detection of biomarkers (CHIL31, IL-8, MCP-1, sICAM1, sFLT1, and IL-1β) in neuroinflammation is achieved by nanosensors, capable of identifying cytokines, thereby allowing early detection of neuroinflammation [71]. Researchers have explored nanoparticles (NPs)as potential carriers for delivering drugs targeting microglia and astrocytes, crucial participants in the inflammatory response. An example of the nanostructure is apolipoprotein E3-reconstituted high-density lipoprotein (ApoE3-rHDL) that binds to amyloid-β and mediates its clearance and mitigates microglial activation in AD [72]. Beyond drug delivery, NPs can transport genetic material to the brain, potentially targeting genes associated with inflammatory responses. As an illustration, extracellular vesicles containing long intergenic non-coding RNA-cyclooxygenase-2 (LincRNA-Cox2) have been reported to show the ability to regulate lipopolysaccharide (LPS)-induced microglial proliferation in BV-2 cells and C57BL/6 N mice [73].

Some researchers have raised concerns regarding the potential toxicity of NPs, necessitating careful consideration of the neurotoxic effect of NPs in treating neurological diseases. Ongoing research examines limitations such as profiling immune responses to diverse non-materials for immunotoxicity, cost-effectiveness, the feasibility of scaling up, and stability. The potential neurotoxicity of NPs is closely associated with several factors, including their shape, size, dosage, and surface area [74]. Despite the advantages of NPs in precise therapeutic targeting, they may face challenges like particle aggregation limiting drug loading. The dosage and drug delivery process of NPs could potentially contribute to increased neuroinflammation, resulting in oxidative stress and DNA damage [75]. It is essential to have a comprehensive understanding of the biocompatibility of nanomedicine that ensures the safety efficacy and minimizes adverse effects. Commonly utilized NPs have undergone scrutiny for potential neurotoxic effects. For instance, prolonged exposure to engineered silver NPs (AgNPs) worsens symptoms such as elevated body temperature, excessive salivation, and lethargy and demonstrates increased disruption of BBB. It also leads to the formation of brain edema, deterioration of cognitive and motor functions, and results in brain damage [76]. Research findings also reported that the oral administration of high-dose TiO2NPs in Wistar rats resulted in the elevation of IL-6 levels in both plasma and brain, showing the potential to induce neuroinflammation [77]. Studies have shown that introducing zinc oxide (ZnO) and titanium dioxide (TiO2) NPs into the brain affects mitochondria. Specifically, this includes mitochondrial swelling and fragmentation, indicating a potential alteration in mitochondrial structure and function [78]. In recent years, biomimetic NPs have garnered interest as a strategy to mitigate the immunotoxicity associated with traditional NPs This innovative technology entails enveloping NPs with cell membranes to prevent phagocytosis and minimize other potentially harmful immune responses. By mimicking natural cellular structure, biomimetic NPs seek to enhance biocompatibility and reduce adverse effects, presenting a promising avenue for safer and more effective nanomedical application [79]. The majority of NPs display anti-inflammatory effect, alone or in combination with drug. However, certain NPs have the potential to trigger neuroinflammation.

## 5.1 NPs Delivering Drug to Brain

The predominant focus of nanomedicine revolves around NPs with a size of less than 1 μm, carrying a variety of drugs designed for controlled release. Minorities of NPs are generated in contrast to free drugs, namely organic nanocarriers (polymeric NPs with drugs, micelles, dendrimers, liposomes, solid-lipid NPs (SLNs), and nanoemulsion (NE) and inorganic nanocarriers (gold nanoparticle, iron-oxide NPs, silica NPs, silver NPs, and carbon nano-tubes) [80]. In Tables 2 and 3, we have detailed a diverse range of NPs, highlighting their distinct function and compositions tailored for targeting the CNS. This comprehensive overview provides insight into the nuanced characteristics of each nanoparticle type, shedding light on their specific roles and structures designed to address CNS-related applications.

**Table 2** Organic NPs targeting CNS

| NPs | Structure | Size range | Biological model | Drug combination | Therapeutic effect/mechanism | Ref. |
|---|---|---|---|---|---|---|
| Liposomes | They are made up of hydrophilic components at the center surrounded by lipid bilayers | 10–100 nm | LPS-induced inflammation of HMC3 cells | Lipo-curcumin | Downregulation of glial cell proliferation, suppression of IL-6, IL-1β, TNF-α, and TGF-β | [81] [82] |
| | | | MCAO stroke model in rats | Lipo-cyclosporine A | Inhibits TNF-α level, MPO activity, and suppresses inflammation | [83] |
| Polymeric NPs (PNPs) | The central part consists of a compact polymer matrix that encases lipophilic drugs | 10–250 nm | LPS-induced BV2 cells | Alginate-derived oligosaccharide (AdO) | Reduces NO, PGE2, TLR4, and c expression, thereby attenuating inflammation | [84] [85] |
| | | | OGD-induced primary neuronal cells<br>MCAO stroke model in C57BL/6 mice | PLGA with superoxide dismutase (SOD) | Reduces apoptosis and inflammatory markers IL-1β, TNF-α, and TGF-β | [86] |
| | | | LPS-induced BV2 cell<br>Neuropathic pain by spinal-nerve ligation in rat | Foxp3 plasmid encapsulates PLGA NPs | Alleviates microglia activity by reducing proinflammatory cytokines | [87] |
| Solid-lipid NPs (SLNs) | Aqueous colloidal mixture encompassing triglycerides/fatty acids | | LPS-induced BV2 microglial cells | Curcumin-loaded SLN | Inhibits NO and proinflammatory cytokines IL-1β, TNF-α, and IL-6 | [88] [89] |
| | | | 5xFAD mouse model of AD | Curcumin-loaded SLN | Anti-amyloid, anti-inflammatory, and neuroprotective | [90] |

(continued)

**Table 2** (continued)

| NPs | Structure | Size range | Biological model | Drug combination | Therapeutic effect/mechanism | Ref. |
|---|---|---|---|---|---|---|
| Micelles | Amphiphilic<br>The outer core comprises PEG and the inner core of polypropylene glycol, polycaprolactone, and fatty acids | 5–50 nm | Ischemic stroke model | Rapamycin in self-assembled micelles with ROS-responsive and fibrin-binding polymers | M2 microglia polarization and BBB preservation | [91]<br>[92] |
| NE | Colloidal system comprised of edible oil, surfactant and water | 100–500 nm | LPS-induced rat astrocyte primary culture | Chitosan-coated rosmarinic acid NE | Increases astrocyte viability and reduces necrosis, antioxidant, anti-inflammatory action | [93] |
| | | | LPS-induced J774A.1 murin macrophage | Anti-TNFα siRNA-NE | Reduced the level of TNFα | [94] |
| Dendrimers | Polymeric structure with branches<br>Example: Polypropylene imine, polyamidoamine (PAMAM) and poly-L-lysine | 1–15 nm | MOG35–55-induced EAE in mice | ABP dendrimers | Promotes IL-10 producing CD4 + T cells | [95]<br>[96]<br>[97] |
| | | | CCI model of TBI on mice | ShCCL20-CCR6 nanodendriplexes with MSC | Downregulates CCL20-CCR6 expression and IL-6<br>Promotes neurogenesis | [98] |
| | | | LPS-induced BV2 murin microglia | PEGOL-60 | Decreases TNFα, IL-6, IL-10 and iNOS.<br>Upregulates CD206, Arg1and IL-4 level | [99] |
| | | | LPS-activated murin macrophages and Rabbit model of pediatric TBI by CCI | Sinomenine with hydroxyl terminated generation-4 PAMAM dendrimer | Suppression of TNFα, IL-6, IL-1β and CCL-3.<br>Inhibition of NFκβ activation and nuclear translocation | [100] |

**Table 3** Inorganic NPs targeting CNS

| NPs | Structure | Size range | Biological model | Drug combination | Therapeutic effect/mechanism | Ref. |
|---|---|---|---|---|---|---|
| AuNPs | Colloidal gold particles | 1 nm–100 nm | Ex-vivo brain slice stroke model | AuNPs | Inhibits TNF-α and NFκβ signaling and induces M2 microglia polarization | [101] |
| | | | Okadaic acid-induced AD | AuNPs | Increases anti-inflammatory IL-4 level | [102] |
| | | | LPS-induced BV-2 cells | Au-quercetin NPs | Decreases COX2 and iNOS expression | [103] |
| | | | BV-2 cells | Anthocyanin with AuNPs | Inhibits p-JNK/NFκβ/p-GSK3β-mediated apoptosis | [104] |
| IONPs | Magnetic iron-oxide core magnetite/ hematite/maghemite with biocompatible surface | 1 nm–100 nm | Murine microglial cells | Resovist | Inhibits LPS-induced IL-1β generation | [105] [106] |
| SiO2NPs | Conjugates with PEG that enhance BBB penetration | 25–100 nm | Rat hippocampus | Sub-acute exposure of SiNPs | Increases mRNA expression of TNF-α, IL-β, IL-6, and COX2 | [107] [108] |
| Carbon nanotubes (CNTs) | A tubular structure formed with folding layer of graphene | SWCNT 0.4–3 nm, MWCNT 2–30 nm | LPS-induced HAPI cells | PPy/CNTb/ Dexamethasone film-coated electrode | Attenuates LPS-induced microglial activation | [109] [110] |
| Graphene quantum dots (GQDs) | Graphene nano-sheets with ten graphene layer | <100 nm | Chronic relapsing EAE in Dark Agouti (DA) rats | GQD as treatment | Activation of MAPK/Akt signaling and suppression of IL-10, IL-17, IFN-γ, and Th1 immune response | [111] |

## 6 Conclusion

Cytokine mediates various cellular mechanisms crucial for the brain's normal physiological function. Despite this, persistent and intense neuroinflammation drives the progression of neurological disorders marked by the excessive release of proinflammatory cytokines. Some drugs and biological agents can potentially reduce inflammation, but obstacles like crossing BBB impede their efficacy. NPs serve as a promising delivery system, overcoming enzymatic degradation, improving targeting, and reducing side effects. However, researchers have expressed concern about the potential toxicity of NPs and challenges such as particle aggregation limiting drug loading. The dosage and drug delivery processes of NPs may potentiate neuroinflammation, leading to oxidative stress and DNA damage followed by neuronal cell death. Despite progress, understanding the drug kinetics and dynamics of these NPs is limited. A comprehensive exploration of their side effects is necessary.

## References


1. Liu, C., et al. (2021). Cytokines: From clinical significance to quantification. *Advancement of Science, 8*, 2004433.
2. Yang, L., Huh, J. R., & Choi, G. B. (2023). One messenger shared by two systems: How cytokines directly modulate neurons. *Current Opinion in Neurobiology, 80*, 102708.
3. Miller, A. H., Haroon, E., Raison, C. L., & Felger, J. C. (2013). Cytokine targets in the brain: Impact on neurotransmitters and neurocircuits. *Depression and Anxiety, 30*, 297–306.
4. Stolp, H. B. (2013). Neuropoietic cytokines in normal brain development and neurodevelopmental disorders. *Molecular and Cellular Neurosciences, 53*, 63–68.
5. Godukhin, O. V., & Levin, S. G. (2017). Modulating effect of cytokines on mechanisms of synaptic plasticity in the brain. *Biochemistry (Moscow), 82*, 264–274.
6. Vainchtein, I. D., et al. (2018). Astrocyte-derived interleukin-33 promotes microglial synapse engulfment and neural circuit development. *Science, 359*, 1269–1273.
7. Almutairi, M. M. A., Gong, C., Xu, Y. G., Chang, Y., & Shi, H. (2016). Factors controlling permeability of the blood–brain barrier. *Cellular and Molecular Life Sciences, 73*, 57–77.
8. Tsai, S.-J. (2017). Effects of interleukin-1beta polymorphisms on brain function and behavior in healthy and psychiatric disease conditions. *Cytokine & Growth Factor Reviews, 37*, 89–97.
9. Thomson, A. W., & Lotze, M. T. (2003). *The cytokine handbook, two-volume set*. Elsevier.
10. Vilotić, A., et al. (2022). IL-6 and IL-8: An overview of their roles in healthy and pathological pregnancies. *International Journal of Molecular Sciences, 23*, 14574.
11. O'Reilly, S. (2023). Interleukin-11 and its eminent role in tissue fibrosis: A possible therapeutic target. *Clinical and Experimental Immunology, 214*(2), 154–161.
12. Beizavi, Z., Zohouri, M., Asadipour, M., & Ghaderi, A. (2021). IL-27, a pleiotropic cytokine for fine-tuning the immune response in cancer. *International Reviews of Immunology, 40*, 319–329.
13. Varricchi, G., Poto, R., Marone, G., & Schroeder, J. T. (2021). IL-3 in the development and function of basophils. *Seminars in Immunology, 54*, 101510.
14. Bachert, C., et al. (2023). The unified airway hypothesis: Evidence from specific intervention with anti-interleukin-5 biologic therapy. *The Journal of Allergy and Clinical Immunology: In Practice, 11*(9), 2630–2641.
15. Wang, M., et al. (2023). A GM-CSF and DOX co-delivery nanoplatform modulates macrophage polarization to promote tumor suppression. *JCIS Open, 9*, 100081.

16. Beck, C., et al. (2023). Development of a TGFβ—IL-2/15 switch receptor for use in adoptive cell therapy. *Biomedicine, 11*, 459.
17. Elisseeff, J. H., Baker, D., & Spangler, J. B. (2023). Design of cell-type-specific hyperstable IL-4 mimetics via modular de novo scaffolds. *Nature Chemical Biology, 19*, 1127.
18. Marton, C., et al. (2023). IL-7 producing immunotherapy improves ex vivo T cell functions of immunosenescent patients, especially post hip fracture. *Human Vaccines & Immunotherapeutics, 19*, 2232247.
19. Tan, C., & Gery, I. (2012). The unique features of Th9 cells and their products. *Critical Reviews in Immunology, 32*, 1.
20. Yoshida, S., et al. (2022). Interleukin-15 receptor subunit alpha regulates interleukin-15 localization and protein expression in skeletal muscle cells. *Experimental Physiology, 107*, 222–232.
21. Schroderus, A.-M., et al. (2023). Evaluation of plasma IL-21 as a potential biomarker for type 1 diabetes progression. *Frontiers in Immunology, 14*, 1157265.
22. Ullrich, K. A.-M., et al. (2020). Immunology of IL-12: An update on functional activities and implications for disease. *EXCLI Journal, 19*, 1563.
23. Feng, B., et al. (2023). PDE4D/cAMP/IL-23 axis determines the immunotherapy efficacy of lung adenocarcinoma via activating the IL-9 autocrine loop of cytotoxic T lymphocytes. *Cancer Letters, 565*, 216224.
24. Li, Y., Yao, L., & Lu, J. (2023). IL-35 inhibits adipogenesis via PPARγ-Wnt/β-catenin signaling pathway by targeting Axin2. *International Immunopharmacology, 122*, 110615.
25. Junttila, I. S. (2018). Tuning the cytokine responses: An update on interleukin (IL)-4 and IL-13 receptor complexes. *Frontiers in Immunology, 9*, 888.
26. Zhang, Q., Putheti, P., Zhou, Q., Liu, Q., & Gao, W. (2008). Structures and biological functions of IL-31 and IL-31 receptors. *Cytokine & Growth Factor Reviews, 19*, 347–356.
27. Li, S., et al. (2018). Type I interferons: Distinct biological activities and current applications for viral infection. *Cellular Physiology and Biochemistry, 51*, 2377–2396.
28. Sabat, R., et al. (2010). Biology of interleukin-10. *Cytokine & Growth Factor Reviews, 21*, 331–344.
29. Autieri, M. V. (2018). IL-19 and other IL-20 family member cytokines in vascular inflammatory diseases. *Frontiers in Immunology, 9*, 700.
30. Witte, K., Witte, E., Sabat, R., & Wolk, K. (2010). IL-28A, IL-28B, and IL-29: Promising cytokines with type I interferon-like properties. *Cytokine & Growth Factor Reviews, 21*, 237–251.
31. Sims, J. E., & Smith, D. E. (2010). The IL-1 family: Regulators of immunity. *Nature Reviews. Immunology, 10*, 89–102.
32. McGeachy, M. J., Cua, D. J., & Gaffen, S. L. (2019). The IL-17 family of cytokines in health and disease. *Immunity, 50*, 892–906.
33. Travis, M. A., & Sheppard, D. (2014). TGF-β activation and function in immunity. *Annual Review of Immunology, 32*, 51–82.
34. Dinarello, C. A., & Kim, S. (2006). IL-32, a novel cytokine with a possible role in disease. *Annals of the Rheumatic Diseases, 65*, iii61–iii64.
35. Richmond, J., Tuzova, M., Cruikshank, W., & Center, D. (2014). Regulation of cellular processes by interleukin-16 in homeostasis and cancer. *Journal of Cellular Physiology, 229*, 139–147.
36. Palomino, D. C. T., & Marti, L. C. (2015). Chemokines and immunity. *Einstein (São Paulo), 13*, 469–473.
37. Gilhus, N. E., & Deuschl, G. (2019). Neuroinflammation—A common thread in neurological disorders. *Nature Reviews. Neurology, 15*, 429–430.
38. Carson, M. J., Thrash, J. C., & Walter, B. (2006). The cellular response in neuroinflammation: The role of leukocytes, microglia and astrocytes in neuronal death and survival. *Clinical Neuroscience Research, 6*, 237–245.

39. Brown, C. M., Mulcahey, T. A., Filipek, N. C., & Wise, P. M. (2010). Production of proinflammatory cytokines and chemokines during neuroinflammation: Novel roles for estrogen receptors α and β. *Endocrinology, 151*, 4916–4925.
40. Takata, F., Nakagawa, S., Matsumoto, J., & Dohgu, S. (2021). Blood-brain barrier dysfunction amplifies the development of neuroinflammation: Understanding of cellular events in brain microvascular endothelial cells for prevention and treatment of BBB dysfunction. *Frontiers in Cellular Neuroscience, 15*, 661838.
41. Bowman, G. L., et al. (2007). Blood–brain barrier impairment in Alzheimer disease: Stability and functional significance. *Neurology, 68*, 1809–1814.
42. Waubant, E. (2006). Biomarkers indicative of blood-brain barrier disruption in multiple sclerosis. *Disease Markers, 22*, 235–244.
43. Gray, M. T., & Woulfe, J. M. (2015). Striatal blood–brain barrier permeability in Parkinson's disease. *Journal of Cerebral Blood Flow and Metabolism, 35*, 747–750.
44. Zlokovic, B. V. (2008). The blood-brain barrier in health and chronic neurodegenerative disorders. *Neuron, 57*, 178–201.
45. Konsman, J. P. (2022). Cytokines in the brain and Neuroinflammation: We didn't starve the fire! *Pharmaceuticals, 15*, 140.
46. López-López, A. L., et al. (2016). Chronic unpredictable mild stress generates oxidative stress and systemic inflammation in rats. *Physiology & Behavior, 161*, 15–23.
47. Hommet, C., et al. (2014). Neuroinflammation and β amyloid deposition in Alzheimer's disease: In vivo quantification with molecular imaging. *Dementia and Geriatric Cognitive Disorders, 37*, 1–18.
48. Thal, D. R. (2012). The role of astrocytes in amyloid β-protein toxicity and clearance. *Experimental Neurology, 236*, 1–5.
49. Patel, N. S., et al. (2005). Inflammatory cytokine levels correlate with amyloid load in transgenic mouse models of Alzheimer's disease. *Journal of Neuroinflammation, 2*, 1–10.
50. Liu, S., et al. (2012). TLR2 is a primary receptor for Alzheimer's amyloid β peptide to trigger neuroinflammatory activation. *Journal of Immunology, 188*, 1098–1107.
51. Lai, K. S. P., et al. (2017). Peripheral inflammatory markers in Alzheimer's disease: A systematic review and meta-analysis of 175 studies. *Journal of Neurology, Neurosurgery, and Psychiatry, 88*, 876–882.
52. Chen, Y., & Yu, Y. (2023). Tau and neuroinflammation in Alzheimer's disease: Interplay mechanisms and clinical translation. *Journal of Neuroinflammation, 20*, 165.
53. Liu, S., et al. (2023). Asiaticoside mitigates Alzheimer's disease pathology by attenuating inflammation and enhancing synaptic function. *International Journal of Molecular Sciences, 24*, 11976.
54. More, S. V., Kumar, H., Kim, I. S., Song, S.-Y., & Choi, D.-K. (2013). Cellular and molecular mediators of neuroinflammation in the pathogenesis of Parkinson's disease. *Mediators of Inflammation, 2013*, 1.
55. Reale, M., et al. (2009). Peripheral cytokines profile in Parkinson's disease. *Brain, Behavior, and Immunity, 23*, 55–63.
56. Marogianni, C., et al. (2020). Neurodegeneration and inflammation—An interesting interplay in Parkinson's disease. *International Journal of Molecular Sciences, 21*, 8421.
57. Zhao, M., Tuo, H., Wang, S., & Zhao, L. (2020). The roles of monocyte and monocyte-derived macrophages in common brain disorders. *BioMed Research International, 2020*, 9396021.
58. Zhao, H., Li, Y., Zhang, Y., He, W.-Y., & Jin, W.-N. (2022). Role of immune and inflammatory mechanisms in stroke: A review of current advances. *Neuroimmunomodulation, 29*, 255–268.
59. Li, C., et al. (2021). Macrophage-disguised manganese dioxide nanoparticles for neuroprotection by reducing oxidative stress and modulating inflammatory microenvironment in acute ischemic stroke. *Advancement of Science, 8*, 2101526.
60. Kolaczkowska, E., & Kubes, P. (2013). Neutrophil recruitment and function in health and inflammation. *Nature Reviews. Immunology, 13*, 159–175.

61. Rayasam, A., et al. (2018). Immune responses in stroke: How the immune system contributes to damage and healing after stroke and how this knowledge could be translated to better cures? *Immunology, 154*, 363–376.
62. Dobson, R., & Giovannoni, G. (2019). Multiple sclerosis–a review. *European Journal of Neurology, 26*, 27–40.
63. Codarri, L., Fontana, A., & Becher, B. (2010). Cytokine networks in multiple sclerosis: Lost in translation. *Current Opinion in Neurology, 23*, 205–211.
64. Palle, P., Monaghan, K. L., Milne, S. M., & Wan, E. C. K. (2017). Cytokine signaling in multiple sclerosis and its therapeutic applications. *Medical Science, 5*, 23.
65. Rasouli, J., et al. (2015). Expression of GM-CSF in T cells is increased in multiple sclerosis and suppressed by IFN-β therapy. *Journal of Immunology, 194*, 5085–5093.
66. Limmroth, V., Putzki, N., & Kachuck, N. J. (2011). The interferon beta therapies for treatment of relapsing–remitting multiple sclerosis: Are they equally efficacious? A comparative review of open-label studies evaluating the efficacy, safety, or dosing of different interferon beta formulations alone or in combination. *Therapeutic Advances in Neurological Disorders, 4*, 281–296.
67. Guzman-Martinez, L., et al. (2019). Neuroinflammation as a common feature of neurodegenerative disorders. *Frontiers in Pharmacology, 10*, 1008.
68. Saraiva, C., et al. (2016). Nanoparticle-mediated brain drug delivery: Overcoming blood–brain barrier to treat neurodegenerative diseases. *Journal of Controlled Release, 235*, 34–47.
69. Feng, X., et al. (2019). Immunomodulatory nanosystems. *Advancement of Science, 6*, 1900101.
70. Kaur, I. P., Bhandari, R., Bhandari, S., & Kakkar, V. (2008). Potential of solid lipid nanoparticles in brain targeting. *Journal of Controlled Release, 127*, 97–109.
71. Abreu, C. M., et al. (2018). Emerging biosensing technologies for neuroinflammatory and neurodegenerative disease diagnostics. *Frontiers in Molecular Neuroscience, 11*, 164.
72. Song, Q., et al. (2014). Lipoprotein-based nanoparticles rescue the memory loss of mice with Alzheimer's disease by accelerating the clearance of amyloid-beta. *ACS Nano, 8*, 2345–2359.
73. Liao, K., et al. (2020). Intranasal delivery of lincRNA-Cox2 siRNA loaded extracellular vesicles decreases lipopolysaccharide-induced microglial proliferation in mice. *Journal of Neuroimmune Pharmacology, 15*, 390–399.
74. Saifi, M. A., Khan, W., & Godugu, C. (2018). Cytotoxicity of nanomaterials: Using nanotoxicology to address the safety concerns of nanoparticles. *Pharmaceutical Nanotechnology, 6*, 3–16.
75. Vega-Villa, K. R., et al. (2008). Clinical toxicities of nanocarrier systems. *Advanced Drug Delivery Reviews, 60*, 929–938.
76. Sharma, H. S., et al. (2009). Chronic treatment with nanoparticles exacerbate hyperthermia induced blood-brain barrier breakdown, cognitive dysfunction and brain pathology in the rat. Neuroprotective effects of nanowired-antioxidant compound H-290/51. *Journal of Nanoscience and Nanotechnology, 9*, 5073–5090.
77. Grissa, I., et al. (2016). The effect of titanium dioxide nanoparticles on neuroinflammation response in rat brain. *Environmental Science and Pollution Research, 23*, 20205–20213.
78. Aijie, C., et al. (2017). Central neurotoxicity induced by the instillation of ZnO and TiO2 nanoparticles through the taste nerve pathway. *Nanomedicine, 12*, 2453–2470.
79. Yaman, S., Chintapula, U., Rodriguez, E., Ramachandramoorthy, H., & Nguyen, K. T. (2020). Cell-mediated and cell membrane-coated nanoparticles for drug delivery and cancer therapy. *Cancer Drug Resist., 3*, 879.
80. Chintapula, U., et al. (2023). Immunomodulation in age-related disorders and nanotechnology interventions. *Wiley Interdisciplinary Reviews. Nanomedicine and Nanobiotechnology, 15*, e1840.
81. Ramos-Cabrer, P., & Campos, F. (2013). Liposomes and nanotechnology in drug development: Focus on neurological targets. *International Journal of Nanomedicine, 8*, 951–960.

# Integration of Peripheral and Central Immune Alteration in Neurodegeneration

Zitin Wali, Pinky, Neha, and Suhel Parvez

**Abstract**

Neurodegenerative disorders are characterized by increasing deterioration and destruction of neurons in the CNS. Immune system failure is the hallmark of neurodegenerative disorders. There is substantial evidence that pathogenic alterations occur gradually in both the central and peripheral immune responses. It has been commonly assumed that the BBB establishes an immunologically favoured surrounding in the CNS via preventing peripheral immune cells and humoral immune components from entering the CNS. Peripheral-produced cytokines can penetrate the BBB, give rise to the stimulation of microglia and astrocytes, and induce direct neurotoxicity. Additionally, peripheral immune cells can invade the brain and contribute in evaluation of neuroinflammation, leading to neurodegenerative disorders. In recent years, neuroinflammation has attracted considerable interest as a therapeutic target for a variety of such conditions. This book chapter focuses on the growing roles of PIS and CIS in neurodegenerative disorders, as well as interconnections. A deeper awareness of the immune system's increasing significance in neurological disorders could enhance therapeutic options. Through this chapter, an effort has been made to emphasize the interplay between peripheral and central immunological alterations to combat neurodegenerative disorders.

Z. Wali · Pinky · Neha · S. Parvez (✉)
Department of Medical Elementology and Toxicology, School of Chemical and Life Sciences, New Delhi, India
e-mail: zitinwali@gmail.com; pinkyatwal_sch@jamiahamdard.ac.in; neha.bhardwaj_sch@jamiahamdard.ac.in; sparvez@jamiahamdard.ac.in



R. Shukla et al. (eds.), *Neuroendocrine-Immune Axis*,
https://doi.org/10.1007/978-981-95-5088-3_4

central immune system (CIS). Because the blood-brain barrier (BBB) prevents peripheral immune cells and humoral immunological components, researchers have generally regarded both peripheral and central immune frameworks to be distinct mechanisms [90]. Traditionally, it was believed that immunological dysfunction was an indication of the formation of neurodegenerative disease, and its progression was restricted to the immunological process in the CNS, but emerging data reveals that the PIS also plays a significant role. Although there is mounting proof that the peripheral immune system (PIS) is essential in neurological pathology. By invading the brain, peripheral immune cells can contribute in advancement of neuroinflammation and neurodegenerative disorders [75]. Furthermore, peripherally produced cytokines can penetrate the BBB, causing immediate neurotoxicity and contributing to glial cell stimulation [65]. Active glial cells secrete more inflammatory chemokines along with cytokines, attracting more immune system cells to CNS (Fig. 1) [146, 189].

Examining neurological pathogenesis through the lens of a systemic immunological mechanism raises fundamental questions regarding how interaction occurs between the peripheral and central sections, whether this interaction changes functionally over time, and whether this interaction can serve as a therapeutic target [101, 140, 184, 194]. AD, PD, and ALS are examples of classic neurodegenerative disorders in which degeneration is the primary feature [58]. Other disorders, subsequently, can be classified as degenerative when a primary insult, such as demyelination, ischemia, or trauma, causes neuronal death. Secondary neurodegenerative illnesses include multiple sclerosis (MS), stroke, and traumatic brain injury [3]. The body's immune system is important in CNS regulation and health issues. The innate immune system (IIS) is the primary line of protection against infectious diseases, and microglia, CNS-resident macrophages, are critical not just as early responders to CNS abnormalities like injury or infection, but additionally for growth and homeostasis [9, 35]. Through the identification of non-self-antigens, adaptive immune system (AIS) displays a vital contribution in the host's defence mechanism. This defence process is controlled by T and B cells, which exhibit a wide spectrum of unique receptors for antigens throughout humoral and cellular immunity [117]. The BBB is disturbed during both aging and neurodegenerative disorders, resulting in a greater influx of PIS cells across the CNS, where they can promote additional deterioration or promote the growth of new tissue [117].

**Fig. 1** (continued) CD36 impacts inflammatory remodelling through the absorption of palmitate, a Toll-like receptor (TLR) ligand. The activation of TLR causes phagocytosis and the generation of ROS. TREM2 (triggering receptor expressed on myeloid cells 2) is an additional lipid sensor produced in microglia that recognizes APOE, Aβ, and aminophospholipid for phagocytosis and degradation in microglial cells. TREM2 in its soluble form is discovered to promote inflammatory cytokine generation by stimulating phosphoinositide 3 kinases/protein kinase B

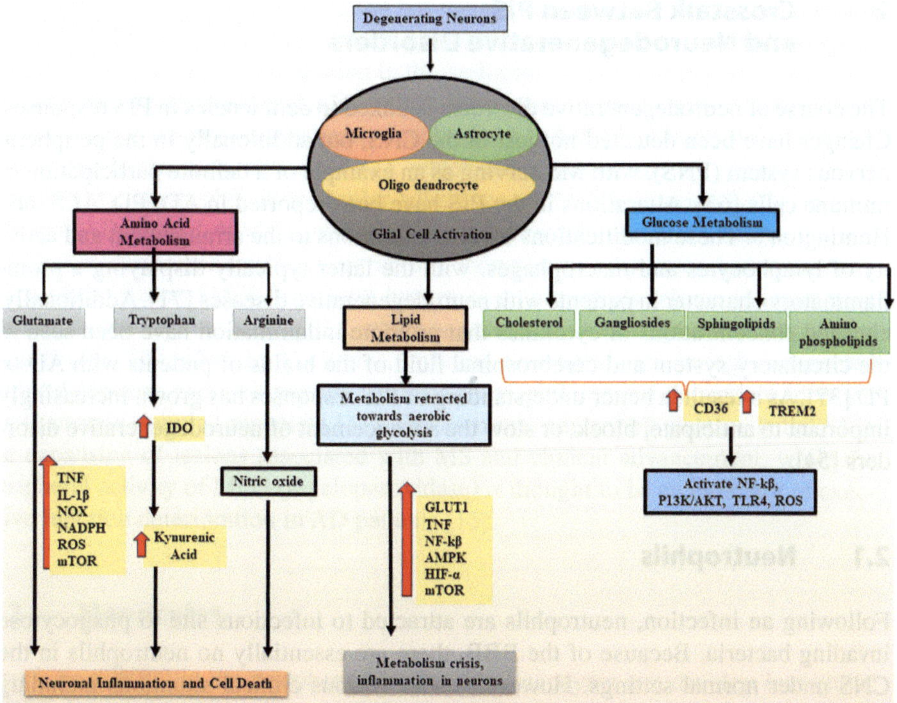


**Fig. 1** Representative figure shows the glial activation response mediated neuronal toxicity: Neuroimmunometabolic sensors are primarily made up of energy components like sugars, amino acids, and lipids, as well as transporters or receptors which detect variations within these components and initiate biochemical and metabolic pathways that employ those substrates. Sensors like this, which are found in neurons and glia, send out signals that change metabolic and immunological health according to the general level of energy requirement. Glucose metabolism in microglia is influenced by their activation status. Microglia generally scan their surroundings for signs of impairment or illness. OXPHOS drives this monitoring process. When impairment is recognized, it causes microglial activation, which entails transformation to inflammatory substances or phagocytic forms. To fulfil this inflammatory demand, stimulated microglia stimulate glycolysis metabolism and shift away from OXPHOS. The transition from OXPHOS to glycolysis in microglia is due to enhanced production of hexokinase and PFKFB3 (limitation enzymes of the glycolytic pathway) and the stimulation of mammalian target of rapamycin (mTOR), which governs glycolysis transcriptional regulation through a mechanism involving hypoxia-inducible factor-1α (HIF-1α). Microglia and astrocytes exhibit tryptophan. Here, the major kynurenine intermediates quinolinic acid, 3-hydroxykynurenine, and kynurenic acid cause distinct inflammatory reactions. Kynurenic acid shields neurons in astrocytes by eliminating excess glutamate transfer, but quinolinic acid and 3-hydroxykynurenine cause neurotoxicity by triggering the NMDAR NR2B component, resulting in a significant calcium inflow. As microglial inflammatory cytokines TNF and IL-1β hinder the activity of astrocytic EAATs, glutamatergic equilibrium is compromised. Furthermore, glutamine-to-glutamate synthesis (glutaminolysis) in microglia promotes mTOR signalling, inhibits autophagy, and produces an inflammatory mediators milieu. Inhibiting glutamine synthetase, which catalyses the contrary conversion of glutamate to glutamine, is found to improve microglia's response to inflammation. Under stressful circumstances, microglia produce numerous lipid-sensing sensors to modify the microglial lipidome. The pattern identification sensor cluster of differentiation 36 (CD36) acts as an important lipid sensor that recognizes LDL and Aβ. When activated,

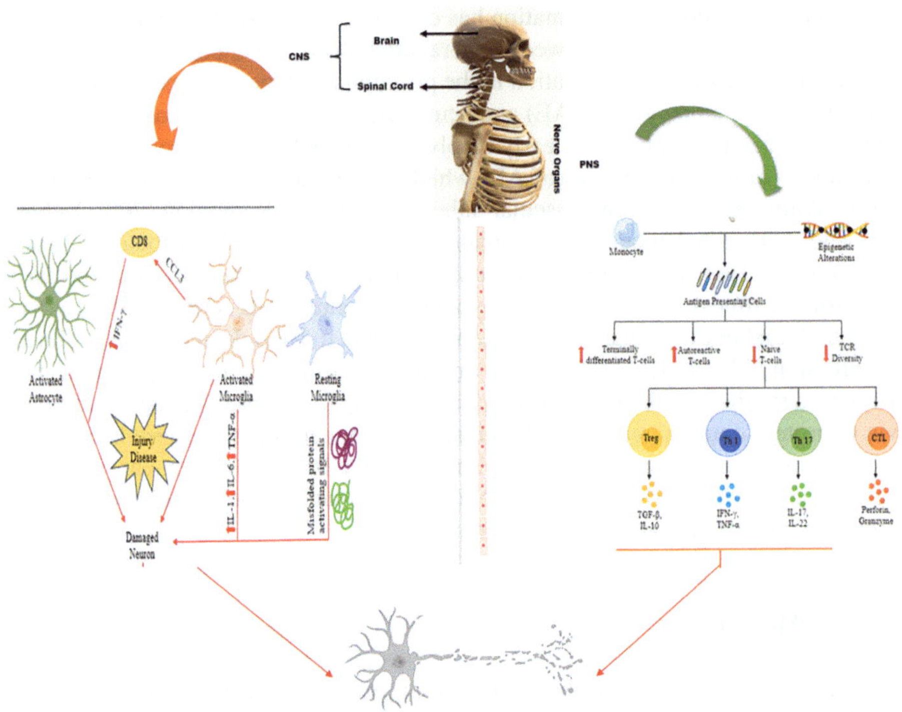


**Fig. 2** Immune processes regulate CNS injury and healing, involving T and B cells possessing harmful and therapeutic effects. T effector lymphocytes such as Th1, Th17, and Th9 increase microglial cell activity by releasing cytokines that promote inflammation (IFN-γ, TNF-α, IL-9, and IL-17) in response to APC-T lymphocyte communication. This increases the death of neurons and demyelination, resulting in a reduction in cognitive ability. TRAIL is widely secreted by activated glia, infiltrating peripheral monocytes, and damaged neurons during brain inflammation. TRAIL functions as a possible killing trigger by communicating with receptors found in neurons, microglia, monocytes, lymphocytes, astrocytes, and oligodendrocytes. Naive T cells mature into antigen-specific T effector cells (Th1, Th2, Th17), and cytotoxic T lymphocytes (CTL) or regulatory T (Treg) cells in response to antigen-presenting cellular stimuli. Th1 and Th17 cells, in particular, breach BBB and immediately promote neuroinflammation by producing neurotoxic and inflammatory substances that have an impact on glial cells in the brain. As a result, activated microglia and astrocytes react by secreting enormous quantities of chemokines, which aid in the recruitment of another set of effector T cells into the brain

related to improved motor neuron preservation in ALS, however, this penetration may be limited [141]. Furthermore, in ALS, monocyte-derived macrophages are stimulated. During the illness, stimulated macrophages perform neuroprotective roles via misfolding protein elimination [175]. The data suggest that the accumulated monocytes in the CNS were caused by infiltrated proliferating cells rather than the infiltration of accumulated circulating monocytes [39]. Circulating monocytes are diverse and malleable innate immune cells that may respond quickly to changes in their surroundings [155]. Humans with neurodegenerative diseases have shown systemic changes in monocyte subsets. Monocyte subset instability in recurrent neuroinflammation in relapsing–remitting multiple sclerosis (RRMS) patients entails the increase of non-classical and transitional monocytes and the decrease of classical monocytes [80]. Monocytes can move into cells and convert into macrophages, resulting in them being key participants in brain homeostasis. Monocyte invasion commences at the onset of the disease and increases until the disease reaches its peak, implying that these cells have a role in disease induction and progression [95]. The monocytes that circulate have been found in AD mice to enter the brain and lower Aβ load, implying a positive function for these cells in AD [113]. Patrolling monocytes were additionally observed crawling along the luminal surfaces of plaque-positive veins, implying their capacity to focus on and eliminate Aβ [121]. However, while monocytes invade the brain and eliminate debris, iron, and Aβ to some extent, these cells are ineffective in eliminating amyloid β (Aβ) deposits in the infected brain in mice with AD-like disease, and their peripheral characteristics alter to have a proinflammatory description that has restricted phagocytic capacity [67].

## 2.3 Natural Killer Cells (NK Cells)

NK cells are hematopoietic cells produced from bone marrow that are broadly dispersed in peripheral lymphoid tissues and bloodstream [81]. NK cells can generate cytokines and chemokines that influence organisms' immune systems, also discharge granules from cytoplasm including perforin and granzyme, which promote death in target cells [192, 212]. Furthermore, for killing target cells, death receptor mechanism is used through Fas ligand and tumour necrosis factor-related apoptosis-inducing ligand (TRAIL) [77, 202]. Chemokines attract these to CNS in diseased situations. During neuroinflammation, for instance, neuronal synthesis of CX3C chemokine ligand 1 (CX3XL1) is required to guide CX3CR1-expressing NK cells towards the brain [83, 86]. By communicating with invading Treg cells (regulatory T cells) and resident microglia, NK cells can modulate Treg recruitment and microglial phenotype. Furthermore, NK cells influence CNS function by destroying glia and releasing IFN-γ [123].

NK cells control beginning and curse of motor neuron degeneration in ALS. They can destroy motor neurons in spinal cord easily. Furthermore, these can communicate with local microglia and invade Treg cells, enhance ALS advancement in gender/age-specific manner, implying that while designing immunotherapy for ALS, we should take gender and age into account [68, 126]. According to one study, accumulation of these cells in the aging brain suppresses neurogenesis, whereas the removal of NK cells decreases neurogenesis and neurological inflammation in aging brain and AD sufferers [94]. Despite major variations in phenotype and function of NK cells, their killing and destruction are unaffected by healthy aging and the onset AD [100]. NK cells additionally communicate with microglia and cause cytotoxicity against hyperactive microglial cells [59]. IFN-derived from NK cells generates an inflammatory phenotype in microglia, regulates the production of CCL2, a chemokine that can control CNS invasion, from MNs, and affects Treg cell migration

[68]. NK cells are essential components of the innate immunity, serving as either primary-line defence or as non-specific effector [20]. The amount of NK cells is reported to rise with age, while alterations to NK cell activity are less obvious. All evidence pointed to impaired NK cell activity in the aged. Clinical symptoms of immunofluorescence could be results of age-related changes in NK cell quantity and functioning [8, 203]. The involvement of NK cells in the pathophysiology of neurodegenerative conditions is currently unknown. For instance, alterations in peripheral NK cell functions have been found in AD patients and healthy aged adults, whereas NK cell killing efficiency and breakdown (CD107 expression) remained intact. The fundamental mechanisms remained unknown [100]. Furthermore, compared to healthy participants, AD patients had circular RNA alterations, and these RNAs were linked to NK cell-mediated cytotoxicity [104]. The percentage of these cells is enhanced in PD patients, yet their activity was unaltered [92]. Amounts of data suggested that NK cell concentrations may be related to the severity of PD [73]. Furthermore, these cells influenced α-synuclein pathology and motor signs in α-synuclein transgenic mouse model [60]. Moreover, the number of NK cells in the peripheral blood of MS patients was enhanced [33]. The function of NK cells in MS is still debatable, with research revealing both protective and detrimental functions in MS mouse models. The potential significance of these cells in neurodegeneration needs further investigation [22].

## 2.4 Dendritic Cells (DCs)

The most common antigen-presenting cells (APCs) are dendritic cells. DCs are found in the CNS in a state of equilibrium, exhibiting significant histocompatibility complex class II (MHCII) with leukocyte integrin CD11c [164]. MHCII/CD11c activation is commonly used to determine presence and location of DCs in the brain [44]. Various investigations have established that under physiological settings, DCs are found in meninges, choroid plexus, CSF, and perivascular spaces in CNS by identifying cell shape and surface markers [167]. All of these regions are highly vascularized, providing credibility to the theory that suggests brain DCs originate from blood rather than the brain. Endogenous dendritic cells in the steady-state brain's meninges and choroidal plexus are probably a result of native precursor DCs that invaded perivascular part of brain initially in life, which implies that these precursor DCs originate from bone marrow instead of monocytes [5]. Chemokines and adhesion molecules are released by peripheral DCs, allowing them to move through meninges or choroid plexus, recognize antigens, and present antigens for T cells [160]. DCs deliver stimulating T cell co-stimulatory signals, boosting the growth and stimulation of CD8+ cytotoxic T cells (CTLs) and CD4+ helper T cells alongside varied proinflammatory cytokine profiles [44]. Under certain conditions, these proinflammatory actions can possibly result in tissue disruption (Fig. 2). However, new research found that bone marrow-derived DCs produce Tregs that provide neurological protection in Parkinson's disease (PD) model, offering neuroprotection through adaptive immune modulation [168]. DCs served a variety of functions, including acting as a communicator for adaptive as well as innate immunity. They processed antigens to be delivered to secondary lymphoid organs after integrating impulses through tissue infection or injury. Furthermore, DCs supplied an array of soluble and surface-coated impulses to aid in T cell development [122]. Overall, immunosenescence is distinguished via reduction in the count of DCs. Modifications in DCs in patients with neurodegenerative disorders thus closely reflect typical immunosenescence, and it can't be ruled out that neurological disorders are simply differentiated via hastened immune system aging stimulated DCs moved to cervical lymph nodes following CNS inflammation, where they stimulated naive lymphocytes before migrating to the point of inflammation, according to multiple investigations [213]. Although DCs can be readily detected in CSF, their ability to transverse the BBB and accumulate within the perivascular compartment remains highly restricted. Furthermore, resident DCs have anti-inflammatory properties. Intranasal administration of vesicular stomatitis virus (VSV) might result in CNS infection. However, the IFN-γ reaction generated via VSV in CNS remained viable in DCs-deficient mice. As a result, inflammation and particular elements of adaptive initial anti-viral immune reaction in CNS are dependent on body's peripheral DCs [25].

In AD, the existence of DCs remained debatable. However, a remarkable decrease in the overall number of DC precursors in the peripheral bloodstreams of AD patients was discovered, along with a decrease in blood myeloid DCs (MDC) [23]. Depletion of DCs in AD animal models was caused by systematic administration of diphtheria toxin, which preferentially focused, hence destroying bone marrow-derived DCs. This revealed peripheral DCs were gathered in the brain and helped remove amyloid plaques [26]. This phenomenon seemed to be primarily related to the course of AD and was modified by the use of acetylcholinesterase inhibitors [41]. The stimulation of DCs preceding the steady loss of neighbouring sensory neurons revealed that immune cells were involved in tau-associated disease from the start in the CNS [93]. Vaccinating Aβ-sensitized DCs using a T cell antigen mutant resulted in antibody reactions in AD mice [26]. Functional alterations of DCs in PIS in PD patients have been thoroughly researched. Modifications in circulating MDC and LDC in PD patients' serum were discovered. MDC moved across tissues that were inflamed and lymph nodes, recognizing pathogens. LDCs evolved into conventional DCs, mostly producing interferon [40]. It was recently proposed that peripheral DCs in PD animal model could penetrate the brain and settle in plexus of choroid or meninges before reacting with diverse antigens to enhance inflammation mechanism [132]. DCF1, a protein found in the membrane that serves critical functions in nerve maturation in mice, may be able to counteract α-syn-induced dopaminergic neuron deletion via accumulation of α-syn in dorsomedial of Drosophila [209]. Recent research has shown that vaccination using DCs can bridge the gap across both AIS and IIS responses [26]. The host immune system is strengthened by employing DCs as organic adjuvants, and the creation of particular antibodies aids in the clearance of abnormal intracellular accumulation of proteins. The vaccine based on DCs keeps the immune response balanced and could offer benefits over regular protein-based immunization [160]

## 2.5 T Cells

T cells possess an essential part in AIS. Multiple investigations have found that the interaction of IIS and AIS plays a role in the pathogenesis of neurodegenerative disorders. T cells in CNS have been frequently seen as harmful, particularly in neuroinflammatory illnesses, and exorbitant inflammatory feedback is assumed to be regulator of neurodegenerative development [61]. Anomalous T cells induce neuroinflammation by interacting directly with microglia in the brain, secreting proinflammatory chemicals [50]. T cells have been demonstrated in recent research to have a vital function in controlling inflammation, CNS damage, infection, and neurodegeneration. The equilibrium between T cells that can be harmful or protective in the CNS has to be studied further [204]. By releasing a range of cytokines, attracting immune cells to area of affliction, and beginning development of CD4+ T cells subgroups alongside various effector roles, CD4+ T cells are primary controllers of immunological reaction [124]. T helper 1 (Th1), T helper 2 (Th2), and T helper 17 (Th17) cells, as well as regulatory T cells (Tregs), develop from naive CD4+ T cells (Th0). Cytotoxic T cells (CTLs) are CD8+ T cells that primarily regulate cellular immunological responses. Th1/Th17 cells are thought to function as proinflammatory cytokine makers that actively encourage neurological inflammation via releasing inflammatory cytokines like IL-1, IL-6, IL-17, TNF-α, and IFN-γ [51]. By increasing the expulsion of ROS and nitric oxide, these cells exacerbate microglia-mediated neurotoxicity (Fig. 1) [110]. Th2 cells, alternately, are believed to release anti-inflammatory cytokines (IL-4). Th2 and Treg cells boost microglia-mediated neuroprotection [117]. The proportion of Th1-Th2 cytokines could be utilized for indication of nature of immune feedback and changes in this proportion are hypothesized as a stimulant for neurodegeneration [29]. Because of their crucial involvement in immunological tolerance and the regulation of inflammatory responses, Tregs are prospective remedial targets for numerous illnesses [174]. Tregs defend the brain by controlling glial cell stimulation. During ALS initial phases, this modulation emerges as an increase in Tregs and an increase in M2-type neuroprotective microglia [17]. Although alongside progression of the disorder, Treg numbers fall, proinflammatory cytokine concentrations rise, glial cells transform into toxic M1 cells [174]. γ/δ T cells contribute to a minor fraction of T lymphocytes in humans [201]. T cells are IIS cells that have lately been revealed to dispense several fundamental characteristics of AIS [52, 53]. Previous research has shown that T cells can modulate immunological responses related to inflammation and serve a proinflammatory function in CNS. T cells have the ability to stimulate regional magnification of immune feedback in CNS, changing inflamed brain's interstitial milieu and eventually ending up in BBB problems [19, 197]. Using Fas or Fas ligand-induced brain-native T cell death, T cells may be able to manage CNS inflammation and disease recovery [144]. The various roles of T cells are determined by their timing and position, in addition to their subtype. Additional research on diverse T cell subtypes must be conducted, though, for understanding their different functions inflammatory feedback of CNS [197]. CD8+ T cells are found to play a role in the pathogenesis of neurodegenerative disorders. CD8+ T lymphocytes promote CNS degeneration of axons in normal senescent rats via TCR and granzyme B, and this impact is amplified in an inflammatory state [76]. Auto-responsive CD8+ T cells engaged with motor neurons and caused them to die. However, there has recently been a surge in demand for CD8+ T cells, as they are now proven to possess activities apart from neurotoxicity [45]. In the brain, peripheral ailment creates a form of CNS tissue-resident cognitive CD8+ T cells. These cells help to regulate CNS ailments by activating quickly, producing more cytokines, and providing mediated protection following brain infection [188]. CD8+ T cell and macrophage invasion stimulates myelin repair, slows muscle desecration, and increases existence duration [130]. In post-mortem samples from PD sufferers as well as MPTP-induced animals, CD4+ and CD8+ T lymphocytes were detected within the SN region, indicating that such immune feedback enhances dopaminergic neurons (DN) degradation via Fas/FasL cytotoxic mechanism [27]. Microglia that have been activated release inflammatory mediators facilitating antigen delivery to CD4+ T cells through MHCII route, leading to the expansion of cells, gradual deterioration, hence dopaminergic death of neurons [116]. T lymphocytes could enter the CNS after being injured, influencing glial cell stimulation and neurodegeneration [125]. T cell stimulation is dependent on antigen-presenting cells (APCs) (Fig. 2). APCs are not present in blood-CSF. APCs are gathered in neurodegenerative disease CSF. Following that, APCs deliver antigen peptides to invading T cells by MHCII molecules, and its surface T cells receptor (TCR) binds these provided antigen peptides. TCR-MHCII association stimulates development of CD4+/CD8+ T lymphocytes [38]. Under different mediators, undistinguishable CD4+ T cells tiggered by APCs developed in diverse operational phenotypes. CD4+ T cells inclined to develop Th1 and Th17 inflammatory phenotypes, associated with neurological inflammation and neuronal injury [142]. CD4+ T cells, whereas, could develop into effective Th2 and Treg cells. These characteristics were critical in the suppression of T cell inflammatory activities and decrease in neurological inflammation [136]. Th1/Th17 cells generated proinflammatory factors like TNF-$\alpha$ and IFN-$\gamma$ to activate glial cells, causing production of additional proinflammatory factors, while Th2/Tregs generated IL-4 and IL-10 for encouragement of transitioning glial cells from proinflammatory to anti-inflammatory phenotypes [71, 145]. T cells boosted development of hippocampal neurons in AD mouse model, but T cell depletion inhibited the regeneration of neurons in the hippocampus. The processes enabling T cell-mediated neuronal regeneration are regulated by increased peripheral T cell activity and production of brain microglial neurotrophic substances [66, 109].

## 2.6 B Cells

B lymphocytes serve a significant function in the development of numerous CNS illnesses via operating peripherally or sectionalizing within CNS. The count of B lymphocytes in CSF rose multiple-fold in patients with CNS inflammation, and these cells expanded by a minimum of several orders of magnitude in parenchyma and perivascular regions of CNS [112]. Several individuals with PD have a reduced number of B lymphocytes subgroups located in peripheral blood, connected to modifies B cell-related gene regulation [98]. Even though B cells are absent in postmortem brain samples of PD sufferers, research has revealed IgG accumulation neurons that are dopaminergic in SN and Lewy bodies in CNS [139]. Few studies discovered higher concentrations of α-syn—particular antibodies in blood and CSF of PD sufferers [173]. This discovery demonstrates importance of B cell antibodies in PD. Regardless of rising proof of B cell participation in neurological disorders, it remains unclear if adaptive immunological alterations are primary or after CNS injury [161]. Pharmacological reduction of B lymphocytes at the outset of the disorder greatly reduced the course of AD in animal model. Additionally, an elderly PSAPP mice model of operational T/B cell elimination was found to have lower brain Aβ concentrations and more glial cell proliferation [183]. B cells participate in AIS and are expected to possess an essential position in the pathological phase of neurodegenerative disorders [161].

B cells were responsible for a variety of tasks, involving providing antigens to T cells, synthesis of proinflammatory proteins, and production of anti-inflammatory cytokines [198]. Auto-reactive B cells were crucial in autoimmune neurological diseases. Cytokines that are secreted by B cells like IL-6 and TNF-α to enhance inflammatory responses, as well as granulocyte-macrophage colony-stimulating factor (GM-CSF) for encouraging development of proinflammatory factors [190]. B cells, on the other hand, may release IL-10 and IL-35 to combat inflammation. The count of these cells is demonstrated to diminish with growing age, which contributes significantly to immunosenescence [31]. Decreased antibody selectivity is caused by B cell immunosenescence. Aging also changes the sensitivity of antibodies. B cell deficits associated with aging include reduced B cell count and antigen sensitivity. Thus, B cell senescence influences the pathogenic mechanism of neurodegenerative disorders. For instance, lower numbers of peripheral B-cell subgroups have been identified in individuals with AD, which might be linked to gene alterations [91]. There were also changes in anti-Aβ antibody concentrations in serum and CSF between adults with AD and robust controls. Such variability could attribute to enhanced anti-Aβ in AD patients [180, 193]. Increasing evidence shows that APP-mutant AD models lacking functional B cells exhibit reduced Aβ pathology, indicating that B lymphocytes play a significant role in modulating Aβ dynamics.

## 3 Central Immune System/Central Nervous System

The CNS is immune-privileged, which means that non-self-antigens that acquire accessibility to the CNS cannot elicit an AIS response. The CSF-blood barrier, BBB, and blood-spinal cord barrier all help to protect the CNS from AIS responses. Furthermore, immune regulatory molecules expressed by CNS cells suppress T cell function and, in conjunction with inadequate amounts of MHC molecules, operate to restrict immunological activation and immune-mediated harm in the CNS (Fig. 2) [2].

### 3.1 Microglia

Microglia are durable macrophages that exist in CNS. Microglia, as a component of CNS equilibrium, persist in being dormant under physiological settings, carry out exceptionally robust immune monitoring via greatly motile mechanisms [134]. This constant mobility allows microglia to react quickly to neuronal pathological alterations. These types of cells perform phagocytic functions, swallowing microbes and cell detritus that enter the brain. Microglia, whereas, can transition into an activated phenotype in response to particular stimuli, along with transcriptional modifications that allow them to undertake inflammatory tasks [4]. The stimulation of glial cells is separated into two phases, historically labelled as M1 (classic activation) and M2 (alternative activating). Persistent stimulation of the M1 phenotype leads to the release of a significant quantity of cytokines that are proinflammatory and neurotoxic chemicals, which harm the organism. M2 microglia, on the other hand, stimulate regeneration and repair of tissues by producing anti-inflammatory cytokines and neurotrophic factors, possessing protective benefits. However, it has recently been demonstrated that stimulated microglia produce canonical product genes linked to M1-M2 phenotypes [149]. Microglia stimulation occurs on a spectrum in neuropathological circumstances, having an ambivalent phenotype regulated via complicated cascade of environmental impulses. Microglial classification according to M1-M2 polarization phases is sufficient to represent many levels of activation of microglial cells [152]. Recent research has demonstrated that microglia can form immediate links with distinct areas of neurons, allowing for improved control of neuronal reactions and cell density [47, 48]. Microglia can detect ATP generated through neuronal activation and convert it to adenosine, suppressing adenosine receptors on active neuron surface, preventing outrageous neuron stimulation, triggering opposing response regulation of neuron activity [10]. However, in few neurological disorders, microglia lose out their capability to detect ATP, creating adenosine. As a result, further research into the process of microglia-neuronal interaction may yield novel therapeutic approaches for specific neurodegenerative illnesses. Indeed, it is widely assumed that microglia can interact not just with immunological constituents of CNS, but with PIS elements penetrating the CNS as well [110]. Furthermore, microglia play a variety of critical but poorly known functions for disease prevention in the brain, opening the door to the development of focused molecular therapeutics. However, stimulated microglia can be harmful under pathological settings, and microglia may regulate the beginning and progression of inflammatory feedback via various transcription factors and cellular signalling mechanisms [30]. Addressing microglia inflammation signalling system could thus act as promising technique for treating neurodegenerative disorders. Microglia are prevalent in the brain and serve as the primary IIS cells and early respondents to pathological abuse [14, 82]. Microglia make up 5-12% of the entire number of cells in the mouse brain, based on site, and they exhibit a variety of morphologies, including compressed spherical, longitudinally branching, and radially branched [99]. They contribute to equilibrium and host defence processes by performing three critical tasks [85]. The first purpose of their senses, which are encoded by numerous genes, is to detect changes in their surroundings [84]. The second purpose is physiological housekeeping, such as migration to wounded locations, synaptic remodelling, and preserving myelin homeostasis [208]. The third step is to defend against potentially detrimental impulse, for instance pathogen-associated molecular patterns (PAMPs) and damage-associated molecular patterns (DAMPs). Microglia demonstrate cellular receptors that recognize PAMPs and DAMPs, including toll-like receptors, nuclear oligomerization domain, and viral receptors [184]. Depending on activation, microglia in CNS can be proinflammatory or neuroprotective. Proinflammatory cytokines are pathogen or damaged cell debris that trigger inactive microglia to generate proinflammatory factors like IL-1β, IL-6, TNF-α, nitric oxide, and proteases (Fig. 1) [105].

IL-4, IL-10, IL-13, and TGF-β stimulate neuroprotective microglia, resulting in production of various factors like, FIZZ1, Ym1, CD206, Arginase, Chitinase-3-like-3 (Chi3l3) and Frizzled class receptor 1 (Fzd1) [70, 105, 186]. These factors are associated with neurological protection and tissue restoration. For instance, IL-4 is expected to hamper synthesis of proinflammatory cytokines like IL-6, NO, or TNF-α (Fig. 2) [210].

## 3.2 Astrocytes

Astrocytes offer nutritional support for neurons, manage nervous system's metabolism, possessing a vital part in encouraging synaptic development and operation, as well as preserving the framework of the brain and BBB [90]. Furthermore, a growing count of investigations have demonstrated that astrocytes are double-edged sword. Murine-responsive astrocytes are classified as A1-A2 in the same way that macrophages are classified as M1 and M2 [106]. A1 astrocytes are proinflammatory and neurotoxic, and their presence has been linked to the advancement of neurodegenerative disorders. A2 astrocytes, whereas, possess a protective role. In an animal model, microglia express IL-1α, TNF, and complement component 1q (C1q) in reaction towards LPS activation, and a combination of actions of these cytokines is required for A1-type astrocyte stimulation (Fig. 2) [153]. A1 astrocytes are implicated in neurological inflammation-mediated neurotoxicity in a variety of mechanisms after activation, producing neuronal and mature distinguished oligodendrocytes death, contributing to development of neurodegeneration [106]. Furthermore, neurodegeneration is characterized via idiopathic neuron damage in various areas of CNS, which is unable to compensate for tissue resurrection. Alternatively, extracellular matrix elements are increasingly replacing these cells, which are primarily generated via endothelial cells, stimulated fibroblasts, and astrocytes [49]. This microglial reaction serves two functions: tissue protection and repair inhibition. Scar-forming astrocyte is typically found near lesion and helps to limit dissemination of inflammatory cytokines and cellular detritus [16].

It has been found that such a fibrotic reaction hastens onset of neurodegeneration. As a result, it was proposed that transforming astrocytes and myofibroblasts from a fibrosis-supporting matrix accumulation state to a fibrosis-supporting retrogressive matrix breakdown state and reducing the development of scars could serve to enhance the neurological processes of ALS and AD [49]. Astrocytes are among a prevalent type of glial cells found in the brain [43]. Although astrocytes were previously thought to simply have passive functions, new research has revealed that they perform active and critical parts in brain equilibrium [138]. They govern blood flow and BBB, deliver energy substrates to neurons, modify synaptic activity, regulate neurotrophin production, eliminate dead cells, manage extracellular equilibrium of ions, transmitters, and fluids [43, 138, 179]. Variations in GFAP-measured astrocyte molecular expression and shape may suggest the extent of reactive astrogliosis, an indicator of CNS disease [99]. Defects in astrocytes in initial phases of damage, such as in spinal cord damage and EAE, frequently linked to worse medical results, neuroinflammation, disruption of BBB, and death of neurons [43]. Research conducted in a chronic investigational model found that astrocytes create lactosylceramide (LacCer) during chronic CNS inflammation, which increases inflammation and neurodegeneration [118]. This explains that impact of astrogliosis might be positive/negative, dependent on time frame, specific disease, and different microenvironmental cues, like microglia. Astrocytes, like microglia, possess proinflammatory and immunoregulatory subgroups. Proinflammatory responsive astrocytes stimulate numerous genes (e.g., complement cascade gene), producing proinflammatory substances. In comparison, neuroprotective reactive astrocytes increase the expression of numerous neurotrophic factors and thrombospondins [105]. Inflammatory mediators released via proinflammatory microglia, like IL-1α/β, TNF-α, and C1q, have been shown to stimulate proinflammatory astrocytes, causing an inflammatory reaction eventually (Fig. 1) [106]. Various cytokines, neurotrophins, or sphingolipids additionally activate detrimental astrocytic signalling mechanisms [43].

## 3.3 Oligodendrocytes

Mature oligodendrocytes (OLs) can be seen across the CNS's grey and white matter [24]. OL is myelin-forming cells found throughout the CNS that generate myelin sheaths encircling axons to offer metabolic support to neurons [159]. Recent research has shown that OL depletion and demyelination are hallmarks of neurodegenerative disorders (Fig. 1). The progenitors of terminally developed OLs are oligodendrocyte precursor cells (OPCs). When demyelination develops, the rate of growth and differentiation of OPCs rises resulting in new OLs as a repair strategy [131, 178]. This could be when it comes to the pathological repair of such disorders. Recent research has indicated that OLs might possess immune-related roles, with several immunomodulatory substances, including cytokines or chemokines with their receptors, expressed in OLs [148, 205]. Ols and OPCs don't serve as passive targets in MS rather regulate active immunity [64]. The injury can result in extensive impairment of OLs, leading to degeneration of the axons they support [87]. Oligodendrocytes are CNS myelin-producing glial cells. Myelin plays an important function in neural transmission by protecting the axon, improving action potential transmission, and promoting high-frequency firing [87]. OL dysfunction causes severe neurodegeneration. PLP is a myelin-related gene that is expressed specifically in OLs and Schwann cells. This gene's deletion or overexpression resulted in significant late-onset neurodegeneration in several CNS areas (Y. [108]).

Other protein aggregation-related neurodegenerative disorders, including tauopathies and synucleinopathy, are additionally linked to OLs. OLs and myelin disruption, as well as their putative parts in onset and enhancement of AD as well as Huntington's disease (HD), have piqued our science community's interest in the past few years [55]. Animals with targeted altered tau expression in OLs established insoluble filament containing tau aggregates as well as axon disruption, illustrating that mutant tau-induced OL malfunction provides partially, if not entirely, to neurodegeneration of tauopathies or instance, AD [108].

## 3.4 Endothelial Cells

The neurovascular unit (NVU) connects neurons, glial cells, and vascular cells structurally as well as functionally [34]. NVU helps to form and maintain BBB, ion equilibrium, and nutrient transfer as well [191]. Endothelial cells serve as a critical constituent of NVU. Brain endothelial cells express more connexins at intercellular junctions than non-neural tissue endothelial cells, conferring barrier limitation features for paracellular permeability [147]. Peripheral cytokines act on endothelial cells during the inflammatory phase, resulting in decreased hindrance. Simultaneously, inflammatory endothelium cells increase expression of adhesion molecules, allowing circulating PIS cells and antibodies to penetrate the barrier [116]. Research lately has discovered an up-to-date method for microvascular endothelial cells for digestion of myelin detritus through autophagy-lysosome mechanism, which encourages advancement of demyelinating disorders via encouraging inflammation, angiogenesis, production of fibrotic scars as well [211]. Endothelial cells in AD brains may be the focus of the harmful effects of severely accumulated proteins, glia-derived cytokines, and stimuli generating oxidative and metabolic stress [163]. Endothelial cells in the brain enhance the ability to survive and multiply OL precursor cells [6]. Endothelial cells act as a structural and physiological barrier between neurons/glia and circulating blood, and alterations in the endothelium have been related to the development of most degenerative disorders [120].

## 3.5 Pericytes

Pericytes are vascular cell walls that reside in the lower membrane and the microvasculature around it. Pericytes, as previously stated, are key biological parts of NVU [187]. They communicate with endothelial cells, which include neurons, glial cells, perivascular macrophages, and it is via such a connection that various roles are carried out [56]. Pericytes have been implicated in both neuro-inflammatory and neurodegenerative illnesses in recent years [137]. Pleiotrophin, generated from pericytes, is hypothesized as a neurotrophic factor essential for existence of neurons. The depletion of pericytes causes BBB, circulation, and neurons to die in mouse brain [133]. Furthermore, CNS pericytes play a role in AD pathogenesis. These are engaged in Aβ-induced capillary narrowing in the brain in an AD animal model [135]. Pericyte disruption hastens amyloid angiopathy and brain amyloidosis via decreasing Aβ eviction from brain fluid [162]. CNS pericytes perform critical regulatory functions in BBB formation and maintenance, microvascular stability, capillary density and angiogenesis, capillary width, bloodstream modulation, macromolecule removal from brain fluid [196]. Pericytes contain a huge count of intracellular inclusions, pinocytotic vesicles, lipid molecules, mitochondrial anomalies indicating malfunction or disruption. Capillary depletion, general dilatation, and tortuosity of existing vessels are all associated with degradation of pericytes [13]. Individually, deficits in pericyte coverage correspond negatively, indicating BBB destruction, like the leaking of proteins from the plasma such as immunoglobulin G and fibrin [171].

## 4 Innate Immunity

Innate immunity, the primary line of defence against infection, yet it additionally serves a significant part in tissue regeneration, clearing cell debris, and responding to malignancies. While microglia and astrocytes, the most important IIS cells in CNS, macrophages, NK cells, mast cells, Ols, and neurons, also help maintain IIS feedback in CNS. PAMPs and DAMPs comprise misfolded and accumulated proteins, as seen in AD, HD, and ALS [107]. Microglia, the CNS's primary resident IIS, perform a variety of roles. They shape neuronal circuits during formation via pruning synapses, as well as by pruning synapses as well as governing cellular death and excretion after inflammation or CNS disruption. According to chemokine and cytokine production in vivo, microglial stimulation is characterized as M1 and M2 [176]. Changing between such polarization is necessary for remyelination and is impacted by age, that was demonstrated via application of parabiosis in mouse [156]. Microglia secrete both proinflammatory and anti-inflammatory substances that can be useful or deleterious in neurodegenerative illnesses. For instance, in AD animal model, microglia depletion decreased neuron reduction without changing pathology of Aβ [182, 195]. Microglia-derived substances, like BDNF, are critical for cognitive operations, which might be impacted by maternal inflammation, resulting in later-life behaviour and learning [166]. Microglia seem to have distinct transcriptome patterns depending on brain location, ageing neuropathological condition, and microbiota [62, 72]. Monocytes are blood-borne progenitors to macrophages and DCs, possessing a significant part in innate immunity. Meanwhile, microglia like other tissue-resident macrophages, develop via yolk sacs of primitive macrophages, with unique involvement in CNS illnesses that is mostly unclear. Creation of CCr2—red fluorescent protein knock—in mice has enabled scientists to distinguish between invading monocytes/macrophages and native microglia in clinical ailments [184]. The new markers TMEM119 and P2Y12 have been shown to aid in the differentiation of microglia and macrophages, implying that the respective role of these cells in neuroinflammatory illnesses might be investigated [214]. Subpopulations of astrocytes were identified producing proinflammatory mediators (A1) and immunoregulatory mediators (A2). This is alike to M1-M2 polarization of macrophages and microglia. The A1 astrocytes releasing IL-1α, TNF, and C1q are thought to be neuroinflammatory, triggering apoptosis, inhibiting Th cell stimulation, development, and functioning of stimulated T cells. A2 astrocytes are also neuroprotective, as they promote neuron development, survival, and synaptic regeneration. Astrocytes are hypertrophic in various neurodegenerative conditions like stroke, ALS, MS, TBI, etc., and other inflammatory conditions, and they respond to a variety of stressors [105].

## 5 Adaptive Immunity

Alterations in T- and B-cell subsets, along with corresponding antibody levels in the blood, CSF, and brain, indicate a substantive involvement of the adaptive immune system in neurodegenerative disorders. The cytokines and chemokine profiles, for instance, determine if cell subsets play a harmful (Th1/Th17) or anti-inflammatory (Th2/Tregs) function. The spectrum of autoimmune encephalopathy disorders, comprising paraneoplastic neuronal disorders (PND), best exemplifies the involvement of AIS responses in neurological diseases [12, 96]. The administration of IVIG or excision of tumour producing abnormal antigen in OND is typically adequate to treat several ailments like this. Natively, MS is defined as the intrusion of CNS via AIS cells—T/B lymphocytes. Although functioning of AIS in ALS and PD is progressively getting attention, in AD, inflammation is predominantly caused via CNS-resident microglial. For instance, in initial phases of PD, there is an increase in the count of Th17 cells in blood, some of these identify α-syn [185]. T and B lymphocytes additionally exist in CNS throughout X-ALD, which might be an additional occurrence because immunosuppressive medications have minimal effect on the disease's progression. Lower Treg counts are associated with motor neuron loss and shorter survival durations in ALS mice models, whereas Treg transfer lowers neuroinflammation and extends survival. In keeping with these findings, Tregs are malfunctioning in ALS patients [18].

## 6 Alzheimer's Disease

AD is a very frequent neurodegenerative condition and a major cause of mental decline in aged individuals. Pathological alterations in AD comprise accumulation of Aβ in brain, which forms senile plaques, hyperphosphorylation of tau protein, which causes NFTs (X. Q. [36]). Microglia may play two functions in the aetiology of AD. Excess Aβ plaques and cell detritus can be removed by activated microglia. Mutations that cause functional decline in genes encoding microglial transmembrane proteins, including TREM2 and CD33, are linked to decrease phagocytosis and are thought to be a genetic risk in AD. The role of microglia alters as the disease proceeds [32, 103]. Astrocytes take up and internalize Aβ from external surroundings as well as assist in its breakdown. Recently, it was found that IL-3 generated by astrocytes in the brains of mice was neuroprotective in an AD animal model and acted as an essential facilitator for astrocyte-microglia crosstalk, suggesting that it could be used as a therapeutic intervention in AD. However, the pathogenic significance of responsive astrocytes in AD is established [119]. Furthermore, CNS pericytes play a role in AD pathogenesis. Pericytes are engaged in Aβ-induced capillary restriction in the brain in a rat model of AD. Pericyte loss hastens amyloid angiopathy along with brain amyloidosis by decreasing Aβ excretion from brain ISF [135].

According to the chemokine sensor and Ly6C expression levels, murine monocytes are reorganized at two primary subpopulations: proinflammatory and anti-inflammatory. On the contrary, Ly6Clow monocyte subpopulation patrolled arterial lumen, promoting tissue regeneration, phagocytosing toxic substances such as Aβ, and alleviating neurodegeneration [129]. Ly6Chigh monocytes, on the contrary, could enter brain parenchyma and release proinflammatory cytokines that encouraged microglial stimulation [65]. In human AD and 3xTg-AD mouse, peripheral NK cell infiltration of the brain and the associated neuroinflammatory alterations have been documented [111]. Immune cells can flow to and from the brain due to increased permeability of BBB in AD sufferers. Crosstalk across microglia and astrocytes has been revealed in recent research to be crucial for T cells to migrate to CNS [29]. In the hippocampus of mice, there was an influx of CD8+ T cells linked to initial chemokine reactions, especially those including CCL3. This discovery highlights the essential function of adaptive immunity in the aetiology of AD (Fig. 2).

## 7 Parkinson's Disease

PD, additionally referred to as tremor palsy, is the second most prevalent neurodegenerative disorder after AD. It is distinguished via substantial decrease of neurons that produce in dopamine in SN and abnormal intraneuronal accumulation of α-syn in Lewy bodies [172]. Here, microglia help eliminate misfolded α-syn clumps. However, through interacting with Toll-like receptors (TLR), α-syn can stimulate NLRP3 inflammasome in microglia. This causes NF-κB translocation, which results in elevated production of proinflammatory cytokines and also defective mitochondria, causing dopaminergic neurons to be damaged [28]. Immuno-responsive astrocytes with increased density and phenotype alterations are seen in postmortem PD samples; nevertheless, the precise role of astrocytes in PD pathology is unknown. On the contrary, astrocytes can help prevent disease development by isolating and digesting proinflammatory extracellular α-syn. High levels of extracellular α-syn promote astrocytes inflammatory feedback in TLR4-dependent way, which might worsen pressurized situations in synucleinopathy brains [151]. Monocytes may play an important role in OD aetiology, and high expression of expressed quantitative trail loci (eQTL) defined to monocytes is linked to the disease [150]. The involvement of NK cells in the aetiology of PD is yet unknown. It was found that NK cells protect against Lewy body-related neurodegenerative disorders. Human NK cells, on the other hand, can efficiently internalize and destroy α-syn accumulates via endosomal or lysosomal mechanism. NK cells help to resolve extracellular α-syn burden via generating IFN-γ, stimulation APCs like microglia. Systemic NK cell depletion exacerbated motor signs and nuclear protein pathology in an experimental animal model of PD [60]. CD4+ and CD8+ T-lymphocytes are found in SN region, indicating that this immune reaction enhances DN degradation via Fas/FasL cytotoxic mechanism in postmortem PD samples and MPTP-induced mice [181].

## 8 Future Prospects

In this chapter, we examine how the interaction between the PIS and CNS regulates CNS homeostasis and disease. We discuss the growing roles of the PIS and CIS in neurodegenerative illnesses, along with their interconnections. In the past few years, the burden of neurodegenerative ailments has steadily risen, placing a tremendous financial and emotional drain on family and sufferers. However, there are no effective treatments to halt or cure neurodegenerative disease development. Despite recent advancements, clinical studies have thus far proven unsatisfactory. In this chapter, we suggest that for neurodegenerative immunotherapies and treatments, it is crucial to consider not just CNS or peripheral immunity, but also their interactions.

**Acknowledgments** The authors would like to thank Central Council for Research in Unani Medicine (File. No.3-51/2022-CCRUM/Tech) for the EMR project awarded to Prof. Suhel Parvez. Ms. Neha received the Senior Research Fellowship from DST-INSPIRE (File No. DST/INSPIRE/03/ 2019/001769. IVR No. 201900031405]. The authors would like to thank DST for FIST (SR/FST/LS-I/2017/05[C]) and PURSE grants (SR/PURSE Phase 2/39[C]).

**Authors Contribution** Zitin, Pinky, and Neha: Data creation; Conceptualization; Roles/Writing-original draft, Resources, Graphical abstract, Figures, Tables. Suhel Parvez: supervision, Writing-review & editing.

**Conflict Statement** The author declares no conflict of interest.

## References

1. Abdel-Nour, M., Carneiro, L. A. M., Downey, J., Tsalikis, J., Outlioua, A., Prescott, D., Da Costa, L. S., Hovingh, E. S., Farahvash, A., Gaudet, R. G., Molinaro, R., van Dalen, R., Lau, C. C. Y., Azimi, F. C., Escalante, N. K., Trotman-Grant, A., Lee, J. E., Gray-Owen, S. D., Divangahi, M., et al. (2019). The heme-regulated inhibitor is a cytosolic sensor of protein misfolding that controls innate immune signaling. *Science, 364*(6448). https://doi.org/10.1126/science.aaw4144
2. Amor, S., & Woodroofe, M. N. (2014). Innate and adaptive immune responses in neurodegeneration and repair. *Immunology, 141*(3), 287–291. https://doi.org/10.1111/imm.12134

3. Amor, S., Puentes, F., Baker, D., & Van Der Valk, P. (2010). Inflammation in neurodegenerative diseases. *Immunology, 129*(2), 154–169. https://doi.org/10.1111/j.1365-2567.2009.03225.x
4. Amor, S., McNamara, N. B., Gerrits, E., Marzin, M. C., Kooistra, S. M., Miron, V. E., & Nutma, E. (2022). White matter microglia heterogeneity in the CNS. *Acta Neuropathologica, 143*(2), 125–141. https://doi.org/10.1007/s00401-021-02389-x
5. Anandasabapathy, N., Victora, G. D., Meredith, M., Feder, R., Dong, B., Kluger, C., Yao, K., Dustin, M. L., Nussenzweig, M. C., Steinman, R. M., & Liu, K. (2011). Flt3L controls the development of radiosensitive dendritic cells in the meninges and choroid plexus of the steady-state mouse brain. *Journal of Experimental Medicine, 208*(18), 1695–1705. https://doi.org/10.1084/jem.20102657
6. Arai, K., & Lo, E. H. (2009). An oligovascular niche: Cerebral endothelial cells promote the survival and proliferation of oligodendrocyte precursor cells. *Journal of Neuroscience, 29*(14), 4351–4355. https://doi.org/10.1523/JNEUROSCI.0035-09.2009
7. Aubé, B., Lévesque, S. A., Paré, A., Chamma, É., Kébir, H., Gorina, R., Lécuyer, M.-A., Alvarez, J. I., De Koninck, Y., Engelhardt, B., Prat, A., Côté, D., & Lacroix, S. (2014). Neutrophils mediate blood–spinal cord barrier disruption in demyelinating neuroinflammatory diseases. *The Journal of Immunology, 193*(5), 2438–2454. https://doi.org/10.4049/jimmunol.1400401
8. Aw, D., Silva, A. B., & Palmer, D. B. (2007). Immunosenescence: Emerging challenges for an aging population. *Immunology, 120*(4), 435–446. https://doi.org/10.1111/j.1365-2567.2007.02555.x
9. Bachiller, S., Jiménez-Ferrer, I., Paulus, A., Yang, Y., Swanberg, M., Deierborg, T., & Boza-Serrano, A. (2018). Microglia in neurological diseases: A road map to brain-disease dependent-inflammatory response. *Frontiers in Cellular Neuroscience, 12*, 488. https://doi.org/10.3389/fncel.2018.00488
10. Badimon, A., Strasburger, H. J., Ayata, P., Chen, X., Nair, A., Ikegami, A., Hwang, P., Chan, A. T., Graves, S. M., Uweru, J. O., Ledderose, C., Kutlu, M. G., Wheeler, M. A., Kahan, A., Ishikawa, M., Wang, Y. C., Loh, Y. H. E., Jiang, J. X., Surmeier, D. J., et al. (2020). Negative feedback control of neuronal activity by microglia. *Nature, 586*(7829), 417–423. https://doi.org/10.1038/s41586-020-2777-8
11. Baik, S. H., Cha, M. Y., Hyun, Y. M., Cho, H., Hamza, B., Kim, D. K., Han, S. H., Choi, H., Kim, K. H., Moon, M., Lee, J., Kim, M., Irimia, D., & Mook-Jung, I. (2014). Migration of neutrophils targeting amyloid plaques in Alzheimer's disease mouse model. *Neurobiology of Aging, 35*(6), 1286–1292. https://doi.org/10.1016/j.neurobiolaging.2014.01.003
12. Balint, B., Vincent, A., Meinck, H. M., Irani, S. R., & Bhatia, K. P. (2018). Movement disorders with neuronal antibodies: Syndromic approach, genetic parallels and pathophysiology. *Brain, 141*(1), 13–36. https://doi.org/10.1093/brain/awx189
13. Baloyannis, S. J., & Baloyannis, I. S. (2012). The vascular factor in Alzheimer's disease: A study in Golgi technique and electron microscopy. *Journal of the Neurological Sciences, 322*(1–2), 117–121. https://doi.org/10.1016/j.jns.2012.07.010
14. Baufeld, C., O'Loughlin, E., Calcagno, N., Madore, C., & Butovsky, O. (2018). Differential contribution of microglia and monocytes in neurodegenerative diseases. *Journal of Neural Transmission, 125*(5), 809–826. https://doi.org/10.1007/s00702-017-1795-7
15. Bawa, K. K., Krance, S. H., Herrmann, N., Cogo-Moreira, H., Ouk, M., Yu, D., Wu, C. Y., Black, S. E., Lanctôt, K. L., & Swardfager, W. (2020). A peripheral neutrophil-related inflammatory factor predicts a decline in executive function in mild Alzheimer's disease. *Journal of Neuroinflammation, 17*(1). https://doi.org/10.1186/s12974-020-01750-3
16. Becerra-Calixto, A., & Cardona-Gómez, G. P. (2017). The role of astrocytes in neuroprotection after brain stroke: Potential in cell therapy. *Frontiers in Molecular Neuroscience, 10*. https://doi.org/10.3389/FNMOL.2017.00088
17. Beers, D. R., Henkel, J. S., Zhao, W., Wang, J., Huang, A., Wen, S., Liao, B., & Appel, S. H. (2011). Endogenous regulatory T lymphocytes ameliorate amyotrophic lateral sclerosis in mice and correlate with disease progression in patients with amyotrophic lateral sclerosis. *Brain, 134*(5), 1293–1314. https://doi.org/10.1093/brain/awr074

18. Beers, D. R., Zhao, W., Wang, J., Zhang, X., Wen, S., Neal, D., Thonhoff, J. R., Alsuliman, A. S., Shpall, E. J., Rezvani, K., & Appel, S. H. (2017). ALS patients' regulatory T lymphocytes are dysfunctional, and correlate with disease progression rate and severity. *JCI Insight, 2*(5). https://doi.org/10.1172/jci.insight.89530
19. Benakis, C., Brea, D., Caballero, S., Faraco, G., Moore, J., Murphy, M., Sita, G., Racchumi, G., Ling, L., Pamer, E. G., Iadecola, C., & Anrather, J. (2016). Commensal microbiota affects ischemic stroke outcome by regulating intestinal γδ T cells. *Nature Medicine, 22*(5), 516–523. https://doi.org/10.1038/nm.4068
20. Berrou, J., Fougeray, S., Venot, M., Chardiny, V., Gautier, J. F., Dulphy, N., Toubert, A., & Peraldi, M. N. (2013). Natural killer cell function, an important target for infection and tumor protection, is impaired in type 2 diabetes. *PLoS One, 8*(4). https://doi.org/10.1371/journal.pone.0062418
21. Bettcher, B. M., Tansey, M. G., Dorothée, G., & Heneka, M. T. (2021). Peripheral and central immune system crosstalk in Alzheimer disease — A research prospectus. *Nature Reviews Neurology, 17*(11), 689–701. https://doi.org/10.1038/s41582-021-00549-x
22. Bielekova, B. (2013). Daclizumab therapy for multiple sclerosis. *Neurotherapeutics, 10*(1), 55–67. https://doi.org/10.1007/s13311-012-0147-4
23. Bossù, P., Spalletta, G., Caltagirone, C., & Ciaramella, A. (2015). Myeloid dendritic cells are potential players in human neurodegenerative diseases. *Frontiers in Immunology, 6*(DEC). https://doi.org/10.3389/fimmu.2015.00632
24. Boulanger, J. J., & Messier, C. (2014). From precursors to myelinating oligodendrocytes: Contribution of intrinsic and extrinsic factors to white matter plasticity in the adult brain. *Neuroscience, 269*, 343–366. https://doi.org/10.1016/j.neuroscience.2014.03.063
25. Bourgeois-Daigneault, M. C., Roy, D. G., Falls, T., Twumasi-Boateng, K., St-Germain, L. E., Marguerie, M., Garcia, V., Selman, M., Jennings, V. A., Pettigrew, J., Amos, S., Diallo, J. S., Nelson, B., & Bell, J. C. (2016). Oncolytic vesicular stomatitis virus expressing interferon-γ has enhanced therapeutic activity. *Molecular Therapy - Oncolytics, 3*. https://doi.org/10.1038/mto.2016.1
26. Brezovakova, V., Valachova, B., Hanes, J., Novak, M., & Jadhav, S. (2018). Dendritic cells as an alternate approach for treatment of neurodegenerative disorders. *Cellular and Molecular Neurobiology, 38*(6), 1207–1214. https://doi.org/10.1007/s10571-018-0598-1
27. Brochard, V., Combadière, B., Prigent, A., Laouar, Y., Perrin, A., Beray-Berthat, V., Bonduelle, O., Alvarez-Fischer, D., Callebert, J., Launay, J. M., Duyckaerts, C., Flavell, R. A., Hirsch, E. C., & Hunot, S. (2009). Infiltration of CD4+ lymphocytes into the brain contributes to neurodegeneration in a mouse model of Parkinson disease. *Journal of Clinical Investigation, 119*(1), 182–192. https://doi.org/10.1172/JCI36470
28. Brück, D., Wenning, G. K., Stefanova, N., & Fellner, L. (2016). Glia and alpha-synuclein in neurodegeneration: A complex interaction. *Neurobiology of Disease, 85*, 262–274. https://doi.org/10.1016/j.nbd.2015.03.003
29. Burgaletto, C., Munafò, A., Di Benedetto, G., De Francisci, C., Caraci, F., Di Mauro, R., Bucolo, C., Bernardini, R., & Cantarella, G. (2020). The immune system on the TRAIL of Alzheimer's disease. *Journal of Neuroinflammation, 17*(1). https://doi.org/10.1186/s12974-020-01968-1
30. Cai, B., Seong, K. J., Bae, S. W., Chun, C., Kim, W. J., & Jung, J. Y. (2018). A synthetic diosgenin primary amine derivative attenuates LPS-stimulated inflammation via inhibition of NF-κB and JNK MAPK signaling in microglial BV2 cells. *International Immunopharmacology, 61*, 204–214. https://doi.org/10.1016/j.intimp.2018.05.021
31. Cancro, M. P., Hao, Y., Scholz, J. L., Riley, R. L., Frasca, D., Dunn-Walters, D. K., & Blomberg, B. B. (2009). B cells and aging: Molecules and mechanisms. *Trends in Immunology, 30*(7), 313–318. https://doi.org/10.1016/j.it.2009.04.005
32. Cao, S., Fisher, D. W., Rodriguez, G., Yu, T., & Dong, H. (2021). Comparisons of neuroinflammation, microglial activation, and degeneration of the locus coeruleus-norepinephrine system in APP/PS1 and aging mice. *Journal of Neuroinflammation, 18*(1). https://doi.org/10.1186/s12974-020-02054-2

33. Caruana, P., Lemmert, K., Ribbons, K., Lea, R., & Lechner-Scott, J. (2017). Natural killer cell subpopulations are associated with MRI activity in a relapsing-remitting multiple sclerosis patient cohort from Australia. *Multiple Sclerosis, 23*(11), 1479–1487. https://doi.org/10.1177/1352458516679267
34. Castellani, G., & Schwartz, M. (2020). Immunological features of non-neuronal brain cells: Implications for Alzheimer's disease immunotherapy. *Trends in Immunology, 41*(9), 794–804. https://doi.org/10.1016/j.it.2020.07.005
35. Chaplin, D. D. (2010). Overview of the immune response. *Journal of Allergy and Clinical Immunology, 125*(2 Suppl. 2). https://doi.org/10.1016/j.jaci.2009.12.980
36. Chen, X. Q., & Mobley, W. C. (2019). Alzheimer disease pathogenesis: Insights from molecular and cellular biology studies of oligomeric Aβ and tau species. *Frontiers in Neuroscience, 13*(Jun). https://doi.org/10.3389/fnins.2019.00659
37. Chen, X., Hu, Y., Cao, Z., Liu, Q., & Cheng, Y. (2018a). Cerebrospinal fluid inflammatory cytokine aberrations in Alzheimer's disease, Parkinson's disease and amyotrophic lateral sclerosis: A systematic review and meta-analysis. *Frontiers in Immunology, 9*(Sept). https://doi.org/10.3389/fimmu.2018.02122
38. Chen, Z., Chen, S., & Liu, J. (2018b). The role of T cells in the pathogenesis of Parkinson's disease. *Progress in Neurobiology, 169*, 1–23. https://doi.org/10.1016/j.pneurobio.2018.08.002
39. Chiot, A., Zaïdi, S., Iltis, C., Ribon, M., Berriat, F., Schiaffino, L., Jolly, A., De La Grange, P., Mallat, M., & Bohl, D. (2020). Modifying macrophages at the periphery has the capacity to change microglial reactivity and to extend ALS survival. *Nature Neuroscience, 23*(11), 1339–1351. https://doi.org/10.1038/s41593-020-00718-z
40. Ciaramella, A., Salani, F., Bizzoni, F., Pontieri, F. E., Stefani, A., Pierantozzi, M., Assogna, F., Caltagirone, C., Spalletta, G., & Bossù, P. (2013). Blood dendritic cell frequency declines in idiopathic Parkinson's disease and is associated with motor symptom severity. *PLoS One, 8*(6). https://doi.org/10.1371/journal.pone.0065352
41. Ciaramella, A., Salani, F., Bizzoni, F., Orfei, M. D., Caltagirone, C., Spalletta, G., & Bossù, P. (2016). Myeloid dendritic cells are decreased in peripheral blood of Alzheimer's disease patients in association with disease progression and severity of depressive symptoms. *Journal of Neuroinflammation, 13*(1). https://doi.org/10.1186/s12974-016-0483-0
42. Ciccocioppo, F., Bologna, G., Ercolino, E., Pierdomenico, L., Simeone, P., Lanuti, P., Pieragostino, D., Del Boccio, P., Marchisio, M., & Miscia, S. (2020). Neurodegenerative diseases as proteinopathies-driven immune disorders. *Neural Regeneration Research, 15*(5), 850–856. https://doi.org/10.4103/1673-5374.268971
43. Colombo, E., & Farina, C. (2016). Astrocytes: Key regulators of neuroinflammation. *Trends in Immunology, 37*(9), 608–620. https://doi.org/10.1016/j.it.2016.06.006
44. Colton, C. A. (2013). Immune heterogeneity in neuroinflammation: Dendritic cells in the brain. *Journal of Neuroimmune Pharmacology, 8*(1), 145–162. https://doi.org/10.1007/s11481-012-9414-8
45. Coque, E., Salsac, C., Espinosa-Carrasco, G., Varga, B., Degauque, N., Cadoux, M., Crabé, R., Virenque, A., Soulard, C., Fierle, J. K., Brodovitch, A., Libralato, M., Végh, A. G., Venteo, S., Scamps, F., Boucraut, J., Laplaud, D., Hernandez, J., Gergely, C., et al. (2019). Cytotoxic CD8 + T lymphocytes expressing ALS-causing SOD1 mutant selectively trigger death of spinal motoneurons. *Proceedings of the National Academy of Sciences of the United States of America, 116*(6), 2312–2317. https://doi.org/10.1073/pnas.1815961116
46. Cova, I., Markova, A., Campini, I., Grande, G., Mariani, C., & Pomati, S. (2017). Worldwide trends in the prevalence of dementia. *Journal of the Neurological Sciences, 379*. https://doi.org/10.1016/j.jns.2017.06.030
47. Cserép, C., Pósfai, B., Lénárt, N., Fekete, R., László, Z. I., Lele, Z., Orsolits, B., Molnár, G., Heindl, S., Schwarcz, A. D., Ujvári, K., Környei, Z., Tóth, K., Szabadits, E., Sperlágh, B., Baranyi, M., Csiba, L., Hortobágyi, T., Maglóczky, Z., et al. (2020). Microglia monitor and protect neuronal function through specialized somatic purinergic junctions. *Science, 367*(6477), 528–537. https://doi.org/10.1126/science.aax6752

48. Cserép, C., Pósfai, B., & Dénes, Á. (2021). Shaping neuronal fate: Functional heterogeneity of direct microglia-neuron interactions. *Neuron, 109*(2), 222–240. https://doi.org/10.1016/j.neuron.2020.11.007
49. D'Ambrosi, N., & Apolloni, S. (2020). Fibrotic scar in neurodegenerative diseases. *Frontiers in Immunology, 11*. https://doi.org/10.3389/fimmu.2020.01394
50. Dai, L., & Shen, Y. (2021). Insights into T-cell dysfunction in Alzheimer's disease. *Aging Cell, 20*(12). https://doi.org/10.1111/acel.13511
51. Dardalhon, V., Korn, T., Kuchroo, V. K., & Anderson, A. C. (2008). Role of Th1 and Th17 cells in organ-specific autoimmunity. *Journal of Autoimmunity, 31*(3). https://doi.org/10.1016/j.jaut.2008.04.017
52. Davey, M. S., Willcox, C. R., Joyce, S. P., Ladell, K., Kasatskaya, S. A., McLaren, J. E., Hunter, S., Salim, M., Mohammed, F., Price, D. A., Chudakov, D. M., & Willcox, B. E. (2017). Clonal selection in the human Vδ1 T cell repertoire indicates γδ TCR-dependent adaptive immune surveillance. *Nature Communications, 8*. https://doi.org/10.1038/ncomms14760
53. Davey, M. S., Willcox, C. R., Hunter, S., Kasatskaya, S. A., Remmerswaal, E. B. M., Salim, M., Mohammed, F., Bemelman, F. J., Chudakov, D. M., Oo, Y. H., & Willcox, B. E. (2018). The human Vδ2+ T-cell compartment comprises distinct innate-like Vγ9+ and adaptive Vγ9-subsets. *Nature Communications, 9*(1). https://doi.org/10.1038/s41467-018-04076-0
54. del Pilar, C., Lebrón-Galán, R., Pérez-Martín, E., Pérez-Revuelta, L., Ávila-Zarza, C. A., Alonso, J. R., Clemente, D., Weruaga, E., & Díaz, D. (2021). The selective loss of Purkinje cells induces specific peripheral immune alterations. *Frontiers in Cellular Neuroscience, 15*. https://doi.org/10.3389/fncel.2021.773696
55. Desai, M. K., Mastrangelo, M. A., Ryan, D. A., Sudol, K. L., Narrow, W. C., & Bowers, W. J. (2010). Early oligodendrocyte/myelin pathology in Alzheimer's disease mice constitutes a novel therapeutic target. *American Journal of Pathology, 177*(3). https://doi.org/10.2353/ajpath.2010.100087
56. Ding, X., Gu, R., Zhang, M., Ren, H., Shu, Q., Xu, G., & Wu, H. (2018). Microglia enhanced the angiogenesis, migration and proliferation of co-cultured RMECs. *BMC Ophthalmology, 18*(1). https://doi.org/10.1186/s12886-018-0886-z
57. Du, Y., Zhao, W., Thonhoff, J. R., Wang, J., Wen, S., & Appel, S. H. (2020). Increased activation ability of monocytes from ALS patients. *Experimental Neurology, 328*. https://doi.org/10.1016/j.expneurol.2020.113259
58. Dugger, B. N., & Dickson, D. W. (2017). Pathology of neurodegenerative diseases. *Cold Spring Harbor Perspectives in Biology, 9*(7). https://doi.org/10.1101/cshperspect.a028035
59. Earls, R. H., & Lee, J. K. (2020). The role of natural killer cells in Parkinson's disease. *Experimental and Molecular Medicine, 52*(9). https://doi.org/10.1038/s12276-020-00505-7
60. Earls, R. H., Menees, K. B., Chung, J., Gutekunst, C. A., Lee, H. J., Hazim, M. G., Rada, B., Wood, L. B., & Lee, J. K. (2020). NK cells clear α-synuclein and the depletion of NK cells exacerbates synuclein pathology in a mouse model of α-synucleinopathy. *Proceedings of the National Academy of Sciences of the United States of America, 117*(3). https://doi.org/10.1073/pnas.1909110117
61. Ellwardt, E., Walsh, J. T., Kipnis, J., & Zipp, F. (2016). Understanding the role of T cells in CNS homeostasis. *Trends in Immunology, 37*(2). https://doi.org/10.1016/j.it.2015.12.008
62. Erny, D., De Angelis, A. L. H., Jaitin, D., Wieghofer, P., Staszewski, O., David, E., Keren-Shaul, H., Mahlakoiv, T., Jakobshagen, K., Buch, T., Schwierzeck, V., Utermöhlen, O., Chun, E., Garrett, W. S., Mccoy, K. D., Diefenbach, A., Staeheli, P., Stecher, B., Amit, I., & Prinz, M. (2015). Host microbiota constantly control maturation and function of microglia in the CNS. *Nature Neuroscience, 18*(7). https://doi.org/10.1038/nn.4030
63. Fakhoury, M. (2016). Immune-mediated processes in neurodegeneration: Where do we stand? *Journal of Neurology, 263*(9). https://doi.org/10.1007/s00415-016-8052-0
64. Falcão, A. M., van Bruggen, D., Marques, S., Meijer, M., Jäkel, S., Agirre, E., Samudyata, F., & E. M., Vanichkina, D. P., ffrench-Constant, C., Williams, A., Guerreiro-Cacais, A. O., & Castelo-Branco, G. (2018). Disease-specific oligodendrocyte lineage cells arise in multiple sclerosis. *Nature Medicine, 24*(12). https://doi.org/10.1038/s41591-018-0236-y

65. Fani Maleki, A., & Rivest, S. (2019). Innate immune cells: Monocytes, monocyte-derived macrophages and microglia as therapeutic targets for Alzheimer's disease and multiple sclerosis. *Frontiers in Cellular Neuroscience, 13*. https://doi.org/10.3389/fncel.2019.00355
66. Ferretti, M. T., Merlini, M., Späni, C., Gericke, C., Schweizer, N., Enzmann, G., Engelhardt, B., Kulic, L., Suter, T., & Nitsch, R. M. (2016). T-cell brain infiltration and immature antigen-presenting cells in transgenic models of Alzheimer's disease-like cerebral amyloidosis. *Brain, Behavior, and Immunity, 54*. https://doi.org/10.1016/j.bbi.2016.02.009
67. Fiala, M., Lin, J., Ringman, J., Kermani-Arab, V., Tsao, G., Patel, A., Lossinsky, A. S., Graves, M. C., Gustavson, A., Sayre, J., Sofroni, E., Suarez, T., Chiappelli, F., & Bernard, G. (2005). Ineffective phagocytosis of amyloid-β by macrophages of Alzheimer's disease patients. *Journal of Alzheimer's Disease, 7*(3). https://doi.org/10.3233/JAD-2005-7304
68. Garofalo, S., Cocozza, G., Porzia, A., Inghilleri, M., Raspa, M., Scavizzi, F., Aronica, E., Bernardini, G., Peng, L., Ransohoff, R. M., Santoni, A., & Limatola, C. (2020). Natural killer cells modulate motor neuron-immune cell cross talk in models of amyotrophic lateral sclerosis. *Nature Communications, 11*(1). https://doi.org/10.1038/s41467-020-15644-8
69. Giannoni, P., Claeysen, S., Noe, F., & Marchi, N. (2020). Peripheral routes to neurodegeneration: passing through the blood–brain barrier. *Frontiers in Aging Neuroscience, 12*. https://doi.org/10.3389/fnagi.2020.00003
70. Glass, C. K., Saijo, K., Winner, B., Marchetto, M. C., & Gage, F. H. (2010). Mechanisms underlying inflammation in neurodegeneration. *Cell, 140*(6). https://doi.org/10.1016/j.cell.2010.02.016
71. González, H., & Pacheco, R. (2014). T-cell-mediated regulation of neuroinflammation involved in neurodegenerative diseases. *Journal of Neuroinflammation, 11*(1). https://doi.org/10.1186/s12974-014-0201-8
72. Grabert, K., Michoel, T., Karavolos, M. H., Clohisey, S., Kenneth Baillie, J., Stevens, M. P., Freeman, T. C., Summers, K. M., & McColl, B. W. (2016). Microglial brain regionâ 'dependent diversity and selective regional sensitivities to aging. *Nature Neuroscience, 19*(3). https://doi.org/10.1038/nn.4222
73. Green, H., Zhang, X., Tiklova, K., Volakakis, N., Brodin, L., Berg, L., Greengard, P., Perlmann, T., & Svenningsson, P. (2017). Alterations of p11 in brain tissue and peripheral blood leukocytes in Parkinson's disease. *Proceedings of the National Academy of Sciences of the United States of America, 114*(10). https://doi.org/10.1073/pnas.1621218114
74. Greenhalgh, A. D., Zarruk, J. G., Healy, L. M., Baskar Jesudasan, S. J., Jhelum, P., Salmon, C. K., Formanek, A., Russo, M. V., Antel, J. P., McGavern, D. B., McColl, B. W., & David, S. (2018). Peripherally derived macrophages modulate microglial function to reduce inflammation after CNS injury. *PLoS Biology, 16*(10). https://doi.org/10.1371/journal.pbio.2005264
75. Greenhalgh, A. D., David, S., & Bennett, F. C. (2020). Immune cell regulation of glia during CNS injury and disease. *Nature Reviews Neuroscience, 21*(3). https://doi.org/10.1038/s41583-020-0263-9
76. Groh, J., Knöpper, K., Arampatzi, P., Yuan, X., Lößlein, L., Saliba, A. E., Kastenmüller, W., & Martini, R. (2021). Accumulation of cytotoxic T cells in the aged CNS leads to axon degeneration and contributes to cognitive and motor decline. *Nature Aging, 1*(4). https://doi.org/10.1038/s43587-021-00049-z
77. Guillerey, C., Huntington, N. D., & Smyth, M. J. (2016). Targeting natural killer cells in cancer immunotherapy. *Nature Immunology, 17*(9). https://doi.org/10.1038/ni.3518
78. Guo, Y., Zeng, H., & Gao, C. (2021). The role of neutrophil extracellular traps in central nervous system diseases and prospects for clinical application. *Oxidative Medicine and Cellular Longevity, 2021*. https://doi.org/10.1155/2021/9931742
79. Hartl, F. U. (2017). Protein misfolding diseases. *Annual Review of Biochemistry, 86*. https://doi.org/10.1146/annurev-biochem-061516-044518
80. Haschka, D., Tymoszuk, P., Bsteh, G., Petzer, V., Berek, K., Theurl, I., Berger, T., & Weiss, G. (2020). Expansion of neutrophils and classical and nonclassical monocytes as a Hallmark in relapsing-remitting multiple sclerosis. *Frontiers in Immunology, 11*. https://doi.org/10.3389/fimmu.2020.00594

81. Hazenberg, M. D., & Spits, H. (2014). Human innate lymphoid cells. *Blood, 124*(5). https://doi.org/10.1182/blood-2013-11-427781
82. Heneka, M. T., Carson, M. J., El Khoury, J., Landreth, G. E., Brosseron, F., Feinstein, D. L., Jacobs, A. H., Wyss-Coray, T., Vitorica, J., Ransohoff, R. M., Herrup, K., Frautschy, S. A., Finsen, B., Brown, G. C., Verkhratsky, A., Yamanaka, K., Koistinaho, J., Latz, E., Halle, A., et al. (2015). Neuroinflammation in Alzheimer's disease. *The Lancet Neurology, 14*(4), 388–405. https://doi.org/10.1016/S1474-4422(15)70016-5
83. Hertwig, L., Hamann, I., Romero-Suarez, S., Millward, J. M., Pietrek, R., Chanvillard, C., Stuis, H., Pollok, K., Ransohoff, R. M., Cardona, A. E., & Infante-Duarte, C. (2016). CX3CR1-dependent recruitment of mature NK cells into the central nervous system contributes to control autoimmune neuroinflammation. *European Journal of Immunology, 46*(8), 1984–1996. https://doi.org/10.1002/eji.201546194
84. Hickman, S. E., Kingery, N. D., Ohsumi, T. K., Borowsky, M. L., Wang, L. C., Means, T. K., & El Khoury, J. (2013). The microglial sensome revealed by direct RNA sequencing. *Nature Neuroscience, 16*(12). https://doi.org/10.1038/nn.3554
85. Hickman, S., Izzy, S., Sen, P., Morsett, L., & El Khoury, J. (2018). Microglia in neurodegeneration. *Nature Neuroscience, 21*(10), 1359–1369. https://doi.org/10.1038/s41593-018-0242-x
86. Huang, D., Shi, F., Jung, S., Pien, G. C., Wang, J., Salazar-Mather, T. P., He, T. T., Weaver, J. T., Ljunggren, H., Biron, C. A., Littman, D. R., & Ransohoff, R. M. (2006). The neuronal chemokine CX3CL1/fractalkine selectively recruits NK cells that modify experimental autoimmune encephalomyelitis within the central nervous system. *The FASEB Journal, 20*(7). https://doi.org/10.1096/fj.05-5465com
87. Huntemer-Silveira, A., Patil, N., Brickner, M. A., & Parr, A. M. (2021). Strategies for oligodendrocyte and myelin repair in traumatic CNS injury. *Frontiers in Cellular Neuroscience, 14*. https://doi.org/10.3389/fncel.2020.619707
88. Ishii, M., & Iadecola, C. (2020). Risk factor for Alzheimer's disease breaks the blood-brain barrier. *Nature, 581*(7806). https://doi.org/10.1038/d41586-020-01152-8
89. Jellinger, K. A. (2010). Basic mechanisms of neurodegeneration: A critical update. *Journal of Cellular and Molecular Medicine, 14*(3). https://doi.org/10.1111/j.1582-4934.2010.01010.x
90. Jeon, M. T., Kim, K. S., Kim, E. S., Lee, S., Kim, J., Hoe, H. S., & Kim, D. G. (2021). Emerging pathogenic role of peripheral blood factors following BBB disruption in neurodegenerative disease. *Ageing Research Reviews, 68*. https://doi.org/10.1016/j.arr.2021.101333
91. Jiang, Q., Jin, S., Jiang, Y., Liao, M., Feng, R., Zhang, L., Liu, G., & Hao, J. (2017a). Alzheimer's disease variants with the genome-wide significance are significantly enriched in immune pathways and active in immune cells. *Molecular Neurobiology, 54*(1), 594–600. https://doi.org/10.1007/s12035-015-9670-8
92. Jiang, S., Gao, H., Luo, Q., Wang, P., & Yang, X. (2017b). The correlation of lymphocyte subsets, natural killer cell, and Parkinson's disease: A meta-analysis. *Neurological Sciences, 38*(8). https://doi.org/10.1007/s10072-017-2988-4
93. Jiao, H., Downie, L. E., Huang, X., Wu, M., Oberrauch, S., Keenan, R. J., Jacobson, L. H., & Chinnery, H. R. (2020). Novel alterations in corneal neuroimmune phenotypes in mice with central nervous system tauopathy. *Journal of Neuroinflammation, 17*(1). https://doi.org/10.1186/s12974-020-01803-7
94. Jin, W. N., Shi, K., He, W., Sun, J. H., Van Kaer, L., Shi, F. D., & Liu, Q. (2021). Neuroblast senescence in the aged brain augments natural killer cell cytotoxicity leading to impaired neurogenesis and cognition. *Nature Neuroscience, 24*(1). https://doi.org/10.1038/s41593-020-00745-w
95. Jordão, M. J. C., Sankowski, R., Brendecke, S. M., Sagar, L., & G., Tai, Y. H., Tay, T. L., Schramm, E., Armbruster, S., Hagemeyer, N., Groß, O., Mai, D., Çiçek, Ö., Falk, T., Kerschensteiner, M., Grün, D., & Prinz, M. (2019). Neuroimmunology: Single-cell profiling identifies myeloid cell subsets with distinct fates during neuroinflammation. *Science, 363*(6425). https://doi.org/10.1126/science.aat7554
96. Kalman, B. (2017). Autoimmune encephalitides: A broadening field of treatable conditions. *The Neurologist, 22*(1). https://doi.org/10.1097/NRL.0000000000000087

97. Kempuraj, D., Thangavel, R., Natteru, P. A., Selvakumar, G. P., Saeed, D., Zahoor, H., Zaheer, S., Iyer, S. S., & Zaheer, A. (2016). Neuroinflammation induces neurodegeneration. *Journal of Neurology, Neurosurgery and Spine, 1*(1).
98. Kobo, H., Bar-Shira, A., Dahary, D., Gan-Or, Z., Mirelman, A., Goldstein, O., Giladi, N., & Orr-Urtreger, A. (2016). Down-regulation of B cell-related genes in peripheral blood leukocytes of Parkinson's disease patients with and without GBA mutations. *Molecular Genetics and Metabolism, 117*(2). https://doi.org/10.1016/j.ymgme.2015.09.005
99. Kwon, H. S., & Koh, S. H. (2020). Neuroinflammation in neurodegenerative disorders: the roles of microglia and astrocytes. *Translational Neurodegeneration, 9*(1). https://doi.org/10.1186/s40035-020-00221-2
100. Le Page, A., Bourgade, K., Lamoureux, J., Frost, E., Pawelec, G., Larbi, A., Witkowski, J. M., Dupuis, G., & Fülöp, T. (2015). NK cells are activated in amnestic mild cognitive impairment but not in mild Alzheimer's disease patients. *Journal of Alzheimer's Disease, 46*(1). https://doi.org/10.3233/JAD-143054
101. Le Page, A., Dupuis, G., Frost, E. H., Larbi, A., Pawelec, G., Witkowski, J. M., & Fulop, T. (2018). Role of the peripheral innate immune system in the development of Alzheimer's disease. *Experimental Gerontology, 107*. https://doi.org/10.1016/j.exger.2017.12.019
102. Leone, M. A., Mandrioli, J., Russo, S., Cucovici, A., Gianferrari, G., Lisnic, V., Muresanu, D. F., Giuliani, F., Copetti, M., & Fontana, A. (2022). Neutrophils-to-lymphocyte ratio is associated with progression and overall survival in amyotrophic lateral sclerosis. *Biomedicine, 10*(2). https://doi.org/10.3390/biomedicines10020354
103. Li, C., Chen, Y. H., & Zhang, K. (2020a). Neuroprotective properties and therapeutic potential of bone marrow–derived microglia in Alzheimer's disease. *American Journal of Alzheimer's Disease and Other Dementias, 35*. https://doi.org/10.1177/1533317520927169
104. Li, Y., Fan, H., Sun, J., Ni, M., Zhang, L., Chen, C., Hong, X., Fang, F., Zhang, W., & Ma, P. (2020b). Circular RNA expression profile of Alzheimer's disease and its clinical significance as biomarkers for the disease risk and progression. *International Journal of Biochemistry and Cell Biology, 123*. https://doi.org/10.1016/j.biocel.2020.105747
105. Liddelow, S. A., & Barres, B. A. (2017). Reactive astrocytes: Production, function, and therapeutic potential. *Immunity, 46*(6). https://doi.org/10.1016/j.immuni.2017.06.006
106. Liddelow, S. A., Guttenplan, K. A., Clarke, L. E., Bennett, F. C., Bohlen, C. J., Schirmer, L., Bennett, M. L., Münch, A. E., Chung, W. S., Peterson, T. C., Wilton, D. K., Frouin, A., Napier, B. A., Panicker, N., Kumar, M., Buckwalter, M. S., Rowitch, D. H., Dawson, V. L., Dawson, T. M., et al. (2017). Neurotoxic reactive astrocytes are induced by activated microglia. *Nature, 541*(7638). https://doi.org/10.1038/nature21029
107. Lill, C. M., Rengmark, A., Pihlstrøm, L., Fogh, I., Shatunov, A., Sleiman, P. M., Wang, L. S., Liu, T., Lassen, C. F., Meissner, E., Alexopoulos, P., Calvo, A., Chio, A., Dizdar, N., Faltraco, F., Forsgren, L., Kirchheiner, J., Kurz, A., Larsen, J. P., et al. (2015). The role of TREM2 R47H as a risk factor for Alzheimer's disease, frontotemporal lobar degeneration, amyotrophic lateral sclerosis, and Parkinson's disease. *Alzheimer's & Dementia, 11*(12), 1407–1416. https://doi.org/10.1016/j.jalz.2014.12.009
108. Liu, Y., & Zhou, J. (2013). Oligodendrocytes in neurodegenerative diseases. *Frontiers in Biology, 8*(2). https://doi.org/10.1007/s11515-013-1260-4
109. Liu, J., Ma, Y. X., Tian, S. M., Zhang, L., Zhao, M. M., Zhang, Y. Q., & Xu, D. C. (2014). T cells promote the regeneration of neural precursor cells in the hippocampus of alzheimer's disease mice. *Neural Regeneration Research, 9*(16). https://doi.org/10.4103/1673-5374.139481
110. Liu, Z., Cheng, X., Zhong, S., Zhang, X., Liu, C., Liu, F., & Zhao, C. (2020). Peripheral and central nervous system immune response crosstalk in amyotrophic lateral sclerosis. *Frontiers in Neuroscience, 14*. https://doi.org/10.3389/fnins.2020.00575
111. Lu, Y., Li, K., Hu, Y., & Wang, X. (2021). Expression of immune related genes and possible regulatory mechanisms in Alzheimer's disease. *Frontiers in Immunology, 12*. https://doi.org/10.3389/fimmu.2021.768966
112. Machado-Santos, J., Saji, E., Tröscher, A. R., Paunovic, M., Liblau, R., Gabriely, G., Bien, C. G., Bauer, J., & Lassmann, H. (2018). The compartmentalized inflammatory response in

the multiple sclerosis brain is composed of tissue-resident CD8+ T lymphocytes and B cells. *Brain, 141*(7). https://doi.org/10.1093/brain/awy151
113. Malm, T. M., Koistinaho, M., Pärepalo, M., Vatanen, T., Ooka, A., Karlsson, S., & Koistinaho, J. (2005). Bone-marrow-derived cells contribute to the recruitment of microglial cells in response to β-amyloid deposition in APP/PS1 double transgenic Alzheimer mice. *Neurobiology of Disease, 18*(1). https://doi.org/10.1016/j.nbd.2004.09.009
114. Manda-Handzlik, A., & Demkow, U. (2019). The brain entangled: The contribution of neutrophil extracellular traps to the diseases of the central nervous system. *Cells, 8*(12). https://doi.org/10.3390/cells8121477
115. Mantovani, A., Cassatella, M. A., Costantini, C., & Jaillon, S. (2011). Neutrophils in the activation and regulation of innate and adaptive immunity. *Nature Reviews Immunology, 11*(8). https://doi.org/10.1038/nri3024
116. Marogianni, C., Sokratous, M., Dardiotis, E., Hadjigeorgiou, G. M., Bogdanos, D., & Xiromerisiou, G. (2020). Neurodegeneration and inflammation—An interesting interplay in parkinson's disease. *International Journal of Molecular Sciences, 21*(22). https://doi.org/10.3390/ijms21228421
117. Mayne, K., White, J. A., McMurran, C. E., Rivera, F. J., & de la Fuente, A. G. (2020). Aging and neurodegenerative disease: Is the adaptive immune system a friend or foe? *Frontiers in Aging Neuroscience, 12*. https://doi.org/10.3389/fnagi.2020.572090
118. Mayo, L., Trauger, S. A., Blain, M., Nadeau, M., Patel, B., Alvarez, J. I., Mascanfroni, I. D., Yeste, A., Kivisäkk, P., Kallas, K., Ellezam, B., Bakshi, R., Prat, A., Antel, J. P., Weiner, H. L., & Quintana, F. J. (2014). Regulation of astrocyte activation by glycolipids drives chronic CNS inflammation. *Nature Medicine, 20*(10). https://doi.org/10.1038/nm.3681
119. McAlpine, C. S., Park, J., Griciuc, A., Kim, E., Choi, S. H., Iwamoto, Y., Kiss, M. G., Christie, K. A., Vinegoni, C., Poller, W. C., Mindur, J. E., Chan, C. T., He, S., Janssen, H., Wong, L. P., Downey, J., Singh, S., Anzai, A., Kahles, F., et al. (2021). Astrocytic interleukin-3 programs microglia and limits Alzheimer's disease. *Nature, 595*(7869). https://doi.org/10.1038/s41586-021-03734-6
120. Michael, O. O., Gbolahan, B. W., Ansa, C. E., Abdulbasitand, A., & Azeez, I. O. (2015). Endothelial proliferation modulates neuron-glia survival and differentiation in ischemic stress. *Annals of Neurosciences, 22*(3). https://doi.org/10.5214/ans.0972.7531.220305
121. Michaud, J. P., Bellavance, M. A., Préfontaine, P., & Rivest, S. (2013). Real-time in vivo imaging reveals the ability of monocytes to clear vascular amyloid beta. *Cell Reports, 5*(3). https://doi.org/10.1016/j.celrep.2013.10.010
122. Mildner, A., & Jung, S. (2014). Development and function of dendritic cell subsets. *Immunity, 40*(5). https://doi.org/10.1016/j.immuni.2014.04.016
123. Moodley, K., Angel, C. E., Glass, M., & Graham, E. S. (2011). Real-time profiling of NK cell killing of human astrocytes using xCELLigence technology. *Journal of Neuroscience Methods, 200*(2). https://doi.org/10.1016/j.jneumeth.2011.07.005
124. Morgan, J., Muskat, K., Tippalagama, R., Sette, A., Burel, J., & Lindestam Arlehamn, C. S. (2021). Classical CD4 T cells as the cornerstone of antimycobacterial immunity. *Immunological Reviews, 301*(1). https://doi.org/10.1111/imr.12963
125. Mundt, S., Greter, M., Flügel, A., & Becher, B. (2019). The CNS immune landscape from the viewpoint of a T cell. *Trends in Neurosciences, 42*(10). https://doi.org/10.1016/j.tins.2019.07.008
126. Murdock, B. J., Famie, J. P., Piecuch, C. E., Raue, K. D., Mendelson, F. E., Pieroni, C. H., Iniguez, S. D., Zhao, L., Goutman, S. A., & Feldman, E. L. (2021). NK cells associate with ALS in a sex – And age-dependent manner. *JCI Insight, 6*(11). https://doi.org/10.1172/jci.insight.147129
127. Naegele, M., Tillack, K., Reinhardt, S., Schippling, S., Martin, R., & Sospedra, M. (2012). Neutrophils in multiple sclerosis are characterized by a primed phenotype. *Journal of Neuroimmunology, 242*(1–2). https://doi.org/10.1016/j.jneuroim.2011.11.009

128. Naert, G., & Rivest, S. (2011). CC chemokine receptor 2 deficiency aggravates cognitive impairments and amyloid pathology in a transgenic mouse model of Alzheimer's disease. *Journal of Neuroscience, 31*(16). https://doi.org/10.1523/JNEUROSCI.0299-11.2011
129. Naert, G., & Rivest, S. (2013). A deficiency in CCR2+ monocytes: The hidden side of Alzheimer's disease. *Journal of Molecular Cell Biology, 5*(5). https://doi.org/10.1093/jmcb/mjt028
130. Nardo, G., Trolese, M. C., de Vito, G., Cecchi, R., Riva, N., Dina, G., Heath, P. R., Quattrini, A., Shaw, P. J., Piazza, V., & Bendotti, C. (2016). Immune response in peripheral axons delays disease progression in SOD1G93A mice. *Journal of Neuroinflammation, 13*(1). https://doi.org/10.1186/s12974-016-0732-2
131. Neely, S. A., Williamson, J. M., Klingseisen, A., Zoupi, L., Early, J. J., Williams, A., & Lyons, D. A. (2022). New oligodendrocytes exhibit more abundant and accurate myelin regeneration than those that survive demyelination. *Nature Neuroscience, 25*(4). https://doi.org/10.1038/s41593-021-01009-x
132. Negi, N., & Das, B. K. (2018). CNS: Not an immunoprivilaged site anymore but a virtual secondary lymphoid organ. *International Reviews of Immunology, 37*(1). https://doi.org/10.1080/08830185.2017.1357719
133. Nikolakopoulou, A. M., Montagne, A., Kisler, K., Dai, Z., Wang, Y., Huuskonen, M. T., Sagare, A. P., Lazic, D., Sweeney, M. D., Kong, P., Wang, M., Owens, N. C., Lawson, E. J., Xie, X., Zhao, Z., & Zlokovic, B. V. (2019). Pericyte loss leads to circulatory failure and pleiotrophin depletion causing neuron loss. *Nature Neuroscience, 22*(7). https://doi.org/10.1038/s41593-019-0434-z
134. Nimmerjahn, A., Kirchhoff, F., & Helmchen, F. (2005). Neuroscience: Resting microglial cells are highly dynamic surveillants of brain parenchyma in vivo. *Science, 308*(5726). https://doi.org/10.1126/science.1110647
135. Nortley, R., Korte, N., Izquierdo, P., Hirunpattarasilp, C., Mishra, A., Jaunmuktane, Z., Kyrargyri, V., Pfeiffer, T., Khennouf, L., Madry, C., Gong, H., Richard-Loendt, A., Huang, W., Saito, T., Saido, T. C., Brandner, S., Sethi, H., & Attwell, D. (2019). Amyloid b oligomers constrict human capillaries in Alzheimer's disease via signaling to pericytes. *Science, 365*(6450). https://doi.org/10.1126/science.aav9518
136. Nyström, S. N., Bourges, D., Garry, S., Ross, E. M., van Driel, I. R., & Gleeson, P. A. (2014). Transient Treg-cell depletion in adult mice results in persistent self-reactive CD4+ T-cell responses. *European Journal of Immunology, 44*(12). https://doi.org/10.1002/eji.201344432
137. Nyúl-Tóth, Á., Kozma, M., Nagyőszi, P., Nagy, K., Fazakas, C., Haskó, J., Molnár, K., Farkas, A. E., Végh, A. G., Váró, G., Galajda, P., Wilhelm, I., & Krizbai, I. A. (2017). Expression of pattern recognition receptors and activation of the non-canonical inflammasome pathway in brain pericytes. *Brain, Behavior, and Immunity, 64*. https://doi.org/10.1016/j.bbi.2017.04.010
138. Oksanen, M., Lehtonen, S., Jaronen, M., Goldsteins, G., Hämäläinen, R. H., & Koistinaho, J. (2019). Astrocyte alterations in neurodegenerative pathologies and their modeling in human induced pluripotent stem cell platforms. *Cellular and Molecular Life Sciences, 76*(14). https://doi.org/10.1007/s00018-019-03111-7
139. Orr, C. F., Rowe, D. B., Mizuno, Y., Mori, H., & Halliday, G. M. (2005). A possible role for humoral immunity in the pathogenesis of Parkinson's disease. *Brain, 128*(11). https://doi.org/10.1093/brain/awh625
140. Paouri, E., & Georgopoulos, S. (2019). Systemic and CNS inflammation crosstalk: Implications for Alzheimer's disease. *Current Alzheimer Research, 16*(6). https://doi.org/10.2174/1567205016666190321154618
141. Peake, K., Manning, J., Lewis, C. A., Tran, K., Rossi, F., & Krieger, C. (2017). Bone marrow-derived cell accumulation in the spinal cord is independent of peripheral mobilization in a mouse model of amyotrophic lateral sclerosis. *Frontiers in Neurology, 8*(Mar). https://doi.org/10.3389/fneur.2017.00075
142. Peck, A., & Mellins, E. D. (2010). Plasticity of T-cell phenotype and function: The T helper type 17 example. *Immunology, 129*(2). https://doi.org/10.1111/j.1365-2567.2009.03189.x

143. Pieper, C., Pieloch, P., & Galla, H. J. (2013). Pericytes support neutrophil transmigration via interleukin-8 across a porcine co-culture model of the blood-brain barrier. *Brain Research, 1524*. https://doi.org/10.1016/j.brainres.2013.05.047
144. Ponomarev, E. D., & Dittel, B. N. (2005). Γδ T cells regulate the extent and duration of inflammation in the central nervous system by a Fas ligand-dependent mechanism. *The Journal of Immunology, 174*(8). https://doi.org/10.4049/jimmunol.174.8.4678
145. Prajeeth, C. K., Kronisch, J., Khorooshi, R., Knier, B., Toft-Hansen, H., Gudi, V., Floess, S., Huehn, J., Owens, T., Korn, T., & Stangel, M. (2017). Effectors of Th1 and Th17 cells act on astrocytes and augment their neuroinflammatory properties. *Journal of Neuroinflammation, 14*(1). https://doi.org/10.1186/s12974-017-0978-3
146. Prinz, M., & Priller, J. (2017). The role of peripheral immune cells in the CNS in steady state and disease. *Nature Neuroscience, 20*(2). https://doi.org/10.1038/nn.4475
147. Procter, T. V., Williams, A., & Montagne, A. (2021). Interplay between brain pericytes and endothelial cells in dementia. *American Journal of Pathology, 191*(11). https://doi.org/10.1016/j.ajpath.2021.07.003
148. Raffaele, S., Boccazzi, M., & Fumagalli, M. (2021). Oligodendrocyte dysfunction in amyotrophic lateral sclerosis: Mechanisms and therapeutic perspectives. *Cells, 10*(3). https://doi.org/10.3390/cells10030565
149. Rahimian, R., Wakid, M., O'Leary, L. A., & Mechawar, N. (2021). The emerging tale of microglia in psychiatric disorders. *Neuroscience and Biobehavioral Reviews, 131*. https://doi.org/10.1016/j.neubiorev.2021.09.023
150. Raj, T., Rothamel, K., Mostafavi, S., Ye, C., Lee, M. N., Replogle, J. M., Feng, T., Lee, M., Asinovski, N., Frohlich, I., Imboywa, S., Von Korff, A., Okada, Y., Patsopoulos, N. A., Davis, S., McCabe, C., Paik, H. I., Srivastava, G. P., Raychaudhuri, S., et al. (2014). Polarization of the effects of autoimmune and neurodegenerative risk alleles in leukocytes. *Science, 344*(6183). https://doi.org/10.1126/science.1249547
151. Rannikko, E. H., Weber, S. S., & Kahle, P. J. (2015). Exogenous α-synuclein induces toll-like receptor 4 dependent inflammatory responses in astrocytes. *BMC Neuroscience, 16*(1). https://doi.org/10.1186/s12868-015-0192-0
152. Ransohoff, R. M. (2016). A polarizing question: Do M1 and M2 microglia exist. *Nature Neuroscience, 19*(8). https://doi.org/10.1038/nn.4338
153. Ridler, C. (2017). Neuroimmunology: Microglia-induced reactive astrocytes-toxic players in neurological disease? *Nature Reviews Neurology, 13*(3). https://doi.org/10.1038/nrneurol.2017.17
154. Rossi, B., Constantin, G., & Zenaro, E. (2020). The emerging role of neutrophils in neurodegeneration. *Immunobiology, 225*(1). https://doi.org/10.1016/j.imbio.2019.10.014
155. Rossi, B., Santos-Lima, B., Terrabuio, E., Zenaro, E., & Constantin, G. (2021). Common peripheral immunity mechanisms in multiple sclerosis and Alzheimer's disease. *Frontiers in Immunology, 12*. https://doi.org/10.3389/fimmu.2021.639369
156. Ruckh, J. M., Zhao, J. W., Shadrach, J. L., Van Wijngaarden, P., Rao, T. N., Wagers, A. J., & Franklin, R. J. M. (2012). Rejuvenation of regeneration in the aging central nervous system. *Cell Stem Cell, 10*(1). https://doi.org/10.1016/j.stem.2011.11.019
157. Rumble, J. M., Huber, A. K., Krishnamoorthy, G., Srinivasan, A., Giles, D. A., Zhang, X., Wang, L., & Segal, B. M. (2015). Neutrophil-related factors as biomarkers in EAE and MS. *Journal of Experimental Medicine, 212*(1). https://doi.org/10.1084/jem.20141015
158. Russo, M. V., & McGavern, D. B. (2016). Inflammatory neuroprotection following traumatic brain injury. *Science, 353*(6301). https://doi.org/10.1126/science.aaf6260
159. Saab, A. S., Tzvetavona, I. D., Trevisiol, A., Baltan, S., Dibaj, P., Kusch, K., Möbius, W., Goetze, B., Jahn, H. M., Huang, W., Steffens, H., Schomburg, E. D., Pérez-Samartín, A., Pérez-Cerdá, F., Bakhtiari, D., Matute, C., Löwel, S., Griesinger, C., Hirrlinger, J., et al. (2016). Oligodendroglial NMDA receptors regulate glucose import and axonal energy metabolism. *Neuron, 91*(1). https://doi.org/10.1016/j.neuron.2016.05.016
160. Sabahi, M., Joshaghanian, A., Dolatshahi, M., Jabbari, P., Rahmani, F., & Rezaei, N. (2021). Modification of glial cell activation through dendritic cell vaccination: Promises for treat-

ment of neurodegenerative diseases. *Journal of Molecular Neuroscience, 71*(7). https://doi.org/10.1007/s12031-021-01818-6

161. Sabatino, J. J., Pröbstel, A. K., & Zamvil, S. S. (2019). B cells in autoimmune and neurodegenerative central nervous system diseases. *Nature Reviews Neuroscience, 20*(12). https://doi.org/10.1038/s41583-019-0233-2
162. Sagare, A. P., Bell, R. D., Zhao, Z., Ma, Q., Winkler, E. A., Ramanathan, A., & Zlokovic, B. V. (2013). Pericyte loss influences Alzheimer-like neurodegeneration in mice. *Nature Communications, 4*. https://doi.org/10.1038/ncomms3932
163. Salmina, A. B., Inzhutova, A. I., Malinovskaya, N. A., & Petrova, M. M. (2010). Endothelial dysfunction and repair in Alzheimer-type neurodegeneration: Neuronal and glial control. *Journal of Alzheimer's Disease, 22*(1). https://doi.org/10.3233/JAD-2010-091690
164. Santos, E. d. S., Aragão-França, L. S. d., Meira, C. S., Cerqueira, J. V., Vasconcelos, J. F., Nonaka, C. K. V., Pontes-de-Carvalho, L. C., & Soares, M. B. P. (2020). Tolerogenic dendritic cells reduce cardiac inflammation and fibrosis in chronic Chagas disease. *Frontiers in Immunology, 11*. https://doi.org/10.3389/fimmu.2020.00488
165. Saresella, M., Marventano, I., Piancone, F., La Rosa, F., Galimberti, D., Fenoglio, C., Scarpini, E., & Clerici, M. (2020). IL-33 and its decoy sST2 in patients with Alzheimer's disease and mild cognitive impairment. *Journal of Neuroinflammation, 17*(1). https://doi.org/10.1186/s12974-020-01806-4
166. Schaafsma, W., Basterra, L. B., Jacobs, S., Brouwer, N., Meerlo, P., Schaafsma, A., Boddeke, E. W. G. M., & Eggen, B. J. L. (2017). Maternal inflammation induces immune activation of fetal microglia and leads to disrupted microglia immune responses, behavior, and learning performance in adulthood. *Neurobiology of Disease, 106*. https://doi.org/10.1016/j.nbd.2017.07.017
167. Schain, A. J., Melo-Carrillo, A., Borsook, D., Grutzendler, J., Strassman, A. M., & Burstein, R. (2018). Activation of pial and dural macrophages and dendritic cells by cortical spreading depression. *Annals of Neurology, 83*(3). https://doi.org/10.1002/ana.25169
168. Schutt, C. R., Gendelman, H. E., & Mosley, R. L. (2018). Tolerogenic bone marrow-derived dendritic cells induce neuroprotective regulatory T cells in a model of Parkinson's disease. *Molecular Neurodegeneration, 13*(1). https://doi.org/10.1186/s13024-018-0255-7
169. Schwartz, M. (2017). Can immunotherapy treat neurodegeneration? *Science, 357*(6348). https://doi.org/10.1126/science.aai8231
170. Schwartz, M., & Deczkowska, A. (2016). Neurological disease as a failure of brain–Immune crosstalk: The multiple faces of neuroinflammation. *Trends in Immunology, 37*(10). https://doi.org/10.1016/j.it.2016.08.001
171. Sengillo, J. D., Winkler, E. A., Walker, C. T., Sullivan, J. S., Johnson, M., & Zlokovic, B. V. (2013). Deficiency in mural vascular cells coincides with blood-brain barrier disruption in alzheimer's disease. *Brain Pathology, 23*(3). https://doi.org/10.1111/bpa.12004
172. Shahnawaz, M., Mukherjee, A., Pritzkow, S., Mendez, N., Rabadia, P., Liu, X., Hu, B., Schmeichel, A., Singer, W., Wu, G., Tsai, A. L., Shirani, H., Nilsson, K. P. R., Low, P. A., & Soto, C. (2020). Discriminating α-synuclein strains in Parkinson's disease and multiple system atrophy. *Nature, 578*(7794). https://doi.org/10.1038/s41586-020-1984-7
173. Shalash, A., Salama, M., Makar, M., Roushdy, T., Elrassas, H. H., Mohamed, W., El-Balkimy, M., & Donia, M. A. (2017). Elevated serum α-synuclein autoantibodies in patients with parkinson's disease relative to Alzheimer's disease and controls. *Frontiers in Neurology, 8*(Dec). https://doi.org/10.3389/fneur.2017.00720
174. Sheean, R. K., McKay, F. C., Cretney, E., Bye, C. R., Perera, N. D., Tomas, D., Weston, R. A., Scheller, K. J., Djouma, E., Menon, P., Schibeci, S. D., Marmash, N., Yerbury, J. J., Nutt, S. L., Booth, D. R., Stewart, G. J., Kiernan, M. C., Vucic, S., & Turner, B. J. (2018). Association of regulatory T-cell expansion with progression of amyotrophic lateral sclerosis a study of humans and a transgenic mouse model. *JAMA Neurology, 75*(6). https://doi.org/10.1001/jamaneurol.2018.0035
175. Shiraishi, W., Yamasaki, R., Hashimoto, Y., Ko, S., Kobayakawa, Y., Isobe, N., Matsushita, T., Kira, J., & ichi. (2021). Clearance of peripheral nerve misfolded mutant protein by infiltrated

macrophages correlates with motor neuron disease progression. *Scientific Reports, 11*(1). https://doi.org/10.1038/s41598-021-96064-6
176. Sica, A., & Mantovani, A. (2012). Macrophage plasticity and polarization: In vivo veritas. *Journal of Clinical Investigation, 122*(3). https://doi.org/10.1172/JCI59643
177. Silvestre-Roig, C., Hidalgo, A., & Soehnlein, O. (2016). Neutrophil heterogeneity: Implications for homeostasis and pathogenesis. *Blood, 127*(18). https://doi.org/10.1182/blood-2016-01-688887
178. Snaidero, N., Schifferer, M., Mezydlo, A., Zalc, B., Kerschensteiner, M., & Misgeld, T. (2020). Myelin replacement triggered by single-cell demyelination in mouse cortex. *Nature Communications, 11*(1). https://doi.org/10.1038/s41467-020-18632-0
179. Sofroniew, M. V. (2009). Molecular dissection of reactive astrogliosis and glial scar formation. *Trends in Neurosciences, 32*(12). https://doi.org/10.1016/j.tins.2009.08.002
180. Söllvander, S., Ekholm-Pettersson, F., Brundin, R. M., Westman, G., Kilander, L., Paulie, S., Lannfelt, L., & Sehlin, D. (2015). Increased number of plasma B cells producing autoantibodies against Aβ 42 protofibrils in Alzheimer's disease. *Journal of Alzheimer's Disease, 48*(1). https://doi.org/10.3233/JAD-150236
181. Sommer, A., Maxreiter, F., Krach, F., Fadler, T., Grosch, J., Maroni, M., Graef, D., Eberhardt, E., Riemenschneider, M. J., Yeo, G. W., Kohl, Z., Xiang, W., Gage, F. H., Winkler, J., Prots, I., & Winner, B. (2018). Th17 lymphocytes induce neuronal cell death in a human iPSC-based model of Parkinson's disease. *Cell Stem Cell, 23*(1). https://doi.org/10.1016/j.stem.2018.06.015
182. Spangenberg, E. E., Lee, R. J., Najafi, A. R., Rice, R. A., Elmore, M. R. P., Blurton-Jones, M., West, B. L., & Green, K. N. (2016). Eliminating microglia in Alzheimer's mice prevents neuronal loss without modulating amyloid-β pathology. *Brain, 139*(4). https://doi.org/10.1093/brain/aww016
183. Späni, C., Suter, T., Derungs, R., Ferretti, M. T., Welt, T., Wirth, F., Gericke, C., Nitsch, R. M., & Kulic, L. (2015). Reduced β-amyloid pathology in an APP transgenic mouse model of Alzheimer's disease lacking functional B and T cells. *Acta Neuropathologica Communications, 3*. https://doi.org/10.1186/s40478-015-0251-x
184. Stephenson, J., Nutma, E., van der Valk, P., & Amor, S. (2018). Inflammation in CNS neurodegenerative diseases. *Immunology, 154*(2). https://doi.org/10.1111/imm.12922
185. Sulzer, D., Alcalay, R. N., Garretti, F., Cote, L., Kanter, E., Agin-Liebes, J., Liong, C., McMurtrey, C., Hildebrand, W. H., Mao, X., Dawson, V. L., Dawson, T. M., Oseroff, C., Pham, J., Sidney, J., Dillon, M. B., Carpenter, C., Weiskopf, D., Phillips, E., et al. (2017). T cells from patients with Parkinson's disease recognize α-synuclein peptides. *Nature, 546*(7660). https://doi.org/10.1038/nature22815
186. Tang, Y., & Le, W. (2016). Differential roles of M1 and M2 microglia in neurodegenerative diseases. *Molecular Neurobiology, 53*(2). https://doi.org/10.1007/s12035-014-9070-5
187. Uemura, M. T., Maki, T., Ihara, M., Lee, V. M. Y., & Trojanowski, J. Q. (2020). Brain microvascular pericytes in vascular cognitive impairment and dementia. *Frontiers in Aging Neuroscience, 12*. https://doi.org/10.3389/fnagi.2020.00080
188. Urban, S. L., Jensen, I. J., Shan, Q., Pewe, L. L., Xue, H. H., Badovinac, V. P., & Harty, J. T. (2020). Peripherally induced brain tissue–resident memory CD8+ T cells mediate protection against CNS infection. *Nature Immunology, 21*(8). https://doi.org/10.1038/s41590-020-0711-8
189. Vainchtein, I. D., & Molofsky, A. V. (2020). Astrocytes and microglia: In sickness and in health. *Trends in Neurosciences, 43*(3). https://doi.org/10.1016/j.tins.2020.01.003
190. Vazquez, M. I., Catalan-Dibene, J., & Zlotnik, A. (2015). B cells responses and cytokine production are regulated by their immune microenvironment. *Cytokine, 74*(2). https://doi.org/10.1016/j.cyto.2015.02.007
191. Vedam-Mai, V. (2021). Harnessing the immune system for the treatment of Parkinson's disease. *Brain Research, 1758*. https://doi.org/10.1016/j.brainres.2021.147308

192. Voskoboinik, I., Whisstock, J. C., & Trapani, J. A. (2015). Perforin and granzymes: Function, dysfunction and human pathology. *Nature Reviews Immunology, 15*(6). https://doi.org/10.1038/nri3839
193. Wang, C., Hao, J., Liu, X., Li, C., Yuan, X., Lee, R. J., Bai, T., & Wang, D. (2020). Isoforsythiaside attenuates alzheimer's disease via regulating mitochondrial function through the PI3K/AKT pathway. *International Journal of Molecular Sciences, 21*(16). https://doi.org/10.3390/ijms21165687
194. Webers, A., Heneka, M. T., & Gleeson, P. A. (2020). The role of innate immune responses and neuroinflammation in amyloid accumulation and progression of Alzheimer's disease. *Immunology and Cell Biology, 98*(1). https://doi.org/10.1111/imcb.12301
195. Wes, P. D., Holtman, I. R., Boddeke, E. W. G. M., Möller, T., & Eggen, B. J. L. (2016). Next generation transcriptomics and genomics elucidate the biological complexity of microglia in health and disease. *GLIA, 64*(2). https://doi.org/10.1002/glia.22866
196. Winkler, E. A., Sagare, A. P., & Zlokovic, B. V. (2014). The pericyte: A forgotten cell type with important implications for Alzheimer's disease? *Brain Pathology, 24*(4). https://doi.org/10.1111/bpa.12152
197. Wo, J., Zhang, F., Li, Z., Sun, C., Zhang, W., & Sun, G. (2020). The role of gamma-delta T cells in diseases of the central nervous system. *Frontiers in Immunology, 11*. https://doi.org/10.3389/fimmu.2020.580304
198. Wortel, C. M., & Heidt, S. (2017). Regulatory B cells: Phenotype, function and role in transplantation. *Transplant Immunology, 41*. https://doi.org/10.1016/j.trim.2017.02.004
199. Wyss-Coray, T. (2016). Ageing, neurodegeneration and brain rejuvenation. *Nature, 539*(7628). https://doi.org/10.1038/nature20411
200. Wyss-Coray, T., & Mucke, L. (2002). Inflammation in neurodegenerative disease - A double-edged sword. *Neuron, 35*(3). https://doi.org/10.1016/S0896-6273(02)00794-8
201. Xu, W., Lau, Z. W. X., Fulop, T., & Larbi, A. (2020). The aging of $\gamma\delta$ T cells. *Cells, 9*(5). https://doi.org/10.3390/cells9051181
202. Yang, J., Kumar, A., Vilgelm, A. E., Chen, S. C., Ayers, G. D., Novitskiy, S. V., Joyce, S., & Richmond, A. (2018). Loss of CXCR4 in myeloid cells enhances antitumor immunity and reduces melanoma growth through NK Cell and FASL mechanisms. *Cancer Immunology Research, 6*(10). https://doi.org/10.1158/2326-6066.CIR-18-0045
203. Yang, Q., Wang, G., & Zhang, F. (2020). Role of peripheral immune cells-mediated inflammation on the process of neurodegenerative diseases. *Frontiers in Immunology, 11*. https://doi.org/10.3389/fimmu.2020.582825
204. Zang, X., Chen, S., Zhu, J. Y., Ma, J., & Zhai, Y. (2022). The emerging role of central and peripheral immune systems in neurodegenerative diseases. *Frontiers in Aging Neuroscience, 14*. https://doi.org/10.3389/fnagi.2022.872134
205. Zeis, T., Enz, L., & Schaeren-Wiemers, N. (2016). The immunomodulatory oligodendrocyte. *Brain Research, 1641*. https://doi.org/10.1016/j.brainres.2015.09.021
206. Zenaro, E., Pietronigro, E., Bianca, V. D., Piacentino, G., Marongiu, L., Budui, S., Turano, E., Rossi, B., Angiari, S., Dusi, S., Montresor, A., Carlucci, T., Nanì, S., Tosadori, G., Calciano, L., Catalucci, D., Berton, G., Bonetti, B., & Constantin, G. (2015). Neutrophils promote Alzheimer's disease-like pathology and cognitive decline via LFA-1 integrin. *Nature Medicine, 21*(8). https://doi.org/10.1038/nm.3913
207. Zenaro, E., Piacentino, G., & Constantin, G. (2017). The blood-brain barrier in Alzheimer's disease. *Neurobiology of Disease, 107*. https://doi.org/10.1016/j.nbd.2016.07.007
208. Zhan, Y., Paolicelli, R. C., Sforazzini, F., Weinhard, L., Bolasco, G., Pagani, F., Vyssotski, A. L., Bifone, A., Gozzi, A., Ragozzino, D., & Gross, C. T. (2014). Deficient neuron-microglia signaling results in impaired functional brain connectivity and social behavior. *Nature Neuroscience, 17*(3). https://doi.org/10.1038/nn.3641
209. Zhang, S., Feng, R., Li, Y., Gan, L., Zhou, F., Meng, S., Li, Q., & Wen, T. (2018). Degradation of alpha-synuclein by dendritic cell factor 1 delays neurodegeneration and extends lifespan in Drosophila. *Neurobiology of Aging, 67*. https://doi.org/10.1016/j.neurobiolaging.2018.03.010

210. Zhao, W., Xie, W., Xiao, Q., Beers, D. R., & Appel, S. H. (2006). Protective effects of an anti-inflammatory cytokine, interleukin-4, on motoneuron toxicity induced by activated microglia. *Journal of Neurochemistry, 99*(4). https://doi.org/10.1111/j.1471-4159.2006.04172.x
211. Zhou, T., Zheng, Y., Sun, L., Badea, S. R., Jin, Y., Liu, Y., Rolfe, A. J., Sun, H., Wang, X., Cheng, Z., Huang, Z., Zhao, N., Sun, X., Li, J., Fan, J., Lee, C., Megraw, T. L., Wu, W., Wang, G., & Ren, Y. (2019). Microvascular endothelial cells engulf myelin debris and promote macrophage recruitment and fibrosis after neural injury. *Nature Neuroscience, 22*(3). https://doi.org/10.1038/s41593-018-0324-9
212. Zhou, Z., He, H., Wang, K., Shi, X., Wang, Y., Su, Y., Wang, Y., Li, D., Liu, W., Zhang, Y., Shen, L., Han, W., Shen, L., Ding, J., & Shao, F. (2020). Granzyme A from cytotoxic lymphocytes cleaves GSDMB to trigger pyroptosis in target cells. *Science, 368*(6494). https://doi.org/10.1126/science.aaz7548
213. Zozulya, A. L., Clarkson, B. D., Ortler, S., Fabry, Z., & Wiendl, H. (2010). The role of dendritic cells in CNS autoimmunity. *Journal of Molecular Medicine, 88*(6). https://doi.org/10.1007/s00109-010-0607-4
214. Zrzavy, T., Hametner, S., Wimmer, I., Butovsky, O., Weiner, H. L., & Lassmann, H. (2017). Loss of "homeostatic" microglia and patterns of their activation in active multiple sclerosis. *Brain, 140*(7). https://doi.org/10.1093/brain/awx113

# Cytokine Alterations and Targeting in Neurodegeneration: Role of Nanomaterials

Abdul Naeem and Syed Shadab Raza

**Abstract**

Neuroinflammation, a hallmark of neurodegenerative diseases including Alzheimer's disease (AD), Parkinson's disease (PD), and amyotrophic lateral sclerosis (ALS), arises from an inflammatory response within the central nervous system (CNS). Over-activated microglia, the resident immune cells in the CNS, release pro-inflammatory cytokines, leading to neuronal damage and accelerated neurodegeneration. However, effective drug delivery to the CNS faces a significant challenge due to the blood-brain barrier (BBB). This natural protective barrier restricts harmful plasma components from accessing brain tissue. Nanomaterials have emerged as a promising therapeutic strategy. Moreover, biodegradability and biocompatibility of nanomaterials offer potential as minimally invasive carriers for anti-inflammatory drugs. By leveraging nanomaterials, emerging research aims to overcome the BBB and inhibit microglial over-activation, thereby modulating neuroinflammation in neurodegenerative diseases. Further research is essential to optimize nanomaterials-based therapeutic protocols and enhance drug delivery across the BBB. In this chapter, we have discussed and compiled information on different types of nanoparticles utilized to treat neuroinflammatory diseases. Here, we explain their mechanisms and effects in modulating inflammatory responses associated with neurodegenerative diseases. Additionally, we provide information and recommendations for using nanoparticles in the treatment of neuroinflammation-related disorders.

**Keywords**

Blood-brain barrier · Central nervous system · Nanoparticles · Nanomaterials · Interleukin · Cytokines · Neuroinflammation

A. Naeem · S. S. Raza (✉)
Laboratory for Stem Cell & Restorative Neurology, Department of Biotechnology, Era's Lucknow Medical College & Hospital, Era University, Lucknow, India



R. Shukla et al. (eds.), *Neuroendocrine-Immune Axis*,
https://doi.org/10.1007/978-981-95-5088-3_5

# 1 Introduction

Neuroinflammation involves a complex interplay between immune cells and signaling molecules within the CNS [83]. It is characterized by the activation of microglia and astrocytes that release inflammatory cytokines and reactive oxygen species (ROS). This inflammatory state can result in synaptic dysfunction, synapse loss, and neuronal damage [137]. Neuroinflammation is a common underlying mechanism in various central nervous system (CNS) diseases. Research has increasingly focused on strategies to alleviate and inhibit this process [145]. Depending on the situation, cytokines can modulate inflammation either by increasing or decreasing it. Pro-inflammatory cytokines such as interleukin (IL)-1β, IL-6, IL-12, IL-18, IL-23, IL-33, and tumor necrosis factor alpha (TNF-α) are actively involved in neurodegenerative diseases and have been observed to be elevated in animal models as well as in humans ([12], [174], [106, 139]). Elevated IL-1β expression has been observed in the microglia around amyloid-β (Aβ) plaques in patients of AD and α-synuclein pathology in PD [9, 104]. However, crossing the selectively permeable BBB is one of the more evident difficulties in delivering anti-inflammatory medications to astrocytes and microglia in the CNS. This barrier is essential for CNS defense by selectively permitting some molecules to pass through while blocking the entry of toxins and infections [17, 108]. Though it poses a considerable obstacle to drug delivery to the CNS, this protective feature of the BBB often reduces the overall effectiveness of therapies for neuroinflammation. However, addressing this difficulty has been fruitful in light of recent advancements in nanotechnology. According to recent research breakthroughs that greatly enhanced drug delivery to its areas of action within the brain were the creation of nanoparticles that could pass the BBB [1, 130, 150].

Nanomaterials for the improvement of CNS neuroinflammation management are potentially useful [150, 164]. A myriad of nanomaterials have emerged in the past two decades, so it is expected that nanomedicine will further evolve and probably change the paradigms of disease pathogenesis, diagnosis, and treatment management [22]. Nanomaterials are materials that are engineered to have at least one dimension in the size range of 1–100 nm. They also enable successful interactions with biological systems at the molecular level [109]. Nanomaterials have the potential to encapsulate high drug loading capacity and allow easy surface modification to achieve targeted drug delivery [158]. Furthermore, nanomaterials can control the site-specific release of drugs at target cells or tissues, enhancing therapeutic efficiency and minimizing drug adverse effects [88]. Under nanomaterials-based delivery systems, drugs that are insoluble or unstable in aqueous solutions can have their solubility improved and their pharmacological effects [125].

Indeed, using nanomaterials to deliver modulation of cytokines straight to the CNS is one promising way to treat neuroinflammation. However, nanoformulations developed from lipid-based nanoparticles like liposomes and solid lipid nanoparticles can encapsulate drugs like dexamethasone or natural plant-based compounds such as curcumin, which have potent anti-inflammatory properties [155]. These lipid-based carriers potentially cross the BBB, whereupon the drug is released in a controlled manner and directly targets inflamed areas to reduce systemic side effects [6]. Moreover, polymer-based nanoparticles are also employed to deliver drugs like minocycline, a tetracycline antibiotic with anti-inflammatory and neuroprotective effects. These nanoparticles ensure a controlled release with sustained drug levels in the CNS, thus improving the therapeutic outcome and reducing dosing frequency [83].

Moreover, inorganic nanomaterials like gold and cerium oxide nanoparticles are reportedly able to alleviate neuroinflammation [97]. AuNPs were shown to decrease the release of pro-inflammatory cytokines, such as IL-1β and TNF-α, which may be effective for neuroprotection against inflammation-mediated injury [32]. In the same way, CeO NPs, which also exerted antioxidant effects related to scavenging ROS and oxidative stress-reducing activity, are one of the major contributors to neuroinflammation. In conclusion, nanomaterials have huge potential for better therapy in neuroinflammatory diseases by modulating cytokines [73, 128]. The continued development and optimization of such nanomaterials hold significant potential for advancing the treatment of neuroinflammatory conditions and improving patient outcomes.

## 2 Inflammation Inducers in Neurodegenerative Diseases

Neuroinflammation refers an inflammatory reaction inside the CNS that resulted in the activation of neuroimmune cells, specifically astrocytes and microglia, leading to a pro-inflammatory state [131, 160]. Neurodegenerative disorders are characterized by various shared and unique pathological and clinical aspects, including the selective susceptibility of specific brain regions and the aggregation of different proteins. In addition to these neuropathological and clinical features, these disorders are associated with insistent and enduring inflammation due to increased cytokine production [52, 162]. Initially, it was believed that inflammation was a consequence of protein aggregation in the CNS [35]. However, accumulating evidence suggests that immune signaling not only results from but also precedes and promotes protein aggregation during the early stages of neurodegenerative diseases [66–68, 120].

The immune system is pivotal in maintaining cellular homeostasis, clearing pathogens, and facilitating tissue repair [51, 116]. Usually, immune activation are advantageous and self-limiting, resolving once the tissue injury is repaired or the infection is exterminated. However, the persistent inflammatory stimulus overwhelmed the normal resolution mechanisms, resulting in chronic inflammation. This chronic state can lead to the release of neurotoxic factors, thereby exacerbating the disease [52].

Persistent inflammation can be driven by endogenous factors such as protein aggregates [66–68], environmental factors including systemic infections, gut microbiota dysbiosis, aging, diet [42, 98, 161], [68], and genetic susceptibilities, such as mutations in progranulin (PGRN) and apolipoprotein E4 (APOE4) [102, 119]. Moreover, specialized pro-resolving lipid mediators (SPMs) are crucial for promoting tissue healing and are also involved in suppressing inflammatory cytokines. A

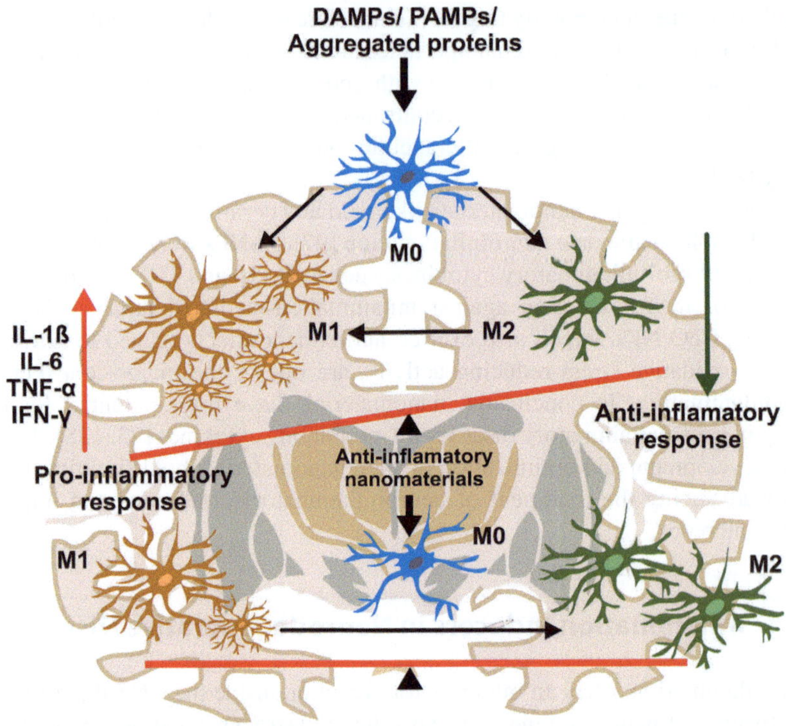


**Fig. 1** DAMPs/PAMPs, and aggregated proteins activate resting microglia (M0) into a pro-inflammatory state (M1), triggers the production of cytokines. To keep the balance of M1/M2 type microglia, anti-inflammatory nanomaterials prevent the over-activation of microglia and encourage them to enter an anti-inflammatory state

deficiency in resolving inflammation due to suppressed level of SPMs is one more contributing factor for brain diseases associated with chronic inflammation [132] (Fig. 1).

Recent studies suggest that the over-activation of resting microglia (M0) in neurodegeneration may be initiated by a range of pathogen-associated molecular patterns (PAMPs) or danger-associated molecular patterns (DAMPs), such as particles, viruses, bacteria, fungi, toxins, lipopolysaccharide (LPS), crystals, silica, and misfolded protein aggregates (Aβ, Tau, α-synuclein, etc.) [3, 5, 8]. The CNS has a specialized neuroimmune system consisting of microglia and astrocytes. Moderately activated microglia (M2) release anti-inflammatory cytokines like Arg-1, TGF-β, and IL-10, which protect the brain by defending against harmful materials [49]. On the other hand, excessive activation of microglia (M1) can lead to neurobehavioral problems such as cognitive impairments and major pathological changes [110]. During the condition of neuroinflammation, inflammatory cytokines such as IL-1β, IL-6, IL-18, and TNF-α are produced, along with chemokines like CCL1, CCL5, and CXCL1. Small molecule messengers such as prostaglandins, nitric oxide (NO), and ROS are also involved [39]. Anti-inflammatory nanomaterial therapy triggers a shift from M1-type to M2-type microglia, evidenced by increased anti-inflammatory cytokines and reduced pro-inflammatory cytokines [2, 21].

In summary, neuroinflammation, characterized by microglial and astrocyte activation and the release of cytokines and ROS, is a common mechanism in various neurodegenerative diseases. The BBB significantly hampers the delivery of many anti-inflammatory drugs to the CNS, thus posing a major challenge in treating neuroinflammation. Nanotechnology, particularly the use of nanoparticles, offers promising solutions for delivering drugs across the BBB, targeting neuroinflammation more effectively, and potentially altering the course of neurodegenerative diseases [1, 130].

## 3 Nanomaterials and Its Therapeutic Option in Neuroinflammatory Diseases

Nanomaterials can broadly be divided into inorganic and organic nanomaterials. Inorganic nanomaterials include gold nanoparticles (AuNPs), titanium dioxide nanoparticles (TiO2 NPs), iron oxide nanoparticles (IONPs), and other metal-based nanoparticles. These nanomaterials show high stability, and because of their magnetic nature and ease of functionalization, they are used as therapeutic agents for image enhancement during drug administration [157]. There can be mainly three types of organic nanomaterials: lipid-based nanoparticles, nanoemulsions, and polymer-based nanoparticles. Lipid nanoparticles represent another vital aspect based on two very important advantages of liposomes and SLNs are biocompatible and can encapsulate both hydrophobic and hydrophilic drugs to enhance their solubility and stability [147]. High surface area and enhanced drug bioavailability are provided by nanoemulsions, which are fine oil-in-water or water-in-oil dispersions stabilized by surfactants. They are polymer nanoparticles, designed in forms such as polymeric nanoparticles, dendrimers, nanogels, and micelles, for controlled drug release and targeted delivery [88]. Carbon nanoparticles (CNPs) are a third class of nanomaterial that is also recognized. CNPs encompass diverse structures, including fullerenes, carbon nanotubes (CNTs), graphene, carbon quantum dots (CQDs), nanodiamonds, and carbon nanohorns [126], [129]. These materials exhibit unique properties, making them valuable in various fields. CNPs find applications in analytical sensing, bioimaging, theranostics, and catalysis [103]. However, challenges like assessing toxicity and achieving scalability remain crucial to fully unlock their potential [84] (Fig. 2).

The application of these nanomaterials in treating neuroinflammation through modulating cytokines holds significant promise. Inorganic nanomaterials, such as AuNPs and TiO2 NPs, have already demonstrated very high efficiency regarding neuroinflammation by reducing oxidative stress and modulating immune responses [135]. IONPs have already been used due to their magnetic properties in magnetic resonance imaging and as carriers for targeted drug delivery into inflamed brain areas [6].

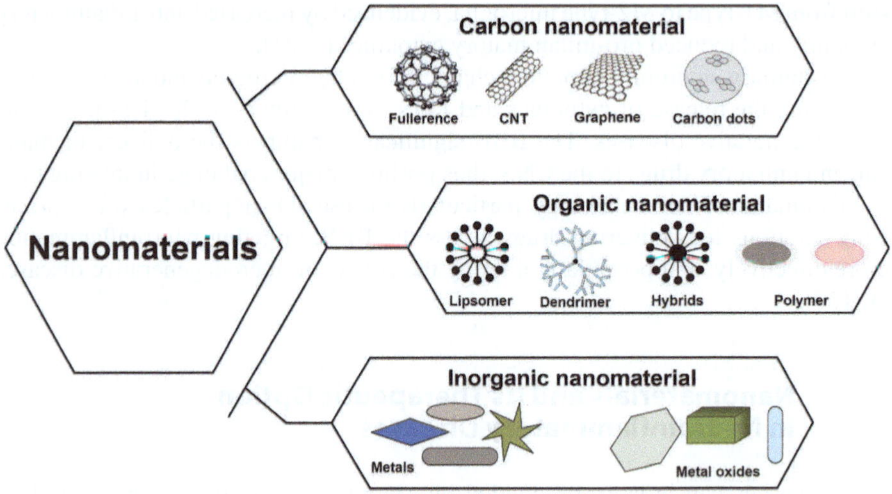


**Fig. 2** Examples of different types of nanomaterials include carbon-based, organic, and inorganic nanomaterials

In that way, organic nanomaterials also offer very considerable advantages in neuroinflammation treatment. Lipid nanoparticles, such as liposomes, may go through the BBB and provide anti-inflammatory responses precisely to the site of inflammation, therefore increasing their in vivo therapeutic efficiency while reducing systemic side effects [105]. Polymeric nanoparticles can be engineered to release drugs in a controlled manner, ensuring sustained therapeutic effects and reducing the frequency of drug administration [135].

NPs offer a wide range of choices for drug administration routes, including nasal, oral, intravenous, parenteral, intra-ocular, and dermal topical applications [58]. This versatility makes them particularly useful for targeting the CNS, where the BBB presents a significant challenge for drug delivery. In recent years, a variety of nanomaterials have been created as efficient and non-invasive transporters that facilitate crossing of the BBB, either by assisting other drugs or by acting as therapeutic agents for neurodegenerative diseases [10].

In conclusion, the development of nanomaterials has been one of the most significant advances made in the field of neuroinflammation treatment. There exist unique properties spanning both inorganic and organic nanomaterials that could potentially be harnessed to improve the pattern of drug delivery, enhance therapeutic efficacies, and reduce adverse effects. With the progress of research in nanomedicine, these materials are, therefore, going to assume a very important place in the future pattern of healthcare as far as the prevention, diagnosis, and treatment of neuroinflammatory diseases are concerned.

## 4 Cytokine Production in Neurodegenerative Diseases

Cytokines serve as critical messengers within the immune system, bridging innate and adaptive immunity [18]. Several cell types within the CNS, including glial cells, blood components, and peripheral tissues, are capable of synthesizing cytokines [84]. These cytokines exert their effects by binding to their respective receptors and activating downstream signaling through the Janus kinases (JAK), Signal Transducer and Activator of Transcription (STAT), or MAPK signaling pathways [123]. Cytokines can identify target cells, prompting the migration of additional immune components and the production of more cytokines [113, 123]. Cytokine expression is typically modest in the CNS and peripheral organs. However, immune cells like T cells and peripheral cytokines can use active transport pathways to breach the blood-brain barrier in response to immunological stressors. Additionally, peripheral cytokines can affect CNS processes by diffusing across the BBB [16, 149].

Moreover, microglia and mast cells residing in the CNS contribute to neuroinflammation by releasing various cytokines. Research has demonstrated that these pro-inflammatory cytokines are essential mediators in initiating and sustaining leukocyte-driven immune responses within the CNS. Among the most crucial cytokines involved are IL-1β, IL-6, IL-23, interferon-gamma (IFNγ), TNF, and granulocyte macrophage colony stimulating factor (GM-CSF) [61, 101]. The presence of these cytokines suggests that all cells within the CNS can produce IL-1β and IL-6, contributing to the inflammatory process. IL-23, primarily produced by T cells, enhances the pro-inflammatory functions of immune cells, contributing to disease pathology, while GM-CSF facilitates the action of inflammatory monocyte-derived cells, leading to tissue damage [61].

Cytokines initiate their effects through intricate signaling pathways. The JAK-STAT pathway involves the binding of cytokines to their receptors, resulting in the activation of JAKs. These kinases then phosphorylate STAT proteins, which translocate to the nucleus to modulate gene expression. Similarly, cytokine-receptor interactions activate the MAPK pathway, leading to a cascade of kinase activations, ending in the regulation of transcription factors that trigger inflammatory responses [122, 123].

In neuroinflammation, cytokines play dual roles, acting as both mediators and regulators. During an immune response, cytokines help recruit immune cells to the site of inflammation, enhancing the immune response. However, their overproduction or prolonged presence can exacerbate inflammation, causative to the pathological progression of neurodegenerative disorders [66, 67]. For instance, IL-6 is involved in both acute and chronic inflammatory responses and has been implicated in the progression of AD and multiple sclerosis [43]. The ability of peripheral cytokines to influence CNS function underscores the importance of the BBB in maintaining CNS homeostasis. However, during neuroinflammation, the integrity of the BBB can be compromised, allowing cytokines and immune cells to infiltrate the CNS, thereby perpetuating the inflammatory response [16].

Moreover, cytokines like IL-1β and TNF-α have been shown to alter BBB permeability, further facilitating the entry of immune cells into the CNS. This can lead to a vicious cycle of inflammation, where cytokine-induced BBB disruption leads to increased CNS infiltration of immune cells and subsequent cytokine production [36]. Understanding the role of cytokines in neuroinflammation is crucial for developing therapeutic strategies for neurodegenerative diseases. Targeting cytokine signaling pathways offers the potential for modulating the inflammatory response in the CNS, thereby mitigating the progression of diseases such as AD, PD, and multiple sclerosis [52].

## 5 Key Cytokines Involved in Neurodegenerative Diseases and Their Modulation by Nanomaterials

Several cytokines have been identified as key players in the pathogenesis of neurodegenerative diseases. Among these, IL-1β, IL-6, TNF-α, and IFN-γ are prominently involved in the inflammatory responses associated with neurodegenerative diseases [121, 156]. Understanding their roles and mechanisms of action is essential for developing targeted therapeutic strategies through nanomaterials (Fig. 3).

## 6 IL-1β

IL-1β is considered to be a pro-inflammatory cytokine that is produced by activated microglia. It is also involved in the elevation of ROS and NO, leading to the process of neurodegeneration through oxidative stress and neuronal injury [64]. Nanomaterials might represent a potential tool for modulating the effects of IL-1β on neurodegenerative diseases. For instance, the anti-inflammatory properties of AuNPs and cerium oxide nanoparticles (CeONPs) have been investigated in many studies [46, 69, 82]. AuNPs showed a reduction in IL-1β levels, attenuated the oxidative stress in neuronal BV2 cells, and provided neuroprotection against neuroinflammation [4]. A research study conducted by Ling et al. demonstrated that the administration of AuNPs mitigates oxidative stress and motor abnormalities induced by MPTP in PD rats. Furthermore, the analysis also revealed that AuNPs treatment significantly reduces the production of IL-1β, TNF-α, and IL-6 [93].

Similarly, CeONPs show catalytic antioxidant activity for the reduction in ROS production and modulation of IL-1β activity, resulting in protection from inflammation-induced neurotoxicity [37]. Das et al. also demonstrated that CeONPs were able to attenuate IL-1β-mediated neurotoxicity by scavenging free radicals and reducing oxidative stress in rat model of spinal cord injury [37]. These results highlight the potential use of CeONPs in lowering neuroinflammatory responses and thus maintaining neuronal integrity.

Furthermore, carbon-based nanomaterials, like graphene oxide (GO), also have significant potential in suppressing IL-1β-induced neuroinflammation [96]. Liao et al. reported that GO would suppress the activation of microglia and astrocytes that in turn down-regulated the production of IL-1β and other pro-inflammatory cytokines in a rat spinal cord injury model [91]. Moreover, the unique properties of

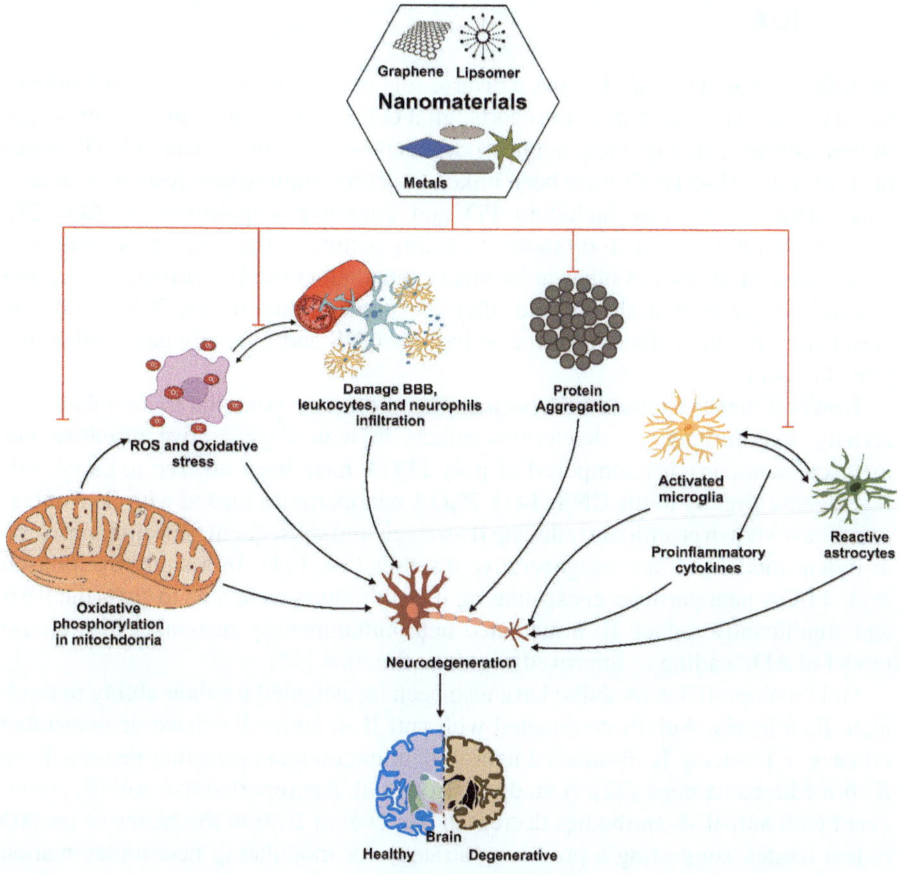


**Fig. 3** Decreased mitochondrial activity causes an imbalance between ROS/RNS and antioxidants, which in turn activates microglia and astrocytes. NPs delivery systems facilitate medication distribution across the BBB to suppress hyperactivated microglia and the ensuing neuroinflammatory response

GO, including its large surface area and high mechanical strength, make it an attractive candidate for delivering anti-inflammatory agents to specific sites of neuroinflammation [15].

Another auspicious nanomaterial is poly lactic-co-glycolic acid (PLGA) nanoparticles, which have already been used for IL-1β antagonist delivery directly into the CNS. For example, PLGA nanoparticles loaded with IL-1Ra crossed the BBB and decreased the concentration of IL-1β, reducing neuroinflammation and attenuating neurological outcomes [19]. Targeted delivery was achieved with this system, enhancing therapeutic efficacy while reducing systemic side effects [117]. These nanomaterials, therefore, hold a potential avenue to modulate the activity of IL-1β and attenuate neuroinflammation in neurodegenerative diseases.

# 7 IL-6

Multifunctional cytokine IL-6 has a diverse role in immune modulation and inflammatory processes. According to reports, glial cell-produced IL-6 may be the cause of both chronic inflammation and neurodegeneration in neurodegenerative illnesses [43]. Elevated IL-6 levels have been linked to neuroinflammatory responses in neurodegenerative diseases including PD and cognitive impairment in AD [24]. Inducing brain injury, IL-6 promotes signaling pathways that alter the synthesis of acute-phase proteins and other inflammatory mediators [124]. Furthermore, recent research suggests that IL-6 could alter the permeability of the BBB, allowing peripheral immune cells to translocate into the CNS and intensify neuroinflammation [17, 141].

However, nanomaterials have been explored for their potential to modulate IL-6 activity and mitigate its deleterious effects in neurodegenerative diseases. For instance, nanoparticles composed of poly PLGA have been utilized to carry anti-IL-6 agents directly to the CNS [141]. PLGA nanoparticles loaded with IL-6 antagonists have shown promise in reducing IL-6 levels and subsequent neuroinflammation in rodent models of neurodegenerative diseases [34, 141]. In a study by Johnson et al., PLGA nanoparticles encapsulating IL-6 inhibitors were able to cross the BBB and significantly reduce IL-6-mediated neuroinflammatory responses in a mouse model of AD, leading to improved cognitive function [72].

Gold nanoparticles (AuNPs) have also been investigated for their ability to modulate IL-6 levels. AuNPs conjugated with anti-IL-6 antibodies have demonstrated efficacy in reducing IL-6-induced neuroinflammation and protecting neurons from IL-6-mediated damage [32]. A study by Hou et al. has reported that AuNPs conjugated with anti-IL-6 antibodies decreased the level of IL-6 in the brains of the AD rodent model, suggesting a promising strategy for modulating neuroinflammation [70]. Similarly, CeONPs modulated the activity of IL-6. Having antioxidant properties, CeONPs can reduce oxidative stress and inflammation in the CNS [14]. In a study done by Li et al., $CeO_2$NPs attenuated neuroinflammation caused by IL-6 via ROS-scavenging mechanism and NF-κB pathway inhibition [89]. Silica nanoparticles are currently under study for their potential for intervention with IL-6 expressions in the context of neurodegenerative diseases [30]. In one of the studies by Lee et al., silica nanoparticles were used as delivery agents for siRNA that would target the IL-6 mRNA molecules. There is a considerable reduction in the resultant IL-6 amount and neuroinflammatory responses while, simultaneously, resulting in high mRNA and protein levels of angiogenic factors [103].

Graphene oxide and other carbon-based nanomaterials have also been investigated for their efficacy in regulating the activity of IL-6 [28, 81]. Moreover, a recent study conducted by Jiang et al. found that chiral metal-organic frameworks have demonstrated a prospective therapy for neuroinflammation by lowering IL-6 and ROS levels, which in turn restores motor function in a mouse model of PD [75].

These studies show the potential of nanomaterials to modulate IL-6 activity and reduce neuroinflammation in neurodegenerative diseases. In this respect, the nanotechnologies targeting IL-6, reducing its pro-inflammatory effects, can be proposed as new therapeutic strategies to protect neuronal health and improve the clinical outcomes of patients with neurodegenerative disorders.

## 8 TNF-α

TNF-α is another critical pro-inflammatory cytokine involved in neurodegenerative diseases [53]. TNF-α is produced by activated microglia and astrocytes in the CNS that play a crucial role in the pathogenesis of AD, PD, and ALS [118, 138]. This cytokine induces the expression of adhesion molecules, chemokines, and other cytokines, thus facilitating inflammation and amplifying the inflammatory response. Moreover, TNF-α directly induces neuronal apoptosis through the activation of death receptors on neurons, leading to neurodegeneration [52, 115]. Very recently, the role of TNF-α in synaptic dysfunction and loss has been identified, further implicating it in cognitive deficits occurring in neurodegenerative diseases [45].

Furthermore, nanomaterials have the potential to modulate TNF-α activity to reduce their deleterious effects in neurodegenerative diseases [57]. For example, AuNPs conjugated with TNF-α inhibitors have been reported to decrease the levels of TNF-α and attenuate neuroinflammation [85]. In another instance, Ou et al. showed that AuNPs conjugated with anti-TNF-α antibodies crossed the BBB and significantly reduced TNF-α-mediated neuroinflammatory responses within the AD mouse models, leading to improved cognitive function [111]. Furthermore, the neuroprotective activity of CeONPs against TNF-α-induced neurotoxicity has also been investigated. In light of their speed-related capabilities for antioxidant activity, CeONPs can scavenge ROS, resulting in reduced oxidative stress in the CNS. Heckman et al. also reported how CeONPs dampened TNF-α-induced neuroinflammation and reduced neuronal apoptosis in a rat model of PD [65].

In addition, silica nanoparticles were also tested for modulation of TNF-α activity [78]. According to the Díaz-García et al. study, silica nanoparticles were used that were loaded with TNF-α inhibitors, targeted against neuroinflammatory responses in the mouse model of ALS, and significantly reduced TNF-α levels and neuroinflammatory markers, improving motor function in treated mice [38].

Furthermore, polymer-based nanoparticles have also been used to facilitate the proper delivery of anti-TNF-α agents into the CNS, such as PLGA. PLGA NPs loaded with TNF-α inhibitors were found to be able to decrease neuroinflammation mediated by TNF-α and to exert neuroprotection on dopaminergic neurons in a mouse model of PD [117].

These studies underline the potential of nanomaterials in modulating TNF-α activity and mitigating neuroinflammation in neurodegenerative diseases. Specifically, such nanotechnologies, targeting TNF-α and reducing its pro-inflammatory effects, might hold some new therapeutic strategies for protecting neuronal health and improving clinical outcomes in patients with neurodegenerative disorders.

## 9 IFN-γ

IFN-γ is a cytokine mainly produced by T cells and NK cells, which exerts very strong immunoregulatory and pro-inflammatory activities in the CNS [151]. IFN-γ triggers activation in microglia and astrocytes, stimulating the generation of several other pro-inflammatory cytokines and chemokines responsible for neuroinflammation [92]. Higher IFN-γ levels were shown during the process of different neurodegenerative diseases, which means a direct relationship with more neuronal damage and inflammation ([70], 67). Besides, IFN-γ could enhance the antigen-presenting ability of microglia, promoting peripheral immune cells to infiltrate CNS more easily, leading to the deterioration of certain diseases such as multiple sclerosis and PD [29, 77, 79].

It has been reported that inhibiting IFN-γ signaling pathways at different levels mitigated neuroinflammation and reduced neuronal damage in mouse models of AD [163]. Furthermore, some strategies based on nanotechnology targeting IFN-γ have shown encouraging results in attenuating its effects on neurodegenerative disorders [146]. Gold nanoparticles, functionalized with inhibitors of IFN-γ, have been engineered to inhibit selectively the IFN-γ-mediated signaling pathways and thus reduce neuroinflammation but preserve neuronal function [44, 136]. Indeed, IFN-γ and its interactions with other cytokines further underline the multifactorial involvement that characterizes neuroinflammation in neurodegenerative diseases. Knowing the mechanisms may help in the design of targeted therapies by modulation of cytokine activity to halt disease progression (Tables 1 and 2).

## 10 Neurotoxicity of Nanomaterials

Nanomaterials hold great promise for treating neuroinflammation, offering notable therapeutic benefits. However, their potential neurotoxicity has raised valid concerns among researchers. At the nanoscale, atoms can exhibit increased reactivity, potentially leading to harmful effects [5]. Thus, it is crucial to assess the neurotoxicity of nanoparticles used for neuroinflammation therapy.

Neurotoxicity caused by physical or chemical agents involves adverse effects on the nervous system's structure, function, or chemistry [142, 143]. The neurotoxicity of nanoparticles is influenced by various parameters, including shape, dosage, size, and surface area, with size and surface area being particularly critical [126]. However, toxic effects are generally more pronounced with inorganic nanoparticles when compared to organic ones [23]. Key mechanisms underlying neurotoxicity involve an overproduction of ROS, which triggers oxidative stress; increased cytokine release that promotes neuroinflammatory responses; and disrupted apoptotic pathways that ultimately lead to neuronal cell death [144].

Various types of nanoparticles have been investigated for their neurotoxic potential. For instance, gold nanoparticles (AuNPs) have been connected to increase in both the number and size of astrocytes that induces cognitive impairments like attention and memory problems in AD patients and is also associated with other

**Table 1** The involvement of nanoparticles in neuroinflammation by cytokine modulation within in vitro models

| S. no. | NPs | Cell line | Treatment time | Dose | Mechanism and detected markers | References |
|---|---|---|---|---|---|---|
| 1 | MWCNTs | iPS cells | 24 h | 16 and 64 μg/ml | NF-κB-KLF4 pathway, and nNOS | [76] |
| 2 | ZnO NPs | PC12 cells | 6 h or 12 h | 0–20 μg/ml | ROS and inflammatory cytokines | [94, 95] |
| 3 | IONP | SH-SY5Y cells | 24 h or 48 h | 200 μg/ml | AMPK pathway | [71] |
| 4 | Silica NPs | bEnd.3, N9, and BV2 cells | 24 h | 25–200 μg/ml | ROS and inflammatory cytokines | [40] |
| 5 | Mn3O4 NPs | PC12 cells | 24 h | 5, 10, and 20 μg/ml | ROS and apoptosis | [31] |
| 6 | Co. NPs | SH-SY5Y cells | 24 h | 1–100 μg/ml | Oxytosis | [60] |
| 7 | Ag NPs | Dopaminergic neuronal cell line N27 | 24 h | 50 μg/ml | ROS and TNF-α | [54] |
| 8 | AuNCs | BV2 and N2a cells | | <5 μg/ml | Inflammatory cytokines and iNOS | [154] |
| 9 | D-mino | BV2 cells | 24 h | Concentration range of 50–500 μM | NO and TNF-α | [134] |
| 10 | PEGOL-60 | BV2 cells | 24 h | 500 μg/ml | Inflammatory cytokines and iNOS | [133] |
| 11 | dPGS | Primary neuroglia | 1 h | 1 M | Aβ and LCN2 | [100] |
| 12 | PAMAM-(COOH)46-(NAC)18 | BV2 cells | 3 h | 0.5, 2, and 8 mM | ROS, NO, and TNF-α | [152] |

brain disorders such as hypoxia, ischemia, and seizures [32, 127]. High doses of anatase titanium dioxide nanoparticles (TiO2NPs) have significantly elevated IL-6 levels in blood and brain tissues. This implies that the ingestion of such nanoparticles may trigger neuroinflammatory responses and exert neurotoxic effects within the central nervous system [55].

Exposure to iron oxide nanoparticles (IONPs) has been shown to disrupt synaptic signaling and impair nerve conduction, which may lead to immune cell infiltration, inflammation in neural tissue, and programmed cell death [153]. Notably, IONPs appear to induce oxidative stress specifically in the striatum, with minimal effects observed in the hippocampus [81]. Studies have also shown that drug-free liposomes can cause neuropathological changes, including neuroinflammation and

**Table 2** The involvement of nanoparticles in neuroinflammation by cytokine modulation within in vivo models

| S. no. | NPs | Animals | Administration route | Dose | Mechanism and detected markers | References |
|---|---|---|---|---|---|---|
| 1 | ZnO NPs | Wistar rats | Tongue instillation | 50 mg/kg | NF-κB, MAPK pathway, and inflammatory cytokines | [90] |
| 2 | $Al_2O_3$ NPs | Wistar rats | Intranasal instillation | 20 μg/g | ROS, inflammatory cytokines | [94, 95] |
| 3 | $CeO_2$-NPs | Rainbow trout | Fish-tank exposure | 0.001–0.1 μg/L | Oxidative stress | [33] |
| 4 | CNTs | C57Bl/6 mice | Stereotactic injection | 0.5 mg/mouse | Inflammatory cytokines | [20] |
| 5 | ABP dendrimer | Mouse model of multiple sclerosis | Intravenous injection | 0.001 to 10 mg/kg | IL-1 β, TNF-a, IL-6, and IL-17 | [63] |
| 6 | D-NAC | Rabbit model of cerebral palsy | Intravenous injection | 10–100 mg/kg | NF-κB, GSH, and TNF-α | [78] |
| 7 | shCCL20-CCR6 | Mouse model of traumatic brain injury | Intranasal and intravenous administration | – | CCL20 | [99] |
| 8 | PEGOL-60 | Rabbit model of cerebral palsy | Intravenous injection | 55 mg/kg | TNF-α | [133] |
| 9 | PEG-PCL | Cerebral artery occlusion rat model | Intranasal administration | 100 mg/kg | TNF-α | [56] |
| 10 | AuNP | Rat model of Parkinson's disease | – | 5–10 mg/kg | IL-1β, IL-6, and TNF-α | [93] |

necrosis [41]. Additionally, the buildup of polysorbate 80-modified chitosan nanoparticles has been associated with neuronal death, significant oxidative damage, and a slight inflammatory reaction [164].

Although many nanoparticles have been shown to have positive anti-neuroinflammatory effects, either directly or by acting as vehicles for anti-inflammatory drugs, some of them may paradoxically cause neuroinflammatory reactions. For example, zinc oxide nanoparticles (ZnO NPs) have been found to induce apoptosis and oxidative stress in neuronal cells, suggesting a potential risk of neurotoxicity [129]. Similarly, silver nanoparticles (AgNPs) have been shown to cause neurotoxicity through oxidative stress and inflammation pathways, leading to cell death [74]. Surface modifications of nanoparticles can also influence their toxicity. Functionalization with polyethylene glycol (PEG) can reduce immunogenicity and increase biocompatibility [112]. However, even PEGylated nanoparticles can induce neuroinflammatory responses under certain conditions [47].

The route of administration is another critical factor in determining the neurotoxicity of nanoparticles. Inhalation and intravenous administration are more likely to result in systemic distribution and potential accumulation in the brain compared to oral administration [50]. Consequently, understanding the biodistribution and long-term effects of nanoparticles is essential for assessing their safety.

Furthermore, the long-term impact of chronic exposure to nanoparticles remains a significant concern. Repeated exposure may lead to the accumulation of nanoparticles in neural tissues, exacerbating neuroinflammation and contributing to neurodegenerative processes [27]. Therefore, meticulous investigation of both acute and chronic toxicity is required to ensure the safe use of nanoparticles in cytokine modulation and treatment of neuroinflammation.

## 11 Conclusion and Future Perspectives

Inflammatory cytokines, such as IL-1β, IL-6, TNF-α, and IFN-γ, play a crucial role in mediating and regulating neurodegeneration in neurodegenerative diseases [48, 52]. Nanomaterials have shown promising potential in modulating these cytokine activities. Future research in the field of nanomaterials should concentrate on several key areas to fully utilize their potential in the treatment of neurodegenerative diseases and the regulation of cytokine activity. Developing more advanced delivery systems that can precisely target affected areas within the brain is crucial. These targeted delivery systems not only enhance the effectiveness of treatments but also minimize adverse effects, ultimately leading to improved treatment outcomes. Nanomaterials specifically designed to target cells producing inflammatory cytokines like IL-1β, IL-6, TNF-α, or IFN-γ are more efficient at reducing inflammation.

In this line, nanomaterials, including inorganic, organic, and CNPs have shown significant promise in treating neuroinflammatory diseases. Inorganic nanomaterials like AuNPs, TiO2 NPs, and IONPs are valued for their stability, magnetic properties, and ease of functionalization, making them suitable for imaging, drug delivery, and therapeutic interventions [13, 59, 142, 143]. AuNPs, CeONPs, and other nanoparticles loaded with cytokine inhibitors demonstrate anti-inflammatory properties by reducing IL-1β levels and oxidative stress. Additionally, reduces IL-6 and TNF-α-mediated neuroinflammation, improving cognitive function, and reducing neuronal apoptosis [80, 87, 164]. Furthermore, CNPs, including fullerenes, CNTs, and graphene, offer unique properties useful in analytical sensing, bioimaging, and catalysis [26]. However, their toxicity and scalability challenges necessitate multidisciplinary collaboration [107].

However, the potential neurotoxicity of nanomaterials is a concern. Inorganic nanoparticles often exhibit more severe toxicity than organic ones. The primary neurotoxicity mechanisms include ROS production, cytokine release, and dysregulated apoptosis ([25], [60]. The toxicity of nanoparticles can be influenced by factors such as their shape, dosage, size, and surface area ([148], [83]. Organic nanomaterials, such as lipid-based nanoparticles, nanoemulsions, and polymer-based nanoparticles, offer significant advantages in neuroinflammation treatment. Lipid nanoparticles like liposomes can cross the BBB to deliver drugs directly to inflammation sites, enhancing therapeutic efficiency while minimizing systemic side effects [86, 159]. Polymeric nanoparticles enable controlled drug release, ensuring sustained therapeutic effects and reducing administration frequency. Their versatility in drug administration routes, including intravenous, nasal, oral, and dermal applications, makes them particularly useful for targeting the CNS ([83], [11]).

Future studies should emphasize the identification and utilization of distinct biomarkers and receptors found only in diseased brain regions. This approach will enable precise targeting while reducing unintended effects on healthy tissues. To reduce toxicity and ensure safe removal from the body, novel nanomaterials that are biocompatible and biodegradable must be developed.

Research efforts should focus on designing nanomaterials that can be easily metabolized and expelled without causing any harmful reactions. The integration of green chemistry principles and sustainable materials science is crucial in achieving this objective [114].

Furthermore, nanomaterials that can decompose safely while also modulating cytokine levels offer a safer alternative for treating chronic neuroinflammatory conditions. Moreover, developing nanomaterials capable of simultaneously performing multiple functions, such as imaging and therapy, opens up promising prospects for comprehensive treatment strategies. These multifunctional nanomaterials not only allow real-time monitoring of treatment progress and disease advancement but also facilitate the delivery of therapeutic agents. For instance, nanomaterials that can track and suppress cytokines like TNF-α and IL-6 may play a crucial role in managing neurodegenerative diseases [62].

Moreover, customizing nanomedicine approaches to individual patients based on their genetic, biochemical, and physiological profiles can lead to more personalized and effective treatments. Future investigations should explore the utility of advanced genomic, proteomic, and metabolomic technologies in identifying patient-specific markers. This tailored approach will enable the adaptation of nanomaterial-based cytokine modulation to address each patient's unique requirements, particularly those with distinct cytokine profiles [7, 140].

In conclusion, it is crucial to establish a robust regulatory framework to oversee the development, evaluation, and approval of nanomedicines to ensure compliance with safety and efficacy standards. Research endeavors should focus on formulating standardized procedures for the preclinical and clinical assessment of nanomaterials and cytokines modulation. By addressing these critical aspects, future research could maximize and standardize the use of nanomaterials in the treatment of neurodegenerative diseases and the modulation of cytokine activity, presenting novel and effective therapeutic options to patients suffering from neurodegenerative diseases.

## References


1. Abbott, N. J., Patabendige, A. A., Dolman, D. E., Yusof, S. R., & Begley, D. J. (2010). Structure and function of the blood-brain barrier. *Neurobiology of Disease, 37*(1), 13–25.
2. Agarwal, H., Nakara, A., & Shanmugam, V. K. (2019). Anti-inflammatory mechanism of various metal and metal oxide nanoparticles synthesized using plant extracts: A review. *Biomedicine & Pharmacotherapy, 109*, 2561–2572.
3. Agostinho, P., Cunha, R. A., & Oliveira, C. (2010). Neuroinflammation, oxidative stress and the pathogenesis of Alzheimer's disease. *Current Pharmaceutical Design, 16*(25), 2766–2778.
4. Aili, M., Zhou, K., Zhan, J., Zheng, H., & Luo, F. (2023). Anti-inflammatory role of gold nanoparticles in the prevention and treatment of Alzheimer's disease. *Journal of Materials Chemistry B, 11*(36), 8605–8621.
5. Akçan, R., Aydogan, H. C., Yildirim, M. Ş., Taştekin, B., & Sağlam, N. (2020) Nanotoxicity: a challenge for future medicine *TURKISH JOURNAL OF MEDICAL SCIENCES 50*(4) 1180-1196. https://doi.org/10.3906/sag-1912-209
6. Akbarzadeh, A., Rezaei-Sadabady, R., Davaran, S., Joo, S. W., Zarghami, N., Hanifehpour, Y., Nejati-Koshki, K. (2013) Liposome: *classification preparation and applications Nanoscale Research Letters 8*(1). https://doi.org/10.1186/1556-276X-8-102
7. Alcendor, D. J., Charest, A. M., Zhu, W. Q., Vigil, H. E., & Knobel, S. M. (2012). Infection and upregulation of proinflammatory cytokines in human brain vascular pericytes by human cytomegalovirus. *Journal of Neuroinflammation, 9*, 95.
8. Alexiou, C., Schmid, R. J., Jurgons, R., Kremer, M., Wanner, G., Bergemann, C., Huenges, E., Nawroth, T., Arnold, W., & Parak, F. G. (2006). Targeting cancer cells: Magnetic nanoparticles as drug carriers. *European Biophysics Journal, 35*(5), 446–450.
9. Alghamdi, M. A., Fallica, A. N., Virzì, N., Kesharwani, P., Pittalà, V., & Greish, K. (2022). The promise of nanotechnology in personalized medicine. *Journal of Personalized Medicine, 12*(5), 673.
10. Allaman, I., Bélanger, M., & Magistretti, P. J. (2011). Astrocyte-neuron metabolic relationships: For better and for worse. *Trends in Neurosciences, 34*(2), 76–87.
11. Althafar, Z. M. (2022). Targeting microglia in Alzheimer's disease: From molecular mechanisms to potential therapeutic targets for small molecules. *Molecules, 27*(13), 4124.
12. Asimakidou, E., Tan, J. K. S., Zeng, J., & Lo, C. H. (2024). Blood-brain barrier-targeting nanoparticles: Biomaterial properties and biomedical applications in translational neuroscience. *Pharmaceuticals (Basel), 17*(5), 612.
13. Ayub, A., & Wettig, S. (2022). An overview of nanotechnologies for drug delivery to the brain. *Pharmaceutics, 14*(2), 224.
14. Bachiller, S., Jiménez-Ferrer, I., Paulus, A., Yang, Y., Swanberg, M., Deierborg, T., & Boza-Serrano, A. (2018). Microglia in neurological diseases: A road map to brain-disease dependent-inflammatory response. *Frontiers in Cellular Neuroscience, 12*, 488.
15. Baetke, S. C., Lammers, T., & Kiessling, F. (2015). Applications of nanoparticles for diagnosis and therapy of cancer. *The British Journal of Radiology, 88*(1054), 20150207.
16. Bailey, Z. S., Nilson, E., Bates, J. A., Oyalowo, A., Hockey, K. S., Sajja, V. S. S. S., Thorpe, C., Rogers, H., Dunn, B., Frey, A. S., Billings, M. J., Sholar, C. A., Hermundstad, A., Kumar, C., VandeVord, P. J., & Rzigalinski, B. A. (2020). Cerium oxide nanoparticles improve outcome after in vitro and in vivo mild traumatic brain injury. *Journal of Neurotrauma, 37*(12), 1452–1462.

17. Ban, G., Hou, Y., Shen, Z., Jia, J., Chai, L., & Ma, C. (2023). Potential biomedical limitations of graphene nanomaterials. *International Journal of Nanomedicine, 18*, 1695–1708.
18. Banks, W. A. (2005). Blood-brain barrier transport of cytokines: A mechanism for neuropathology. *Current Pharmaceutical Design, 11*(8), 973–984.
19. Banks, W. A. (2016). From blood-brain barrier to blood-brain interface: New opportunities for CNS drug delivery. *Nature Reviews Drug Discovery, 15*(4), 275–292.
20. Banyer, J. L., Hamilton, N. H., Ramshaw, I. A., & Ramsay, A. J. (2000). Cytokines in innate and adaptive immunity. *Reviews in Immunogenetics, 2*(3), 359–373.
21. Barcia, E., Boeva, L., García-García, L., Slowing, K., Fernández-Carballido, A., Casanova, Y., & Negro, S. (2017). Nanotechnology-based drug delivery of ropinirole for Parkinson's disease. *Drug Delivery, 24*(1), 1112–1123.
22. Bardi, G., Nunes, A., Gherardini, L., Bates, K., Al-Jamal, K. T., Gaillard, C., Prato, M., Bianco, A., Pizzorusso, T., & Kostarelos, K. (2013). Functionalized carbon nanotubes in the brain: Cellular internalization and neuroinflammatory responses. *PLoS One, 8*(11), e80964.
23. Battaglini, M., Marino, A., Montorsi, M., Carmignani, A., Ceccarelli, M. C., & Ciofani, G. (2024). Nanomaterials as microglia modulators in the treatment of central nervous system disorders. *Advanced Healthcare Materials, 13*(12), e2304180.
24. Bhaskar, S., Tian, F., Stoeger, T., Kreyling, W., de la Fuente, J. M., Grazú, V., Borm, P., Estrada, G., Ntziachristos, V., & Razansky, D. (2010). Multifunctional nanocarriers for diagnostics, drug delivery and targeted treatment across blood-brain barrier: Perspectives on tracking and neuroimaging. *Particle and Fibre Toxicology, 7*, 3.
25. Bhatti, R., Shakeel, H., Malik, K., Qasim, M., Khan, M. A., Ahmed, N., & Jabeen, S. (2022). Inorganic nanoparticles: Toxic effects, mechanisms of cytotoxicity and phytochemical interactions. *Advanced Pharmaceutical Bulletin, 12*(4), 757–762.
26. Boche, D., Perry, V. H., & Nicoll, J. A. (2013). Review: Activation patterns of microglia and their identification in the human brain. *Neuropathology and Applied Neurobiology, 39*(1), 3–18.
27. Boyes, W. K., & van Thriel, C. (2020). Neurotoxicology of nanomaterials. *Chemical Research in Toxicology, 33*(5), 1121–1144.
28. Brito, C. L., Silva, J. V., Gonzaga, R. V., La-Scalea, M. A., Giarolla, J., & Ferreira, E. I. (2024). A review on carbon nanotubes family of nanomaterials and their health field. *ACS Omega, 9*(8), 8687–8708.
29. Buzea, C., Pacheco, I. I., & Robbie, K. (2007). Nanomaterials and nanoparticles: Sources and toxicity. *Biointerphases, 2*(4), MR17.
30. Cellot, G., Franceschi Biagioni, A., & Ballerini, L. (2022). Nanomedicine and graphene-based materials: Advanced technologies for potential treatments of diseases in the developing nervous system. *Pediatric Research, 92*(1), 71–79.
31. Chastain, E. M., Duncan, D. S., Rodgers, J. M., & Miller, S. D. (2011). The role of antigen presenting cells in multiple sclerosis. *Biochimica et Biophysica Acta, 1812*(2), 265–274.
32. Chen, J., Jin, J., Li, K., Shi, L., Wen, X., & Fang, F. (2022). Progresses and prospects of neuroprotective agents-loaded nanoparticles and biomimetic material in ischemic stroke. *Frontiers in Cellular Neuroscience, 16*, 868323.
33. Chen, X., Wu, G., Zhang, Z., Ma, X., & Liu, L. (2020). Neurotoxicity of Mn3O4 nanoparticles: Apoptosis and dopaminergic neurons damage pathway. *Ecotoxicology and Environmental Safety, 188*, 109909.
34. Chiang, M. C., Yang, Y. P., Nicol, C. J. B., & Wang, C. J. (2024). Gold nanoparticles in neurological diseases: A review of neuroprotection. *International Journal of Molecular Sciences, 25*(4), 2360.
35. Correia, A. T., Rebelo, D., Marques, J., & Nunes, B. (2019). Effects of the chronic exposure to cerium dioxide nanoparticles in Oncorhynchus mykiss: Assessment of oxidative stress, neurotoxicity and histological alterations. *Environmental Toxicology and Pharmacology, 68*, 27–36.
36. Cunha, A., Gaubert, A., Latxague, L., & Dehay, B. (2021). PLGA-based nanoparticles for neuroprotective drug delivery in neurodegenerative diseases. *Pharmaceutics, 13*(7), 1042.

37. Currais, A., Fischer, W., Maher, P., & Schubert, D. (2017). Intraneuronal protein aggregation as a trigger for inflammation and neurodegeneration in the aging brain. *The FASEB Journal, 31*(1), 5–10.
38. Dantzer, R., O'Connor, J. C., Freund, G. G., Johnson, R. W., & Kelley, K. W. (2008). From inflammation to sickness and depression: When the immune system subjugates the brain. *Nature Reviews Neuroscience, 9*(1), 46–56.
39. Das, M., Patil, S., Bhargava, N., Kang, J. F., Riedel, L. M., Seal, S., & Hickman, J. J. (2007). Auto-catalytic ceria nanoparticles offer neuroprotection to adult rat spinal cord neurons. *Biomaterials, 28*(10), 1918–1925.
40. Díaz-García, D., Ferrer-Donato, Á., Méndez-Arriaga, J. M., Cabrera-Pinto, M., Díaz-Sánchez, M., Prashar, S., Fernandez-Martos, C. M., & Gómez-Ruiz, S. (2022). Design of mesoporous silica nanoparticles for the treatment of amyotrophic lateral sclerosis (ALS) with a therapeutic cocktail based on leptin and pioglitazone. *ACS Biomaterials Science & Engineering, 8*(11), 4838–4849.
41. DiSabato, D. J., Quan, N., & Godbout, J. P. (2016). Neuroinflammation: The devil is in the details. *Journal of Neurochemistry, 139*(Suppl 2), 136–153.
42. Du, Q., Ge, D., Mirshafiee, V., Chen, C., Li, M., Xue, C., Ma, X., & Sun, B. (2019). Assessment of neurotoxicity induced by different-sized Stöber silica nanoparticles: Induction of pyroptosis in microglia. *Nanoscale, 11*(27), 12965–12972.
43. El-Shamarka, M. E., Abdel-Salam, O. M., Shafee, N., & Zeidan, H. M. (2023). Curcumin modulation of L-dopa and rasagiline-induced neuroprotection in rotenone model of Parkinson's disease. *Iranian Journal of Basic Medical Sciences, 26*(2), 139–147.
44. Erny, D., Hrabě de Angelis, A. L., Jaitin, D., Wieghofer, P., Staszewski, O., David, E., Keren-Shaul, H., Mahlakoiv, T., Jakobshagen, K., Buch, T., Schwierzeck, V., Utermöhlen, O., Chun, E., Garrett, W. S., McCoy, K. D., Diefenbach, A., Staeheli, P., Stecher, B., Amit, I., & Prinz, M. (2015). Host microbiota constantly control maturation and function of microglia in the CNS. *Nature Neuroscience, 18*(7), 965–977.
45. Erta, M., Quintana, A., & Hidalgo, J. (2012). Interleukin-6, a major cytokine in the central nervous system. *International Journal of Biological Sciences, 8*(9), 1254–1266.
46. Feng, C., Deng, L., Yong, Y. Y., Wu, J. M., Qin, D. L., Yu, L., Zhou, X. G., & Wu, A. G. (2023). The application of biomaterials in spinal cord injury. *International Journal of Molecular Sciences, 24*(1), 816.
47. Frankola, K. A., Greig, N. H., Luo, W., & Tweedie, D. (2011). Targeting TNF-α to elucidate and ameliorate neuroinflammation in neurodegenerative diseases. *CNS & Neurological Disorders Drug Targets, 10*(3), 391–403.
48. Fujita, T., Zysman, M., Elgrabli, D., Murayama, T., Haruta, M., Lanone, S., Ishida, T., & Boczkowski, J. (2021). Anti-inflammatory effect of gold nanoparticles supported on metal oxides. *Scientific Reports, 11*(1), 23129.
49. Ganson, N. J., Povsic, T. J., Sullenger, B. A., Alexander, J. H., Zelenkofske, S. L., Sailstad, J. M., Rusconi, C. P., & Hershfield, M. S. (2016). Pre-existing anti-polyethylene glycol antibody linked to first-exposure allergic reactions to pegnivacogin, a PEGylated RNA aptamer. *The Journal of Allergy and Clinical Immunology, 137*(5), 1610–1613.e7.
50. Gao, C., Jiang, J., Tan, Y., & Chen, S. (2023). Microglia in neurodegenerative diseases: Mechanism and potential therapeutic targets. *Signal Transduction and Targeted Therapy, 8*(1), 359.
51. Gąssowska-Dobrowolska, M., Chlubek, M., Kolasa, A., Tomasiak, P., Korbecki, J., Skowrońska, K., Tarnowski, M., Masztalewicz, M., & Baranowska-Bosiacka, I. (2023). Microglia and Astroglia-the potential role in neuroinflammation induced by pre- and neonatal exposure to Lead (Pb). *International Journal of Molecular Sciences, 24*(12), 9903.
52. Geiser, M., & Kreyling, W. G. (2010). Deposition and biokinetics of inhaled nanoparticles. *Particle and Fibre Toxicology, 7*, 2.
53. Ginhoux, F., Lim, S., Hoeffel, G., Low, D., & Huber, T. (2013). Origin and differentiation of microglia. *Frontiers in Cellular Neuroscience, 7*, 45.

54. Glass, C. K., Saijo, K., Winner, B., Marchetto, M. C., & Gage, F. H. (2010). Mechanisms underlying inflammation in neurodegeneration. *Cell, 140*(6), 918–934.
55. Gonzalez, C. N. (2023). Role of tumor necrosis factor-alpha in the central nervous system: A focus on autoimmune disorders. *Frontiers in Immunology, 14*, 1213448.
56. Gonzalez-Carter, D. A., Leo, B. F., Ruenraroengsak, P., Chen, S., Goode, A. E., Theodorou, I. G., Chung, K. F., Carzaniga, R., Shaffer, M. S., Dexter, D. T., Ryan, M. P., & Porter, A. E. (2017). Silver nanoparticles reduce brain inflammation and related neurotoxicity through induction of H2S-synthesizing enzymes. *Scientific Reports, 7*, 42871.
57. Grissa, I., Guezguez, S., Ezzi, L., Chakroun, S., Sallem, A., Kerkeni, E., Elghoul, J., El Mir, L., Mehdi, M., Cheikh, H. B., & Haouas, Z. (2016). The effect of titanium dioxide nanoparticles on neuroinflammation response in rat brain. *Environmental Science and Pollution Research International, 23*(20), 20205–20213.
58. Grossen, P., Witzigmann, D., Sieber, S., & Huwyler, J. (2017). PEG-PCL-based nanomedicines: A biodegradable drug delivery system and its application. *Journal of Controlled Release, 260*, 46–60.
59. Guo, Z. H., Khattak, S., Rauf, M. A., Ansari, M. A., Alomary, M. N., Razak, S., Yang, C. Y., Wu, D. D., & Ji, X. Y. (2023). Role of nanomedicine-based therapeutics in the treatment of CNS disorders. *Molecules, 28*(3), 1283.
60. Günter, Oberdörster Eva, Oberdörster Jan, Oberdörster (2005) Nanotoxicology: *An Emerging Discipline Evolving from Studies of Ultrafine Particles Environmental Health Perspectives 113*(7) 823-839. https://doi.org/10.1289/ehp.7339
61. Gupta, A., Eral, H. B., Hatton, T. A., & Doyle, P. S. (2016). Nanoemulsions: Formation, properties and applications. *Soft Matter, 12*(11), 2826–2841.
62. Gupta, A. K., & Gupta, M. (2005). Synthesis and surface engineering of iron oxide nanoparticles for biomedical applications. *Biomaterials, 26*(18), 3995–4021.
63. Gupta, G., Gliga, A., Hedberg, J., Serra, A., Greco, D., Odnevall Wallinder, I., & Fadeel, B. (2020). Cobalt nanoparticles trigger ferroptosis-like cell death (oxytosis) in neuronal cells: Potential implications for neurodegenerative disease. *The FASEB Journal, 34*(4), 5262–5281.
64. Hamilton, J. A. (2020). GM-CSF-dependent inflammatory pathways. *Frontiers in Immunology, 11*, 2055.
65. Hannah D., Holscher (2017) *Dietary fiber and prebiotics and the gastrointestinal microbiota Gut Microbes 8*(2) 172-184. https://doi.org/10.1080/19490976.2017.1290756
66. Harun-Or-Rashid, M., Aktar, M. N., Hossain, M. S., Sarkar, N., Islam, M. R., Arafat, M. E., Bhowmik, S., & Yusa, S. I. (2023). Recent advances in micro- and nano-drug delivery systems based on natural and synthetic biomaterials. *Polymers (Basel), 15*(23), 4563.
67. Hayder, M., Varilh, M., Turrin, C. O., Saoudi, A., Caminade, A. M., Poupot, R., & Liblau, R. S. (2015). Phosphorus-based dendrimer ABP treats neuroinflammation by promoting IL-10-producing CD4(+) T cells. *Biomacromolecules, 16*(11), 3425–3433.
68. He, P., Zhong, Z., Lindholm, K., et al. (2007). Deletion of tumor necrosis factor death receptor inhibits amyloid beta generation and prevents learning and memory deficits in Alzheimer's mice. *The Journal of Cell Biology, 178*(5), 829–841.
69. Heckman, K. L., DeCoteau, W., Estevez, A., Reed, K. J., Costanzo, W., Sanford, D., Leiter, J. C., Clauss, J., Knapp, K., Gomez, C., Mullen, P., Rathbun, E., Prime, K., Marini, J., Patchefsky, J., Patchefsky, A. S., Hailstone, R. K., & Erlichman, J. S. (2013). Custom cerium oxide nanoparticles protect against a free radical mediated autoimmune degenerative disease in the brain. *ACS Nano, 7*(12), 10582–10596.
70. Heneka, M. T., Golenbock, D. T., & Latz, E. (2014). Innate immunity in Alzheimer's disease. *Nature Immunology, 16*(3), 229–236.
71. Heneka, M. T., Kummer, M. P., & Latz, E. (2014). Innate immune activation in neurodegenerative disease. *Nature Reviews Immunology, 14*(7), 463–477.
72. Heppner, F. L., Ransohoff, R. M., & Becher, B. (2015). Immune attack: The role of inflammation in Alzheimer disease. *Nature Reviews Neuroscience, 16*(6), 358–372.
73. Hirst, S. M., Karakoti, A. S., Tyler, R. D., Sriranganathan, N., Seal, S., & Reilly, C. M. (2009). Anti-inflammatory properties of cerium oxide nanoparticles. *Small, 5*(24), 2848–2856.

74. Hou, K., Zhao, J., Wang, H., Li, B., Li, K., Shi, X., Wan, K., Ai, J., Lv, J., Wang, D., Huang, Q., Wang, H., Cao, Q., Liu, S., & Tang, Z. (2020). Chiral gold nanoparticles enantioselectively rescue memory deficits in a mouse model of Alzheimer's disease. *Nature Communications, 11*(1), 4790.
75. Huang, H., Zhou, M., Ruan, L., Wang, D., Lu, H., Zhang, J., Chen, J., Hu, Y., & Chai, Z. (2019). AMPK mediates the neurotoxicity of iron oxide nanoparticles retained in mitochondria or lysosomes. *Metallomics, 11*(7), 1200–1206.
76. Huang, N., Lu, S., Liu, X. G., Zhu, J., Wang, Y. J., & Liu, R. T. (2017). PLGA nanoparticles modified with a BBB-penetrating peptide co-delivering Aβ generation inhibitor and curcumin attenuate memory deficits and neuropathology in Alzheimer's disease mice. *Oncotarget, 8*(46), 81001–81013.
77. Huang, X., He, D., Pan, Z., Luo, G., & Deng, J. (2021). Reactive-oxygen-species-scavenging nanomaterials for resolving inflammation. *Materials Today Bio, 11*, 100124.
78. Huang, Y., Li, P., Zhao, R., Zhao, L., Liu, J., Peng, S., & Zhang, Z. (2022) *Silica nanoparticles: Biomedical applications and toxicity Biomedicine & Pharmacotherapy* 151113053. https://doi.org/10.1016/j.biopha.2022.113053
79. Hussain, S. M., Javorina, A. K., Schrand, A. M., Duhart, H. M., Ali, S. F., & Schlager, J. J. (2006). The interaction of manganese nanoparticles with PC-12 cells induces dopamine depletion. *Toxicological Sciences, 92*(2), 456–463.
80. Jiang, W., Li, Q., Zhang, R., Li, J., Lin, Q., Li, J., Zhou, X., Yan, X., & Fan, K. (2023). Chiral metal-organic frameworks incorporating nanozymes as neuroinflammation inhibitors for managing Parkinson's disease. *Nature Communications, 14*(1), 8137.
81. Jiang, Y., Gong, H., Jiang, S., She, C., & Cao, Y. (2020). Multi-walled carbon nanotubes decrease neuronal NO synthase in 3D brain organoids. *Science of the Total Environment, 748*, 141384.
82. Jiang, Z., Jiang, J. X., & Zhang, G. X. (2014). Macrophages: A double-edged sword in experimental autoimmune encephalomyelitis. *Immunology Letters, 160*(1), 17–22.
83. Jörg, Kreuter (2014) Drug delivery to the central nervous system by polymeric nanoparticles: *What do we know? Advanced Drug Delivery Reviews* 712-14. https://doi.org/10.1016/j.addr.2013.08.008
84. Jović, D., Jaćević, V., Kuča, K., Borišev, I., Mrdjanovic, J., Petrovic, D., & Djordjevic, A (2020) The Puzzling Potential of Carbon Nanomaterials: *General Properties Application and Toxicity Nanomaterials 10*(8)1508. https://doi.org/10.3390/nano10081508
85. Kannan, S., Dai, H., Navath, R. S., Balakrishnan, B., Jyoti, A., Janisse, J., Romero, R., & Kannan, R. M. (2012). Dendrimer-based postnatal therapy for neuroinflammation and cerebral palsy in a rabbit model. *Science Translational Medicine, 4*(130), 130ra46.
86. Kanninen, K. M., & White, A. R. (2017). Type-I interferons in Parkinson's disease: Innate inflammatory response drives fate of neurons in model of degenerative brain disorder: An editorial comment on 'Type-I interferons mediate the neuroinflammatory response and neurotoxicity induced by rotenone'. *Journal of Neurochemistry, 141*(1), 9–11.
87. Khan, H. A., Ibrahim, K. E., Alrashood, S. T., Alamery, S., Alrokayan, S. H., Al-Harbi, N., Al-Mutary, M. G., Sobki, S. H., & Khan, I. (2020). Immunohistochemistry of IL-1β, IL-6 and TNF-α in spleens of mice treated with gold nanoparticles. *Saudi Journal of Biological Sciences, 27*(4), 1163–1168.
88. Kim, H. Y., Yoon, H. H., Seong, H., Seo, D. K., Choi, S. W., Kang, K. S., & Jeon, S. R. (2023). Preventive effects of nano-graphene oxide against Parkinson's disease via reactive oxygen species scavenging and anti-inflammation. *BMB Reports, 56*(2), 202–207.
89. Kim, J. S., Yoon, T. J., Yu, K. N., Kim, B. G., Park, S. J., Kim, H. W., Lee, K. H., Park, S. B., Lee, J. K., & Cho, M. H. (2006). Toxicity and tissue distribution of magnetic nanoparticles in mice. *Toxicological Sciences, 89*(1), 338–347.
90. Kölliker-Frers, R., Udovin, L., Otero-Losada, M., Kobiec, T., Herrera, M. I., Palacios, J., Razzitte, G., & Capani, F. (2021). Neuroinflammation: An integrating overview of reactive-Neuroimmune cell interactions in health and disease. *Mediators of Inflammation, 2021*, 9999146.

91. Konsman, J. P. (2022). Cytokines in the brain and neuroinflammation: We didn't starve the fire! *Pharmaceuticals (Basel), 15*(2), 140. https://doi.org/10.3390/ph15020140
92. Koushki, K., Keshavarz Shahbaz, S., Keshavarz, M., Bezsonov, E. E., Sathyapalan, T., & Sahebkar, A. (2021). Gold nanoparticles: Multifaceted roles in the management of autoimmune disorders. *Biomolecules, 11*(9), 1289.
93. Kulkarni, M., Patel, K., Patel, A., Patel, S., Desai, J., Patel, M., Shah, U., Patel, A., & Solanki, N. (2023). Nanomaterials as drug delivery agents for overcoming the blood-brain barrier: A comprehensive review. *ADMET DMPK, 12*(1), 63–105.
94. Kumar, A., Sudevan, S. T., Nair, A. S., Singh, A. K., Kumar, S., Jose, J., Behl, T., Mangalathillam, S., Mathew, B., & Kim, H. (2023). Current and future nano-carrier-based approaches in the treatment of Alzheimer's disease. *Brain Sciences, 13*(2), 213.
95. Kumari, A., Yadav, S. K., & Yadav, S. C. (2010). Biodegradable polymeric nanoparticles based drug delivery systems. *Colloids and Surfaces. B, Biointerfaces, 75*(1), 1–18.
96. Li, Y., Li, Y., Bai, Y., Lin, L., & Sun, Y. (2020). High catalytic efficiency from Er3+-doped CeO2-x nanoprobes for in vivo acute oxidative damage and inflammation therapy. *Journal of Materials Chemistry B, 8*(37), 8634–8643.
97. Liang, H., Chen, A., Lai, X., Liu, J., Wu, J., Kang, Y., Wang, X., & Shao, L. (2018). Neuroinflammation is induced by tongue-instilled ZnO nanoparticles via the Ca2+-dependent NF-κB and MAPK pathways. *Particle and Fibre Toxicology, 15*(1), 39.
98. Liao, X., Liu, Y., Zheng, J., Zhao, X., Cui, L., Hu, S., Xia, T., & Si, S. (2022). Diverse pathways of engineered nanoparticle-induced NLRP3 inflammasome activation. *Nanomaterials (Basel), 12*(21), 3908.
99. Liddelow, S. A., Guttenplan, K. A., Clarke, L. E., Bennett, F. C., Bohlen, C. J., Schirmer, L., et al. (2017). Neurotoxic reactive astrocytes are induced by activated microglia. *Nature, 541*(7638), 481–487.
100. Ling, L., Jiang, Y., Liu, Y., Li, H., Bari, A., Ullah, R., & Xue, J. (2019). Role of gold nanoparticle from Cinnamomum verum against 1-methyl-4-phenyl-1, 2, 3, 6-tetrahydropyridine (MPTP) induced mice model. *Journal of Photochemistry and Photobiology. B, 201*, 111657.
101. Liu, H., Yang, H., Fang, Y., Li, K., Tian, L., Liu, X., Zhang, W., Tan, Y., Lai, W., Bian, L., Lin, B., & Xi, Z. (2020). Neurotoxicity and biomarkers of zinc oxide nanoparticles in main functional brain regions and dopaminergic neurons. *Science of the Total Environment, 705*, 135809.
102. Liu, H., Zhang, W., Fang, Y., Yang, H., Tian, L., Li, K., Lai, W., Bian, L., Lin, B., Liu, X., & Xi, Z. (2020). Neurotoxicity of aluminum oxide nanoparticles and their mechanistic role in dopaminergic neuron injury involving p53-related pathways. *Journal of Hazardous Materials, 392*, 122312.
103. Manoharan, D., Wang, L. C., Chen, Y. C., Li, W. P., & Yeh, C. S. (2024) Catalytic Nanoparticles in Biomedical Applications: *Exploiting Advanced Nanozymes for Therapeutics and Diagnostics Abstract Advanced Healthcare Materials 13*(22) https://doi.org/10.1002/adhm.202400746
104. Marin, D., & Marchesan, S. (2022). Carbon graphitization: Towards greener alternatives to develop nanomaterials for targeted drug delivery. *Biomedicine, 10*(6), 1320.
105. Martano, S., De Matteis, V., Cascione, M., & Rinaldi, R. (2022). Inorganic nanomaterials versus polymer-based nanoparticles for overcoming neurodegeneration. *Nanomaterials (Basel), 12*(14), 2337. https://doi.org/10.3390/nano12142337. PMID: 35889562; PMCID: PMC9317100.
106. Marques, F. Z., Mackay, C. R., & Kaye, D. M. (2020). Beyond gut feelings: How the gut microbiota regulates blood pressure. *Nature Reviews Cardiology, 17*(1), 20–32.
107. Mayilsamy, K., Markoutsa, E., Das, M., Chopade, P., Puro, D., Kumar, A., Gulick, D., Willing, A. E., Mohapatra, S. S., & Mohapatra, S. (2020). Treatment with shCCL20-CCR6 nanodendriplexes and human mesenchymal stem cell therapy improves pathology in mice with repeated traumatic brain injury. *Nanomedicine, 29*, 102247.
108. Maysinger, D., Ji, J., Moquin, A., Hossain, S., Hancock, M. A., Zhang, I., Chang, P. K. Y., Rigby, M., Anthonisen, M., Grütter, P., Breitner, J., McKinney, R. A., Reimann, S., Haag, R.,

& Multhaup, G. (2018). Dendritic polyglycerol sulfates in the prevention of synaptic loss and mechanism of action on glia. *ACS Chemical Neuroscience, 9*(2), 260–271.

109. McCoy, M. K., & Tansey, M. G. (2008). TNF signaling inhibition in the CNS: Implications for normal brain function and neurodegenerative disease. *Journal of Neuroinflammation, 5*, 45.
110. Michaelson, D. M. (2014). APOE epsilon4: The most prevalent yet understudied risk factor for Alzheimer's disease. *Alzheimer's & Dementia, 10*(6), 861–868.
111. Miyashita, R., Chen, L., Oshiro, H., Uchino, H., & Shibasaki, F. (2012). Int6 silencing causes induction of angiogenic factors in neuronal cells via accumulation of hypoxia-inducible factor 2α and decreases brain damage in rats. *Neuroscience Letters, 528*(1), 83–88.
112. Muhammad, B., Li, H., Gu, Y., Xue, S., Gao, Y., Xu, Z., Fang, X., Ding, H., Wu, F., Geng, D., & Niu, H. (2023). IL-1β/IL-1R1 signaling is involved in the propagation of α-synuclein pathology of the gastrointestinal tract to the brain. *Journal of Neurochemistry, 166*(5), 830–846.
113. Müller, R. H., Mäder, K., & Gohla, S. (2000). Solid lipid nanoparticles (SLN) for controlled drug delivery – A review of the state of the art. *European Journal of Pharmaceutics and Biopharmaceutics, 50*(1), 161–177.
114. Naeem, A., Prakash, R., Kumari, N., Ali Khan, M., Quaiyoom Khan, A., Uddin, S., Verma, S., Ab Robertson, A., Boltze, J., & Shadab, R. S. (2024). MCC950 reduces autophagy and improves cognitive function by inhibiting NLRP3-dependent neuroinflammation in a rat model of Alzheimer's disease. *Brain, Behavior, and Immunity, 116*, 70–84.
115. Nel, A., Xia, T., Mädler, L., & Li, N. (2006). Toxic potential of materials at the nanolevel. *Science, 311*(5761), 622–627.
116. Neuwelt, E. A., Bauer, B., Fahlke, C., Fricker, G., Iadecola, C., Janigro, D., Leybaert, L., Molnár, Z., O'Donnell, M. E., Povlishock, J. T., Saunders, N. R., Sharp, F., Stanimirovic, D., Watts, R. J., & Drewes, L. R. (2011). Engaging neuroscience to advance translational research in brain barrier biology. *Nature Reviews Neuroscience, 12*(3), 169–182.
117. Niemeyer, C. M. (2001). Nanoparticles, proteins, and nucleic acids: Biotechnology meets materials science. *Angewandte Chemie (International Ed. in English), 40*(22), 4128–4158.
118. Norden, D. M., & Godbout, J. P. (2013). Review: Microglia of the aged brain: Primed to be activated and resistant to regulation. *Neuropathology and Applied Neurobiology, 39*(1), 19–34.
119. Ou, W., Ohno, Y., Yang, J., Chandrashekar, D. V., Abdullah, T., Sun, J., Murphy, R., Roules, C., Jagadeesan, N., Cribbs, D. H., & Sumbria, R. K. (2022). Efficacy and safety of a brain-penetrant biologic TNF-α inhibitor in aged APP/PS1 mice. *Pharmaceutics, 14*(10), 2200.
120. Owens, D. E., 3rd, & Peppas, N. A. (2006). Opsonization, biodistribution, and pharmacokinetics of polymeric nanoparticles. *International Journal of Pharmaceutics, 307*(1), 93–102.
121. Pranzatelli, M. R. (2018). Advances in biomarker-guided therapy for pediatric- and adult-onset neuroinflammatory disorders: Targeting chemokines/cytokines. *Frontiers in Immunology, 9*, 557.
122. Patra, J. K., Das, G., Fraceto, L. F., Campos, E. V. R., Rodriguez-Torres, M. D. P., Acosta-Torres, L. S., Diaz-Torres, L. A., Grillo, R., Swamy, M. K., Sharma, S., Habtemariam, S., & Shin, H. S. (2018). Nano based drug delivery systems: Recent developments and future prospects. *Journal of Nanobiotechnology, 16*(1), 71.
123. Perry, V. H., & Holmes, C. (2014). Microglial priming in neurodegenerative disease. *Nature Reviews Neurology, 10*(4), 217–224.
124. Prinz, M., Priller, J., Sisodia, S. S., & Ransohoff, R. M. (2014). Heterogeneity of CNS myeloid cells and their roles in neurodegeneration. *Nature Neuroscience, 17*(10), 1338–1344.
125. Puricelli, C., Gigliotti, C. L., Stoppa, I., Sacchetti, S., Pantham, D., Scomparin, A., Rolla, R., Pizzimenti, S., Dianzani, U., Boggio, E., & Sutti, S. (2023). Use of poly lactic-co-glycolic acid nano and micro particles in the delivery of drugs modulating different phases of inflammation. *Pharmaceutics, 15*(6), 1772.
126. Qasim, M., Clarkson, A. N., & Hinkley, S. F. (2023) *Green Synthesis of Carbon Nanoparticles (CNPs) from Biomass for Biomedical Applications International Journal of Molecular Sciences 24*(2)1023. https://doi.org/10.3390/ijms24021023

127. Qin, J., Ma, Z., Chen, X., & Shu, S. (2023). Microglia activation in central nervous system disorders: A review of recent mechanistic investigations and development efforts. *Frontiers in Neurology, 14*, 1103416.
128. Rademakers, R., Neumann, M., & Mackenzie, I. R. (2012). Advances in understanding the molecular basis of frontotemporal dementia. *Nature Reviews Neurology, 8*(8), 423–434.
129. Rahmati, M., & Mozafari, M. (2019) Biological Response to Carbon-Family Nanomaterials: Interactions at the Nano-Bio Interface. *Frontiers in Bioengineering and Biotechnology* 7. https://doi.org/10.3389/fbioe.2019.00004
130. Ransohoff, R. M. (2016). How neuroinflammation contributes to neurodegeneration. *Science, 353*(6301), 777–783.
131. Rauf, A., Badoni, H., Abu-Izneid, T., Olatunde, A., Rahman, M. M., Painuli, S., Semwal, P., Wilairatana, P., & Mubarak, M. S. (2022). Neuroinflammatory markers: Key indicators in the pathology of neurodegenerative diseases. *Molecules, 27*(10), 3194.
132. Rawlings, J. S., Rosler, K. M., & Harrison, D. A. (2004). The JAK/STAT signaling pathway. *Journal of Cell Science, 117*(Pt 8), 1281–1283.
133. Roberto, M., Patel, R. R., & Bajo, M. (2018). Ethanol and cytokines in the central nervous system. *Handbook of Experimental Pharmacology, 248*, 397–431.
134. Rose-John, S., Jenkins, B. J., Garbers, C., Moll, J. M., & Scheller, J. (2023). Targeting IL-6 trans-signalling: Past, present and prospects. *Nature Reviews Immunology, 23*(10), 666–681.
135. Sahoo, S. K., & Labhasetwar, V. (2003). Nanotech approaches to drug delivery and imaging. *Drug Discovery Today, 8*(24), 1112–1120. https://doi.org/10.1016/S1359-6446(03)02903-9
136. Saifi, M. A., Khan, W., & Godugu, C. (2018). Cytotoxicity of nanomaterials: Using nanotoxicology to address the safety concerns of nanoparticles. *Pharmaceutical Nanotechnology, 6*(1), 3–16.
137. Saijo, K., Crotti, A., & Glass, C. K. (2013). Regulation of microglia activation and deactivation by nuclear receptors. *Glia, 61*(1), 104–111.
138. Salatin, S., Farhoudi, M., Farjami, A., Maleki Dizaj, S., Sharifi, S., & Shahi, S. (2023). Nanoparticle formulations of antioxidants for the management of oxidative stress in stroke: A review. *Biomedicine, 11*(11), 3010.
139. Sanjeeb K., Sahoo Vinod, Labhasetwar (2003) *Nanotech approaches to drug delivery and imaging Drug Discovery Today* 8(24) 1112-1120. https://doi.org/10.1016/S1359-6446(03)02903-9
140. Saquib, Q., Al-Khedhairy, A. A., Siddiqui, M. A., Abou-Tarboush, F. M., Azam, A., & Musarrat, J. (2012). Titanium dioxide nanoparticles induced cytotoxicity, oxidative stress and DNA damage in human amnion epithelial (WISH) cells. *Toxicology In Vitro, 26*(2), 351–361.
141. Saraiva, C., Praça, C., Ferreira, R., Santos, T., Ferreira, L., & Bernardino, L. (2016). Nanoparticle-mediated brain drug delivery: Overcoming blood-brain barrier to treat neurodegenerative diseases. *Journal of Controlled Release, 235*, 34–47.
142. Schain, M., & Kreisl, W. C. (2017). Neuroinflammation in neurodegenerative disorders-a review. *Current Neurology and Neuroscience Reports, 17*(3), 25.
143. Serhan, C. N., Chiang, N., & Dalli, J. (2014). The resolution code of acute inflammation: Novel pro-resolving lipid mediators in resolution. *Seminars in Immunology, 26*(4), 237–245.
144. Sharma, A., Sharma, R., Zhang, Z., et al. (2020). Dense hydroxyl polyethylene glycol dendrimer targets activated glia in multiple CNS disorders. *Science Advances, 6*(4), eaay8514.
145. Sharma, R., Kim, S. Y., Sharma, A., Zhang, Z., Kambhampati, S. P., Kannan, S., & Kannan, R. M. (2017). Activated microglia targeting dendrimer-minocycline conjugate as therapeutics for neuroinflammation. *Bioconjugate Chemistry, 28*(11), 2874–2886.
146. Shukla, R., Bansal, V., Chaudhary, M., Basu, A., Bhonde, R. R., & Sastry, M. (2005). Biocompatibility of gold nanoparticles and their endocytotic fate inside the cellular compartment: A microscopic overview. *Langmuir, 21*(23), 10644–10654.
147. Siddiqi, N. J., Abdelhalim, M. A., El-Ansary, A. K., Alhomida, A. S., & Ong, W. Y. (2012). Identification of potential biomarkers of gold nanoparticle toxicity in rat brains. *Journal of Neuroinflammation, 9*, 123.

148. Sofroniew, M. V., & Vinters, H. V. (2010). Astrocytes: Biology and pathology. *Acta Neuropathologica, 119*(1), 7–35.
149. Sriram, K., & O'Callaghan, J. P. (2007). Divergent roles for tumor necrosis factor-alpha in the brain. *Journal of Neuroimmune Pharmacology, 2*(2), 140–153.
150. Stephenson, J., Nutma, E., van der Valk, P., & Amor, S. (2018). Inflammation in CNS neurodegenerative diseases. *Immunology, 154*(2), 204–219.
151. Svensson, E., von Mentzer, U., & Stubelius, A. (2023). Achieving precision healthcare through nanomedicine and enhanced model systems. *ACS Materials Au, 4*(2), 162–173.
152. Takata, F., Nakagawa, S., Matsumoto, J., & Dohgu, S. (2021). Blood-brain barrier dysfunction amplifies the development of neuroinflammation: Understanding of cellular events in brain microvascular endothelial cells for prevention and treatment of BBB dysfunction. *Frontiers in Cellular Neuroscience, 15*, 661838.
153. Teleanu, D. M., Chircov, C., Grumezescu, A. M., & Teleanu, R. I. (2019). Neurotoxicity of nanomaterials: An up-to-date overview. *Nanomaterials (Basel), 9*(1), 96.
154. Teleanu, D. M., Negut, I., Grumezescu, V., Grumezescu, A. M., & Teleanu, R. I. (2019). Nanomaterials for drug delivery to the central nervous system. *Nanomaterials (Basel), 9*(3), 371.
155. Teleanu, R. I., Negut, I., Grumezescu, V., Grumezescu, A. M., & Teleanu, D. M. (2018). Nanomaterials for drug delivery to the central nervous system. *Nanomaterials (Basel), 9*(3), 371.
156. Tohidpour, A., Morgun, A. V., Boitsova, E. B., Malinovskaya, N. A., Martynova, G. P., Khilazheva, E. D., Kopylevich, N. V., Gertsog, G. E., & Salmina, A. B. (2017). Neuroinflammation and infection: Molecular mechanisms associated with dysfunction of neurovascular unit. *Frontiers in Cellular and Infection Microbiology, 7*, 276.
157. Tong, S., Cinelli, M. A., El-Sayed, N. S., Huang, H., Patel, A., Silverman, R. B., & Yang, S. (2022). Inhibition of interferon-gamma-stimulated melanoma progression by targeting neuronal nitric oxide synthase (nNOS). *Scientific Reports, 12*(1), 1701.
158. Torchilin, V. P. (2005). Recent advances with liposomes as pharmaceutical carriers. *Nature Reviews Drug Discovery, 4*(2), 145–160.
159. Trouiller, B., Reliene, R., Westbrook, A., Solaimani, P., & Schiestl, R. H. (2009). Titanium dioxide nanoparticles induce DNA damage and genetic instability in vivo in mice. *Cancer Research, 69*(22), 8784–8789.
160. Varatharaj, A., & Galea, I. (2017). The blood-brain barrier in systemic inflammation. *Brain, Behavior, and Immunity, 60*, 1–12.
161. Vashist, A., Manickam, P., Raymond, A. D., Arias, A. Y., Kolishetti, N., Vashist, A., Arias, E., & Nair, M. (2023). Recent advances in nanotherapeutics for neurological disorders. *ACS Applied Bio Materials, 6*(7), 2614–2621.
162. Vogel, D. Y., Vereyken, E. J., Glim, J. E., Heijnen, P. D., Moeton, M., van der Valk, P., Amor, S., Teunissen, C. E., van Horssen, J., & Dijkstra, C. D. (2013). Macrophages in inflammatory multiple sclerosis lesions have an intermediate activation status. *Journal of Neuroinflammation, 10*, 35.
163. Wang, B., Navath, R. S., Romero, R., Kannan, S., & Kannan, R. (2009). Anti-inflammatory and anti-oxidant activity of anionic dendrimer-N-acetyl cysteine conjugates in activated microglial cells. *International Journal of Pharmaceutics, 377*(1–2), 159–168.
164. Weifeng, Zhang Dan, Xiao Qinwen, Mao Haibin, Xia (2023) *Role of neuroinflammation in neurodegeneration development Abstract Signal Transduction and Targeted Therapy 8*(1) https://doi.org/10.1038/s41392-023-01486-5
165. Wu, J., Wang, C., Sun, J., & Xue, Y. (2011). Neurotoxicity of silica nanoparticles: Brain localization and dopaminergic neurons damage pathways. *ACS Nano, 5*(6), 4476–4489.
166. Xiao, L., Wei, F., Zhou, Y., Anderson, G. J., Frazer, D. M., Lim, Y. C., Liu, T., & Xiao, Y. (2020). Dihydrolipoic acid-gold nanoclusters regulate microglial polarization and have the potential to alter neurogenesis. *Nano Letters, 20*(1), 478–495.
167. Yallapu, M. M., Nagesh, P. K., Jaggi, M., & Chauhan, S. C. (2015). Therapeutic applications of curcumin Nanoformulations. *The AAPS Journal, 17*(6), 1341–1356.

168. Yin, Z., Chen, J., Xia, M., Zhang, X., Li, Y., Chen, Z., Bao, Q., Zhong, W., Yao, J., Wu, K., Zhao, L., & Liang, F. (2023). Assessing causal relationship between circulating cytokines and age-related neurodegenerative diseases: A bidirectional two-sample Mendelian randomization analysis. *Scientific Reports, 13*(1), 12325.
169. Yeh, Y. C., Creran, B., & Rotello, V. M. (2012). Gold nanoparticles: Preparation, properties, and applications in bionanotechnology. *Nanoscale, 4*(6), 1871–1880.
170. Yetisgin, A. A., Cetinel, S., Zuvin, M., Kosar, A., & Kutlu, O. (2020). Therapeutic nanoparticles and their targeted delivery applications. *Molecules, 25*(9), 2193.
171. Zha, S., Liu, H., Li, H., Li, H., Wong, K. L., & All, A. H. (2024). Functionalized nanomaterials capable of crossing the blood-brain barrier. *ACS Nano, 18*(3), 1820–1845.
172. Zhang, F., Nance, E., Alnasser, Y., Kannan, R., & Kannan, S. (2016). Microglial migration and interactions with dendrimer nanoparticles are altered in the presence of neuroinflammation. *Journal of Neuroinflammation, 13*(1), 65.
173. Zhang, H., Chen, Y., Wang, Z., Xie, G., Liu, M., Yuan, B., Chai, H., Wang, W., & Cheng, P. (2022). Implications of gut microbiota in neurodegenerative diseases. *Frontiers in Immunology, 13*, 785644.
174. Zhang, W., Xiao, D., Mao, Q., & Xia, H. (2023). Role of neuroinflammation in neurodegeneration development. *Signal Transduction and Targeted Therapy, 8*(1), 267. https://doi.org/10.1038/s41392-023-01486-5
175. Zheng, Y., Zhang, X., Zhang, R., Wang, Z., Gan, J., Gao, Q., Yang, L., Xu, P., & Jiang, X. (2023). Inflammatory signaling pathways in the treatment of Alzheimer's disease with inhibitors, natural products and metabolites (review). *International Journal of Molecular Medicine, 52*(5), 111.
176. Zhu, F. D., Hu, Y. J., Yu, L., Zhou, X. G., Wu, J. M., Tang, Y., Qin, D. L., Fan, Q. Z., & Wu, A. G. (2021). Nanoparticles: A Hope for the treatment of inflammation in CNS. *Frontiers in Pharmacology, 12*, 6839.

# Natural Product Assembly in Nanoformulation: Importance in Neurodegenerative Diseases

Kanika Verma, Ritu Singh, Smita Jain, and Swapnil Sharma

**Abstract**

Nanotechnology is one of the innovative approaches that utilizes engineered materials to assist the management of multitudes of ailments, including neurodegenerative disorders (NDDs). Diverse nanocarriers such as nanomicelles, nanoliposomes, lipid nanoparticles, polymeric nanoparticles, and carbon nanotubes offer various advantages, including improved retention impact and penetration, precise administering and targeting, enhanced bioavailability, stability, and reliability. In recent years, plant as a whole or their derived compounds/secondary metabolites are gaining the utmost attention of investigators for their utility in the therapeutic management of various NDDs like Alzheimer's disease (AD), Parkinson's disease (PD), Huntington's disease (HD), etc. This chapter briefs the therapeutic efficacy of different natural product-based nanoformulations for NDDs management with special emphasis on their possible mode of action, merits, and limitations evident in preclinical and/or clinical studies.

**Keywords**

Neurodegenerative disorder · Nanocarrier · Therapeutics · Natural product · Neuroinflammation

K. Verma · R. Singh · S. Sharma (✉)
Department of Pharmacy, Banasthali Vidyapith, Banasthali, Rajasthan, India

S. Jain
Department of Pharmacy, Banasthali Vidyapith, Banasthali, Rajasthan, India

Department of Pharmacy, SVKM's NMIMS, Shirpur Campus, Rajasthan, India

 
R. Shukla et al. (eds.), *Neuroendocrine-Immune Axis*,
https://doi.org/10.1007/978-981-95-5088-3_6

## 1 Introduction

In a developed and advanced medical era, technological enrichments have proven helpful for curing many challenging conditions and disorders. The preliminary thing for diagnosis and treatment of any disorder is to uptake, transport, and targeted response of any imaging/detecting agents and therapeutic drugs to the desired site. In NDD cases, the treatment and diagnosis are even more challenging due to the involvement of the blood-brain barrier (BBB) for drugs to enter the brain. The neurodegeneration term derives from two words, 'neuro' and 'degeneration' means degeneration of neurons. NDD is a diverse condition characterized by dysfunction and continuous loss of neurons, glia, synapse, brain, and spinal cord. Various neurological diseases like AD, PD, MS, ALS, HD, etc., have complex pathophysiology and are still under research. NDDs are incurable and debilitating diseases, increasingly spreading globally due to increased population aging. NDD is the most common illness in geriatric populations, surpasses cancer, myocardial infarction, and stroke, and accounts for 4th leading cause of mortality. In 2019, approximately 50 million people had an NDD globally, often resulting in dementia, which postulates to rise to 152 million by 2060. Dementia is the most common NDD known. The 1.6% of males and 1% of females in the age range of 65–69 suffering from dementia in Europe will increase to 11% and 12.6% respectively, at the age of 85–89 years. This increasing neurological burden poses a global threat. The WHO 2012 report states that >35.6 million people were suffering from dementia worldwide, of which 60%–70% of cases are AD [116].

In contrast to prevailing NDD, we only have seven USFDA-approved drugs for AD treatment, namely Memantine, Galantamine, Donepezil, Rivastigmine, combination of Donepezil and Memantine, Aduhelm, and Leqembi. However, none of the drugs can slow down/reverse or even stop progressing neuronal damage. Additionally, the prime treatment for the second most common NDD type, i.e., PD includes dopamine agonists and levodopa. Another autosomal inherited NDD known as HD, associated with dystonia, chorea, jerks, incoordination, and behavioural changes, has only symptomatic therapy as well. Notably, dopamine antagonists are the chief pharmaceutical option for HD treatment. The unmet needs and healthcare, societal, emotional, and physical dependency-related impacts of NDD forced regulatory authority and society to understand and develop more effective methods to treat these conditions. Thus, enormous efforts have been invested in identifying the pathophysiological path and developing targeted therapies for NDDs management [24, 88]. The prime hurdle in drug development is to overcome the cerebral physiological barriers. The BBB is a prime physical barrier between the bloodstream and CNS, inhibits outflow of peripheral substance to the brain parenchyma. The BBB consists of an endothelial cell layer and on outer side surrounds with astrocytes and pericytes layer which form a tighter junction that is ~50–100 times more firm than the normal circulatory endothelial cell layer. The BBB has specialized ion channels and transporters to maintain the optimal ionic environment for neuronal and synaptic signalling cascade that remain unaffected even in fluctuated disease condition. Likewise, BBB also prevents the traverse of some therapeutic agents that are important for the NDD management [30].

Looking toward the pathophysiology of NDDs, the development and pathogenesis are due to the disturbance of various pathophysiological phenomena like mitochondrial dysfunctions (midysfunction), oxidative stress (OS), metabolic syndrome, proteinopathy formation and accumulation, and neuroinflammation (Fig. 1) [60, 109]. Moreover, some external stimulants like environmental factors, aging, and genetic factors also play a significant role in the NDD progression. Researchers illustrate that these three factors (i.e., OS, midysfunction and metabolic syndrome) are interlinked, and enhancement of one phenomenon is responsible to increase another process. This trio functions to augment the proteinopathy accumulation and neuroinflammation or vice versa. The proteinopathy accumulation on synapse impairs the $Ca^{2+}$ sensitive receptors (NMDA, AMPA, PMCA, MCX), are responsible for long-term depression (LTD) formation, and hinder long-term potentiation (LTP) formation of memory. The trauma or pathologic insult induces the formation of distorted protein in the endoplasmic reticulum (ER) that leads to ER stress condition. The stress condition impairs the mitochondrial energy production via oxidative phosphorylation (OXPHOS), increasesOXPHOS derived by-product including, reactive oxygen species (ROS), and hinder ROS mitigation system. This overall results inproduction of OS which triggers cellular damage, midysfunction, and impairment of DNA repair system, culminating in activationof apoptotic pathways. Neuroinflammation is also a peculiar pathophysiological event in NDD. Generally, the glial cells are responsible for pro-inflammatory markers (cytokine, chemokine,

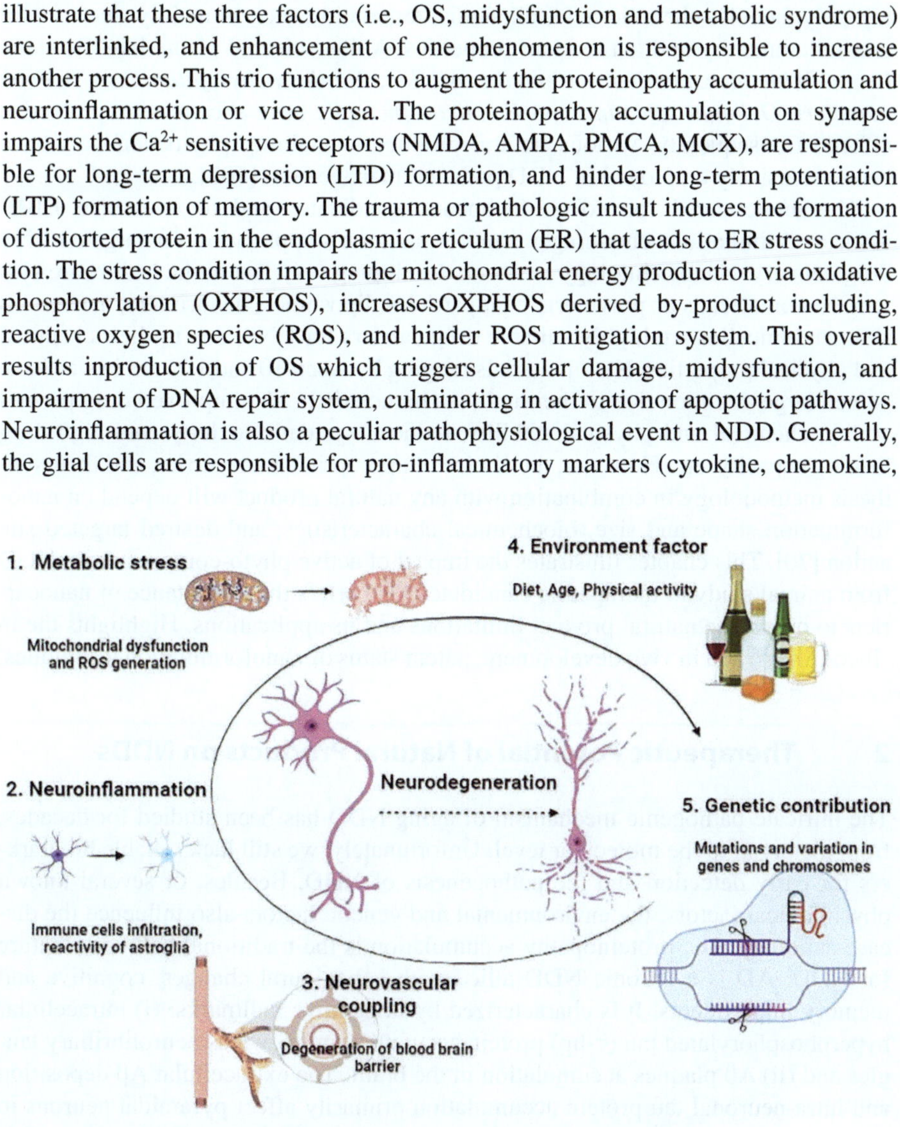


**Fig. 1** Various factors trigger the development and pathogenesis of neurodegeneration

leukotriene, and cytotoxic molecules) released as CNS self-defence mechanism [21, 83]. However, pathologic conditions overwhelm the glia and cause excessive pro-inflammatory markers release, which worsens the NDD case.

The natural bioactive compounds derived from different sources, like microbes, plants, animals, and the marine world, have been extensively investigated for decades in NDD management. The repeated failure of pharmaceutical moieties increases researcher interest in the phyto-compounds as an alternative or integrative treatment for NDD. The beneficial effect of neuroprotective phytochemicals including polyphenols, phenolic acid, alkaloids, terpenoids, etc., is mainly due to their anti-inflammatory and antioxidant properties. A few selected plants like *Curcuma longa*, *Withania somnifera*, *Centella asiatica*, *Nardostachys jatamansi*, *Bacopa monnieri*, *Ocimum sanctum*, and *Emblica officinalis* have shown neuroprotective effects. Consequently, various natural products containing phytochemicals have been extensively employed in NDD patients' management [107].

The unmet need for novel therapeutic candidates and to overcome BBB-associated limitations encourages nanoformulation's application for targeted CNS drug delivery. Nanotechnology has been developing very fast in the healthcare system. Nanomaterials as polymeric nanoparticles, liposomes, dendrimers, nanoemulsion, and micelles are highly suitable drug carriers for CNS, due to their chemical and physical properties. Nanomaterials ranging between 10 and 100 nm in size have gained large attention due to their unique characteristic like improved drug permeability, chemical and physical stability, reduced systemic toxicity, and high drug loading capacity, etc. However, the choice of appropriate nanocarrier, and their synthesis methodology in combination with any natural product will depend on nanoformulation shape and size, biochemical characteristics, and desired targeted site action [70]. This chapter illustrates the impact of active phyto-compounds on NDD from animal study to therapeutic candidate. Also briefs the importance of nanocarriers to overcome natural product limitations and its applications. Highlights the in silico/in vitro and in vivo development, patent status of nanoformulation techniques.

## 2 Therapeutic Potential of Natural Products on NDDs

The intricate pathogenic mechanism of aging NDD has been studied for decades, from the organ to the molecular level. Unfortunately, we still lack reliable biomarkers for early detection and the pathogenesis of NDD. Besides, of several known physiological factors, the environmental and genetic factors also influence the disease aetiology. The proteinopathy accumulation is the traditional hallmark feature for NDD. AD is a chronic NDD allied with behavioural changes, cognitive and memory impairments. It is characterized by imperative hallmarks: (i) intracellular hyperphosphorylated tau (τ-hp) protein accumulation known as neurofibrillary tangles and (ii) Aβ plaques accumulation in the brain. The extracellular Aβ deposition and intra-neuronal tau protein accumulation primarily affect pyramidal neurons in entorhinal cortex (layer-II) progressed toward the hippocampus-CA1 region. This proteinopathy also affects cholinergic neurons (basal forebrain) and noradrenergic

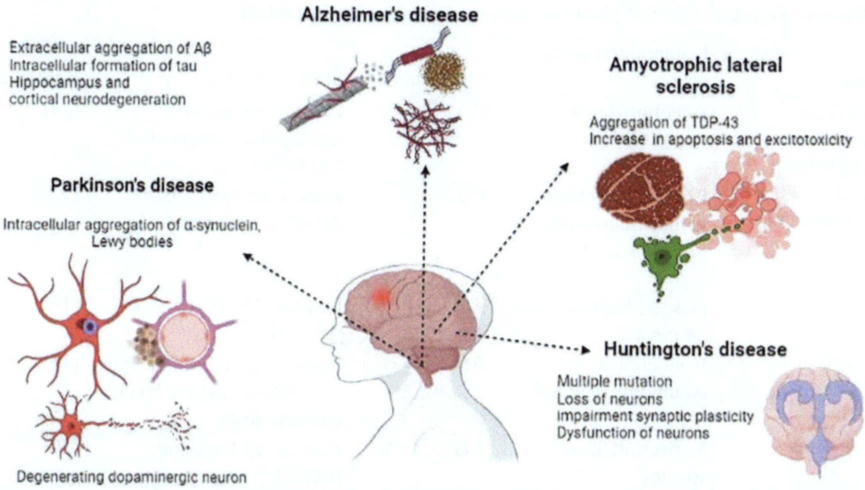


**Fig. 2** Neurodegeneration-associated pathological conditions

neurons (locus coeruleus) as well. PD is also a form of progressive and chronic NDD that generally have dopaminergic nigrostriatal neuronal loss symptomized by reduced motor functions including muscular rigidity, bradykinesia, postural imbalance, and resting tremors. PD mainly affects dopaminergic neurons of the substantia nigra pars compacta, olfactory bulb (OB), and DMV region of the brain. Similar to AD, PD also has proteinopathy accumulation of α-synuclein known as Lewy neurites, Lewy bodies. The HD is a hereditary autosomal dominant NDD characterized by cognitive loss, mental disorders, and motor dysfunction. Pathologically, it is associated with reduced gamma-aminobutyric acid (GABA) and increased dopaminergic activity in basal ganglia. Additionally, an unstable nucleotides (like cytosine–adenine–guanine) expansion is also seen in the HD. Amyotrophic lateral sclerosis (ALS), also a form of NDD, is a disorder linked with superoxide dismutase-1 (SOD1) enzyme-coded gene mutations [24, 39, 54, 68, 71, 91, 96] (Fig. 2). Recently, the natural products have shown their ability of reducing neuroinflammation, proteinopathy and protein misfolding induced–stress revealing their potential as an integrative or alternative therapeutic agent against AD and other NDDs. For example, flavonoids (baicalein, apigenin, narirutin, hesperidin, quercetin, and isoquercetin), alkaloid (berberine), carotenoid (curcumin), EGCG, stilbenoid (resveratrol), linalool, etc. (Table 1).

Different biological processes, including midysfunction, neuroinflammation, and OS, have been evidenced in the development and pathogenesis of NDD. Mitochondria are responsible for cellular respiration and energy generation – OXPHOS (oxidative phosphorylation). Mitochondria are also linked with the ER via $Ca^{2+}$ channel for ease transverse of $Ca^{2+}$ ion and have an important role in managing cellular redox environment. The $Ca^{2+}$ overloading triggers stress condition and increases ROS level resulting in the opening of mitochondrial pores, midysfunction and apoptotic caspase cascade. The AD brains have disturbed OXPHOS

**Table 1** Potential effect of plant secondary metabolites against NDD

| Neuroprotective activity | Compound/natural source | Study model | Study outcome | Reference |
|---|---|---|---|---|
| Reduce protein misfolding and aggregation | Aegeline/*Aegle marmelos* | PD | Reduce α-synuclein aggregation, apoptosis and ROS level | [34] |
| | Epigallocatechin gallate/*Scutellaria baicalensis*, *Angelica gigas* | PD | Reduce Aβ load, OS, AChE, α-synuclein | [124] |
| | Hesperidin/*Valeriana officinalis* | AD | Reduce BACE1 level, OS, Aβ load | [19] |
| | Isoquercitin/commonly available | AD | Reduce Aβ, τ-hp, BACE1, γ-secretase enzyme level, and cell death | [15] |
| | Narirutin/Citrus species | AD | Reduce Aβ level and BACE1 | [20] |
| | Silibinin/*Silybinisus laborinum* | AD/PD/stroke | Reduce ROS, Aβ, AChE, hypoxia, ischemic injury, OS, LC-3, and beclin-1 level | [32] |
| Antioxidant activity | Luteolin/commonly available | Stroke | Have antioxidant and anti-inflammatory properties, reduce OS, MDA, and induce Nrf2 level | [110] |
| | Naringenin/*Citrus* paradise | AD | Reduce OS, cytokines, NFkB, NO, and induce Nrf2 | [100] |
| | Rutin/Citrus species | AD | Reduce cytokines, ROS, OS, IL-1 and 6 level and have neuroprotective effects | [104] |
| Anti-inflammatory activity | Kolaviron/*Garcinia kola* | Stroke | Reduce MPO, cell death and restore Na-K- ATPase pump | [3] |
| Reduce protein misfolding/antioxidant activity/anti-inflammatory activity | Curcumin/*Curcuma longa* | NDD & PD | Show neuroprotective efficacy, reduce OS, infarct volumes, Aβ load, NFKB and α-synuclein level and induce GSH level in hippocampus and cortex region, and microglia M2 polarization | [61] |
| | Linalool/Lavandula species | AD | Reduce inflammatory condition, COX2, Aβ, hp-τ, and some $Ca^{2+}$ channels | [95] |
| | Resveratrol/*Vitis vinifera* | PD | Reduce COX2, TNFα, IL-6, Aβ, NO, and induce BDNF, IL-10, GDNF, SIRT-1 | [117] |

(continued)

**Table 1** (continued)

| Neuroprotective activity | Compound/natural source | Study model | Study outcome | Reference |
|---|---|---|---|---|
| Antioxidant activity/ anti-inflammatory activity | Sulforaphane/ Cruciferous species | AD | Reduce IL-1β, Aβ, aluminium load and induce Nrf2, and neurogenesis factors | [5] |
| Anti-apoptotic & Neurotrophic activity | Withanolide A/*Withania somnifera* | AD | Enhance neurogenesis | [120] |
| Reduce protein misfolding/ antioxidant activity/ anti-apoptotic activity | Withanamide A & C/*Withania somnifera* | AD | Reduce Aβ load, OS, ROS | [35] |
| | Apigenin/flavonoid content-commonly in plants | AD & neurogenesis | Reduce cytokines & cortisol level, Aβ load, oxidative stress, neurotoxicity. And augment BDNF synthesis and mitochondria protection | |
| | Baicalein/*Scutellaria baicalensis* | PD | Reduce Aβ load, hypoxic condition, NO level, OS, ROS, TNFα, IL-6, NFKB, and dopaminergic neuron loss | [105] |
| Reduce protein misfolding/ anti-inflammatory activity/ anti-apoptotic activity | Quercetin/Citrus species and tea | AD/PD | Reduce OS, ROS, TNFα, IL-6, GFAP, MDA, apoptosis, GSK3β, τ-hp, Aβ, BACE1, PKC-ε | [113] |
| Antioxidant activity/ anti-apoptotic activity | Acacetin/Calamintha species | PD | Reduce cell death, caspase cascade, ROS, midysfunction | [52] |
| | Asiatic acid/*Centella asiatica* | PD | Reduce ROS, dopamine reduction, apoptosis, and induce ERK, AKT pathway | [77] |
| | Ginsenoside Rd/*Panax ginseng* | Stroke | Reduce ROS, $Ca^{2+}$ uptake, excitotoxicity | [76] |

(continued)

**Table 1** (continued)

| Neuroprotective activity | Compound/natural source | Study model | Study outcome | Reference |
|---|---|---|---|---|
| Reduce protein misfolding/ antioxidant/ anti-inflammatory/ anti-apoptotic activity | Ginsenoside/*Panax notoginseng* | AD | Reduce Aβ load, NO, ROS level, interleukins, TNFα, β and γ-secretase enzyme level, and apoptosis | [51] |
| Reduce protein misfolding/ anti-apoptotic activity | Morin/commonly available | AD | Reduce Aβ, hp-τ, BACE1, γ-secretase, apoptosis factors | [15] |
| Reduce protein misfolding/ AChEI activity | Crocin/*Crocus sativus* and *Gardenia jasminoides* | AD | Reduce OS, MDA, AChE, TNFα, COX2 and τ-hyperphosphorylation and induce SOD, Ach activity | [37] |
| Reduce protein misfolding/ anti-inflammatory/ anti-apoptotic/ AChEI activity | Naringin/Citrus species | AD/PD | Reduce AChE, GFAP, and induce neurotrophic factors and memory recognition | [27] |
| MAO-inhibitor | Chrysin/*Cytisus villosus* | In silico | Reduce MAO-A | [56] |
| | Myricetin | | | |
| Reduce protein misfolding/ MAO- I | Berberine/Berberis genus | AD | Reduce AChE, BACE1, TNFα, IL-6, NFKB, COX2, Aβ plaques and induce p38 kinase ERK1/2 pathway and AKT | [123] |
| Reduce protein misfolding/ anti-oxidant/ anti-apoptotic/ MAO-I | Genistein/Glycine max | AD | Reduce MAO, NFKB, inflammation, Aβ burden, cell death | [56] |
| Anti-thrombotic activity | Carvacrol/*Cyperus rotundus* | In vitro | Reduce platelet aggregation | [87] |
| | Nootkatone and α-cyperone/*Cyperus rotundus* | | | |
| | Wogonin/*Suctellaria baicalensis* | Stroke | Reduce platelet aggregation, thrombin formation, activated partial thromboplastin, and prothrombin time | [23] |

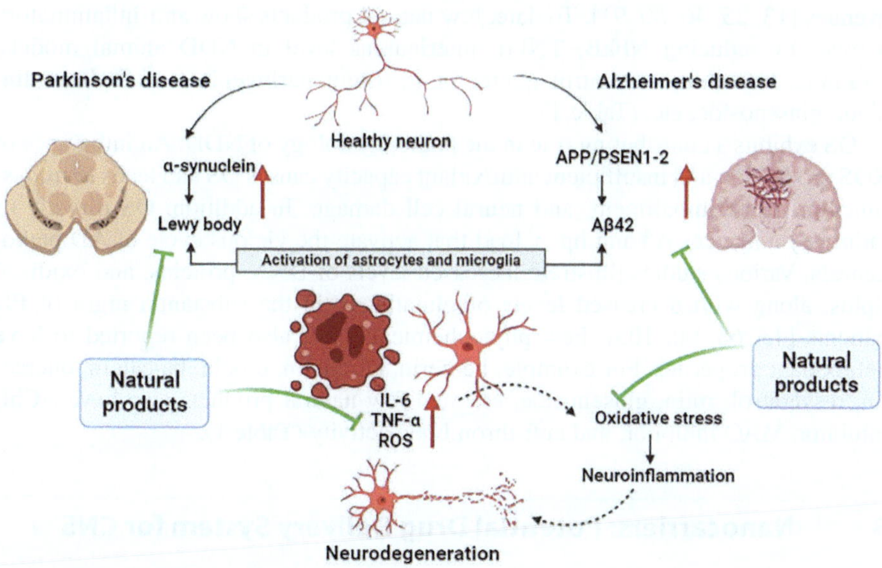


**Fig. 3** Effect of natural products on the pathogenesis of degenerated neurons in AD and PD

involved enzymes like ATP synthase complex, pyruvate dehydrogenase, α-ketoglutarate dehydrogenase, isocitrate dehydrogenase, and complex I-IV involved enzymes, which affect mibiogenesis and lead to abnormal ATP production. This progress toward impaired bioenergetics in the brain and trigger synapse loss led to cognitive impairment [18, 68, 92, 101, 108] (Fig. 3). Biologists illustrate that PD patients also have disturbed complex-I-related enzymes that trigger ROS production, decrease bioenergetics function, and cause mitochondrial instability. In the brain, the apoptotic factors stimulate nucleus that led in condensation of chromatin and fragmentation of DNA. However, complex-I dysfunction is an important step in dopaminergic neuronal death [65, 82]. Few natural product shows beneficial anti-apoptotic effect in NDD cases. For example, ginsenoside, acacetin, baicalein, etc. (Table 1).

Neuroinflammation is an imperative immune response against several factors in CNS and characterized by activated astrocytes and microglia. It's a natural defence system to mitigate triggering factors like misfolded proteins, cell fragments, and other stimuli. Nevertheless, neuroinflammation is a different process that engenders BBB disruption and neuronal degeneration, by releasing chemokines, interleukins, and pro-inflammatory cytokines primarily by astrocytes and activated microglia. Microglia are macrophages that promote neurite formation, strip excess synapse from developing neurons, and regulate programmed neuronal cell death. In case of any micro environmental changes the microglia are activated and polarised into M1 or M2 type, and secrete chemokines and pro-inflammatory cytokines molecules. Overall, increasing evidence of targeting neuroinflammation has been well-documented for understanding the risk of NDD and identifying novel therapeutic avenues [13, 25, 46, 80, 91]. To date, few natural products show anti-inflammatory activity by reducing NFkB, TNFα, interleukins level in NDD animal models. Example, curcumin, resveratrol, quercetin, genistein, naringenin, sulforaphane, linalool, ginsenoside, etc. (Table 1).

OS exhibits a contributing role in the pathophysiology of NDD. An imbalance of ROS production and insufficient antioxidant capacity causes OS and leads to midysfunction, DNA impairment, and neural cell damage. In addition, OS directly or indirectly enhances Aβ and hp- τ load that activate the vicious cycle of AD pathogenesis. Various studies illustrate increased levels of DNA, proteins, and oxidized lipids, along with decreased levels of glutathione in the substantia nigra of PD patients [43, 68, 90, 103]. Few phytochemicals have also been reported to have antioxidant properties. For example, berberin, curcumin, crocin, baicalein, quercetin, resveratrol, rutin, ginsenoside, etc. and few natural products also have AChE inhibitor, MAO inhibitor, and anti-thrombotic activity (Table 1).

## 3 Nanocarriers: Potential Drug Delivery System for CNS

Generally, molecular complexity, functional group reactivity, limited solubility, and instability are rate-limiting factors for natural products that decrease their therapeutic efficacy. Moreover, for CNS-targeted medications, the physiological barriers like BBB permeability, blood-CSF (cerebrospinal fluid) barrier permeability, P-glycoprotein outflow, and the action of plasma proteins in the circulatory system add natural product (NP) inherent limitations. Hence, nanocarriers are potential hope for natural products to aid neurodegeneration therapeutic medicational list (Fig. 4).

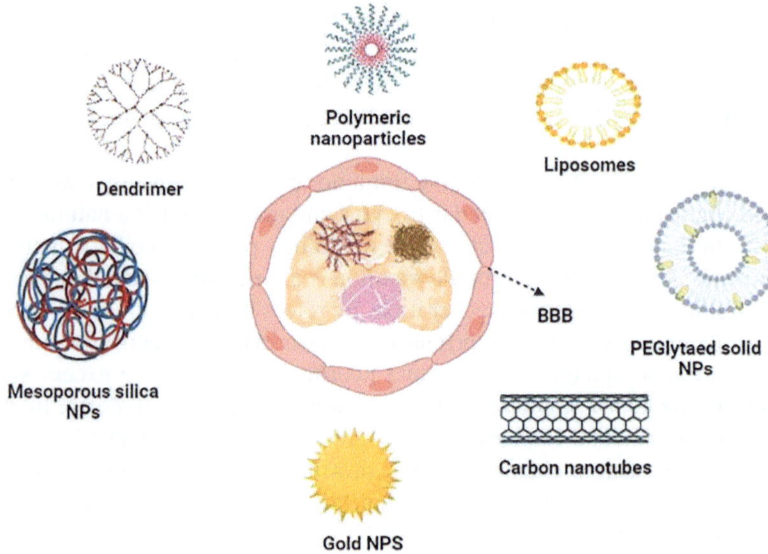


**Fig. 4** Different nanocarriers for targeted CNS drug delivery

## 3.1 Nanomicelles

Micelles are monolayered, nano-sized particles of size range 10–100 nm, self-assembling, and colloidal dispersions of a hydrophobic core covered with hydrophilic shell. Due to its unique features of solubility, size, exposure to the environment and customized surface show novel characteristics to make it multifunctional. At certain temperatures (known as critical micelle temperature (CMT)) and at specific concentrations (critical micelle concentration (CMC)), the amphiphilic molecule aggregates to form a micelle. This micelle aggregation formed due to removal of hydrophobic fragments and formation of hydrogen bonds, which leads to a decrease in free energy of system in order to stabilize the micelle. Nanomicelles have a natural phenomenon to regulate their shape and size, have high drug loading capacity, encapsulation efficiency, and exhibit short circulation time, compared to other nanocarriers. For example, curcumin, a well-known drug for AD, shows different bioavailability with different nanocarriers in a mice study model. The micelle formulation has 45-fold bioavailability compared to nanoparticles (9-fold) and liposomal (5-fold) nanoformulations. Some other drugs like coenzyme Q10 (CoQ10; for PD), flurbiprofen and rivastigmine (for AD), formed nanomicelles to increase bioavailability, brain penetration, and solubility [93, 102].

Polymeric micelles consist of a core-shell and amphiphilic block copolymer, for example, L,D-lactine polycaprolactone and PEGs contain hydrophobic and hydrophilic block polymer, respectively. It incorporates a wide range of drugs (hydrophilic to hydrophobic) released at the target site by diffusion mechanism. This nanosystem has various advantages like (i) long circulating characteristics and significant accumulation on target, (ii) more stable and lower CMC, and (iii) slower dissociation rate. However, the nanoformulation has low encapsulation efficiency of hydrophilic therapeutics, inherent toxicity, and instability over prolonged storage time and will limit its utility [28, 86].

## 3.2 Nanoparticles

Nanoparticles are a colloidal arrangement of inorganic (like gold, iron, cerium, molybdenum, silica) and organic (like trehalose, PLGA, and PLA) materials in core that able this formulation to permeate BBB. The therapeutic moiety is either entrapped within the colloid matrix or surficial coated via adsorption or conjugation methodology. The distinct characteristics of nanoparticles are: (i) they carry a relatively high drug amount with comparatively small size and able to deliver drug moiety at a specific site in a controlled and sustained rate, (ii) they show excellent imaging performance, especially inorganic nanoparticles, (iii) few nanoparticle carrier materials show therapeutic efficacy like antioxidant, reduce ROS level, inhibit Aβ aggregation, etc., (iv) they are efficient in delivering insoluble drugs to CNS. All these factors make them more promising carrier systems for NDD. In addition, several modifiers enhance the stability, for example, PEG, polycyanoacrylate, polystyrene nanospheres (20–200 nm), nanosuspensions (a colloidal system that stabilizes by application of non-ionic surfactants, e.g., PEG, mixture of lipids), etc. Despite this, nanoparticles have few common lacunas such as nanotoxicity, particle aggregation, and organic solvents used during fabrication. The uniform manufacturing of polymeric nanoparticles is also a major limiting factor. Morphologically, nanoparticles remain either nanocapsule or nanosphere forms. The nanocapsules have an oily core that dissolves drug moiety within themself and the outer body cover that controls the drug release rate [1, 4]. While nanospheres have the drug either inside or adsorbed onto the surface to form a continuous polymeric network. To protect the drug moiety from elimination, nanocapsules or nanospheres encapsulated with the nanofibers/synthetic polymer semipermeable membrane.

The iron nanoparticles show promising results in animal and *in vitro* models to reduce the physiological deficit of PD. However, biologists illustrate that iron or metal accumulation in substantia nigra pars compacta causes PD pathophysiological features, which increases oxidative stress and protein accumulation and may cause brain cells toxicity. Thus, neurotoxicity caused by iron nanoparticles was examined carefully at *in vitro* and *in vivo* stages. The gold nanoparticles are prevalent due to their low cytotoxic effect and optical properties. Many biomolecules like anthocyanin (antioxidant) and peptide inhibitors (e.g., VVIA and LPFFD) are laden with gold nanoparticles to prevent Aβ aggregation and cytotoxicity in AD. However, NDD is a multifaceted and complex disease; every nanoparticle formulation cannot be used as nanocarriers for delivery, diagnosis and imaging purpose. Now, the nanotheranostic system has been developed as a promising tool for imaging and therapeutic amalgamation for disease detection and therapy [54].

## 3.3 Lipid Nanoparticles

Lipid nanoparticles are generally lipid-based nanomedicine useful (i) to protect phyto-therapeutic candidate from *in vivo* degradation, (ii) to amend biodistribution by enhancing solubility and bioavailability, (iii) to control drug release, and (iv) to target drug delivery to the desired site of action. They have added advantages over colloidal nanocarrier systems, like improved physical and chemical stability, phototherapeutic entrapment, prolonged drug release, and efficient entrapment of lipophilic drugs in the core. They are generally available in three forms, i.e., SLNs (solid lipid NPs), NLCs (nanostructured lipid carriers), and liposomes. SLNs are nanospheres of size ranging from 50 to 1000 nm, have lipid contain matrix (e.g., glycerides, fatty acids, waxes, etc.). Moreover, this nanoformulation is stabilized by the use of emulsifiers (e.g., polyvinyl alcohol, polyoxyethylene ethers, tween, bile salts, etc.). SLNs include iron oxide microemulsions injected into rats to assess BBB permeation and found that they can be used in MRI imaging techniques due to their paramagnetic features. SLNs also have high ability of BBB permeation thus used as a contrast agent during MRI procedure. Research shows that in combination with PAT (photoacoustic tomography) it enhances the sensitivity of MRI techniques to a picometer concentration level [12, 79].

NLCs are an advanced version of SLN to overcome their limitations. During NLC fabrication, instead of solid lipid, an oil lipid matrix (triglycerides, steroids, fatty acids, waxes, etc.) is used to enhance the drug-loading capacity and to prevent drug leakage during storage. Based on compositional material and synthesis methodology NLCs can be of three types, i.e., imperfect, amorphous, and multiple oils in solid fat-in-water (O/F/W) type. As the name suggests, the imperfect type NLC has an imperfection in their crystal order during mixing of several lipids, to increase their drug-loading capacity. The amorphous type has a structureless amorphous matrix made up by mixing of solid lipids and lipids (e.g., isopropyl myristate, hydroxyoctacosanyl, hydroxyl stearate) to elongate storage time period. O/F/W type NLC contains plentiful nanosized liquid oil scattered in the solid form of matrix [22].

Liposomes are spherical-shaped, lipid-based vesicular systems, a sandwich of hydrophilic and lipophilic bilayers. It provides a versatile drug delivery opportunity for peptides, small molecules, genes, monoclonal antibodies, etc. [99]. Liposomes are the first clinically approved nanoformulation, for example, PLD (pegylated liposomal doxorubicin, use for ovarian cancer, AIDS-related Kaposi's sarcoma, lymphoma, etc.) and nanoparticle-albumin-bound paclitaxel (for non-small cell lung cancer and breast cancer) [94]. The Global Liposomal nanoformulations Market (like, Liposomal form of Paclitaxel, Doxorubicin & Amphotericin B, etc.) captures a wide area of therapeutic indications (like Cancer and Tumor Therapy, Fungal Infection Therapy, Pulmonary and Ocular therapeutic area, etc.), contributes in global pharma market growth. Various market research forecasting agencies postulate a current CAGR of ~8.50%, of USD 4.16 billion in 2021 globally, which increases up to USD 7.99 billion by 2029 [14].

For neurodegeneration, the liposome drug delivery system is a classical dosage form to aid therapeutic moiety (both hydrophilic and lipophilic compounds) penetration to BBB. Despite various favourable outcomes, they have fewer limitations like quick metabolic degradation of the phospholipid layer, moderately efficient for lipophilic drug moieties, fast systemic elimination, stability issues after long storage time period, and not able to provide sustained release. Recently, fewer modifications on the liposomal surface, like attachment of specific ligands (glucose, lactoferrin, transferrin, PEG (polyethylene glycol), glutathione, cationic &/or specific peptides), enhance its brain targeting ability. Before application, the liposome nanoformulations scrutinized via various extensive characterization parameters, like shape and size, zeta potential, phase behaviour, encapsulation efficiency, polydispersity index, lamellarity, and *in vitro* drug release efficiency, etc., later safety and efficacy parameters, are investigated in various animal models [84].

## 3.4 Dendrimer

Dendrimers are highly branched, three-dimensional, synthetic polymeric macromolecules composed of several layers over core, active groups on the surface and repetitive units in the nanoscale size range. Due to their structural orientation, they have multiple drug binding locations to trap high amount of drug thus have a high drug loading capacity. Additionally, drugs distributed uniformly per size have thermodynamic stability, a very low dispersion rate thus shows high performance. Various dendrimers include polyhydroxylamine, polyamidoamine (PAMAM) dendrimers, and polypropylene amine, and generally transport drugs by endocytosis mechanism through cellular internalization. However, the developmental complexity and toxicity limit their applicability. Polyamidoamine (PAMAM) dendrimers attract great attention by paving therapeutics delivery in a controlled release fashion for CNS diseases like Rett syndrome and autism [85] (Fig. 4 and Table 2).

### 3.5 Nanogels

Nanogels are unique drug delivery systems having both hydrogel and nanotechnological features. They are a network of nano-sized polymers with polyethylene amine, PEG, polyacrylic acid, and pluronic, by forming ionic or non-ionic chains. The distinct characteristic feature of this system is having a high drug loading capacity (40–60%). The surface modification of this nanosystem by either transferrin or insulin application enhances the drug distribution through BBB. They are potential transporters for drug delivery to CNS. Scientists developed nano-gel-based methotrexate injection and post-in vitro characterization assessed in animal model. Post injection, these nanogels enter the bloodstream via binding to apolipoproteins and brain endothelial cells and uptake drug by endocytosis mechanism (mediated receptors) [48].

### 3.6 Nanotubes

The promising example of this nanosystem are carbon nanotubes, which are carbon-based cylindrical structures having more than one layer of carbon molecules. Due to its number of layers, this system is categorized as either single or multi-wall tubes. This nanoformulation has characteristic chemical, mechanical, and electrical properties and is valuable in CNS medicine via its neuronal activity and network connectivity. Scientists examined nanotube formulations in animal models using a scanning electron microscope and found that the drug delivery system has energy dependency [119].

### 3.7 Nanoemulsion

Nanoemulsion, as the name suggests, is an either water-in-oil (w/o) or oil-in-water (o/w) formulation system, that is stabilized by incorporating an amphiphilic surfactant. This nanosystem represents an assuring delivery system to protect nanoformulation from degradation, controlled release and enhance bioavailability, suitable for intranasal route delivery systems by surpassing the BBB [78].

**Table 2** Advantages and disadvantages of different nanoformulations

| Nanoformulations | Advantages | Disadvantages |
|---|---|---|
| Polymeric nanoparticles | They are able to capsulate both hydrophilic and/or hydrophobic drug moieties, able to protect encapsulated active phyto-ingredient from metabolic digestion thus extending the nanosystem residence time. They are biocompatible and recyclable candidate | The burst release of active ingredient may cause neurotoxic effect in some cases. The drug loading capacity is limited and the manufacturing cost is also high |
| Liposomes | Able to contain both hydrophilic and/or hydrophobic active drug. Have controlled drug release ability, protect encapsulated active phyto-ingredient from metabolic digestion thus extending the nanosystem residence time. They are biocompatible and recyclable candidate | Have smaller half-life compared to other formulations. *In vivo* the active ingredient might leak from lipid bilayers impairing the sustained drug release. Hydrophobic compounds are less efficient thus hindering drug loading capacity. After prolong storage it has low stability profile. The manufacturing is complicated, has complex fabrication process, is difficult to sterilize the nanosystem, & have high manufacturing cost |
| SLNs | The large-scale production is comparatively easy with relaxed sterilization process. Enhance stability profile, and has controlled release of active ingredient capacity<br>This nanoformulation has the ability to protect drug from degradation, in vivo | The drug entrapment capacity is limited |
| NLCs | The large-scale production is comparatively easy with relaxed sterilization process. In vivo shows better bioavailability profile by oral administration | Burst release of entrapped active drug candidate |
| Nanocrystals | The most advantageous feature of this nanoformulation is that they are carrier-free system, have 100% active drug content, this feature is promising for water-insoluble drugs and shows enhanced bioavailability, in vivo | Due to carrier-free technology very difficult to control the nanosystem morphology, and incorporate high manufacturing cost by complying nanoformulation manufacturing regulations |
| Metallic nanoparticles | Due to incorporation of metal, this nanosystem has unique optical, electrical, and magnetic properties. In influence of external stimuli able to release drug, and have enhanced permeability rate | Due to incorporation of metal the accumulation and aggregation of nanoparticles are generally seen in organ or tissue causing neurotoxicity |

# 4 Application of Numerous Natural Products-Based Nanoformulations in NDDs

Numerous studies have evidenced the natural products' therapeutic potential for NDDs diagnosis, treatment, and management. The phyto-compounds' neuroprotective effect is mainly because of their anti-inflammatory and antioxidant properties. They also have potential in the reduction of proteinopathy aggregation. However, the BBB penetration is a major hindrance for CNS-targeted drug delivery. In this manner, the novel technologies along with multidisciplinary approaches will undoubtedly allow the development of early diagnostic methods and the establishment of therapies for NDD. Pharmaceutical nanotechnology or nanomedicine provides an efficient drug delivery system for the management of NDD employing improved diagnosis, repairing, controlling, and monitoring at a molecular level. Nanoformulations of these natural products, including polyphenols, alkaloids, and terpenoids, are efficient to enhance the bioavailability and stability and overcoming such problems [71, 126]. Herein, we enlist the major phyto-constituents exhibiting neuroprotective effects.

## 4.1 Polyphenols

Polyphenols are major secondary metabolites derived from plants with highly diversified structures. Their structures vary from highly complex polymeric forms in lignins and tannins to simpler phenol candidates (aromatic ring with ≥1 hydroxyl group). They have high antioxidant properties and play an imperative role in suppressing NDD and redox condition. Recently, various side effects have been associated with traditional drugs including NMDA antagonists and cholinesterase inhibitors. The polyphenol mainly presents in food and plants, generally contains high amount of flavonoids (60%), phenolic acids (30%), lignan, stilbene, and lignins. The flavonoids are subclassified into flavanone, isoflavanone, flavan-3-ol, flavonoids, anthocyanin, and others [59].

Flavonoids including epigallocatechin-3-gallate, hesperidin, genistein, and quercetin modulate the inflammatory response and decrease the cytokine expression (COX-2, IL-1β, IL-6 and TNF-α) and reduce neuronal damage. The anti-inflammatory effect is associated with microglial cell regulation via NFkB and MAPK-mediated signalling pathways [106]. Several hypotheses have been proposed for flavonoids neuroprotective effect, which also plays an important role in MAPK and PI3K/Akt signalling pathways to regulate gene expressions. Researchers illustrate its involvement in the proteinopathy, by hindering Aβ aggregation, GSK3β and CDK-5 activity, and prevent tau hyperphosphorylation [8]. Nevertheless, the exact alleyway needs to be explored to state the flavonoid-specific mechanism of action for exerting a neuroprotective effect.

Several pre-clinical studies have investigated flavonoids importance in different NDD cases. Anwar et al. explored the neuroprotective effect of *Harrisonia abyssinica* in aluminium chloride-induced neurotoxic effects in the Wistar rat model. The results revealed that the presence of flavonoids restored the levels of glutamate, monoamine, MDA, TNF-α, IL-1β, caspase-3, ERK1/2, AChE activity, and down-regulated the pro-apoptotic Bax protein, normalized Bcl2 level in the hippocampus region of aluminium-treated animals [6]. Li et al. studied the pharmacokinetic behaviour of Puerarin (*Radix puerariae*) in MCA occlusion rat model and noticed that the intravenous administration of Peurarin (an isoflavone-C-glycoside) affected by BBB. The intranasal administration seems an efficient route for drug delivery by surpassing BBB barrier but has a nasal mucosal damage. Thus, the study demanded new dosage forms like nanoformulations (such as microspheres, microemulsions, liposomes, etc.) with low irritating effects and efficient BBB penetration efficiency [58]. Naringenin is a natural antioxidant and anti-inflammatory flavanone, however, lacks the ability to cross BBB. Owing to its neuroprotective ability, the SLNs nanoformulation was evaluated via in silico and in vivo *Drosophila melanogaster* (*D. melanogaster*) animal model. The naringenin-loaded SLNs with a diameter of 134.5 ± 20 nm showed potential neuroprotective effects possibly through sirtuin, NF-κB, and Nrf-Z/ARE pathway in *D. melanogaster* and rotenone-induced PD Wistar rats study. The formulation enhanced the penetration, bioavailability, and neuroprotective effect of naringenin in the PD study [64].

Another flavonoid, quercetin can induce neuroprotection by preventing tau phosphorylation and improving functional and cognitive symptoms in AD. However, limited bioavailability and targeted site action reduce its clinical application in NDD. Exosomes have gathered enormous attention as a cargo carrier in drug delivery systems and neuronal disease management. Qi et al. developed quercetin-loaded exosomes in a 40–150 nm diameter nanosystem with better targeting, bioavailability, innate stability, high transmission efficiency, low immunogenicity, good compatibility, and ability to improve cognitive dysfunctions in okadaic acid-induced AD mice model study. Quercetin-loaded exosomes might target BBB by specifically targeting endothelial TLR4 and HSP70 in the brain. The developed formulation reduced NFT formation and inhibited CDK5-mediated phosphorylation of Tau, suggesting its neuroprotective behaviour [89].

The inclusion of biocompatible and biodegradable polymers allows rationalization in the design of innovative nanostructures to cross BBB and attenuate the limitations of traditional drugs. Nowadays, due to the promising results of active phyto-compounds obtained in NDD models, curcumin importance can't be ignored. The curcumin efficacy is limited by rapid metabolism, low gastrointestinal tract absorption, and poor aqueous solubility. Recent studies have evidenced the neuroprotective effect of curcumin-loaded PGLA nanoparticles by interfering with the aggregation of Aβ and enhancing neural differentiation. Similarly, curcumin-loaded liposomes efficiently inhibit the deposition and generation of Aβ fibrils in vivo in the brain. Curcumin-loaded glycerol mono-oleate nanoparticles and SLNs inhibit synuclein aggregation, reduce apoptosis, and oxidative stress. Moreover, curcumin-loaded nanoemulsions via intranasal delivery do not show cytotoxic effect. Similarly, resveratrol also has multifaceted effects, it removes Aβ load and induces amyloid protein precursors degradation. Due to its extensive and rapid metabolism and poor bioavailability, it has been under investigation. Resveratrol-loaded PGLA and

**Table 3** Natural-based nanoformulation and their applications

| Natural moiety | Nano formulation | Study model | Properties | Reference |
|---|---|---|---|---|
| Curcumin | Liposomes | AD | Decrease Aβ load | [111] |
| | TAT | | Decrease Aβ load and enhance drug availability in brain | [98] |
| | Apo-E3 mediated poly | | Enhance drug bioavailability and stability | [74] |
| | PLGA-based nanosphere | | Decrease Aβ load | [44] |
| | PLGA-based nanoparticles | | Enhance bioavailability and efficacy | [66] |
| | Nanoliposomes | | specific for $A\beta_{1\text{-}42}$ plaques | [72] |
| | pCB | | Improve PK profile and inhibit $A\beta_{42}$ load | [125] |
| | PEG-PLA | | Enhance bioavailability | [26] |
| | SLNs | | Reverse $AlCl_3$-induced neurotoxicity | [49] |
| | Mucoadhesive liposome | | Shows controlled release, high bioavailability and stability | [97] |
| | Lipid-based nanoparticles | PD | Enhance bioavailability and reduce α-synuclein load | [55] |
| | Nanomicelle | MS | Supress cytokine level, miRNAs, inflammatory condition and mediators | [36] |
| Curcumin | Lactoferrin nanoparticles | Neuro protection | Enhance drug availability in brain | [11] |
| Berberine | Multiwall carbon nanotubes | AD | Reverse memory impairment, reduce AChE activity | [62] |
| Q10 6-coumarin | PLGA | AD | Reduce proteinopathy, increase bioavailability and enhance memory skills | [114] |
| Hespereti n | Nanocrystal | AD | Improve memory skills and antioxidant enzyme activity | [50] |
| Chrysin | SLN | AD | Enhance oral bioavailability, antioxidant activity and reduce neuronal damage | [112] |

PEG-PCL nanoparticles show controlled drug release that is crucial to enhance the antioxidant effects [33, 68]. Unfortunately, the translation of pre-clinical outcome fails to show a similar effect in clinical trials, still the encouraging results open new avenue for futuristic development of new therapeutic approaches (Table 3).

## 4.2 Alkaloids

Alkaloids are the second most secondary metabolites containing nitrogen groups in their structure. Based on their pharmacokinetics, biochemical precursors, and chemical structures, alkaloids are classified into various types. The most common subtype of alkaloid is heterocyclic alkaloids, which are charcaterized by presence of nitrogen group in the cyclic ring structure. An indole alkaloid, harmine (*Peganum harmala*), a methylxanthine derivative, caffeine (*Coffea arabica*), a pyridine alkaloid, nicotine (*Nicotiana tobaccum*), an indole alkaloid, geissospermine (*Geissospermum vellosii*), and a piperidine alkaloid, PIP were shown to possess neuroprotective effects on NDDs. In addition, galantamine (*Galanthus nivalis*), salsolin (*Salsola oppositifolia*), morphine (*Papaver somniferum*), montanine (*Rhodophiala bifida*), and berberine (*Berberis vulgaris)* belong to isoquinolone alkaloids, which have evidenced positive effects in NDDs. Alkaloids produce neuroprotective effects via inhibiting α-synuclein aggregation, AChE and BuChE (butyrylcholinesterase), Aβ, MAO, and modulation of neurotransmitters. In addition, alkaloids also act as NMDA antagonists or nicotinic or dopaminergic agonists [71]. Berberine aids the neuroprotective effect by inhibiting neural cell apoptosis, tau phosphorylation, GSK-3β, and AChE activity, and increasing protein phosphatase 2A activity. In addition, berberine reduces the formation of neurofibrillary tangles, protein carbonyl activity, and enhances cognitive skills in mice AD study model. However, serious adverse effects and poor bioavailability of berberine make it difficult for oral administration and greatly impede its clinical application. To overcome these limitations, [7] prepared a berberine-loaded nanomicelle formulation in a diameter of 9.247 and 18.46 nm using the film hydration method for achieving a better neuroprotective effect and improving encapsulation efficiency and bioavailability of berberine. [115] synthesized lactoferrin-modified berberine nanoliposomes in a diameter range of 120.5 ± 3.5 to 129.5 ± 4.5 nm with high biodegradability, entrapment efficiency, and low toxicity. The prepared nanoformulation reduced tau phosphorylation in cortex-inhibited apoptosis and AChE activity in the hippocampus, and improved mouse behaviour in Aβ injected AD-induced ICR mice. However, they did not explore the prepared nanoformulation dose-dependent nature in the AD mouse model study. Additionally, they did not compare the different methods of administration route as well [7, 115]. Mishra et al. prepared curcumin and berberine- encapsulated transferosomes of a diameter of 165.3 ± 6.5 nm by film hydration method for effective management of AD. The developed transferosomes showed increased stability, greater cellular penetration, and improved cognitive memory in scopolamine-induced-AD Swiss mice. The levels of BACE-1, NO, ROS, and inflammatory cytokines in the hippocampus were decreased [69].

Magnoflorine is an aporphine quaternary alkaloid that possesses good neuroprotective effects in NDD due to its antioxidant and antiradical effect. It also prevents the copper-induced lipid peroxidation and exhibits an anti-amnesic property. However, due to insufficiency to cross BBB magnoflorine has been associated with lateral side effects. Thus, Bashir et al. encapsulated magnoflorine with chitosan–collagen nanocapsules in a diameter of 12 ± 2 nm. Chitosan is a collagen and a polysaccharide, widely used as a drug delivery vehicle for NDD management. The developed formulation showed a good antioxidant effect, encapsulation efficiency, sustained release, and no cytotoxicity via AChE inhibition [9]. Thus, alkaloid nanoformulations also show promising role in NDD management.

### 4.3 Terpenoids

Terpenoids are the organic unsaturated compounds mainly composed of isoprene monomers. Few terpenoids like ginkgolides A/B/C, bilobalide (*Gingko biloba*), thymoquinone (*Nigella sativa*), and thymol (*Thymus vulgaris*) have neuroprotective effects [71, 81]. Thymoquinone possesses crucial characteristics in treating NDDs like multiple sclerosis for its immunomodulatory, antioxidant, and anti-inflammatory effects. It reduces ROS, regulates the immune system, and inhibits pro-inflammatory cytokines. However, its poor hydrophobicity and bioavailability need to be explored in clinical trials. Fahmy et al. formulate thymoquinone-loaded SLNs and investigate the non-enzymatic and enzymatic oxidative stress effect in various brain regions, and evaluate the biodistribution characteristics. The developed nanoformulation revealed a considerable enhancement in the delivery of thymoquinone in the thalamus and a reduction in the cortex. Improved biocompatibility, bioavailability, and permeability achieved along with the biodistribution of thymoquinone at multiple brain regions. These thymoquinone-loaded SLNs may pave the method for targeting specific regions in the brain for diagnostic and clinical applications. Thymoquinone exhibits neuroprotective via JNK protein activation, mir-124 upregulation, and ERK1/2 and NOS enzymes downregulation [29] [129]. Wu et al. developed a ginkgolide B lipid microemulsion having an average diameter of 185.9 ± 52.5 nm; this nanoformulation has phospholipid-covered-oil-based nanosphere that exhibited safe, biocompatible formulation with improved bioavailability, effectiveness, and stability in *in vitro* analysis for the treatment of NDDs [118].

## 5 Regulatory Constraint and Patent Status

As nanoformulation technology arises, there is a requirement for adjoining teamwork between regulatory agencies and pharma industries to develop proper robust regulatory regulations. There is a lack of specific preclinical protocol to characterize nanoformulations' physicochemical characteristics, efficacy, and safety profiles, which challenges the translation of preclinical outcome to clinical stage. The therapeutic efficacy and pre-defined *in vitro* physiochemical characteristics of nanoformulation or nanocarrier might change in biological systems. Hence, their characterization parameters should be well defined under robust quality control specifications to mimic the same characteristic features at a large-scale industrial production. Nanoformulations might interact with the immune cells and plasma proteins. Hence, their immunotoxicity and biocompatibility parameters should be scrutinized under preclinical and clinical developmental programs. Considering the above challenges of nanoformulations, the conventional drug development methods are not robust for nanomedicine regulation. In addition, translating nanomedicine from benchside to bedside is even more challenging with the large-scale and cost-effective production. Hence, risk assessment and appropriate robust quality parameters need to be well defined in the nanomedicine development regulations. There is also a demand to change the current medical standard practices related to environmental safety, ethics, and regulations. Hence, certain regulatory agencies and

**Table 4** Patent granted for nanoformulation drug delivery system for neurodegeneration disease

| Patent No. | Details | Publication year/ granted; country | Reference |
|---|---|---|---|
| EP 2550020 B1 | Reversing micellar microemulsion system by utilizing metal ions | 2015; Europe | [67] |
| US 2015 9192644B2 | Enhance bioavailability of curcuminoid nanoformulation against NDD and AD | 2015; USA | [38] |
| US 2015 0017235A1 | Liposomes with various methods used for lenalidomide form-I anhydrous formulation. | 2015; USA | [53] |
| US 2015 0086616A1 | NSAIDS nanoformulation via nasal route for neuroprotective effects in NDD, AD | 2015; USA | [57] |
| CN10443455 1A | Preparation method of curcumin flexible liposome cream | 2015; China | [31] |
| KR10180418 3B1 | Synthesizing marine product nano-sized liposome for brain disease | 2017; South Korea | [121] |
| WO 2018 081460A1 | Synthesizing novel antibody for AD management | 2018; WIPO | [41] |
| WO 2018 197383A1 | Synthesizing combined therapeutic and their derivatives for AD management | 2018; WIPO | [42] |
| US10485766 B2 | Synthesizing orally administered Bryostatin-1 nanoparticles for reversing cognitive impairment in NDD cases | 2019; USA | [17] |
| EP2994160 (B1) | Aβ antibody and mAb potential towards NDD | 2019; Europe | [40] |
| WO 2020 023530A3 | Aβ antibody treatment to deplete Aβ load in AD | 2019; WIPO | [63] |
| US10,828,276 B2 | Enhanced α-secretase enzyme activity reduces Aβ plaque in NDD | 2020; USA | [16] |
| US201702817 97A1 | Lactoferrin - nanoparticle formulation for brain tumours | 2020, USA | [122] |

legislation committees need to take the initiative to define the regulatory parameters for nanoformulation applications at the commercial stage. The US-FDA, Europe and Japan collaborated to develop a common regulatory guideline (ICH Q8, Q9, and Q10) to define the regulatory requirement of pharmaceutical industries. In the same way, a harmonized strategy and quality control methods need to develop specifically for nanomedicine. Thus, NCL (nanotechnology characterization laboratory), an inter-agency, collaborated with NCI, NIST and FDA to implement the comprehensive resource to evaluate nanomaterials. The quality assurance certificate of nanocarriers' physicochemical characteristics (like, size, morphology, charge, purity, composition, etc.), safety, and efficacy studies via in silico/*in vitro/in vivo* requires by industries in nanomedicine development. NCL also extended to Europe to establish the EUNCL (European Nanomedicine Characterization Laboratory) to encourage the development of nanomedicine under regulatory guidance. The nanoformulation patent generally describes three factors: BBB modulator, nanocarrier loaded with drug moiety, and therapeutic agent encapsulated in nanocarrier. In the last five years, approximately 479 relevant patents filed for neuronal disease management [45, 73, 96]. Some recent nanoformulation patents are listed in Table 4.

# 6 Toxicity and Limitation

Due to nano-size characteristics, the nanoparticles can easily interact with the biological systems and able to cross BBB, thus providing hope for potential therapeutic for NDD. Unfortunately, these unique features which praised in advanced biomedical era, also cursed due to its neurotoxicity. The field that studies the side effects/ toxicity caused by nanoparticle exposure is known as 'Nanotoxicology'. It is a rapidly emerging field due to the increased application of nanoparticles in biomedicine, bio-nanotechnology and nanocarriers drug delivery systems. Thus, a standardized nanotoxicological study protocol is required. The nanoformulation's unique physiochemical properties and approximately similar dimensional characteristics close to biological cells generally interfere in the normal cellular process in vivo conditions. Therefore, it is necessary to assess the pharmacokinetic parameters, cell-to-cell interactions of nanoformulations, and time course of SAE development, which is quite impossible with available *in vitro* models. The nanotoxicity caused by nanoformulations is entirely different from their bulk counterparts. Hence, nanoformulation safety parameters are fundamentally crucial for their application. The natural compound historic safety information does not support the nanosystem safety aspects. In CNS, nanoformulations can cross BBB and may have a tendency to pile up at certain regions of the brain (like in neurons, astrocytes, and microglia), and may cause neuronal nanotoxicity by hindering ROS haemostasis and augmenting oxidative stress condition. This redox environment accelerates neuro-inflammation followed by neuronal apoptosis. The most common neuro nanotoxicity seen with nanoformulations are damage of neurons, glia cells, axons, and spinal cord [127, 128, 75].

The conventional toxicity assessment methods are not reliable for nano-neurotoxicity. Henceforth, there is a prerequisite for a novel technique that deeply evaluate the nanotoxicity and toxicology. Hence, a deep understanding of nanoformulation connectivity to neurons and glial cells is required to define the safety parameters. Another limiting factor is to predict the neurotoxic symptom onset, toxicity interval, and/or duration. The neurotoxic effects might be immediate or symptomatize after some time, leading to either reversible or permanent damage to the either whole brain or some portion of the brain. Along with safety and toxicity profile the manufacturing techniques also need to be well defined to cover the stability, drug delivery, PK profile, and well-accepted quality-related regulations. The challenging parameter for nanoformulation development on an industrial scale is their low therapeutic efficacy and self-aggregation propensity at small concentrations, which affect the drug entrapment variability due to poly-dispersity and overall formulation stability [47]. For example, doxorubicin nanoformulated candidate self-aggregated due to its high ionic strength that enlarge overall size of formulation and affects *in vivo* solubility and efficiency. This problem can be resolved by using small-size nanoparticles (<20 nm), still practical constraints remain challenging.

Another practical challenge is the swelling of nanoformulation that can be a merit or demerit. For targeted delivery, the swelling phenomenon is worthy and used in nanocapsules. However, swelling increases the nanoformulation overall size and further hampers bioavailability and solubility. The swelling mechanism control by applying a proper capping agent or coating with pH-sensitive formulation. Further, to comply US-FDA cGMP (Current Good Manufacturing Practices) approval requirements and cost-effectiveness, formulation development is also a limiting factor. In future, researchers need to focus on the scaling-up manufacturing procedures in compliance with FDA regulations and cost challenges. The characterization of nanoformulations requires an emerging and costly analytical technique (like DLS, SEM, TEM, NMR, NTA, etc.), which is done by a team of trained specialists, ultimately increasing the manufacturing cost. From regulatory perspectives, repeatability, reproducibility, precision, and accuracy of intra- and inter-laboratory facility or analyst variation needs proper evaluation [2, 10].

## 7 Future Prospects and Conclusion

Although there are many advantages of nanoformulations for natural products to better delivery, still a few limitations need to be taken care. The targeted drug delivery and nanoparticle toxicity is a major challenging part that requires careful investigation, especially if these nanosystems are for chronic human disease treatment. For example, 10–15% of patients reported adverse events like hypersensitivity, mild skin reactions, and peripheral neuropathy, with liposomal doxorubicin therapy. Another significant issue with nanoformulation is the lack of targeted therapy. For example, only 0.7% of marketed nanoformulations can reach up to solid tumour site despite massive doses of administration to yield desired therapeutic effects. The high doses ultimately affect safety and undue augmenting treatment cost. Still, nanomedicine technology opens new avenues as a potential approach for drug delivery systems by crossing BBB. It was observed that choosing a suitable synthesis methodology helps nanosystems achieve effective usage. Hence, futuristic development is necessary in nanoformulations synthesis methodologies by regulating the nanocarrier choice, its size and morphology.

## References


1. Agrahari, V., Burnouf, P. A., Burnouf, T., & Agrahari, V. (2019). Nanoformulation properties, characterization, and behavior in complex biological matrices: Challenges and opportunities for brain-targeted drug delivery applications and enhanced translational potential. *Advanced Drug Delivery Reviews, 148*, 146–180.
2. Ahmad, J., Akhter, S., Rizwanullah, M., Md Khan, A., Pigeon, L., Addo, R. T., & Md Kamal, A. (2017). Nanotechnology based theranostic approaches in Alzheimer's disease management: Current status and future perspective. *Current Alzheimer Research, 14*(*11*), 1164–1181. https://doi.org/10.2174/1567205014666170508121031
3. Akinmoladun, A. C., Akinrinola, B. L., Olaleye, M. T., & Farombi, E. O. (2015). Kolaviron, a Garcinia kola biflavonoid complex, protects against ischemia/reperfusion injury: Pertinent mechanistic insights from biochemical and physical evaluations in rat brain. *Neurochemical Research, 40*, 777–787.

4. Amo, L. D., Cano, A., Arriola, M. E., Souto, E. B., Espina, M., Camins, A., Garcia, M. L., & Sanchez-Lopez, E. (2021). Surface functionalization of PLGA nanoparticles to increase transport across the BBB for Alzheimer's disease. *Applied Sciences, 11(9)*, 4305. https://doi.org/10.3390/app11094305
5. An, Y. W., Jhang, K. A., Woo, S. Y., Kang, J. L., & Chong, Y. H. (2016). Sulforaphane exerts its anti-inflammatory effect against amyloid-beta peptide via STAT-1 dephosphorylation and activation of Nrf2/HO-1 cascade in human THP-1 macrophages. *Neurobiology of Aging, 38*, 1–10.
6. Anwar, H. M., Georgy, G. S., Hamad, S. R., Badr, W. K., El Raey, M. A., Abdelfattah, M. A., & Sobeh, M. (2021). A leaf extract of *Harrisonia abyssinica* ameliorates neurobehavioral, histological and biochemical changes in the hippocampus of rats with aluminum chloride-induced alzheimer's disease. *Antioxidants, 10*(6), 947.
7. Azadi, R., Musavi, S. E., Motekef, N., Rezayat, S. M., & Jafari, M. (2020). Preparation and characterization of berberine loaded micelle formulations with approach to oral drug delivery. *Trends in Pharmaceutical Sciences, 6*(4), 255–262.
8. Baptista, F. I., Henriques, A. G., Silva, A. M., Wiltfang, J., & da Cruz e Silva, O. A. (2014). Flavonoids as therapeutic compounds targeting key proteins involved in Alzheimer's disease. *ACS Chemical Neuroscience, 5*(2), 83–92.
9. Bashir, D. J., Manzoor, S., Khan, I. A., Bashir, M., Agarwal, N. B., Rastogi, S., & Samim, M. (2022). Nanonization of magnoflorine-encapsulated novel chitosan– collagen nanocapsules for neurodegenerative diseases: In vitro evaluation. *ACS Omega, 7*(8), 6472–6480.
10. Bhattacharya, T., Soares, G. A. B. E., Chopra, H., Rahman, M. M., Hasan, Z., Swain, S. S., & Cavalu, S. (2022). Applications of phyto-nanotechnology for the treatment of neurodegenerative disorders. *Materials, 15*(3), 804.
11. Bollimpelli, V. S., Kumar, P., Kumari, S., & Kondapi, A. K. J. N. I. (2016). Neuroprotective effect of curcumin-loaded lactoferrin nano particles against rotenone induced neurotoxicity. *Neurochemistry International, 95*, 37–45. https://doi.org/10.1016/j.neuint.2016.01.006
12. Bouchard, L. S., Anwar, M. S., Liu, G. L., Hann, B., Xie, Z. H., Gray, J. W., & Chen, F. F. (2009). Picomolar sensitivity MRI and photoacoustic imaging of cobalt nanoparticles. *Proceedings of the National Academy of Sciences of the United States of America, 106(11)*, 4085–4089. https://doi.org/10.1073/pnas.0813019106
13. Boyd, R. J., Avramopoulos, D., Jantzie, L. L., & McCallion, A. S. (2022). Neuroinflammation represents a common theme amongst genetic and environmental risk factors for Alzheimer and Parkinson diseases. *Journal of Neuroinflammation, 19*(1), 223.
14. Bridge, D. (2021). *Europe loyalty management market report–industry trends and forecast to 2027*. Data Bridge Market Research.
15. Carmona, V., Martin-Aragon, S., Goldberg, J., Schubert, D., & Bermejo-Bescos, P. (2018). Several targets involved in Alzheimer's disease amyloidogenesis are a ected by morin and isoquercitrin. *Nutritional Neuroscience*, 1–16.
16. Castor, T. P. (2020). Combination therapeutics and methods for the treatment of neurodegenerative and other diseases. *US patent*. Patent no. US10828276B2.
17. Castor, T. P., Alexander, J. S., Purdum G., Rios, J. D., Schrott, L. M., Tyler, T. A., & Vizcaino, M. I. (2019). Drug delivery system and method for the treatment of neuro- degenerative disease. *United State patent*. Patent no. US10485766B2.
18. Cenini, G., & Voos, W. (2019). Mitochondria as potential targets in Alzheimer disease therapy: An update. *Frontiers in Pharmacology, 10*, 902.
19. Chakraborty, S., & Basu, S. (2017). Multi-functional activities of citrus flavonoid narirutin in Alzheimer's disease therapeutics: An integrated screening approach and in vitro validation. *International Journal of Biological Macromolecules, 103*, 733–743.
20. Chakraborty, S., Bandyopadhyay, J., Chakraborty, S., & Basu, S. (2016). Multi-target screening mines hesperidin as a multi-potent inhibitor: Implication in Alzheimer's disease therapeutics. *European Journal of Medicinal Chemistry, 121*, 810–822.

21. Chakraborty, B., Mukerjee, N., Maitra, S., Zehravi, M., Mukherjee, D., Ghosh, A., & Rahman, M. (2022). Therapeutic potential of different natural products for the treatment of alzheimer's disease. *Oxidative Medicine and Cellular Longevity, 2022.*
22. Chauhan, I., Yasir, M., Verma, M., & Singh, A. P. (2020). Nanostructured lipid carriers: A groundbreaking approach for transdermal drug delivery. *Advanced Pharmaceutical Bulletin, 10(2)*, 150–165. https://doi.org/10.34172/apb.2020.021
23. Chen, C., Yang, F. Q., Zhang, Q., Wang, F. Q., Hu, Y. J., & Xia, Z. N. (2015). Natural products for antithrombosis. *Evidence-Based Complementary and Alternative Medicine*, 876426.
24. Chen, X., Drew, J., Berney, W., & Lei, W. (2021). Neuroprotective natural products for Alzheimer's disease. *Cells, 10*(6), 1309.
25. Chen, Y., Peng, F., Xing, Z., Chen, J., Peng, C., & Li, D. (2022). Beneficial effects of natural flavonoids on neuroinflammation. *Frontiers in Immunology, 6267.*
26. Cheng, K. K., Yeung, C. F., Ho, S. W., Chow, S. F., Chow, A. H. L., & Baum, L. J. T. (2013). Highly stabilized Curcumin nanoparticles tested in an in vitro blood–brain barrier model and in alzheimer's disease Tg2576 mice. *The AAPS Journal, 15*, 324–336. https://doi.org/10.1208/s12248-012-9444-4
27. Costa, S. L., Silva, V. D. D., Santos Souza, C., Santos, C. C., Paris, I., Munoz, P. S., & Aguilar, J. (2016). Impact of plant-derived flavonoids on neurodegenerative diseases. *Neurotoxicity Research, 30*, 41–52.
28. Croy, S. R., & Kwon, G. S. (2006). Polymeric micelles for drug delivery. *Current Pharmaceutical Design, 12(36)*, 4669–4684. https://doi.org/10.2174/138161206779026245
29. Dalli, T., Beker, M., Terzioglu-Usak, S., Akbas, F., & Elibol, B. (2018). Thymoquinone activates MAPK pathway in hippocampus of streptozotocin-treated rat model. *Biomedicine & Pharmacotherapy, 99*, 391–401.
30. Daneman, R., & Prat, A. (2015). The blood–brain barrier. *Cold Spring Harbor Perspectives in Biology, 1, a020412.* https://doi.org/10.1101/cshperspect.a020412
31. Daquan, C. (2015). Preparation method of curcumin flexible liposome cream. *China patent.* Patent no. CN104434551A.
32. de Andrade Teles, R. B., Diniz, T. C., Costa Pinto, T. C., de Oliveira Junior, R. G., Gama, E. S. M., de Lavor, E. M., & da Silva, A. A. M. (2018). Flavonoids as therapeutic agents in Alzheimer's and Parkinson's diseases: A systematic review of preclinical evidences. *Oxidative Medicine and Cellular Longevity*, 7043213.
33. de Laureto, P. P., Palazzi, L., & Acquasaliente, L. (2019). Polyphenols as potential therapeutic drugs in neurodegeneration. In *Neuroprotection-new approaches and prospects*. IntechOpen.
34. Derf, A., Sharma, A., Bharate, S. B., & Chaudhuri, B. (2018). Aegeline, a natural product from the plant *Aegle marmelos*, mimics the yeast SNARE protein Sec22p in suppressing alpha-synuclein and Bax toxicity in yeast. *Bioorganic & Medicinal Chemistry Letters, 29*, 454–460.
35. Dey, A., Bhattacharya, R., Mukherjee, A., & Pandey, D. K. (2017). Natural products against Alzheimer's disease: pharmaco-therapeutics and biotechnological interventions. *Biotechnology Advances, 35*, 178–216.
36. Dolati, S., Babaloo, Z., Ayromlou, H., Ahmadi, M., Rikhtegar, R., Rostamzadeh, D., et al. (2019). Nanocurcumin improves regulatory T-cell frequency and function in patients with multiple sclerosis. *Journal of Neuroimmunology, 327*, 15–21. https://doi.org/10.1016/j.jneuroim.2019.01.007
37. Finley, J. W., & Gao, S. (2017). A perspective on Crocus sativus L. (Sa ron) constituent crocin: A potent water-soluble antioxidant and potential therapy for Alzheimer's disease. *Journal of Agricultural and Food Chemistry, 65*, 1005–1020.
38. Frautschy, S. A., & Cole, G. M. (2015). Bioavailable curcuminoid formulations for treating Alzheimer's disease and other age-related disorders. *United States Patent.* Patent no. US 2015 9192644B2.

39. Fu, H., Hardy, J., & Duff, K. E. (2018). Selective vulnerability in neurodegenerative diseases. *Nature Neuroscience, 21*(10), 1350–1358.
40. Gelmont, D. M., Singer, J., Fritsch, S., & Schwarz, H. P. (2019). Treatment of Alzheimer's disease subpopulations with pooled immunoglobulin G. *European patent office*. Patent no. EP2994160B1.
41. Hope, M. J., Mui, B., Lin, P. J. C., Barbosa, C., Madden, T., Ansell, S. M., & Du, X. (2018). Lipid nanoparticle formulations. Patent no. WO2018081480.
42. Horn, M. E., Duelund, S. M., & Lundqvist, J. (2018). GLP-1 compositions and uses thereof. Patent no. WO2019038412.
43. Huang, T. I., & Hsieh, C. L. (2021). Effects of acupuncture on oxidative stress amelioration via nrf2/are-related pathways in alzheimer and parkinson diseases. *Evidence-Based Complementary and Alternative Medicine, 2021*.
44. Huo, X., Zhang, Y., Jin, X., Li, Y., & Zhang, L. (2019). A novel synthesis of selenium nanoparticles encapsulated PLGA nanospheres with curcumin molecules for the inhibition of amyloid b aggregation in Alzheimer's disease. *Journal of Photochemistry and Photobiology B: Biology, 190*, 98–102. https://doi.org/10.1016/j.jphotobiol.2018.11.008
45. Jadhav, N., Powar, T., Shinde, S., & Nadaf, S. J. (2014). Herbal nanoparticles: A patent review *Asian. Journal of Pharmaceutics, 8*(1), 58. https://doi.org/10.4103/0973-8398.134101
46. Jain, S., Singh, R., Paliwal, S., & Sharma, S. (2023). Targeting neuroinflammation as disease modifying approach to Alzheimer's disease: Potential and challenges. *Mini Reviews in Medicinal Chemistry*. https://doi.org/10.2174/1389557523666230511122435
47. Jeevanandam, J., San Chan, Y., & Danquah, M. K. (2016). Nano-formulations of drugs: recent developments, impact and challenges. *Biochimie, 128*, 99–112.
48. Kabanov, A. V., & Gendelman, H. E. (2007). Nanomedicine in the diagnosis and therapy of neurodegenerative disorders. *Progress in Polymer Science, 32*(*8–9*), 1054–1082. https://doi.org/10.1016/j.progpolymsci.2007.05.014
49. Kakkar, V., & Kaur, I. P. (2011). Evaluating potential of curcumin loaded solid lipid nanoparticles in aluminium induced behavioural, biochemical and histopathological alterations in mice brain. *Food and Chemical Toxicology, 49*, 2906–2913. https://doi.org/10.1016/j.fct.2011.08.006
50. Kheradmand, E., Hajizadeh Moghaddam, A., & Zare, M. (2018). Neuroprotective effect of hesperetin and nano-hesperetin on recognition memory impairment and the elevated oxygen stress in rat model of Alzheimer's disease. *Biomedicine & Pharmacotherapy, 97*, 1096–1101. https://doi.org/10.1016/j.biopha.2017.11.047
51. Kim, M. H., Kim, S. H., & Yang, W. M. (2014). Mechanisms of action of phytochemicals from medicinal herbs in the treatment of Alzheimer's disease. *Planta Medica, 80*, 1249–1258.
52. Kim, S. M., Park, Y. J., Shin, M. S., Kim, H. R., Kim, M. J., Lee, S. H., & Kwon, S. H. (2017). Acacetin inhibits neuronal cell death induced by 6-hydroxydopamine in cellular Parkinson's disease model. *Bioorganic and Medicinal Chemistry Letters, 27*, 5207–5212.
53. Konakanchi, D. P., Gongalla, B., Sikha, K. B., Kandaswamy, C., Adibhatla, K. S. B. R., & Nannapaneni, V. C. (2015). Anhydrous lenalidomide form-i. *United States patent*. Patent no. US 2015 0017235A1.
54. Kumar, A., Chaudhary, R. K., Singh, R., Singh, S. P., Wang, S. Y., Hoe, Z. Y., & Dai, X. (2020). Nanotheranostic applications for detection and targeting neurodegenerative diseases. *Frontiers in Neuroscience, 14*, 305.
55. Kundu, P., Das, M., Tripathy, K., & Sahoo, S. K. (2016). Delivery of dual drug loaded lipid based nanoparticles across the blood–brain barrier impart enhanced neuroprotection in a rotenone induced mouse model of Parkinson's disease. *ACS Chemical Neuroscience, 7*, 1658–1670. https://doi.org/10.1021/acschemneuro.6b00207
56. Larit, F., Elokely, K. M., Chaurasiya, N. D., Benyahia, S., Nael, M. A., Leon, F., Belouahem-Abed, D., et al. (2018). Inhibition of human monoamine oxidase A and B by flavonoids isolated from two Algerian medicinal plants. *Phytomedicine: International Journal of Phytotherapy and Phytopharmacology, 40*, 27–36.

57. Lehrer, S. (2015). Method for the prevention and treatment of Alzheimer's disease. *US Patent*. Patent No 2015/0086616 A1.
58. Li, P., Zhong, L., Yang, L., Bai, J., Lu, Y., & Du, S. (2020). Coexisting flavonoids and administration route effect on pharmacokinetics of Puerarin in MCAO rats. *Open Life Sciences, 15*(1), 449–457.
59. Li, Z., Zhao, T., Shi, M., Wei, Y., Huang, X., Shen, J., & Qin, D. (2023). Polyphenols: Natural food grade biomolecules for treating neurodegenerative diseases from a multi-target perspective. *Frontiers in Nutrition, 10*, 298.
60. Lin, M. T., & Beal, M. F. (2006). Mitochondrial dysfunction and oxidative stress in neurodegenerative diseases. *Nature, 443(19)*.
61. Liu, Z., Ran, Y., Huang, S., Wen, S., Zhang, W., Liu, X., & Du, H. (2017). Curcumin protects against ischemic stroke by titrating microglia/macrophage polarization. *Frontiers in Aging Neuroscience, 9*, 233.
62. Lohan, S., Raza, K., Mehta, S. K., Bhatti, G. K., Saini, S., & Singh, B. (2017). Anti-Alzheimer's potential of berberine using surface decorated multi-walled carbon nanotubes: A preclinical evidence. *International Journal of Pharmaceutics, 530*, 263–278. https://doi.org/10.1016/j.ijpharm.2017.07.080
63. Luthman, J., Swanson, C. J., Zhang, Y., Dhadda, S., & Wang, J. (2019). Methods of treatment and prevention of alzheimer's disease. WO Patent No 2019/023530A3.
64. Mani, M., Balasubramanian, S., Manikandan, K. R., & Kulandaivel, B. (2021). Neuroprotective potential of Naringenin-loaded solid-lipid nanoparticles against rotenone-induced Parkinson's disease model. *Journal of Applied Pharmaceutical Science, 11*(2), 019–028.
65. Marella, M., Seo, B. B., Yagi, T., & Matsuno-Yagi, A. (2009). Parkinson's disease and mitochondrial complex I: A perspective on the Ndi1 therapy. *Journal of Bioenergetics and Biomembranes, 41*(6), 493.
66. Mathew, A., Fukuda, T., Nagaoka, Y., Hasumura, T., Morimoto, H., Yoshida, Y., et al. (2012). Curcumin loaded-PLGA nanoparticles conjugated with Tet- 1 peptide for potential use in Alzheimer's disease. *PLoS One, 7*, e32616. https://doi.org/10.1371/journal.pone.0032616
67. Maurel, J. C. (2015). Reverse micelle microemulsion comprising metal ions and use thereof. *European Patent office*. Patent no. EP 2550020 B1.
68. Md Sairazi, N. S., & Sirajudeen, K. N. S. (2020). Natural products and their bioactive compounds: Neuroprotective potentials against neurodegenerative diseases. *Evidence-Based Complementary and Alternative Medicine, 2020*.
69. Mishra, G., Awasthi, R., Singh, A. K., Singh, S., Mishra, S. K., Singh, S. K., & Nandi, M. K. (2022). Intranasally co-administered berberine and curcumin loaded in transfersomal vesicles improved inhibition of amyloid formation and BACE-1. *ACS Omega, 7*(47), 43290–43305.
70. Mitchell, M. J., Billingsley, M. M., Haley, R. M., Wechsler, M. E., Peppas, N. A., & Langer, R. (2021). Engineering precision nanoparticles for drug delivery. *Nature Reviews Drug Discovery, 20*, 101.
71. Moradi, S. Z., Momtaz, S., Bayrami, Z., Farzaei, M. H., & Abdollahi, M. (2020). Nanoformulations of herbal extracts in treatment of neurodegenerative disorders. *Frontiers in Bioengineering and Biotechnology, 8*, 238.
72. Mourtas, S., Canovi, M., Zona, C., Aurilia, D., Niarakis, A., La Ferla, B., et al. (2011). Curcumin-decorated nanoliposomes with very high affinity for amyloid- b1-42 peptide. *Biomaterials, 32*, 1635–1645. https://doi.org/10.1016/j.biomaterials.2010.10.027
73. Muheem, A., Jahangir, M. A., Jaiswal, C. P., Jafar, M., Ahmad, M. Z., Ahmad, J., & Warsi, M. H. (2021). Recent patents, regulatory issues, and toxicity of nanoparticles in neuronal disorders. *Current Drug Metabolism, 22*(4), 263–279.
74. Mulik, R. S., Mönkkönen, J., Juvonen, R. O., Mahadik, K. R., & Paradkar, A. R. (2010). ApoE3 mediated Poly(butyl) cyanoacrylate nanoparticles containing curcumin: study of enhanced activity of curcumin against beta amyloid induced cytotoxicity using in vitro cell culture model. *Molecular Pharmaceutics, 7*, 815–825. https://doi.org/10.1021/mp900306x

75. Mushtaq, G., Khan, J. A., Joseph, E., & Md Kamal, A. (2015). Nanoparticles, neurotoxicity and neurodegenerative diseases. *Current Drug Metabolism, 16(8)*, 676–684. https://doi.org/10.2174/1389200216666150812122302
76. Nabavi, S. F., Sureda, A., Habtemariam, S., & Nabavi, S. M. (2015). Ginsenoside Rd and ischemic stroke: A short review of literatures. *Journal of Ginseng Research, 39*, 299–303.
77. Nataraj, J., Manivasagam, T., Justin Thenmozhi, A., & Essa, M. M. (2017). Neuroprotective e ect of asiatic acid on rotenone-induced mitochondrial dysfunction and oxidative stress-mediated apoptosis in dierentiated SH-SYS5Y cells. *Nutritional Neuroscience, 20*, 351–359.
78. Niralc, P., Paul, A., & Yadav, K. S. (2020). Nanoemulsions for targeting the neurodegenerative diseases: Alzheimer's, Parkinson's and Prion's. *Life Sciences, 245*, 117394. https://doi.org/10.1016/j.lfs.2020.117394
79. Niu, X., Chen, J., & Gao, J. (2019). Nanocarriers as a powerful vehicle to overcome blood-brain barrier in treating neurodegenerative diseases: Focus on recent advances. *Asian Journal of Pharmaceutical Sciences, 14*(5), 480–496.
80. Olajide, O. A., & Sarker, S. D. (2020). Alzheimer's disease: Natural products as inhibitors of neuroinflammation. *Inflammopharmacology, 28*(6), 1439–1455.
81. Ovais, M., Zia, N., Ahmad, I., Khalil, A. T., Raza, A., Ayaz, M., & Shinwari, Z. K. (2018). Phyto-therapeutic and nanomedicinal approaches to cure Alzheimer's disease: Present status and future opportunities. *Frontiers in Aging Neuroscience, 10*, 284.
82. Park, J. S., Davis, R. L., & Sue, C. M. (2018). Mitochondrial dysfunction in Parkinson's disease: New mechanistic insights and therapeutic perspectives. *Current Neurology and Neuroscience Reports, 18*, 1–11.
83. Pathak, K., Mishra, S. K., Porwal, A., & Bahadur, S. (2021). Nanocarriers for Alzheimer's disease: Research and patent update. *Journal of Applied Pharmaceutical Science, 11*(3), 001–021.
84. Patra, J. K., Das, G., Fraceto, L. F., Campos, E. V. R., Rodriguez-Torres, M. P., Acosta-Torres, L. S., Diaz-Torres, L. A., & Shin, H. S. (2018). Nano based drug delivery systems: Recent developments and future prospects. *Journal of Nanbiotechnology, 16*, 71. https://doi.org/10.1186/s12951-018-0392-8
85. Perez-Carrion, M., & Posadas, I. (2023). Dendrimers in neurodegenerative diseases. *Processes., 11(2)*, 319. https://doi.org/10.3390/pr11020319
86. Perumal, S., Atchudan, R., & Lee, W. (2022). A review of polymeric micelles and their applications. *Polymers (Basel)., 14(12)*, 2510. https://doi.org/10.3390/polym14122510
87. Pirzada, A. M., Ali, H. H., Naeem, M., Latif, M., Bukhari, A. H., & Tanveer, A. (2015). Cyperus rotundus L.: Traditional uses, phytochemistry, and pharmacological activities. *Journal of Ethnopharmacology, 174*, 540–560.
88. Poovaiah, N., Davoudi, Z., Peng, H., Schlichtmann, B., Mallapragada, S., Narasimhan, B., & Wang, Q. (2018). Treatment of neurodegenerative disorders through the blood– brain barrier using nanocarriers. *Nanoscale, 10*(36), 16962–16983.
89. Qi, Y., Guo, L., Jiang, Y., Shi, Y., Sui, H., & Zhao, L. (2020). Brain delivery of quercetin-loaded exosomes improved cognitive function in AD mice by inhibiting phosphorylated tau-mediated neurofibrillary tangles. *Drug Delivery, 27*(1), 745–755.
90. Rahman, M. H., Bajgai, J., Fadriquela, A., Sharma, S., Trinh, T. T., Akter, R., & Lee, K. J. (2021). Therapeutic potential of natural products in treating neurodegenerative disorders and their future prospects and challenges. *Molecules, 26*(17), 5327.
91. Rahman, M. M., Wang, X., Islam, M. R., Akash, S., Supti, F. A., Mitu, M. I., & Sharma, R. (2022). Multifunctional role of natural products for the treatment of Parkinson's disease: At a glance. *Frontiers in Pharmacology, 13*.
92. Rhein, V., Giese, M., Baysang, G., Meier, F., Rao, S., Schulz, K. L., & Eckert, A. (2010). Ginkgo biloba extract ameliorates oxidative phosphorylation performance and rescues Aβ-induced failure. *PLoS One, 5*(8), e12359.

93. Riccardi, C., Napolitano, F., Montesarchio, D., Sampaolo, S., & Melone, M. A. B. (2021). Nanoparticle-guided brain drug delivery: Expanding the therapeutic approach to neurodegenerative diseases. *Pharmaceutics., 13(11)*, 1897. https://doi.org/10.3390/pharmaceutics13111897
94. Rodriguez, F., Caruana, P., la Fuente, N. D., Espanol, P., Gamez, M., Balart, J., Llurba, E., & Cespedes, M. V. (2022). Nano-based approved pharmaceuticals for cancer treatment: Present and future challenges. *Biomolecules, 12(6)*, 784. https://doi.org/10.3390/biom12060784
95. Sabogal-Guaqueta, A. M., Osorio, E., & Cardona-Gomez, G. P. (2016). Linalool reverses neuropathological and behavioural impairments in old triple transgenic Alzheimer's mice. *Neuropharmacology, 102*, 111–120.
96. Saeedi, M., Eslamifar, M., Khezri, K., & Dizaj, S. M. (2019). Applications of nanotechnology in drug delivery to the central nervous system. *Biomedicine & Pharmacotherapy, 111*, 666–675.
97. Samudre, S., Tekade, A., Thorve, K., Jamodkar, A., Parashar, G., & Chaudhari, N. J. D. D. L. (2015). Xanthan gum coated mucoadhesive liposomes for efficient nose to brain delivery of curcumin. *Drug Delivery Letters, 5*, 201–207. https://doi.org/10.2174/2210303106666160120215857
98. Sancini, G., Gregori, M., Salvati, E., Cambianica, I., Re, F., Ornaghi, F., et al. (2013). Functionalization with TAT-peptide enhances blood-brain barrier crossing in vitro of nanoliposomes carrying a curcumin-derivative to bind amyloid-b peptide. *Science, 4*, 1–8.
99. Shah, S., Dhawan, V., Holm, R., Nagarsenker, M. S., & Perrie, Y. (2020). Liposomes: Advancements and innovation in the manufacturing process. *Advanced Drug Delivery Reviews, 154*, 102–122.
100. Shal, B., Ding, W., Ali, H., Kim, Y. S., & Khan, S. (2018). Anti-neuroinflammatory potential of natural products in attenuation of Alzheimer's disease. *Frontiers in Pharmacology, 9*, 548.
101. Sheng, B., Wang, X., Su, B., Lee, H. G., Casadesus, G., Perry, G., & Zhu, X. (2012). Impaired mitochondrial biogenesis contributes to mitochondrial dysfunction in Alzheimer's disease. *Journal of Neurochemistry, 120*(3), 419–429.
102. Sikorska, M., Lanthier, P., Miller, H., Beyers, M., Sodja, C., Zurakowski, B., Gangarau, S., Pandey, S., & Sandhu, J. K. (2014). Nanomicellar formulation of coenzyme Q10 (Ubisol-Q10) effectively blocks ongoing neurodegeneration in the mouse 1-methyl-4- phenyl-1,2,3,6-tetrahydropyridine model: Potential use as an adjuvant treatment in Parkinson's disease. *Neurobiology of Aging, 35(10)*. https://doi.org/10.1016/j.neurobiolaging.2014.03.032
103. Singh, A., Kukreti, R., Saso, L., & Kukreti, S. (2019). Oxidative stress: A key modulator in neurodegenerative diseases. *Molecules, 24*(8), 1583.
104. Sivanantham, B., Krishnan, U., & Rajendiran, V. (2018). Amelioration of oxidative stress in di erentiated neuronal cells by rutin regulated by a concentration switch. *Biomedicine and Pharmacotherapy, 108*, 15–26.
105. Sowndhararajan, K., Deepa, P., Kim, M., Park, S. J., & Kim, S. (2017). Baicalein as a potent neuroprotective agent: A review. *Biomedicine & Pharmacotherapy, 95*, 1021–1032.
106. Spagnuolo, C., Moccia, S., & Russo, G. L. (2018). Anti-inflammatory effects of flavonoids in neurodegenerative disorders. *European Journal of Medicinal Chemistry, 153*, 105–115.
107. Srivastava, P., & Yadav, R. S. (2016). Efficacy of natural compounds in neurodegenerative disorders. *The Benefits of Natural Products for Neurodegenerative Diseases*, 107–123.
108. Stojakovic, A., Trushin, S., Sheu, A., Khalili, L., Chang, S. Y., Li, X., & Trushina, E. (2021). Partial inhibition of mitochondrial complex I ameliorates Alzheimer's disease pathology and cognition in APP/PS1 female mice. *Communications Biology, 4*(1), 61.
109. Swerdlow, R. H., Koppel, S., Weidling, I., Hayley, C., Ji, Y., & Wilkins, H. M. (2017). Mitochondria, cybrids, aging, and Alzheimer's disease. *Progress in Molecular Biology and Translational Science, 146*, 259–302. https://doi.org/10.1016/bs.pmbts.2016.12.017

110. Tambe, R., Patil, A., Jain, P., Sancheti, J., Somani, G., & Sathaye, S. (2017). Assessment of luteolin isolated from Eclipta alba leaves in animal models of epilepsy. *Pharmaceutical Biology, 55*, 264–268.
111. Taylor, M., Moore, S., Mourtas, S., Niarakis, A., Re, F., Zona, C., et al. (2011). Effect of curcumin-associated and lipid ligand-functionalized nanoliposomes on aggregation of the Alzheimer's Ab peptide. *Nanomedicine: Nanotechnology, Biology and Medicine, 7*, 541–550. https://doi.org/10.1016/j.nano.2011.06.015
112. Vedagiri, A., & Thangarajan, S. (2016). Mitigating effect of chrysin loaded solid lipid nanoparticles against Amyloid b25–35 induced oxidative stress in rat hippocampal region: An efficient formulation approach for Alzheimer's disease. *Neuropeptides, 58*, 111–125. https://doi.org/10.1016/j.npep.2016.03.002
113. Velmurugan, B. K., Rathinasamy, B., Lohanathan, B. P., Thiyagarajan, V., & Weng, C. F. (2018). Neuroprotective role of phytochemicals. *Molecules, 23*, 2485.
114. Wang, Z. H., Wang, Z. Y., Sun, C. S., Wang, C. Y., Jiang, T. Y., & Wang, S. L. J. B. (2010). Trimethylated chitosan-conjugated PLGA nanoparticles for the delivery of drugs to the brain. *Biomaterials, 31*, 908–915. https://doi.org/10.1016/j.biomaterials.2009.09.104
115. Wang, L., Zhou, B. Q., Li, Y. H., Jiang, Q. Q., Cong, W. H., Chen, K. J., & Wu, Z. Z. (2023). Lactoferrin modification of berberine nanoliposomes enhances the neuroprotective effects in a mouse model of Alzheimer's disease. *Neural Regeneration Research, 18*(1), 226.
116. WHO Fact Sheet Report. (2023). *Dementia*. Available from link: https://www.who.int/news-room/fact-sheets/detail/dementia
117. Wightman, E. L. (2017). Potential benefits of phytochemicals against Alzheimer's disease. *The Proceedings of the Nutrition Society, 76*, 106–112.
118. Wu, X., Wang, Z., Zhao, Z., Chen, L., Tao, F., Zhang, Q., & Huo, Q. (2019). The optimization and production of ginkgolide B lipid microemulsion. *Open Chemistry, 17*(1), 357–364.
119. Xiang, C., Zhang, Y., Guo, W., & Lianga, X. J. (2020). Biomimetic carbon nanotubes for neurological disease therapeutics as inherent medication. *Acta Pharmaceutica Sinica B, 10*(2), 239–248. https://doi.org/10.1016/j.apsb.2019.11.003
120. Xu, J., Lacoske, M. H., & Theodorakis, E. A. (2014). Neurotrophic natural products: Chemistry and biology. *Angewandte Chemie (International Ed. in English), 53*, 956–987.
121. Yeoseon, S. G. (2017). A pharmaceutical composition comprising liposome containing protein from marine microorganism as an active ingredient for brain disease. *South Korea Patent*. Patent no. KR101804183B1.
122. Yun, L. D., Jae, L. S., & Shik, K. H. (2017). Lactoferrin-conjugated nanoparticle complex and use thereof. *US Patent 20170,281,*797.
123. Zanforlin, E., Zagotto, G., & Ribaudo, G. (2017). The medicinal chemistry of natural and semisynthetic compounds against Parkinson's and Huntington's diseases. *ACS Chemical Neuroscience, 8*, 2356–2368.
124. Zhang, H., Bai, L., He, J., Zhong, L., Duan, X., Ouyang, L., & Shi, J. (2017). Recent advances in discovery and development of natural products as source for anti- Parkinson's disease lead compounds. *European Journal of Medicinal Chemistry, 141*, 257–272.
125. Zhao, G., Dong, X., & Sun, Y. (2018). Self-assembled curcumin–poly (carboxybetaine methacrylate) conjugates: potent nano-inhibitors against amyloid b- protein fibrillogenesis and cytotoxicity. *Langmuir, 35*, 1846–1857. https://doi.org/10.1021/acs.langmuir.8b01921
126. Zheng, J. C., & Chen, S. (2022). Translational neurodegeneration in the era of fast growing international brain research. *Translational Neurodegeneration, 11*(*1*), 1–2.
127. Zia, S., Islam Aqib, A., Muneer, A., Fatima, M., Atta, K., Kausar, T., Zaheer, C.N.F., Ahmad, I., Saeed, M. and Shafique, A. (2023). Insights into nanoparticles-induced neurotoxicity and cope up strategies. Frontiers in Neuroscience, 17, p.1127460. https://doi.org/10.3389/fnins.2023.1127460

128. ZHu, Y. and Hammarlund-Udenaes, M. (2020). Perspectives on nanodelivery to the brain: prerequisites for successful brain treatment. Molecular Pharmaceutics, 17(11), pp.4029-4039.https://doi.org/10.1021/acs.molpharmaceut.0c00881
129. Fahmy, H.M., Ahmed, M.M., Mohamed, A.S., Shams-Eldin, E., Abd El-Daim, T.M., El-Feky, A.S., Mustafa, A.B., Abd Alrahman, M.W., Mohammed, F.F. and Fathy, M.M. (2022). Novel lipid-coated mesoporous silica nanoparticles loaded with thymoquinone formulation to increase its bioavailability in the brainand organs of Wistar rats. BMC Pharmacology and Toxicology, 23(1), p.71. https://doi.org/10.1186/s40360-022-00616-z

# Nano Probiotics in Gut Health: Link with Neuroprotection

Keshari Kumar Sriwastawa and Ashutosh Kumar

**Abstract**

The gut microbiota or gut flora encompasses diverse microorganisms, including bacteria, viruses, yeasts, archaea, and fungi that regulate gut health and overall body function. These microbes communicate with the body through neural, endocrine, and immunological pathways. Disruption in gut microbiota, known as gut dysbiosis, impairs these systems, contributing to neurodegenerative diseases, predominantly Alzheimer's, Parkinson's, and multiple sclerosis via the gut-brain axis. Gut dysbiosis compromises the intestinal barrier, increasing permeability and allowing bacterial metabolites and proinflammatory mediators to enter circulation, damaging the blood-brain barrier and triggering neuroinflammation. Probiotics such as *Lactobacillus acidophius*, *Bifidobacterium longum*, and *Lactobacillus helveticus R0052* have demonstrated neuroprotective effects in several preclinical models of neurodegenerative diseases. However, their gastric degradation limits their effectiveness. Nanotechnology-based probiotics (nano probiotics), including nanofibers and nanoparticles, have emerged as an effective strategy by enhancing bioavailability, increasing oxidative resistance, site-specific delivery, and nerve regeneration, offering a promising therapeutic strategy for neurodegenerative disease management. This book chapter highlights the potential role of gut microbiota in neurodegenerative diseases and the neuroprotective effects of probiotics enhanced through nanotechnology.

K. K. Sriwastawa · A. Kumar (✉)
Department of Pharmacology and Toxicology, National Institute of Pharmaceutical Education and Research (NIPER) SAS Nagar, Mohali, Punjab, India
e-mail: ashutosh@niper.ac.in



R. Shukla et al. (eds.), *Neuroendocrine-Immune Axis*,
https://doi.org/10.1007/978-981-95-5088-3_7

**Keywords**

Gut microbiota · Gut-brain axis · Neurodegenerative diseases · Neuroprotection · Nano probiotics

## Abbreviations

| | |
|---|---|
| AD | Alzheimer's disease |
| ALS | Amyotrophic lateral sclerosis |
| Aβ | Amyloid beta |
| BBB | Blood-brain barrier |
| BCAA | Branched-chain amino acid |
| BDNF | Brain-derived neurotrophic factor |
| CNS | Central nervous system |
| CREB | cAMP response element binding |
| ENS | Enteric nervous system |
| FOS | Fructo-oligosaccharides |
| GBA | Gut-brain axis |
| GM | Gut microbiota |
| GOS | Galacto-oligosaccharides |
| GSH | Glutathione |
| HPA | Hypothalamic-pituitary-adrenal axis |
| IBD | Inflammatory bowel disease |
| LPS | Lipopolysaccharide |
| MPTP | 1-methyl-4-phenyl-1,2,3,6-tetrahydropyridine |
| MS | Multiple sclerosis |
| MSN | Mesoporous nanoparticles |
| NPs | Nanoparticles |
| NR2A | NMDA receptor subunit 2A |
| PD | Parkinson's disease |
| PRR | Pattern recognition receptor |
| PRR | Pattern recognition receptors |
| PTSD | Post-traumatic stress disorder |
| PVA | Polyvinyl alcohol |
| SCFA | Short-chain fatty acid |
| SNpc | Substantia nigra pars compacta |
| SOD | Superoxide dismutase |
| TLR | Toll-like receptors |
| TOS | Trans-galacto-oligosaccharides |

# 1 Introduction

## 1.1 Gut Microbiota (GM)

A plethora of microorganisms residing in the GIT of our body are collectively known as gut microbiota (GM) or gut flora which communicates with the human host via neural, endocrine, and immunological pathways [1]. The composition of the gut microbiota includes innumerable bacteria, viruses, yeast, archaea, and fungi with bacteria having the highest population in the large intestine as many as $10^{13}$ bacterial cells found in the human colon [2]. It has been reported that gut microbiota has a very important correlation with human health and any disruption in the bacterial balance known as dysbiosis manifests various diseases including neurodegenerative diseases such as Alzheimer's disease (AD), Parkinson's disease (PD), multiple sclerosis (MS), etc. So, it gives a very clear indication that gut health and the brain have a very symbiotic relationship and thus it becomes essential to study more and more about the gut microbiota and its association with the occurrence and treatment of neurological disorders [3]. The adult gut microbiota is comprised majorly of microorganisms belonging to the phyla Firmicutes, Actinobacteria, and Bacteroidetes, and a minority of the population is taken up by Verrucomicrobia and Proteobacteria [4]. Approximately $1 \times 10^{13}$—$1 \times 10^{14}$ microbes reside in the human GIT which is a ten-fold more significant population in comparison to the other human cells and they contain a 150-fold more significant genomic population than the humans and for that reason, due to this fact it is sometimes stated as a forgotten organ. The total number of species of the gut microbiota differs significantly but the estimated population is greater than 1000 with greater than 700 strains. The segregation of the microbes of the gut microbiota into different phyla takes place after the puerperal period as following parturition the infant gets exposed to an intricate population of microbes. The bacterial population varies in an individual-specific manner but the determination of their exact constitution holds a genetic basis. They are divided into 3 major types which are called enterotypes depending on the bacterial constitution as Bacteroides spp., Prevotella spp., and Ruminococcus spp. Upcoming evidence shows that a healthy microbiota is beneficial for the body and various factors tend to modulate it which can lead to the manifestation of various diseases, for instance, diet which significantly affects the gut microbiota; diets that are rich in protein and fat are reported to be related to Bacteroides spp., diets rich in carbohydrates have an association with Prevotella spp. Several other factors such as prevailing infections, disease states, and drugs such as antibiotics also affect the gut microbiota composition which may adversely affect human health [5]. The gut microbiota has a two-way connection with the brain defined as the gut-brain axis (GBA) which displays an interconnected relationship between the cognitive and neuronal functions of the brain and the GIT. Gut dysbiosis is marked by the release of LPS and inflammatory cytokines and impaired intestinal permeability which is one of the most probable causes of intestinal diseases. In context to these neurodegenerative diseases, mood disorders and memory impairment occur due to the release of excessive amounts of LPS into the systemic circulation leading to the disruption of the blood-brain barrier (BBB). According to a study, preferring antibiotics as a mode of treatment in mice has led to the degradation of the GM, whereas bacterial probiotics such as *Lactobacillus* and *Bifidobacterium* coming into the picture have shown ameliorative effects on the CNS-related manifestations caused as a consequence of gut dysbiosis such as improvement in memory function and neuroprotection. The yeast probiotics such as *Saccharomyces boulardii* have also been investigated for their neuroprotective potential in response to gut dysbiosis pertaining to their ameliorative effects against antibiotic-associated diarrhea, replenishment of the function of the GI barrier, resurrection of the proteins of the tight junction in the enterocytes via functionalizing the proteins E-cadherin and p120 catenin and maintenance of the levels of the short-chain fatty acids (SCFA) such as butyrate, acetate and propionate. Moreover, it was also reported that *Saccharomyces boulardii* counteracted the impairment of psychological functions related to gut dysbiosis by exerting neuroprotective effects in the hippocampus [6]. Prebiotics are substantially beneficial to enhance the activity of probiotics and to provide health benefits. They have been reported to have neuroprotective potential by decreasing ROS generation and also to exert positive neuromodulatory effects by the probiotics found endogenously in the gut via affecting the GBA [7]. So, all these facts can be summarized altogether as an interdependent relationship that exists between the GM, GBA, probiotics, and their role in neuroprotection to appear as a promising therapeutic approach to treat neurological disorders.

## 1.2 Neuroprotective Effects of the GM

GM regulates neuronal functioning and working via the formation of neurotransmitters or by altering the catabolism of the neurotransmitters released in the host's body. Various bacteria are found to release neurotransmitters such as catecholamines, GABA, serotonin, and glutamate. The metabolites produced by the microbiota affect the activity of the glial cells in the ENS (enteric nervous system) and CNS (central nervous system) [8]. ENS is also called the second brain of the body and plays a vital role in mediating the interactions and between the GIT and CNS. The connection between the neurons of the ENS and that of CNS is established through the neural (vagus nerve), endocrine, and immune pathways which is regulated by the GM. It has been reported that GM affects the serum protein zonulin which is responsible for the maintenance of the tight junctions present between the vascular endothelial cells (found in the blood-brain barrier) and intestinal cells [9]. Various studies have proved the neuroprotective potential of gut microbiota, reporting that *Lactobacillus buchneri* KU200793 obtained from a Korean fermented food known as kimchi exerted enhanced antioxidant effects and protected SH-SY5Y cells from 1-methyl-4-phenyl pyridinium ($MPP^+$) which also presented its activity as a probiotic [9, 10]. Another study revealed that exopolysaccharides derived from *Lactobacillus delbrueckii* ssp. *Bulgaricus* B3 and *Lactobacillus plantarum GD2* exerted anti-apoptotic effects in SH-SY5Y cells against Aβ (1–42) which indicates

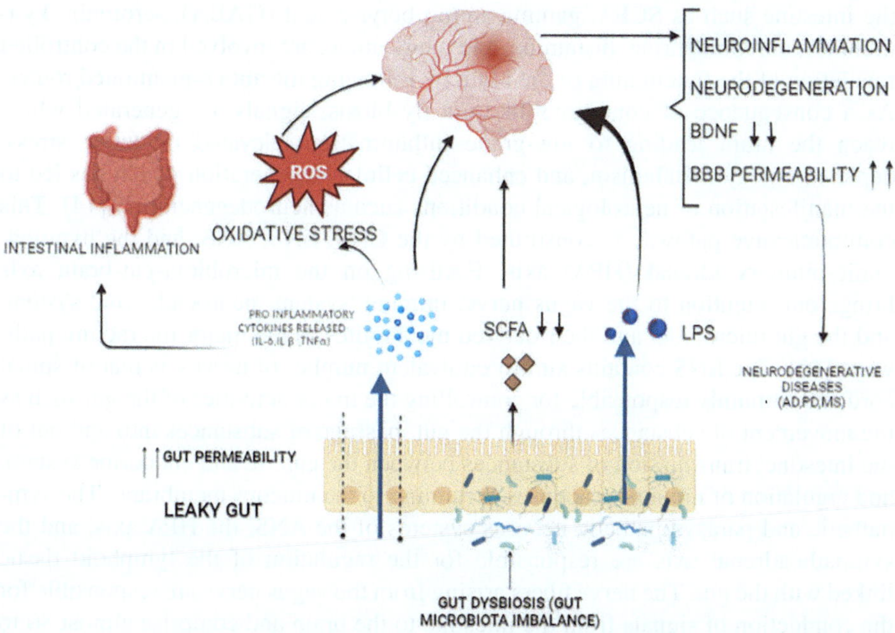


**Fig. 1** Role of gut dysbiosis in pathogenesis of neurodegenerative diseases

their therapeutic potential to treat AD. Moreover, heat-killed strains of *Ruminococcus albus* exerted antioxidant activity via reducing ROS production and enhancing the production of superoxide dismutase (SOD) and glutathione (GSH) in an in vitro (SH-SY5Y cells) and in vivo (rat) model of PD following $H_2O_2$ and sodium arsenate as disease-inducing agents respectively [11]. It has been reported that *Lactobacillus pentosus var plantarum* C29 exerts ameliorative effects against memory dysfunction in mice induced by scopolamine and D-galactose [12]. The upregulation or downregulation of the serotonin receptors ($5HT1_A$), brain-derived neurotrophic factors (BDNF), and NMDA receptor subunit 2 (NR2A) is controlled by the GM. In mice, it facilitates the release of serotonin in the ENS and systemic circulation and has effects on anxiety, cognition, and high blood pressure. The bacteria present in the gut play a significant role in the functionalization of the immune cells present in the intestine, brain, and periphery [13]. All these studies confirm the neuroprotective effects of GM attributed to their anti-inflammatory, antioxidant, and immunoprotective effects (Fig. 1).

## 1.3 Gut-Brain Axis (GBA)

The bidirectional connection between the brain and intestine which mediates its function through neuroendocrine, neural, endocrine, and immune signaling routes is called the gut-brain axis. The metabolites released by the microbiome residing in the intestine such as SCFA, gamma-amino butyric acid (GABA), serotonin, kynurenines, norepinephrine, histamine, and many others are involved in the controlled regulation of the functioning of the brain by following the above-mentioned routes. As a consequence of conditions like gut dysbiosis, signals are generated which reach the brain leading to low-grade inflammation, elevated oxidative stress, depleted energy metabolism, and enhanced cellular degeneration which has led to the manifestation of neurological conditions such as neurodegeneration [14]. This communicative pathway is constituted by the CNS, ENS, ANS, and the hypothalamic-pituitary-adrenal (HPA) axis. Focusing on the microbiota-gut-brain axis brings our attention to the vagus nerve, immune system, neuroendocrine system, and the gut microbiota and their derived metabolites as the major interacting pathways [15]. The ENS contains almost equivalent number of nerves as that of spinal cord and is mainly responsible for controlling the major activities of the gut such as the movement of substances through the gut, passage of substances into and out of the intestine, transmission of substances between the enteric and endocrine system, and regulation of immune responses pertaining to the mucous membrane. The sympathetic and parasympathetic nervous systems of the ANS, the HPA axis, and the sympathoadrenal axis are responsible for the regulation of the lymphoid tissue linked with the gut. The nerve fibers arising from the vagus nerve are responsible for the conduction of signals from the intestine to the brain and comprise almost 80 to 90 percent of all the vagal nerve fibers, in addition to this these nerve fibers interact with the HPA axis and modulate the stress and anti-inflammatory responses [16]. The GBA has been upcoming as a potential target for the treatment of gastrointestinal and psychiatric diseases such as inflammatory bowel disease (IBD), post-traumatic stress disorder (PTSD), and depression and through application of novel methods such as stimulating the vagus nerve the GBA can be regulated and can be helpful in curing disorders associated with anxiety, mood and inflammation [17]. The term microbiota gut-brain axis implies that the GM plays an indispensable role in GBA. The underlying effects of the GM on the components of the GBA explain the interaction between the GM and GBA and its therapeutic significance. It has been found through various studies that the GM interacts and stimulates various receptors known as pattern recognition receptors (PRRs), e.g., TLR 2 and TLR 4 (Toll-like receptors) and leads to the manifestations of diseases associated with the ENS and CNS. Modifications in the composition of the GM arising due to the ingestion of antibiotics significantly influence the physiology of ENS and the chemical processes and phenomena related to the nervous system and also lead to the reduction in the ganglionic population in the glial cells of the ENS. The gut-derived metabolites, such as bile acids, branched-chain amino acids (BCAAs), etc., are responsible for the regulation of various processes such as the formation of neurons, functioning of glial cells and myelin sheath of neurons, elimination of extra synapses by the brain, and proper functioning of the BBB [18] (Fig. 2).

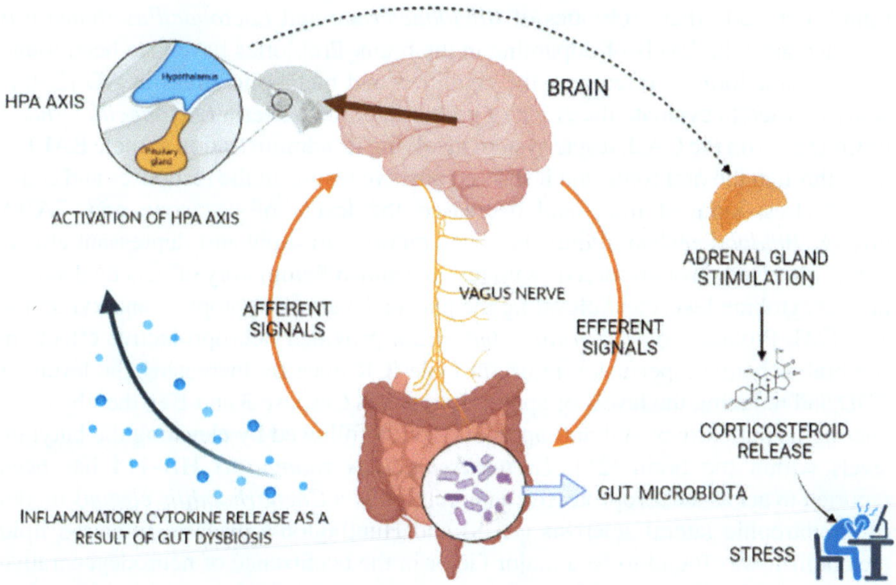


**Fig. 2** Gut-brain interactions through the gut brain axis

## 2 Neuroprotective Effects of Probiotics

The living microbes that, when taken in appropriate amounts, exert health benefits and enhance the activity of the GM are called probiotics [19]. Mostly, the probiotics are identified in Gram-positive bacteria such as *Streptococcus, Bacillus, Lactobacillus*, and *Pediococcus.* Certain probiotics of fungal and yeast origin such as *Sacharomyces cerevisiae* and *Kluyveromyces* are also present [20]. Probiotics function as pH controllers in the body, maintaining the strength and structure of the walls of the intestine, and increasing the levels of BDNF [21]. As we have known the neuroprotective effects of the GM and the involvement of the GBA in enhancing the beneficiary effects of the GM so it becomes very important to maintain the GM composition so as to cure various neurological diseases. One such approach is the use of probiotics which restore the disturbed gut microbiota population in conditions such as gut dysbiosis and exert therapeutic effects in neurodegenerative diseases, cancers, osteoporosis, and several other diseases. Nowadays, probiotics-based therapies are gaining importance and preference because they provide subsequent advantages such as prevention of cell-cell adhesion, killing pathogens, antimicrobial activity, and host immunomodulatory effects. Due to these benefits, they not only serve as good therapeutic agents but can also be used as carriers for delivering drug molecules [22]. It was reported that the synaptic transmission was retrieved and the hippocampal long-term potentiation (LTP) was normalized and regained after administration of probiotics of *Lactobacillus acidophilus, Bifidobacterium lactis,* and *Lactobacillus fermentum*, to rats suffering from diabetes. It has also been found in a study that probiotics of *Bifidobacterium* and *Lactobacillus rhamnosus* GG increased the levels of dopamine in the brain. Probiotics have also been found to show ameliorative effects on the cognition and metabolic stress in AD [23]. A study focused to evaluate the effect of probiotic of the bacteria *Lactobacillus rhamnosus* (JB-1) on the GABAergic system involving its administration to male BALB/c mice through the oral route and it showed positive results in the responses and activity of stress-induced mice and decreased the levels of cortisone and GABA mRNA. *Bifidobacterium infantis* has been reported to show anti-depressant effects in rats observed through forced swim test and anti-inflammatory effects by decreasing the cytokine levels and elevating the plasma levels of tryptophan and kynurenic acid [24]. Probiotic of *Clostridium butyricum* provided neuroprotective effects in cerebral ischemia/reperfusion injury in male ICR mice by increasing the levels of SOD and reducing the levels of apoptotic proteins Caspase 3 and Bax thereby deciphering its antioxidant and anti-apoptotic action followed by elevating the butyrate levels within the brain [25]. *Lacteicaseibacillus rhamnosus* HA-114 has been reported to act as a neuroprotective intervention in a *Caenorhabditis elegans* model of amyotrophic lateral sclerosis (ALS) and Huntington's disease. Impaired lipid metabolism was found to be a major factor in the occurrence of neurodegenerative diseases and *L. rhamnosus* HA-114 was found to repair the impaired lipid metabolism and maintain the energy equilibrium via β oxidation in the mitochondria [26] (Fig. 3 and Table 1).

## 3 Need for Development of Nano Probiotics

Oral drug delivery systems have been developed for probiotics since many years. But their optimum effect is not being observed because of their degradation by the strongly acidic gastric fluid resulting in detrimental effects on their structure and pharmacological actions. As a result, the bacterial probiotics after reaching to their site of action (intestine) are degraded and damaged followed by loss of their metabolic potential and decrease in the production of GABA which is not favorable for proper neuronal activity. Fecal bacteria transplantation and use of antibiotics have some limitations such as in case of fecal bacteria transplantation the procedures are very complex, donor needs are very stringent, and the method is less compliant to patients. In case of antibiotics, adverse effects, development of resistance, and other associated diseases limit their use. So, it becomes the need of the hour to develop effective strategies for the efficient delivery of probiotics [33].

## 4 Nanotechnology-Based Approaches for Probiotics for Neuroprotection (Nano Probiotics)

One such approach is encapsulating probiotics. Dating back to the conventional microencapsulation technology followed by the application of nanofibers and nanoparticles for encapsulating probiotics in huge quantities and the use of nano

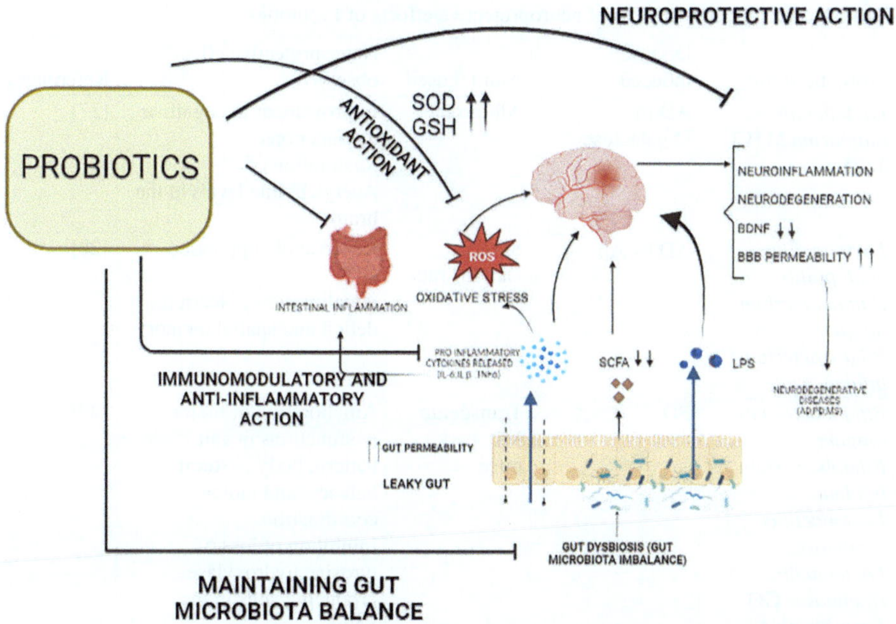


**Fig. 3** Mechanism of action of probiotics

armors for a probiotic by means of biofilm, biological membrane, and nano-coating has been witnessed to gain importance in recent years. These nanoagents of probiotics have provided new insights for the enhanced and efficient delivery of probiotics by methods such as encapsulation. The conventional microcapsule encapsulation technologies used are spray drying, extrusion, freeze drying, and emulsification. The microcapsules prepared using spray drying are prone to high temperature and can trigger stress-like conditions due to loss of water from body which leads to the depletion in the number of probiotics and moreover spray drying technique is expensive and tedious. In case of freeze-drying method, the ice crystals cause severe harm to the cell membrane of the probiotics which precipitates the death of live probiotics. The microcapsules prepared by extrusion technique have greater particle size which is a major drawback for its application. For these reasons incorporation of nanotechnology to probiotics becomes necessary. Nanocarriers enhance the bioavailability of the agent that needs to be encapsulated by virtue of a greater surface volume ratio and increased adherence to the intestine [34].

According to a study, mesoporous silica nanoparticles (MSNs) have been used as nanocarriers for the delivery of probiotic *Bifidobacterium* which produces SCFA in huge amounts and is helpful for the treatment of AD by virtue of its therapeutic benefits in the brain. MSNs are preferred drug carriers because of their safety, remarkable porosity, compatibility with the living tissues of the body, and greater pore volume. Moreover, substantial amounts of bacteria can be incorporated into it, and the administration of conjugated bacterial nanoparticles can increase the entry

**Table 1** Preclinical evidences of neuroprotective effects of Probiotics

| Probiotic strain | Disease induced | Animal used | Neuroprotective effect observed | References |
|---|---|---|---|---|
| *Lactobacillus plantarum* MTCC 1325 | AD by D-galactose | Albino rats | Improvement in cognitive dysfunctions<br>Restoration of Acetylcholine levels in the brain | [27] |
| *Lactobacillus acidophilus*<br>*Bifidobacterium longum*<br>*Bifidobacterium bifidum* | AD by Aβ | Sprague-Dawley rats | Reversal of suppressed LTP<br>Amelioration of learning deficit and spatial memory | [28] |
| *Bifidobacterium longum*<br>*Bifidobacterium bifidum*<br>*Lactobacillus rhamnosis*<br>*Lactobacillus rhamnosus* GG<br>*Lactobacillus plantarum* LP28<br>*Lactococcus lactis* subsp. Lactis | PD | Transgenic Mitopark mice | Amelioration of motor dysfunctions in gait pattern, body postural balance, and motor coordination.<br>Inhibition of loss of tyrosine hydroxylase-producing cells in the SNpc | [29] |
| *Lactobacillus helveticus* R0052<br>*Bifidobacterium longum* | Chronic psychological stress | C57B16 mice | Inhibited the hippocampal neuronal degeneration due to water avoidance stress<br>Prevention of stress-induced aberrant synaptic plasticity | [30] |
| *Clostridium butyricum* | Vascular dementia | Mice | Amelioration of cognitive impairments<br>Elevation in BDNF levels<br>Exerted anti-apoptotic effect in neurons<br>Maintenance of fecal and neural butyrate levels | [31] |
| *Lactobacillus plantarum D189* | PD by MPTP | C57BL/6 mice | Decreased alpha-synuclein aggregation<br>Elevation in SOD and GSH levels<br>Inhibited activation of NLRP3 inflammasome. | [32] |

of substances into the tissues. It was found that the MSN-conjugated Bifidobacterium when administered through the intranasal route to APP/Ps1 mice reduced the increased levels of Aβ and ameliorated the impaired cognition in AD [35]. A study was performed to apply the nanoencapsulation technique to strains of probiotic *Lactobacillus acidophilus, Lactobacillus rhamnosus*, *Bifidobacterium bifidum*, and

*Bifidobacterium animalis* by using corn starch and sodium alginate followed by incorporation into nanofiber mats and it was found that the survival of the probiotics was increased in the acidic environment of the GIT [36]. It was reported that microgels prepared by combining alginate and gelatin brought about an increment in the rate of survival and delivery of the probiotic *Lactobacillus salivarius* Li01 to the gut. But the presence of pores in the microgels allowed the entry of substances such as bile acids, hydrogen ions, and digestive enzymes which caused the degradation of the probiotic. The use of inorganic nanoparticles (NP) was a solution to this issue as they filled the pores. Even certain inorganic NPs, for example, Mg $(OH)_2$ also act as buffers which can counterbalance the acidic pH when the microgels remain in the gastric environment for longer periods. These NPs are able to regulate the entry of unwanted substances into the encapsulated probiotic and maintain a stable pH thereby increasing the survival rate of the probiotics in the site of action [36]. In a study nanosized probiotics of *Lactobacillus acidophilus* and *Bifidobacterium longum* were investigated for their neuroprotective potential and it was found to exert anti-inflammatory effects by decreasing the release of LPS-triggered cytokines and inhibited the microglial activation in the brain [37]. Probiotic bacteria have also been explored for their metallic/non-metallic nanoparticle synthesizing capability, providing various advantages. This ability of probiotics to synthesize NPs can be termed as biosynthesis, and biosynthesis of NPs are gaining great significance. The different types of metallic NPs that can be synthesized by probiotics include NPs of gold, silver, iron, copper, zinc, tungsten oxides, etc., and non-metallic NPs such as selenium [38]. It was reported that the probiotic *Lactobacillus casei* ATCC 393 combined with selenium NPs produced optimum results in AD by reducing the Aβ clustering, ameliorating the memory impairments, normalizing the gut dysbiosis, and regulating the BDNF and cAMP response element binding protein (CREB) pathways which protects from neurodegeneration [39]. In a study, the probiotic bacteria *B animalis Bb12* were encapsulated into PVA nanofibers by the use of co-electrospinning method, and it was reported that the survival of the probiotic was increased even after 40 and 130 days at room temperature and refrigeration temperature of 4 and −20 °C respectively [40]. All these facts and results necessitate the formulation of nano probiotics for neuroprotection (Figs. 4 and 5).

## 5 Neuroprotective Effects of Prebiotics

Prebiotics are defined as non-digestible oligosaccharides whose activity can be altered by the use of probiotics causing specific changes in the composition and/or of the GM, and prove to be healthy for the host [7]. Prebiotics are available in non-digestable oligosaccharides, non-digestable carbohydrates, yacon, unrefined wheat and barley, and many others. Various important constituents present in prebiotics are soya-oligosaccharide, pyrodextrins, lactulose, inulin, isomalto-oligosaccharide, etc. [41]. The commonly used prebiotics are fructo-oligosaccharides (*FOS*), galacto-oligosaccharides (*GOS*), and trans-galacto-oligosaccharides (*TOS*). A substance to be called as a prebiotic should possess the following characteristics: (i) resistance

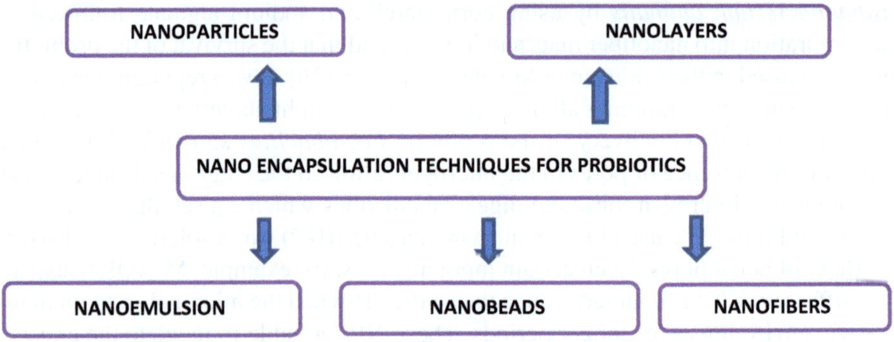


**Fig. 4** Nanoencapsulation strategies for probiotics

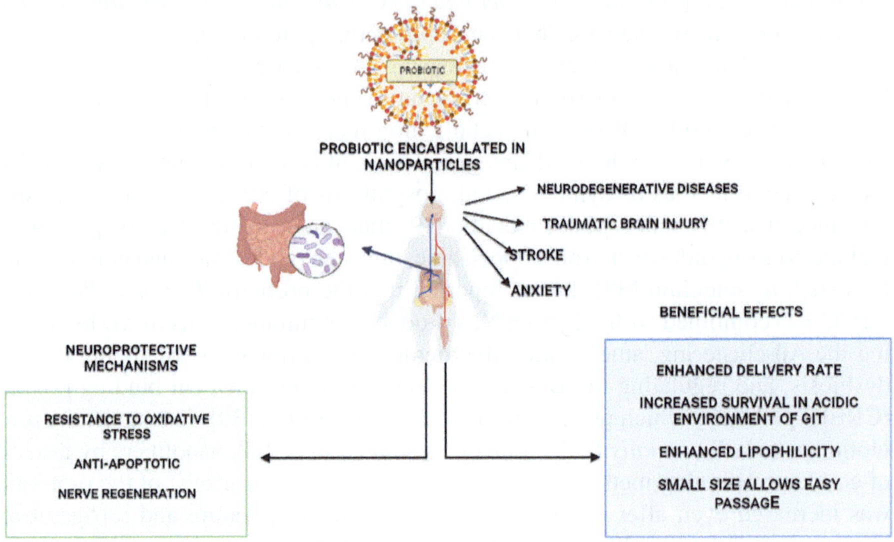


**Fig. 5** Neuroprotective mechanisms and beneficial effects of probiotics encapsulated in nanoparticles

toward the acidic environment and hydrolytic enzymes of GIT, (ii) fermentability by microbes present in the intestine, (iii) ability to modulate the growth of the intestinal gut flora (bacteria) and bringing about improvement in the health condition of the person consuming prebiotics [42]. Various studies on animals have suggested the roles of prebiotics in neuroprotection. They play a vital role in neural development during the early stages of life pertaining to their ability to modulate the expression of BDNF in infants and adults. Preclinical studies of ALS or AD have provided subsequent information regarding the potential roles of prebiotics in neuroprotection such as treatment with 2% GOS through the oral route in $SOD1^{G93A}$ transgenic mice for a time period of 10 weeks induced with ALS slowed down the pathogenesis of the disease and increased the life expectancy by 2 weeks. Chitosan oligosaccharides have been found to provide ameliorative effects in rats suffering from AD due to the administration of $A\beta_{1-42}$ [43]. The prebiotics lactulose and melibiose which are basically the analogs of trehalose have been reported to produce neuroprotective effects in mice against the neurological manifestations of PD caused by the administration of MPTP. The ameliorative effects can be summarized as an increment of the striatal dopamine levels, a decrease in the microglial and astrocyte activation and hyperactivity, and reduction in the levels of 4-hydroxynoneal which is a second messenger of oxidative stress [44]. Similarly, the prebiotic lactulose and trehalose have been used as a therapeutic intervention in murine model of AD, and it was found that they improved the cognitive dysfunctions and memory impairment, reduced the neuroinflammation, and facilitated macroautophagy and chaperone-mediated autophagy [45].

## 6 Conclusion

In this book chapter, we reviewed the neuroprotective roles of the gut-microbiota and interactions between the gut and brain through the gut-brain axis and the neuroprotective effects of probiotics and nano probiotics. The neuroprotective roles of the gut microbiota make it a potentially therapeutic target for treating neurodegenerative diseases such as PD, AD, and other neurological conditions such as stroke, depression, and anxiety. Gut dysbiosis is a major contributor in the pathogenesis of neurodegenerative diseases. So, in order to strengthen the gut microbiota by maintaining its natural composition and balance in the intestine and gut, administration of probiotics is a preferable option. Use of antibiotics limits their use due to the occurrence of resistance and other infections or diseases as a side effect. Oral administration of the drugs has the issue of first pass metabolism, and the inability of the probiotic to sustain in a highly acidic environment raises the need to develop new strategies for the delivery of probiotics to the target site of action. One such approach is the use of nano probiotics or encapsulating probiotics into nanoformulations such as nanoparticles, nanolayers, nanobeads, or nanofibers. Nano probiotics attributed to its small size has the advantage to easily pass through the blood vessels and other areas of the body and its enhanced lipophilicity makes it easy to cross areas such as the blood-brain barrier or GI barrier and deliver the probiotic with an enhanced delivery rate. Various probiotics such as *Lactobacillus* and *Bifidobacterium* have been found to exert neuroprotective effects evident through pre-clinical studies. So, application of nanotechnology to probiotics opens new doors for enhancing the therapeutic efficacy of probiotics to maintain a healthy gut and thereby providing neuroprotective effects in various neurological conditions such as stroke, anxiety, depression, and neurodegenerative diseases through anti-inflammatory, immunomodulatory, anti-oxidant, and anti-apoptotic actions.

## References


1. Toribio-Mateas, M. (2018). Harnessing the power of microbiome assessment tools as part of neuroprotective nutrition and lifestyle medicine interventions. *Microorganisms, 6*(2), 35. https://doi.org/10.3390/microorganisms6020035
2. Leprun, P. M. B., & Clarke, G. (2019). The gut microbiome and pharmacology: A prescription for therapeutic targeting of the gut–brain axis. *Current Opinion in Pharmacology, 49*, 17–23. https://doi.org/10.1016/j.coph.2019.04.007
3. Tetreault, M., & Parker, A. (2020). Probiotic Lacticaseibacillus Rhamnosus HA-114 suppresses age-dependent neurodegeneration via mitochondrial Beta-oxidation. https://doi.org/10.21203/rs.3.rs-59192/v1
4. González-Arancibia, C., Escobar-Luna, J., Barrera-Bugueño, C., et al. (2016). What goes around comes around: Novel pharmacological targets in the gut-brain axis. *Therapeutic Advances in Gastroenterology, 9*(3), 339–353. https://doi.org/10.1177/1756283X16630718
5. Cryan, J. F., & Dinan, T. G. (2012). Mind-altering microorganisms: The impact of the gut microbiota on brain and behaviour. *Nature Reviews Neuroscience, 13*(10), 701–712. https://doi.org/10.1038/nrn3346
6. Roy Sarkar, S., Mitra Mazumder, P., Chatterjee, K., Sarkar, A., Adhikary, M., Mukhopadhyay, K., & Banerjee, S. (2021). Saccharomyces boulardii ameliorates gut dysbiosis associated cognitive decline. *Physiology and Behavior, 236*. https://doi.org/10.1016/j.physbeh.2021.113411
7. Divyashri, G., Sadanandan, B., Chidambara Murthy, K. N., Shetty, K., & Mamta, K. (2021). Neuroprotective potential of non-digestible oligosaccharides: An overview of experimental evidence. *Frontiers in Pharmacology, 12*. Frontiers Media S.A. https://doi.org/10.3389/fphar.2021.712531
8. Giovannini, M. G., Lana, D., Traini, C., & Vannucchi, M. G. (2021). The microbiota–gut–brain axis and alzheimer disease. From dysbiosis to neurodegeneration: Focus on the central nervous system glial cells. *Journal of Clinical Medicine, 10*(11) MDPI. https://doi.org/10.3390/jcm10112358
9. Shandilya, S., Kumar, S., Kumar Jha, N., Kumar Kesari, K., & Ruokolainen, J. (2022). Interplay of gut microbiota and oxidative stress: Perspective on neurodegeneration and neuroprotection. *Journal of Advanced Research, 38*, 223–244. Elsevier B.V. https://doi.org/10.1016/j.jare.2021.09.005
10. Cheon, M. J., Lim, S. M., Lee, N. K., & Paik, H. D. (2020). Probiotic properties and neuroprotective effects of lactobacillus buchneri ku200793 isolated from korean fermented foods. *International Journal of Molecular Sciences, 21*(4). https://doi.org/10.3390/ijms21041227
11. Park, J., Lee, J., Yeom, Z., Heo, D., & Lim, Y. H. (2017). Neuroprotective effect of Ruminococcus albus on oxidatively stressed SH-SY5Y cells and animals. *Scientific Reports, 7*(1). https://doi.org/10.1038/s41598-017-15163-5
12. Woo, J. Y., Gu, W., Kim, K. A., Jang, S. E., Han, M. J., & Kim, D. H. (2014). Lactobacillus pentosus var. plantarum C29 ameliorates memory impairment and inflammaging in a d-galactose-induced accelerated aging mouse model. *Anaerobe, 27*, 22–26. https://doi.org/10.1016/j.anaerobe.2014.03.003
13. Sampson, T. R., Debelius, J. W., Thron, T., Janssen, S., Shastri, G. G., Ilhan, Z. E., Challis, C., Schretter, C. E., Rocha, S., Gradinaru, V., Chesselet, M. F., Keshavarzian, A., Shannon, K. M., Krajmalnik-Brown, R., Wittung-Stafshede, P., Knight, R., & Mazmanian, S. K. (2016). Gut microbiota regulate motor deficits and Neuroinflammation in a model of Parkinson's disease. *Cell, 167*(6), 1469–1480.e12. https://doi.org/10.1016/j.cell.2016.11.018
14. Sun, P., Su, L., Zhu, H., Li, X., Guo, Y., Du, X., Zhang, L., & Qin, C. (2021). Gut microbiota regulation and their implication in the development of neurodegenerative disease. *Microorganisms, 9*(11) MDPI. https://doi.org/10.3390/microorganisms9112281
15. Chakrabarti, A., Geurts, L., Hoyles, L., Iozzo, P., Kraneveld, A. D., la Fata, G., Miani, M., Patterson, E., Pot, B., Shortt, C., & Vauzour, D. (2022). The microbiota–gut–brain axis: Pathways to better brain health. Perspectives on what we know, what we need to investigate and how to

put knowledge into practice. *Cellular and Molecular Life Sciences, 79*(2) Springer Science and Business Media Deutschland GmbH. https://doi.org/10.1007/s00018-021-04060-w
16. Peterson, C. T. (2020). Dysfunction of the microbiota-gut-brain Axis in neurodegenerative disease: The promise of therapeutic modulation with prebiotics, medicinal herbs, probiotics, and Synbiotics. *Journal of Evidence-Based Integrative Medicine, 25*. SAGE Publications Ltd. https://doi.org/10.1177/2515690X20957225
17. Breit, S., Kupferberg, A., Rogler, G., & Hasler, G. (2018). Vagus nerve as modulator of the brain-gut axis in psychiatric and inflammatory disorders. *Frontiers in Psychiatry, 9*(MAR) Frontiers Media S.A. https://doi.org/10.3389/fpsyt.2018.00044
18. Suganya, K., & Koo, B. S. (2020). Gut–brain axis: Role of gut microbiota on neurological disorders and how probiotics/prebiotics beneficially modulate microbial and immune pathways to improve brain functions. *International Journal of Molecular Sciences, 21*(20), 1–29. MDPI AG. https://doi.org/10.3390/ijms21207551
19. Zagórska, A., Marcinkowska, M., Jamrozik, M., Wiśniowska, B., & Paśko, P. (2020). From probiotics to psychobiotics – The gut-brain axis in psychiatric disorders. *Beneficial Microbes, 11*(8), 717–732. Wageningen Academic Publishers. https://doi.org/10.3920/BM2020.0063
20. Durazzo, A., Nazhand, A., Lucarini, M., Atanasov, A. G., Souto, E. B., Novellino, E., Capasso, R., & Santini, A. (2020). An updated overview on nanonutraceuticals: Focus on nanoprebiotics and nanoprobiotics. *International Journal of Molecular Sciences, 21*(7), 2285. https://doi.org/10.3390/ijms21072285
21. Naomi, R., Embong, H., Othman, F., Ghazi, H. F., Maruthey, N., & Bahari, H. (2022). Probiotics for Alzheimer's disease: A systematic review. *Nutrients, 14*(1), 20. https://doi.org/10.3390/nu14010020
22. Baral, K. C., Bajracharya, R., Lee, S. H., & Han, H. K. (2021). Advancements in the pharmaceutical applications of probiotics: Dosage forms and formulation technology. *International Journal of Nanomedicine, 16*, 7535–7556. https://doi.org/10.2147/IJN.S337427
23. Srivastav, S., Neupane, S., Bhurtel, S., Katila, N., Maharjan, S., Choi, H., Hong, J. T., & Choi, D. Y. (2019). Probiotics mixture increases butyrate, and subsequently rescues the nigral dopaminergic neurons from MPTP and rotenone-induced neurotoxicity. *Journal of Nutritional Biochemistry, 69*, 73–86. https://doi.org/10.1016/j.jnutbio.2019.03.021
24. Lim, S. M., Ramli, M. Z., Alwi, N. A. A., Mani, V., Majeed, A. B. A., & Ramasamy, K. (2015). Probiotics and neuroprotection. In *Diet and nutrition in dementia and cognitive decline* (pp. 859–868). Elsevier. https://doi.org/10.1016/B978-0-12-407824-6.00079-3
25. Sun, J., Ling, Z., Wang, F., Chen, W., Li, H., Jin, J., Zhang, H., Pang, M., Yu, J., & Liu, J. (2016). Clostridium butyricum pretreatment attenuates cerebral ischemia/reperfusion injury in mice via anti-oxidation and anti-apoptosis. *Neuroscience Letters, 613*, 30–35. https://doi.org/10.1016/j.neulet.2015.12.047
26. Labarre, A., Guitard, E., Tossing, G., Forest, A., Bareke, E., Labrecque, M., Tétreault, M., Ruiz, M., & Alex Parker, J. (2022). Fatty acids derived from the probiotic Lacticaseibacillus rhamnosus HA-114 suppress age-dependent neurodegeneration. *Communications Biology, 5*(1). https://doi.org/10.1038/s42003-022-04295-8
27. Nimgampalle, M., & Yellamma, K. (2017). Anti-Alzheimer properties of probiotic, Lactobacillus plantarum MTCC 1325 in Alzheimer's disease induced albino rats. *Journal of Clinical and Diagnostic Research, 11*(8), KC01–KC05. https://doi.org/10.7860/JCDR/2017/26106.10428
28. Rezaeiasl, Z., Salami, M., & Sepehri, G. (2019). The effects of probiotic Lactobacillus and Bifidobacterium strains on memory and learning behavior, long-term potentiation (LTP), and some biochemical parameters in β-amyloid-induced rat's model of Alzheimer's disease. *Preventive Nutrition and Food Science, 24*(3), 265–273. https://doi.org/10.3746/pnf.2019.24.3.265
29. Hsieh, T., Kuo, C., Hsieh, K., Shieh, M., Peng, C., Chen, Y., Chang, Y., Huang, Y., Chen, C., Chang, P., Chen, K., & Chen, H. (2020). Probiotics alleviate the progressive deterioration of motor functions in a mouse model of Parkinson's disease. *Brain Sciences, 10*(4), 206. https://doi.org/10.3390/brainsci10040206

30. Ait-Belgnaoui, A., Colom, A., Braniste, V., Ramalho, L., Marrot, A., Cartier, C., Houdeau, E., Theodorou, V., & Tompkins, T. (2014). Probiotic gut effect prevents the chronic psychological stress-induced brain activity abnormality in mice. *Neurogastroenterology and Motility, 26*(4), 510–520. https://doi.org/10.1111/nmo.12295
31. Liu, J., Sun, J., Wang, F., Yu, X., Ling, Z., Li, H., Zhang, H., Jin, J., Chen, W., Pang, M., Yu, J., He, Y., & Xu, J. (2015). Neuroprotective effects of clostridium butyricum against vascular dementia in mice via metabolic butyrate. *BioMed Research International, 2015*. https://doi.org/10.1155/2015/412946
32. Wang, L., Zhao, Z., Zhao, L., Zhao, Y., Yang, G., Wang, C., Gao, L., Niu, C., & Li, S. (2022). Lactobacillus plantarum DP189 reduces α-SYN aggravation in MPTP-induced Parkinson's disease mice via regulating oxidative damage, inflammation, and gut microbiota disorder. *Journal of Agricultural and Food Chemistry, 70*(4), 1163–1173. https://doi.org/10.1021/acs.jafc.1c07711
33. Zhou, X., Guo, Q., Guo, M., Li, B., Peng, W., Wang, D., Ming, D., & Zheng, B. (2021). Nanoarmour-shielded single-cell factory for bacteriotherapy of Parkinson's disease. *Journal of Controlled Release, 338*, 742–753. https://doi.org/10.1016/j.jconrel.2021.09.009
34. Xu, C., Ban, Q., Wang, W., Hou, J., & Jiang, Z. (2022). Novel nano-encapsulated probiotic agents: Encapsulate materials, delivery, and encapsulation systems. *Journal of Controlled Release, 349*, 184–205. https://doi.org/10.1016/j.jconrel.2022.06.061
35. Liu, N., Yang, C., Liang, X., Cao, K., Xie, J., Luo, Q., & Luo, H. (2022). Mesoporous silica nanoparticle-encapsulated Bifidobacterium attenuates brain Aβ burden and improves olfactory dysfunction of APP/PS1 mice by nasal delivery. *Journal of Nanobiotechnology, 20*(1). https://doi.org/10.1186/s12951-022-01642-z
36. Atraki, R., & Azizkhani, M. (2021). Survival of probiotic bacteria nanoencapsulated within biopolymers in a simulated gastrointestinal model. *Innovative Food Science and Emerging Technologies, 72*. https://doi.org/10.1016/j.ifset.2021.102750
37. Feher, F., Pinter, E., Helyes, Z., & Szolcsanyi, J. (2012). Nano-size particles of probiotics for preventing and treating neuroinflammation. *ARVO Annual Meeting. Investigative Ophthalmology & Visual Science, 53*, 331.
38. Akhtar, N., & Pathak, K. (2017). Probiotics as a tool to biosynthesize metallic nanoparticles: Research reports and patents survey. *Recent Patents on Drug Delivery & Formulation, 11*(1), 5–18. https://doi.org/10.2174/1872211311666170313124335
39. Ashraf, H., Cossu, D., Ruberto, S., Noli, M., Jasemi, S., Simula, E. R., & Sechi, L. A. (2023). Latent potential of multifunctional selenium nanoparticles in neurological diseases and altered gut microbiota. *Materials, 16*(2), 699. https://doi.org/10.3390/ma16020699
40. Nagy, Z. K., Wagner, I., Suhajda, Á., Tobak, T., Harasztos, A. H., Vigh, T., Sóti, P. L., Pataki, H., Molnár, K., & Marosi, G. (2014). Nanofibrous solid dosage form of living bacteria prepared by electrospinning. *Express Polymer Letters, 8*(5), 352–361. https://doi.org/10.3144/expresspolymlett.2014.39
41. Davani-Davari, D., Negahdaripour, M., Karimzadeh, I., Seifan, M., Mohkam, M., Masoumi, S. J., Berenjian, A., & Ghasemi, Y. (2019). Prebiotics: Definition, types, sources, mechanisms, and clinical applications. *Foods, 8*(3), 92. https://doi.org/10.3390/foods8030092
42. Kao, A. C. C., Harty, S., & Burnet, P. W. J. (2016). The influence of prebiotics on neurobiology and behavior. *International Review of Neurobiology, 131*, 21–48). Academic Press. https://doi.org/10.1016/bs.irn.2016.08.007
43. Lin, C. H., Wei, P. C., Chen, C. M., Huang, Y. T., Lin, J. L., Lo, Y. S., Lin, J. L., Lin, C. Y., Wu, Y. R., Chang, K. H., & Lee-Chen, G. J. (2020). Lactulose and Melibiose attenuate MPTP-induced Parkinson's disease in mice by inhibition of oxidative stress, reduction of Neuroinflammation and up-regulation of autophagy. *Frontiers in Aging Neuroscience, 12*. https://doi.org/10.3389/fnagi.2020.00226

44. Lee, Y. S., Lai, D. M., Huang, H. J., Lee-Chen, G. J., Chang, C. H., Hsieh-Li, H. M., & Lee, G. C. (2021). Prebiotic lactulose ameliorates the cognitive deficit in Alzheimer's disease mouse model through macroautophagy and chaperone-mediated autophagy pathways. *Journal of Agricultural and Food Chemistry, 69*(8), 2422–2437. https://doi.org/10.1021/acs.jafc.0c07327
45. Song, L., Gao, Y., Zhang, X., & Le, W. (2013). Galactooligosaccharide improves the animal survival and alleviates motor neuron death in SOD1G93A mouse model of amyotrophic lateral sclerosis. *Neuroscience, 246*, 281–290. https://doi.org/10.1016/j.neuroscience.2013.05.002

# Nanocorona, Immunity Switch, and Hide and Seek Interactions of Nanomaterials in Neuroimmune Crosstalk

K. M. Gijith Mohan, Vineet K. Jain, Harvinder Popli, and Keerti Jain

**Abstract**

Nanoparticles (NPs) can be harnessed for a variety of applications in the biomedical field, taking advantage of their size range which is 1–100 nm. Conventional formulations possess many pitfalls as they interact with the biomolecules and their efficiency or activity is largely affected. This points out the importance of the study of the interactions between the drugs and the biomolecules and the possible outcomes. As with any other drug molecules, the NPs' interaction with the body and biomolecules is also to be studied intensively for the same reason, to obtain the maximum therapeutic efficacy and minimize the toxic effects. The changes happening to the NPs can influence their biological fate like absorption, therapeutic action, eliciting an immune response, or their clearance from the body. Nanocorona (NC) formation over the surface of the NPs by biomolecules is one of the major events happening during their course inside the body. This property influences their interaction with the various components of the innate immune system, biological activity, and toxicity as well. The recent advancements in nanotechnology enabled the production of engineered NPs which can not only elicit an immunological reaction but also can be used for immunosuppression. This reflects the immunity-switching capability of the NPs which can be used for the treatment of many immunological diseases. The NPs are capable of interfering with the neuroimmune crosstalk, which is

K. M. Gijith Mohan · K. Jain (✉)
Drug Delivery and Nanomedicine Research Laboratory, Department of Pharmaceutics, National Institute of Pharmaceutical Education and Research (NIPER)-Raebareli, Lucknow, India
e-mail: keertijain.02@niperraebareli.edu.in

V. K. Jain · H. Popli
Department of Pharmaceutics, Delhi Pharmaceutical Sciences and Research University (DPSRU), New Delhi, India

 
R. Shukla et al. (eds.), *Neuroendocrine-Immune Axis*,
https://doi.org/10.1007/978-981-95-5088-3_8

believed to be a governing factor in the development and progression of diseases such as cancer, stroke, mood disorders, multiple sclerosis, and many others. In general, if the NPs are detected by the phagocytes of the immune system, they are rapidly cleared from the body resulting in therapeutic failure. This indicates the significance of understanding the mechanisms of interaction of NPs with the immune system, along with the possibilities of altering the structure of NC to tune their properties. In this chapter, we have reviewed the significance of NC and how it influences neuroimmune communications, which can be harnessed for novel treatment methods for several diseases.

**Keywords**

Nanoparticles · Nanocorona · Immune system · Switch · Interaction · Immunomodulation

## Abbreviations

| | |
|---|---|
| AD | Alzheimer's disease |
| CNT | Carbon nanotube |
| DC | Differential centrifugation |
| IL | Interleukin |
| MPS | Mononuclear phagocyte system |
| NC | Nanocorona |
| NET | Neutrophil extracellular trap |
| NGF | Nerve growth factor |
| NICU | Neuro-immune cell unit |
| NP | Nanoparticle |
| PEG | Polyethylene glycol |
| PRRS | Pattern-recognition receptors |
| ROS | Reactive oxygen species |
| SEC | Size exclusion chromatography |
| Teff | Effector T cells |
| TLR | Toll-like receptor |
| TNF | Tumour necrosis factor |
| Treg | Natural regulatory T cells |

## 1 Introduction

Nanotechnology has become an inevitable element in the healthcare field due to its wide range of applications and potential to become a more efficient tool to challenge emerging health concerns. Nanoparticles (NPs) are defined as particles of size range between 1 and 100 nm and their properties will be different from the bulk samples of the same material [1, 2]. Owing to their thermal conductivity, electron transfer property, small size, and high surface/volume ratio, they possess multifaceted applications, which include bioimaging, targeted delivery of drugs, and theranostics [3, 4]. The main attraction of the NPs is their ability to reach difficult targets like the brain where the other treatments fail. Once the NPs reach the physiological system, they undergo a variety of interactions with various components of the body and these interactions influence their biological activity including immunological reactions, targeting, therapeutic activity, and toxicity. The interactions of NPs with various components of the body lead to the formation of a nanocorona (NC) and this is one of the critical phenomena that decides the fate of NPs inside the body. The NC is mainly composed of proteins, formed by the irreversible adsorption over the surface of the NPs. Apart from protein, several other biomolecules like nucleic acids, sugars, and lipids can also be adsorbed over the NP surface forming the NC, but to a lesser extent. All the pharmacokinetic properties of the NPs depend on the nature of this NC, and the various possibilities to alter this structure may be exploited to obtain maximum therapeutic activity, high safety, better biocompatibility, and a reduced toxicity profile [2]. In association with many diseases including cancer and CNS disorders, it is evident that there is an interaction between the components of the immune system and the nervous system, promoting pathogenesis. For this purpose, it is crucial to understand the NC, its mechanisms of interactions with the immunological system, and interference in the neuroimmune crosstalk.

## 2 Nanomaterials and Nanotechnology

Research technology development by designing and applying functionalized nanomaterials, nanostructures, nanosystems, and nanodevices is nanotechnology by taking benefit of materials' properties at nanoscale. Nanomaterials and nanotechnology are being investigated extensively in medicine and healthcare to manage severe and fatal diseases including infectious diseases such as AIDS, multidrug-resistant tuberculosis, AIDS and non-infectious diseases like cancer, neurodegenerative disorders, diabetes, etc. The ability to use nanostructures as delivery agents to encapsulate or attach therapeutic pharmaceuticals and deliver them to target tissues more accurately with a controlled release has led to the increased appreciation of nanomedicines in recent years [5]. Nanotechnology can be exploited to deliver drugs, peptides, proteins, and nucleic acids using various delivery systems and plays a significant role in drug delivery and targeting of pharmaceutical, therapeutic, and diagnostic agents. Bioimplants coated with nanoparticles were found to have less biocompatibility issues as compared with the conventional implants [6]. Various nanomaterials like liposomes and other vesicular carriers, dendrimers, nanoparticles, quantum dots, carbon nanotubes, nanobubbles, nanoshells, nanoemulsions, etc. are being investigated for various applications in the field of medicine and biotechnology [7]. Many biomedical devices are developed like nanorobots and nanomachines which have huge potential in healthcare industry [8–16].

These nanomaterials have been explored by scientists for in vitro and performance using cell culture and disease-induced in vivo animal models. Before examining the in vitro and in vivo performances of these nanomaterials it is imperative to characterize them for nanoscale size and size distribution as well as for other characterization parameters according to the type of nanomaterials and intended use which may include analysis by Fourier transform infrared (FTIR) spectroscopy, differential scanning calorimetry (DSC), powder X-ray diffraction, and determination of entrapment and loading efficiency, drug release, particle or globule or vesicular size, size distribution, surface charge etc. to evaluate the performance of prepared nanosystem [17–20].

Nanomaterials are extensively being investigated by scientists in delivery of drug molecules, nucleic acids, diagnostic and imaging agents, antibodies, and vaccines. Nanotechnology is applied to develop fluorescent materials which can act as imaging agent, molecular diagnostics kits, and developing novel targeting drugs used in the treatment of cancer. Although these nanomaterials are showing promises in achieving the controlled drug release of payload, reducing the side effects, and improving the therapeutic efficacy in in vitro studies as well as in preclinical investigations in animal models, yet their interactions with biological systems require to be investigated comprehensively [3, 4, 20–26]. How these systems interact with the immune system of the body is very important to be investigated to further explore their potential and as well as confirming safety.

## 3 Nanocorona

When the bare NPs are administered, they encounter many protein components in the physiological fluids of the body. As with any other nanoscale object, NPs possess a large surface area and high surface charge. To reduce the high energy, at the interface between the NP surface and the biological fluid, the proteins get adsorbed leading to the formation of NC resulting in a more stable structure. This NC formation has much influence over the physicochemical properties of the NPs because they will alter the size, charge, and function of the NPs. As a result, most of their properties like biodistribution, diffusion time, cellular uptake, intracellular distribution, bioavailability, clearance, and toxicity are greatly affected [10, 27]. Apart from that, recently it has been found that they can influence so many enzyme systems and can act as immunomodulators, and it is believed that they can play important roles in the treatment of many diseases which are immunological in origin [28]. The proteins will undergo conformational changes during the formation of the NC, and this depends on the properties of both the NPs and the interacting protein. This process can result in either increase, decrease, or complete loss of activity of NPs [29]. The importance of the NC formation can be realized from the fact that only 0.7% of the administered dose of the NPs could reach successfully to the targets [30].

NC formation is a dynamic process depending on the affinity and interactions between the NPs and the protein units which can be explained by the Vroman effect [31]. According to the Vroman effect, the composition of the NC will vary from time to time but the amount of the protein surrounding the NC is relatively constant [29]. Initially, proteins will get adsorbed over the surface of the NPs and with time, those with lesser affinity will get replaced with higher affinity ones, and this continuous adsorption-desorption process results in the formation of NC [32]. It depends also on the surface properties like the charge, the presence of the type of bonding functional groups capable of forming hydrogen bonds, electrostatic interactions, covalent or coordinate bonds, etc. Nevertheless, the size, shape, hydrophobicity, surface smoothness, steric factors, and curvature of the NPs also exert a crucial role in NC formation [31, 33]. Depending on all the variables involved in the process, the NC formation could be a result of physical entanglement or may involve a chemical reaction [34]. After isolation of NC using techniques such as differential centrifugation (DC) or size exclusion chromatography (SEC), numerous methods of analysis of the protein units surrounding the NPs, regarding their conformations and the nature of the interactions, reveal that the NC can fall into 2 classes: hard corona and soft corona.

### 3.1 Hard Corona

Initially, when NPs encounter the protein units in the biological milieu, those with a high affinity toward the NP surface will get tightly bound and form a permanent hard corona layer. This structure may retain for several hours as they are stable due to the stronger attachment. The exchange of the protein units with the surrounding medium will be at a slower pace as the interaction with the NP surface is stronger in the case of hard corona. It is plausible that the hard corona has a major role in the biological response of the NPs and more important entity to be evaluated [35].

### 3.2 Soft Corona

After the formation of the hard NC, some protein units with less affinity will get loosely adsorbed either over the NP surface or over the hard corona to form a temporary soft corona. Being weakly bound to the hard corona, these structures are not stable enough to get retained for a long time and will be restructured in short intervals. The exchange rate of the protein units between the soft corona and the surrounding medium will be higher and there is a continuous replacement that happens at the interphase depending on the affinity involved in the interaction [29, 36] (Fig. 1).

Bioengineering of the NPs is done to give specific functionality, and this depends on the surface properties and the type of NC formed. By altering the conformations of the proteins or other molecules present in the NC, receptor selectivity of the NPs can be manipulated, and this provides the desired biological activity to the NPs [37, 38]. Moreover, the cellular uptake of the NPs is predominantly affected by the nature of the NC, whether it is hard or soft leading to an altered availability to the receptors and intended bioactivity. Nanocoronas with different compositions are identified differently by cell receptors, and this plays an important role in the internalization by various mechanisms [10]. Thus, it is very important to have an in-depth idea about the adsorption of proteins around the NPs, their conformations, the mechanisms involved, and the resultant alterations of the physicochemical properties [39].

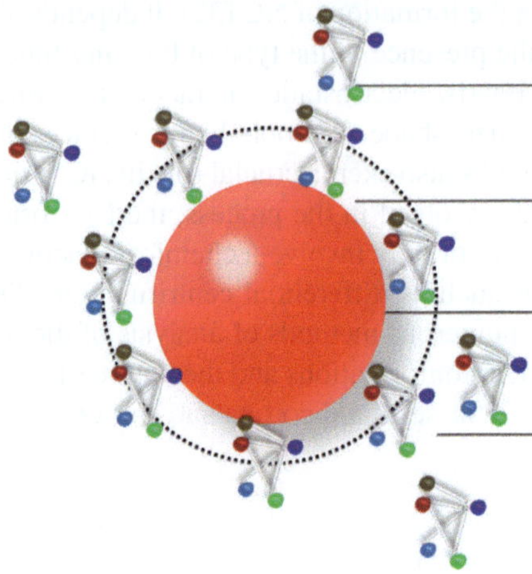


**Fig. 1** Schematic representation of the nanocorona

## 4 NC and Immune System

NPs possess a variety of functional groups, and these can exert a variety of actions when introduced into the physiological environment, capable of activating both innate and adaptive immune systems. Mostly NPs in the biological environment project the properties of a pathogen and this results in the activation of pattern-recognition receptors (PRRS), which can distinguish between 'self' and 'non-self' constituents of the body. This process is further followed by the activation of the innate immune system [40]. Many elements of the immune system, especially the complement system and the mononuclear phagocyte system (MPS), will interact with the NPs resulting in the detection and elicitation of immune reactions that will be either desirable or can lead to toxicity-producing adverse reactions [41–43]. There are situations associated with NPs in the progression of many autoimmune diseases, cancers, and infections, as they hold remarkable immunomodulatory effects including complement activation [44]. The overall behaviour of the NPs will also depend on the properties of the physiological environment they are subjected to, like temperature, ionic strength, pH, oxygen content, and amount of organic matter present in the medium as these factors have a remarkable influence on the composition of the NC and the rate of exchange of the proteins from the NC [34, 45]. NPs can induce the production of several proinflammatory and inflammatory signalling mediators, like cytokines and chemokines leading to several manifestations of inflammation as they can act as adjuvants or haptens [46, 47]. The up-regulation of the expression of the genes for interleukin-1 (IL-1), interleukin-6 (IL-6), and tumour necrosis factor (TNF-alpha) is the key player in the cytotoxicity and inflammation [45].

The MPS constituted by the monocytes and macrophages interacts with the NPs. The primary function of these cells is to induce phagocytosis and remove the

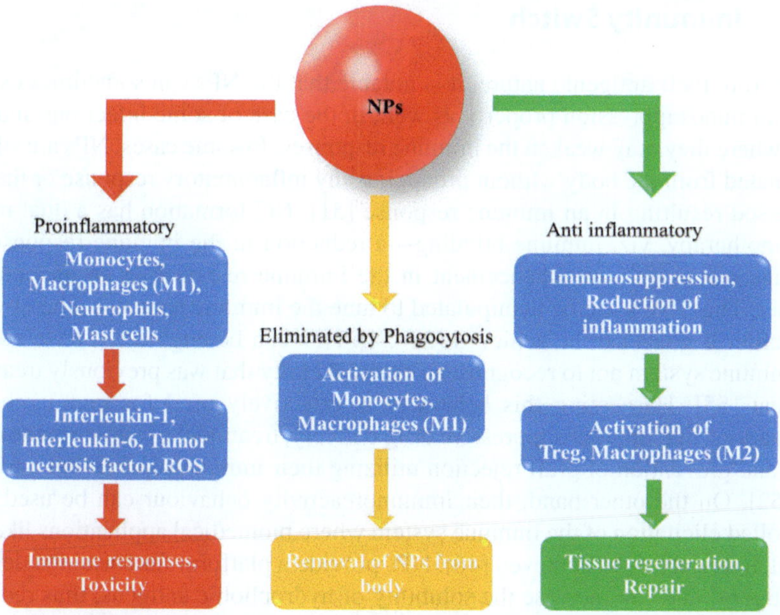


**Fig. 2** Immune response pathways for NPs in the body

external entities including NPs. Macrophages have toll-like receptors (TLRs) over them, which can recognize the NPs and induce phagocytosis by initiating the signal transduction pathway and inflammatory response [46]. Further to add, the uptake of the NPs by the macrophages, generally depends on the nature, the dose, and the charge of the NPs. The "alternative pathway", the binding of the complement C3, to the macrophages, neutrophils as well as the NPs leads to opsonization, the initial process by which phagocytosis will begin and leads to cytotoxic reactions [40, 48, 49].

Neutrophils, the most abundant phagocytes, sequester the NPs and immobilize them resulting in granuloma formation. Activation of the neutrophils by the NPs leads to significant neutrophil infiltration and neutrophilia. Neutrophil extracellular traps (NETs) are specialized structures formed by the neutrophils in response to the formation of reactive oxygen species (ROS) and are involved in the resolution of inflammation and tissue damage. Further, NETs reduce neutrophilic inflammation by degrading the cytokines and chemokines, which causes a reduction in neutrophil activation and recruitment [40, 50]. Similarly, with the dendritic cells also it was found that the interaction with NPs resulted in maturation of the cells and the generation of ROS and inflammatory cytokines [40]. When comes to the case of the mast cells, exposure to NPs tends to activate and release histamine, contributing to the cascade of toxic reactions (Fig. 2) [46]. The initial immune responses carried out by the innate immune system are followed by the activation of the adaptive immune system where a specific reaction against the foreign particle is carried out by the body.

## 5 Immunity Switch

Apart from their antigenic nature, it is notable that the NPs can sometimes express their immunosuppression property as well, in the case of some infections and cancers, where they may weaken the immune responses. In some cases, NPs are silently eliminated from the body without producing any inflammatory response or they are processed resulting in an immune response [51]. NC formation has a dual role in immunotherapy, viz. immune blinding—a reduction in the immune response and immune reactivity—an enhancement in the immune response in an uncontrolled manner. Thus, NCs can be manipulated to tune the immune response whereby specific clinical goals can be achieved [52–54]. NPs can be engineered to re-educate the immune system not to recognize an external entity that was previously treated as a threat [55]. Harnessing this behaviour is effectively used for immunotherapy applications like immunosuppressive drug delivery, treatment of autoimmune disorders, and prevention of graft rejection utilizing their immunosuppression attributes [46, 52]. On the other hand, their immunoreactivity behaviour can be used for a controlled elicitation of the immune system where biomedical applications like vaccine delivery are to be achieved [46]. NPs provide a platform for sustained delivery of antigens, and can increase the solubility of hydrophobic antigens, thus reducing the frequency of dosing for immunizations. When NPs are used as adjuvants in vaccines, they can raise the antibody production several folds higher than conventional adjuvants thereby providing improved protection from the disease [46, 56].

The immune switch behaviour of the NPs can be explained based on the behaviour of the macrophages toward the NPs. Macrophages can be divided into 2 groups, classically activated (M1), which is a pro-inflammatory phenotype or alternatively activated (M2), which is an anti-inflammatory phenotype [57]. Bioengineering paved the way for the functionalization of the NPs, enabling the possibility of reprogramming macrophage polarization, which is a key determinant of the immunomodulation process [40, 43, 58]. It is evident from many studies that the NPs can block the macrophage-related synthesis of mediators like prostaglandins, leading to immunosuppression and a remarkable reduction in the levels of ROS, IL-6, and TNF-$\alpha$ leading to the reduction of inflammation [46, 49].

The role of natural regulatory T cells (Treg) and effector T cells (Teff) can be drawn on the ground of immunomodulation in cases of many immune-related diseases like AIDS. Regulatory T cells are important in bringing immunological self-tolerance by suppressing the activity of self-reactive effector helper and cytotoxic T cells. They also help prevent excessive T cell responses from damaging host tissues. Alternatively, Teff cells hold the responsibility of the exacerbation of the immune response and play a key role in the progression of the diseases [59, 60].

### 5.1 Immunosuppression

In many instances, it is seen that NPs themselves can induce immunosuppressive action without carrying a therapeutic payload. Mitchell et al. proved that in mice when exposed to carbon nanotubes (CNTs), following the interaction with lung cells, cyclooxygenase enzymes were activated in the spleen, leading to immunosuppression [61]. Fullerenes have a negative regulator role and are reported to block the release of many allergic mediators from the mast cells and basophils, in response to an antigen [62]. Bioengineering of the NC to make the NPs escape from the immune system is one of the strategies followed in designing the formulation. In cases of the drug targeting specific parts of the body, usually, the NPs are made to evade the immune system to stay long time in circulation and to reach the target [10, 24]. One of the most common methods involves the usage of polyethylene glycol (PEG) which is grafted over the surface of the NPs and prevents the adsorption of the biomolecules over it. Thus, the opsonization is inhibited to some extent, the NPs' circulation time is prolonged, and the payload is targeted efficiently. Decoration of the NPs with CD47 is another method to prevent the recognition by the phagocytes as it gives 'self' marking to the NPs [63].

## 6 Interactions of Nanomaterials in Neuroimmune Crosstalk

Crosstalk between the nervous system and the immune system has a remarkable influence on the development and progression of many diseases like cancer, stroke, AIDS, chronic inflammatory diseases, and many neurological disorders [20]. Discrete bodies known as neuro–immune cell units (NICUs) communicate with each other through the mediators like neurotransmitters, including catecholamines and acetylcholine, neuropeptides, neurotrophins namely nerve growth factor (NGF), chemokines, and cytokines, where the cells reciprocally express receptors complimenting to the respective mediator. This multidirectional crosstalk is reported to have a direct influence on the macrophages responsible for the maintenance of homeostasis and an important factor for immunomodulatory function in the body. Kennedy JM et al. found that the delayed regeneration of the diabetic nerve is a consequence of abnormal macrophage-mediated nerve repair [64]. As a sequel, aiming NICU as a target site for modulating macrophage function would be a novel therapeutic approach in the management of many diseases [65–69]. Protein misfolding is one of the main causes for many neurological diseases like Alzheimer's disease (AD) with associated neuron inflammation leading to a degeneration cycle inside the cell. Introducing a functionalized NP capable of targeting at the cellular level is found to be effective in the reversal of this process [70].

A hide and seek interaction of NPs in between neuroimmune communication is always a desirable property due to its applications in the biomedical field. The immune blinding can be done in two ways. One approach is to mask fully or partially, the antigens/molecules present on the surface of the NPs with NC. So, the specific stimulation will be prevented, and the immune response will be blocked. Alternatively, the homeostasis of the immune cells also contributes to immune blinding. The phagocytes of the immune system become incapable of detecting the PRRS, and NPs are spared for delivering the payload [52]. So, to summarize, both immune blinding and immune reactivity are highly essential criteria according to the nature of the pathological state and the bioengineered NPs can become a key factor in the future therapeutic strategies [13].

## 7 Conclusion and Future Perspectives

In this chapter, we detailed about the significance of NC and its various implications in the neuroimmune crosstalk. Research is actively done on the utilization of NPs at their best output and needs a clearer understanding of the behaviour of NPs inside the body on a molecular level. One of the dragging factors for the advancements in the field is the unpredictable behaviour of the NPs inside the body. Holding a confined technology, efforts are made to improve the characterization and application of the NPs. One of the most important factors which should be studied is the NC. Almost all properties of NPs depend on the nature of NC and with the help of bioengineering, several applications can be drawn to harness the real potential of the NPs. Further, a deep understanding of the NPs interaction with the NICUs and the possibilities of manipulations of these interactions is required as it can be a valuable solution for many incurable diseases.

**Acknowledgements** The authors Gijith Mohan K M and Keerti Jain acknowledge the Department of Pharmaceuticals (DOP), Ministry of Chemicals and Fertilizers, Government of India for providing facilities for writing this book chapter. The NIPER-R communication number for this chapter is NIPER-R/Communication/540.

## References

1. Pathak, A., Naqvi, S., & Jain, K. (2023). Functionalized dendrimers: Promising nanocarriers for theranostic applications. In *Multifunctional and targeted theranostic nanomedicines: Formulation, design and applications* (pp. 163–185). Springer Nature Singapore.
2. Auffan, M., Rose, J., Bottero, J. Y., Lowry, G. V., Jolivet, J. P., & Wiesner, M. R. (2009). Towards a definition of inorganic nanoparticles from an environmental, health and safety perspective. *Nature Nanotechnology, 4*(10), 634–641.
3. Jain, K., & Zhong, J. (2022). Theranostic applications of nanomaterials. *Current Pharmaceutical Design, 28*(2), 77.
4. Patel, P., Kumar, K., Jain, V. K., Popli, H., Yadav, A. K., & Jain, K. (2023). Nanotheranostics for diagnosis and treatment of breast cancer. *Current Pharmaceutical Design, 29*(10), 732–747.
5. Patra, J. K., Das, G., Fraceto, L. F., Campos, E. V. R., Rodriguez-Torres, M. d. P., Acosta-Torres, L. S., et al. (2018). Nano based drug delivery systems: Recent developments and future prospects. *Journal of Nanobiotechnology, 16*(1), 71.
6. Abaszadeh, F., Ashoub, M. H., Khajouie, G., & Amiri, M. (2023). Nanotechnology development in surgical applications: Recent trends and developments. *European Journal of Medical Research, 28*(1), 537.
7. Mazayen, Z. M., Ghoneim, A. M., Elbatanony, R. S., Basalious, E. B., & Bendas, E. R. (2022). Pharmaceutical nanotechnology: From the bench to the market. *Future Journal of Pharmaceutical Sciences, 8*(1), 12.
8. Haleem, A., Javaid, M., Singh, R. P., Rab, S., & Suman, R. (2023). Applications of nanotechnology in medical field: A brief review. *Global Health Journal, 7*, 70–77.

9. Kaur, S., Mehra, N. K., Jain, K., & Jain, N. K. (2017). Development and evaluation of targeting ligand-anchored CNTs as prospective targeted drug delivery system. *Artificial Cells, Nanomedicine, and Biotechnology, 45*(2), 242–250.
10. Gauro, R., Nandave, M., Jain, V. K., & Jain, K. (2021). Advances in dendrimer-mediated targeted drug delivery to the brain. *Journal of Nanoparticle Research, 23*(3), 76.
11. Jain, K. (2017). Dendrimers: Smart nanoengineered polymers for bioinspired applications in drug delivery. In *Biopolymer-based composites* (pp. 169–220). Woodhead Publishing.
12. Bajwa, N., Kumar Mehra, N., Jain, K., & Kumar Jain, N. (2016). Targeted anticancer drug delivery through anthracycline antibiotic bearing functionalized quantum dots. *Artificial Cells, Nanomedicine, and Biotechnology, 44*(7), 1774–1782.
13. Ojha, B., Jain, V. K., Gupta, S., Talegaonkar, S., & Jain, K. (2022). Nanoemulgel: A promising novel formulation for treatment of skin ailments. *Polymer Bulletin, 79*(7), 4441–4465.
14. Patel, P., Handa, M., Jain, K., & Shukla, R. (2022). Recent pharmaceutical engineered approaches as prophylaxis and management of frostbite. *Journal of Drug Delivery Science and Technology, 76*, 103838.
15. Yadav, S., Jain, V. K., & Jain, K. (2022). Marine biopolymers for gene delivery. In *Marine biomaterials: Therapeutic potential* (pp. 149–172). Springer Nature Singapore.
16. Rambaran, T., & Schirhagl, R. (2022). Nanotechnology from lab to industry-a look at current trends. *Nanoscale Advances, 4*(18), 3664–3675.
17. Sharma, P., Jain, K., Jain, N. K., & Mehra, N. K. (2017). Ex vivo and in vivo performance of anti-cancer drug loaded carbon nanotubes. *Journal of Drug Delivery Science and Technology, 1*(41), 134–143.
18. Chauhan, S., Jain, K., & Naqvi, S. (2022). Dendrimers and its theranostic applications in infectious diseases. In *Nanotheranostics for treatment and diagnosis of infectious diseases* (pp. 199–228). Academic Press.
19. Jain, K., & Ahmad, J. (Eds.). (2022). *Nanotheranostics for treatment and diagnosis of infectious diseases*. Academic Press.
20. Juneja, M., Suthar, T., Pardhi, V. P., Ahmad, J., & Jain, K. (2022). Emerging trends and promises of nanoemulsions in therapeutics of infectious diseases. *Nanomedicine, 17*(11), 793–812.
21. Malik, S., Muhammad, K., & Waheed, Y. (2023). Nanotechnology: A revolution in modern industry. *Molecules, 28*(2), 661.
22. Jain, V., Gupta, A., Pawar, V. K., Asthana, S., Jaiswal, A. K., Dube, A., et al. (2014). Chitosan-assisted immunotherapy for intervention of experimental Leishmaniasis via amphotericin B-loaded solid lipid nanoparticles. *Applied Biochemistry and Biotechnology, 174*(4), 1309–1330.
23. Mehra, N. K., & Jain, K. (2021). *Dendrimers in nanomedicine: Concept, theory and regulatory perspectives*. CRC Press.
24. Jain, K., Mehra, N. K., Jain, V. K., & Jain, N. K. (2020). IPN dendrimers in drug delivery. In *Interpenetrating polymer network: Biomedical applications* (pp. 143–181).
25. Jain, V. K., Jain, K., & Popli, H. (2024). Conjugates of amphotericin B to resolve challenges associated with its delivery. *Expert Opinion on Drug Delivery, 21*, 187–210.
26. Pokrajac, L., Abbas, A., Chrzanowski, W., Dias, G. M., Eggleton, B. J., Maguire, S., et al. (2022). Nanotechnology for a sustainable future: Addressing global challenges with the International Network4Sustainable Nanotechnology. *UCLA School of Law, Public Law Research Paper, 22*(3), 18608–18623.
27. Wang, Y. F., Zhang, C., Yang, K., Wang, Y., Shan, S., Yan, Y., et al. (2021). Transportation of AIE-visualized nanoliposomes is dominated by the protein corona. *National Science Review, 8*(6), nwab068.
28. Park, S. J. (2020). Protein–nanoparticle interaction: Corona formation and conformational changes in proteins on nanoparticles. *International Journal of Nanomedicine, 15*, 5783–5802.
29. Foroozandeh, P., & Aziz, A. A. (2015). Merging worlds of nanomaterials and biological environment: Factors governing protein Corona formation on nanoparticles and its biological consequences. *Nanoscale Research Letters, 10*(1), 1–2.

30. Wilhelm, S., Tavares, A. J., Dai, Q., Ohta, S., Audet, J., Dvorak, H. F., et al. (2016). Analysis of nanoparticle delivery to tumours. *Nature Reviews Materials, 1*(5), 16014.
31. Nguyen, V. H., & Lee, B. J. (2017). Protein corona: A new approach for nanomedicine design. *International Journal of Nanomedicine, 12*, 3137–3151.
32. García-álvarez, R., & Vallet-Regí, M. (2021). Hard and soft protein corona of nanomaterials: Analysis and relevance. *Nanomaterials, 11*(4), 888.
33. Bashiri, G., Padilla, M. S., Swingle, K. L., Shepherd, S. J., Mitchell, M. J., & Wang, K. (2023). Nanoparticle protein corona: From structure and function to therapeutic targeting. *Lab on a Chip, 23*(6), 1432–1466.
34. Abarca-Cabrera, L., Fraga-García, P., & Berensmeier, S. (2021). Bio-nano interactions: Binding proteins, polysaccharides, lipids and nucleic acids onto magnetic nanoparticles. *Biomaterials Research, 25*(1), 1–8.
35. Singh, N., Marets, C., Boudon, J., Millot, N., Saviot, L., Maurizi, L., et al. (2021). In vivo protein corona on nanoparticles: Does the control of all material parameters orient the biological behavior? *Nanoscale Advances, 3*(5), 1209–1229.
36. Corbo, C., Molinaro, R., Parodi, A., Toledano Furman, N. E., Salvatore, F., & Tasciotti, E. (2016). The impact of nanoparticle protein corona on cytotoxicity, immunotoxicity and target drug delivery. *Nanomedicine, 11*(1), 81–100.
37. Suthar, T., Jain, V. K., Popli, H., & Jain, K. (2022). Nanoemulsions as effective carriers for targeting brain tumors. In *Nanocarriers for drug-targeting brain tumors* (pp. 347–363). Elsevier.
38. Cedervall, T., Lynch, I., Lindman, S., Berggård, T., Thulin, E., Nilsson, H., Dawson, K. A., & Linse, S. (2007). Understanding the nanoparticle–protein corona using methods to quantify exchange rates and affinities of proteins for nanoparticles. *National Academy of Sciences of the United States of America, 104*(7), 2050–2055.
39. Khan, S., Sharifi, M., Gleghorn, J. P., Babadaei, M. M. N., Bloukh, S. H., Edis, Z., et al. (2022). Artificial engineering of the protein corona at bio-nano interfaces for improved cancer-targeted nanotherapy. *Journal of Controlled Release, 348*, 127–147.
40. Cronin, J. G., Jones, N., Thornton, C. A., Jenkins, G. J. S., Doak, S. H., & Clift, M. J. D. (2020). Nanomaterials and innate immunity: A perspective of the current status in nanosafety. *Chemical Research in Toxicology, 33*(5), 1061–1073.
41. Wolf, M., Poupardin, R. W., Ebner-Peking, P., Andrade, A. C., Blöchl, C., Obermayer, A., et al. (2022). A functional corona around extracellular vesicles enhances angiogenesis, skin regeneration and immunomodulation. *Journal of Extracellular Vesicles, 11*(4), e12207.
42. Palmieri, V., & Caracciolo, G. (2022). Tuning the immune system by nanoparticle–biomolecular corona. *Nanoscale Advances, 4*(16), 3300–3308.
43. La-Beck, N. M., & Gabizon, A. A. (2017). Nanoparticle interactions with the immune system: Clinical implications for liposome-based cancer chemotherapy. *Frontiers in Immunology, 8*, 416.
44. Moghimi, S. M. (2014). Cancer nanomedicine and the complement system activation paradigm: Anaphylaxis and tumour growth. *Journal of Controlled Release, 190*, 556–562.
45. Dimitriou, N. M., Tsekenis, G., Balanikas, E. C., Pavlopoulou, A., Mitsiogianni, M., Mantso, T., et al. (2017). Gold nanoparticles, radiations and the immune system: Current insights into the physical mechanisms and the biological interactions of this new alliance towards cancer therapy. *Pharmacology & Therapeutics, 178*, 1–17.
46. Kononenko, V., Narat, M., & Drobne, D. (2015). Nanoparticle interaction with the immune system. *Arhiv za Higijenu Rada i Toksikologiju, 66*(2), 97–108.
47. Dwivedi, P. D., Misra, A., Shanker, R., & Das, M. (2009). Are nanomaterials a threat to the immune system? *Nanotoxicology, 3*(1), 19–26.
48. Thau, L., Asuka, E., & Mahajan, K. StatPearls. 2023 May 1 [cited 2024 Feb 27]. *Physiology, opsonization*. Available from: https://www.ncbi.nlm.nih.gov/books/NBK534215/
49. Dobrovolskaia, M. A., Aggarwal, P., Hall, J. B., & McNeil, S. E. (2008). Preclinical studies to understand nanoparticle interaction with the immune system and its potential effects on nanoparticle biodistribution. *Molecular Pharmaceutics, 5*(4), 487–495.

50. Schauer, C., Janko, C., Munoz, L. E., Zhao, Y., Kienhöfer, D., Frey, B., et al. (2014). Aggregated neutrophil extracellular traps limit inflammation by degrading cytokines and chemokines. *Nature Medicine, 20*(5), 511–517.
51. Ernst, L. M., Casals, E., Italiani, P., Boraschi, D., & Puntes, V. (2021). The interactions between nanoparticles and the innate immune system from a nanotechnologist perspective. *Nanomaterials, 11*(11), 2991.
52. Mikelez-Alonso, I., Aires, A., & Cortajarena, A. L. (2020). Cancer nano-immunotherapy from the injection to the target: The role of protein Corona. *International Journal of Molecular Sciences, 21*(2), 519.
53. Giulimondi, F., Digiacomo, L., Vulpis, E., Loconte, L., Ferri, G., Cardarelli, F., et al. (2022). In vitro and ex vivo nano-enabled immunomodulation by the protein corona. *Nanoscale, 14*(29), 10531–10539.
54. Jain, K. (2018). Nanohybrids of dendrimers and carbon nanotubes: A benefaction or forfeit in drug delivery? *Nanoscience & Nanotechnology-Asia, 9*(1), 21–29.
55. Nogrady, B. (2021). How nanotechnology can flick the immunity switch. *Nature, 595*(7865), S18–S19.
56. Dobrovolskaia, M. A., Shurin, M., & Shvedova, A. A. (2016). Current understanding of interactions between nanoparticles and the immune system. *Toxicology and Applied Pharmacology, 299*, 78–89.
57. Mills, C. (2012). M1 and M2 macrophages: Oracles of health and disease. *Critical Reviews in Immunology, 32*(6), 463–488.
58. Guo, M., Zhao, L., Liu, J., Wang, X., Yao, H., Chang, X., et al. (2021). The underlying function and structural organization of the intracellular protein corona on graphdiyne oxide nanosheet for local immunomodulation. *Nano Letters, 21*(14), 6005–6013.
59. Kraft-Terry, S. D., Buch, S. J., Fox, H. S., & Gendelman, H. E. (2009). A coat of many colors: Neuroimmune crosstalk in human immunodeficiency virus infection. *Neuron, 64*(1), 133–145.
60. Alberts, B. (2017). *Molecular biology of the cell* (5th ed., pp. 1572–1575). Garland Science.
61. Mitchell, L. A., Lauer, F. T., Burchiel, S. W., & McDonald, J. D. (2009). Mechanisms for how inhaled multiwalled carbon nanotubes suppress systemic immune function in mice. *Nature Nanotechnology, 4*(7), 451–456.
62. Ryan, J. J., Bateman, H. R., Stover, A., Gomez, G., Norton, S. K., Zhao, W., et al. (2007). Fullerene nanomaterials inhibit the allergic response. *The Journal of Immunology, 179*(1), 665–672.
63. Fadeel, B. (2019). Hide and seek: Nanomaterial interactions with the immune system. *Frontiers in Immunology, 10*, 133.
64. Kennedy, J. M., & Zochodne, D. W. (2000). The regenerative deficit of peripheral nerves in experimental diabetes: Its extent, timing and possible mechanisms. *Brain, 123*(10), 2118–2129.
65. Balogh, M., Janjic, J. M., & Shepherd, A. J. (2022). Targeting neuroimmune interactions in diabetic neuropathy with nanomedicine. *Antioxidants & Redox Signaling, 36*(1–3), 122–143.
66. Choi, Y. H., Laaker, C., Hsu, M., Cismaru, P., Sandor, M., & Fabry, Z. (2021). Molecular mechanisms of neuroimmune crosstalk in the pathogenesis of stroke. *International Journal of Molecular Sciences, 22*(17), 9486.
67. Ordovas-Montanes, J., Rakoff-Nahoum, S., Huang, S., Riol-Blanco, L., Barreiro, O., & von Andrian, U. H. (2015). The regulation of immunological processes by peripheral neurons in homeostasis and disease. *Trends in Immunology, 36*(10), 578–604.
68. Veiga-Fernandes, H., & Mucida, D. (2016). Neuro-immune interactions at barrier surfaces. *Cell, 165*(4), 801–811.
69. Godinho-Silva, C., Cardoso, F., & Veiga-Fernandes, H. (2019). Neuro-immune cell units: A new paradigm in physiology. *Annual Review of Immunology, 37*, 19–46.
70. Ambesh, P., & Angeli, D. G. (2015). Nanotechnology in neurology: Genesis, current status, and future prospects. *Annals of Indian Academy of Neurology, 18*(4), 382–386.

# Neuroimmune Signalling: Novel Emerging Options to Treat Neurodegeneration

Ashwani Sharma, Bhaskaranand Pancholi, Anjali Sharma, and Debapriya Garabadu

**Abstract**

Neuroimmune signalling plays a critical role in maintaining the normal functioning of the central nervous system. Dysregulation of this signalling pathway is now recognised as a critical factor in the development and progression of neurodegenerative diseases. Neurodegenerative disorders such as Alzheimer's disease, Parkinson's disease, and multiple sclerosis are major public health concerns with limited treatment options available. Recent research has highlighted the importance of the interplay between the immune system and the nervous system in the development and progression of these disorders. This has led to the identification of potential therapeutic targets in neuroimmune signalling pathways. This chapter aims to review the current understanding of neuroimmune signalling pathways and their role in neurodegenerative disorders. We will discuss emerging options to modulate these pathways for therapeutic benefit, including immunomodulatory drugs, neurotrophic factors, and stem cell therapy. The potential of targeting neuroimmune signalling pathways is promising, with preclinical and clinical studies showing promising results in improving cognitive function, reducing neuroinflammation, and promoting neuroregeneration. However, further research is needed to fully understand the complexity of these pathways and their specific roles in different neurodegenerative disorders. In summary, targeting neuroimmune signalling pathways represents a promising avenue for the development of novel therapies to treat neurodegenerative disorders.

A. Sharma · A. Sharma
School of Pharmaceutical Sciences, MVN University, Palwal, Haryana, India

B. Pancholi · D. Garabadu (✉)
Department of Pharmacology, Central University of Punjab, Bathinda, India
e-mail: debapriya.garabadu@cup.edu.in

 
R. Shukla et al. (eds.), *Neuroendocrine-Immune Axis*,
https://doi.org/10.1007/978-981-95-5088-3_9

The ongoing research in this field provides hope for improving the lives of individuals affected by these devastating conditions.

**Keywords**

Neuroimmune signalling · Neurodegenerative disorders · Neuroinflammation · Alzheimer's disease · Parkinson's disease · Multiple sclerosis

## 1 Introduction

The immune system and the nervous system are two complex and interconnected systems in the human body that work together to protect neurons against infections. This collaboration of parts and functions is referred to as the neuroimmune system. Through various biochemical and electrophysiological mechanisms, the neuroimmune system plays a crucial role in regulating neuroinflammation, facilitating wound repair in injured neurons, mobilising host defences against pathogens, and maintaining selectively permeable barriers such as the blood-brain barrier and the blood-cerebrospinal fluid barrier [1]. Together, these processes safeguard neurons from illness and maintain their overall well-being. The neurological system and immune system interact biochemically through a network of signalling molecules. Immune cells, such as microglia, which are the central nervous system's resident immune cells, communicate with neurons and other nervous system cells. Immune cells called cytokines and chemokines emit chemical signals when an infection or damage occurs. These immune cell-attracting signalling molecules stimulate inflammation and the immunological responses required to ward against infections and start the healing process. Additionally, these chemical signals have the power to affect neuronal function and encourage neuroprotective reactions [2].

Equally significant are the electrical connections between the immune system and the neurological system. Electrical impulses are used by neurons to communicate, and the immune system can modify these electrical signals. For instance, neurotransmitters released by immune cells can influence synaptic transmission and neuronal excitability. Immune cells can also have direct interactions with neurons, changing their electrical characteristics and impacting their survival and functionality. Neuroimmune signalling is the term for this two-way communication between the immune and neurological systems that permits the coordination of immune responses and neural defence [2, 3]. The control of neuroinflammation is one of the neuroimmune system's key roles. Even while inflammation is a crucial defence mechanism against infections and wounds, it can be harmful to neurons if it persists for an extended period of time. By regulating the intensity and duration of inflammation in the central nervous system, the neuroimmune system contributes to the preservation of equilibrium. Damage to neurons and neurodegenerative disorders can result from excessive neuroinflammation. Through immune cells and signalling chemicals, the immune system aids in tissue healing, the removal of pathogens, and the reduction of excessive inflammation. The neuroimmune system is also involved in wound repair in injured neurons [1]. After an injury, immune cells are recruited to the site to clear debris, promote tissue regeneration, and provide growth factors necessary for neuronal recovery. This immune-mediated wound healing process is essential for repairing damaged neurons and restoring their functionality. The host defences against infections that could infiltrate the central nervous system are also mobilised by the neuroimmune system [2, 4]. The brain and spinal cord are continually being scanned by immune cells for possible infections. Immune cells are stimulated to create an immune response and get rid of the invaders if infections get past the defences. Both the blood-brain barrier and the blood-cerebrospinal fluid barrier serve as selectively permeable barriers that keep infections and dangerous chemicals out of the brain and spinal cord. The neuroimmune system strives to protect neurons from infectious substances and preserve the integrity of these barriers [5].

The peripheral immune system and the neuroimmune system have different anatomical components. The neuroimmune system is mostly made up of glial cells, unlike the peripheral system; only mast cells, out of all the hematopoietic immune system cells, are often present in the neuroimmune system. But in a neuroimmune reaction, certain peripheral immune cells can pass through different blood or fluid-brain barriers to react to infections that have already invaded the brain. For instance, there is proof that immune system cells like macrophages and T cells go into the spinal cord after damage [1, 6]. It has also been established that the central nervous system directly produces immune cells of the complement system.

The neuroimmune system is the result of biochemical and electrophysiological interactions between the immune and neurological systems. This cooperation is essential for keeping neurons healthy and safeguarding them from infections. The neuroimmune system is essential for maintaining neuronal function and avoiding disease by controlling neuroinflammation, encouraging wound healing, mobilising host defences, and protecting selectively permeable barriers. Understanding the complex interaction between these two systems is crucial for improving our understanding of neurological disorders and creating treatment approaches to support neuronal health [2, 3, 5].

## 1.1 Cellular Components of Neuroimmune Signalling

The neuroimmune system consists of various cellular components that play essential roles in maintaining the interaction between the immune system and the nervous system. Glial cells, including astrocytes, microglia, and oligodendrocytes, are key players in the neuroimmune system [7–9]. These cells are present in the central nervous system (CNS) and are involved in immune responses, inflammation regulation, and neuronal support.

- Astrocytes are star-shaped glial cells that provide structural and metabolic support to neurons. They are involved in the regulation of the blood-brain barrier, the clearance of toxins and metabolic waste, and the modulation of synaptic transmission [7]. Astrocytes also participate in immune responses by releasing signalling molecules and cytokines that can influence the activation and recruitment of immune cells [7, 9].

- Microglia are the resident immune cells of the CNS and act as the primary immune defence system in the brain. They constantly survey the brain for signs of infection, injury, or abnormal protein accumulation. Upon detection of such signals, microglia become activated and carry out various functions, including phagocytosis of pathogens and cellular debris, production of inflammatory cytokines, and antigen presentation [8]. Microglia play a critical role in neuroinflammation and the maintenance of neuronal health.
- Oligodendrocytes are glial cells responsible for producing myelin, a protective sheath around neuronal axons that facilitates efficient signal transmission. While their main function is related to neuronal insulation and myelination, oligodendrocytes also contribute to immune responses in the CNS. They can release immune signalling molecules and participate in the regulation of neuroinflammation [10].

In addition to glial cells, mast cells are another component of the neuroimmune system that naturally occur in the brain [9, 10]. Mast cells are immune cells typically associated with allergic reactions and are abundant in peripheral tissues. However, recent research has shown that mast cells also exist in the brain and play a role in mediating interactions between gut microbes, the immune system, and the central nervous system. This involvement is part of the microbiota-gut-brain axis, which refers to the bidirectional communication between the gut microbiota and the brain, mediated by the immune and nervous systems [10].

Neuroimmune signalling involves the activation of specific receptors that are present in both CNS and immune cell types. Some examples of these receptors include:

- Chemokine receptors, such as CXCR4, which are involved in cell migration and immune responses [9].
- Cannabinoid receptors, including CB1, CB2, and GPR55, which play a role in regulating immune and inflammatory processes [9].
- Trace amine-associated receptors, specifically TAAR1, which are involved in modulating neurotransmitter release and immune responses [9].
- μ-Opioid receptors, which are opioid receptors present in both immune cells and neurons and are involved in pain modulation and immune regulation [9].

These receptors allow for communication and coordination between the immune system and the nervous system, contributing to the overall functioning of the neuroimmune system.

## 1.2 Cellular Physiology of Neuroimmune Signalling

The neuroimmune system is a field of study that focuses on understanding the interactions and cross-regulatory impacts between the immune system and the neurological system [11]. It involves investigating how these systems communicate and influence each other's functions.

Cytokines, which are signalling molecules produced by immune cells, play a crucial role in regulating immune responses. It is suggested that cytokines may activate the hypothalamic-pituitary-adrenal (HPA) axis, which is involved in the stress response. Additionally, cytokines have been implicated in the coordination between the nervous system and the immune system [12]. For example, there have been documented instances of cytokines binding to neural receptors, such as the binding of the immune cell-released cytokine IL-1 β to the neural receptor IL-1R [13]. This binding can result in the generation of electrical impulses that transmit pain sensations [12].

Furthermore, there is growing evidence suggesting the involvement of autoimmune T cells in neurogenesis, the process of generating new neurons in the brain. Research has indicated that during adaptive immune system responses, there is an increase in hippocampal neurogenesis [12]. Moreover, studies have shown that autoimmune T cells and microglia, immune cells found in the central nervous system, play a significant role in neurogenesis, memory, and learning in healthy adults [14]. To recognise and react to challenging or damaging stimuli, the neuroimmune system employs complimentary mechanisms of both sensory neurons and immune cells. As an illustration, invading bacteria might activate inflammasomes, which break down IL-1, and depolarise sensory neurons by releasing hemolysins at the same time [15]. Hemolysins produce holes that allow calcium ions to enter the eukaryotic cell and release potassium ions that depolarise the cell [12]. Together, these events cause inflammasomes to activate and sensory neurons to experience an action potential.

A neuroimmune response is also triggered by injury and necrosis. Adenosine triphosphate (ATP), which is released from injured cells, binds to and activates both P2X3 receptors in nociceptors and P2X7 receptors on macrophages in the immune system. Due to the depolarisation brought on by the inflow of calcium and potassium ions as well as the activation of inflammasomes, this results in a response that includes both an action potential and inflammation. The immune system releases IL-1 as a result of the ATP P2X7 receptor binding, and the generated action potential is also responsible for the perception of pain [11, 12]. Whereas inflammation is frequently considered to be an immunological response, the immune system's inflammatory response involves an orchestration of neurological processes. A series of inflammatory reactions, including the release of cytokines and chemokines, along with the release of neuropeptides (like substance P) and neurotransmitters (like serotonin), occur in response to damage or infection. This linked neuroimmune response amplifies inflammation when it works together. These findings highlight the intricate relationship between the immune system and the nervous system, and their impact on various physiological processes. The understanding of the neuro-immune system contributes to our knowledge of immune responses, neuroinflammation, pain perception, and even cognitive functions.

## 2 Neurodegeneration Brought on by Ineffective Neuroimmune Signalling

The gradual loss of structure and function of neurons in the central nervous system is referred to as neurodegeneration, and it is the cause of neurological conditions including Alzheimer's disease (AD), Parkinson's disease (PD), amyotrophic lateral sclerosis (ALS), Huntington's disease (HD), and multiple sclerosis (MS). Despite the fact that neurodegeneration is typically linked to intrinsic nervous system variables, new research indicates that neuroimmune signalling also plays a significant part in these processes.

In the context of neurodegenerative illnesses, the idea of neuroimmune signalling draws attention to the complex relationship between the immune system and the neurological system. Neurodegenerative illnesses have been linked to the pathophysiology of inflammation, a crucial aspect of immunological responses [16]. In reaction to neuronal injury or the development of aberrant protein aggregates, such as beta-amyloid plaques in Alzheimer's disease or Lewy bodies in Parkinson's disease, certain immune cells, including microglia and astrocytes, get activated. Pro-inflammatory cytokines, chemokines, and reactive oxygen species are released by these activated immune cells, and these substances help to maintain a long-term neuroinflammatory state.

Neuroinflammation can affect neurodegenerative processes in both positive and negative ways. The pathogenic proteins or damaged cells are first cleared out as a defence mechanism. However, persistent or severe neuroinflammation may trigger the production of neurotoxic compounds, aggravating the harm to the neurons. This ongoing inflammation has the potential to start a chain reaction that leads to synapse loss, neuronal malfunction, and eventually neurodegeneration [16, 17]. Furthermore, neuroimmune signalling involves the interaction of immune signalling molecules with neuronal receptors. For example, cytokines, such as interleukin-1β (IL-1β) and tumour necrosis factor-alpha (TNF-α), can bind to specific receptors on neurons and initiate intracellular signalling pathways. This activation can impact neuronal survival, synaptic plasticity, and neurotransmitter release. Disruptions in these neuroimmune signalling processes have been implicated in the progression of neurodegenerative diseases [18]. New opportunities for therapeutic approaches are made possible by understanding the function of neuroimmune signalling in neurodegeneration. Neurons may be shielded from further harm by immune system manipulation and targeting of certain inflammatory pathways. Additionally, approaches that encourage immune cells' capacity to perform neuroprotective roles, such as removing harmful protein aggregates, might be investigated [18, 19].

The neurodegeneration involves not only intrinsic factors within the nervous system but also the complex interplay between the immune system and the nervous system. Neuroimmune signalling, including neuroinflammation and the interaction of immune molecules with neuronal receptors, contributes to the progression of neurodegenerative diseases. Further research into these mechanisms holds promise for the development of novel therapeutic strategies aimed at slowing down or preventing neurodegeneration.

## 2.1 Neurodegeneration: The Impact of Microglia

Microglia are specialised immune cells that act as resident macrophages in the central nervous system (CNS). They play a crucial role in defending the CNS and can have both beneficial and detrimental effects on neurons.

Microglia sustain host defence in the CNS when things are normal. They have the capacity to regenerate and are in charge of starting immunological reactions in the CNS. Microglia undergo alterations in their proliferation and shape, a process known as microgliosis, in a variety of neurological diseases. Reactive microglia play two distinct roles: they direct stem cells to the site of damage and activate the production of neurotrophic substances, which promote brain health [20]. On the other hand, pro-inflammatory chemicals including tumour necrosis factor-alpha (TNF-alpha), interleukin-1 beta (IL-1 beta), and reactive oxygen species (ROS) are released by chronically activated microglia, which can harm neurons [21, 22].

Toll-like receptors (TLRs), which cause the production of pro-inflammatory cytokines, are responsible for activating microglia. Additionally, because of this activation, ROS are produced, which fuels neuroinflammation [23]. Particularly in conditions like Alzheimer's disease (AD) and Parkinson's disease (PD), which are characterised by progressive dementia, microgliosis plays a role in synapse loss [24]. After a stroke, M1 microglia, which have a pro-inflammatory nature, produce inflammatory proteins such as IL-6, TNF-, and IL-1, which cause secondary wounds. M2 microglia, on the other hand, are more prevalent in the core following stroke and have anti-inflammatory capabilities [25, 26]. The activation of the anti-inflammatory M2 phenotype is supported by regulatory factors such as IL-4, IL-10, IL-13, and TGF-3 [27, 28].

It is essential for therapeutic treatments to comprehend the equilibrium between pro-inflammatory M1 and anti-inflammatory M2 microglia and to modulate their responses. Neurological illnesses may be treated in novel ways by focusing on the inflammatory pathways linked to microglia activation and enhancing their neuroprotective capabilities [29] as shown in Fig. 1.

## 2.2 Inflammation Causes Neurodegeneration Mediated by Inflammasome

An intricate protein called the inflammasome is in charge of causing inflammation in response to diverse stimuli. Its three primary parts are pro-caspase-1, apoptosis-associated spec-like protein (ASC), and NOD-like receptors (NLRs). The N-terminal pyrin domain (PYD) or caspase recruitment domain (CARD), the NACHT or NOD

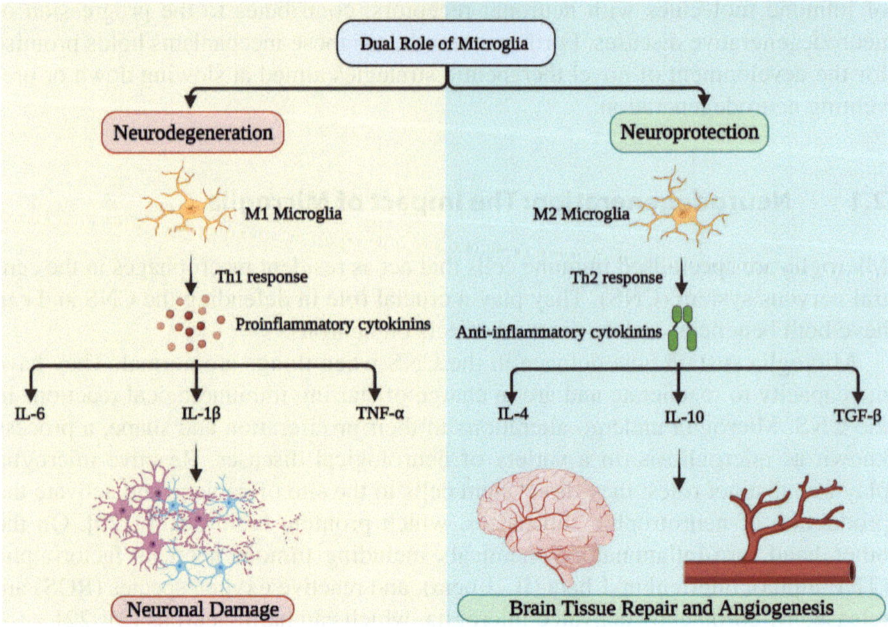


**Fig. 1** Potential bidirectional function in the neurological system for microglia. In response to a Th-1 response, M1 microglia release neurodegenerative cytokines such as IL-1b, TNF-a, and IL6. As a result of the Th2 response, M2 microglia release anti-inflammatory cytokines such as IL-4, IL-10, and TGF-b. IL (interleukin), TNF-a (tumour necrosis factor alpha), TGF-b (transforming growth factor beta), and Th1 (T helper cell 1) are anti-inflammatory cytokines that aid in brain tissue repair, angiogenesis, and, therefore, neuroprotection

domain (nucleotide-binding oligomerisation domain), and the C-terminal leucine-rich repeats (LRRs) make up the tripartite structure of NLRs. Through signalling pathways like MAPK or NF-B, neurodegenerative illnesses activate certain inflammasomes such as the NLRP1 and NLRP3 inflammasomes, along with ASC and pro-caspase-1 [30]. In the central nervous system (CNS) and in neurodegenerative illnesses including Alzheimer's disease (AD), Parkinson's disease (PD), multiple sclerosis (MS), and amyotrophic lateral sclerosis (ALS), NLRP3 plays a key role in inflammation and damage. Additionally contributing to the aetiology of ischemic stroke are NLRP1 and NLRP3 [31].

The NLRP3 protein binds with ASC after activation, and ASC subsequently communicates with the CARD domain through the PYD domain. The NLRP3 inflammasome, also known as the NLRP3-ASC-pro-caspase-1 complex, is created as a result of this recruiting process. The combination activates caspase-1 and changes dormant pro-inflammatory cytokines (pro-IL-1 and pro-IL-18) into active inflammatory cytokines (IL-1 and IL-18), which leads to neuroinflammation and pyroptosis [32, 33]. Gasdermin-D (GSDMD), which creates holes on the plasma membrane, mediates pyroptosis. Numerous causes, including energy depletion, ion flux, cathepsin release, mitochondrial DNA release, PKR activation, and reactive oxygen species (ROS), can cause inflammasome signalling. These include damage-associated molecular patterns (DAMPs) and pathogen-associated molecular patterns (PAMPs), as well as reactive oxygen species (ROS) [33, 34]. Inflammasome-mediated neurodegeneration mechanism is shown in Fig. 2.

The coordinated interplay of the innate and adaptive immune systems constitutes the immunological response. DAMPs and PAMPs are recognised by pattern recognition receptors (PRRs), such as NLRP-1, NLRP-3, NLRC-4, and AIM2, which also have an impact on the adaptive immune response via Toll-like receptors (TLRs). These PRRs are expressed by the immune cell type known as microglia in the CNS, and their enhanced expression causes inflammasome signalling to be activated [33–35].

PAMPs or DAMPs stimulate TLR and control inflammasome non-transcriptional priming. Through NF-jB signalling, TNF activates the TNF receptor and elevates the expression of NLRP3, ASC, and pro-caspase-1. Additionally, it causes the inflammasome to prime, which eventually aids in the formation of the inflammasome. Pro-inflammatory cytokines (pro-IL-1b, pro-IL-18) become inflammatory cytokines (IL-1b, IL-18) after further activation of pro-caspase-1 into caspase-1. Additionally, the P2X7 receptor and mitochondrial stress cause the inflammasome to assemble. It will eventually result in pyroptosis and, as a result, neurodegeneration (PAMPs, or pathogen-associated molecular pattern; DAMPs, or damage-associated molecular pattern; TNF, or tumour necrosis factor; NLRP3, or nucleotide-binding domain; LRD, or leucine-rich repeat-containing receptor (NLR) family pyrin domain; ASC, or apoptosis-associated speck- interleukin, purinergic receptor P2X7-P2, and kappa-light-chain-enhancer of activated B cells).

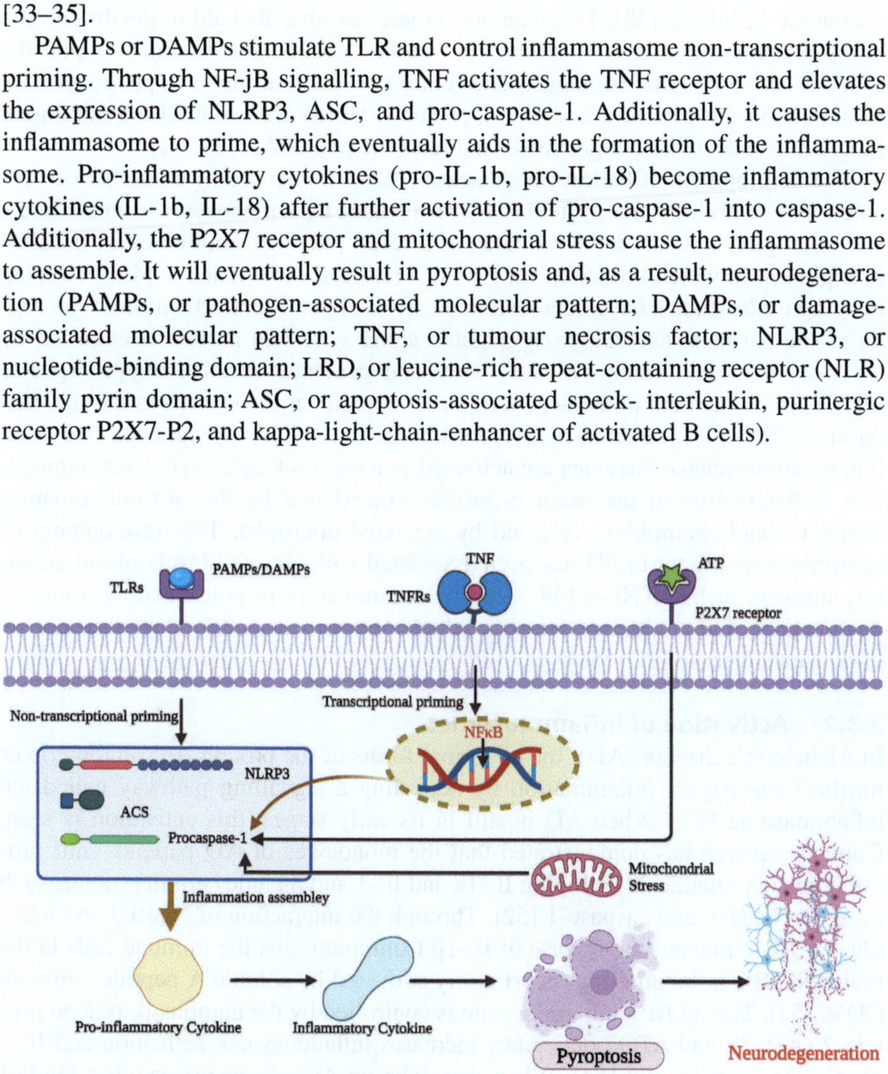


**Fig. 2** Inflasome-mediated neurodegeneration mechanism

## 2.3 The Immune System Machinery Is Impaired in Both Alzheimer's and Parkinson's Diseases

### 2.3.1 Effects of Microglia

Microglia, specialised immune cells in the brain, are able to recognise the Aβ oligomer protein structure in Alzheimer's disease (AD) [36]. The microglia become "reactive" or "primed" as a result of this discovery, altering their appearance and behaviour. Microglia activation in AD has both negative and positive consequences [37]. An adverse effect of active microglia is that they can secrete proinflammatory compounds such as IL-1β, IL-6, and TNF-α that injure cholinergic neurons and encourage cell death [38]. They can also create enzymes that aid in the dissolution of Aβ plaques and offer neuroprotection. Certain genetic abnormalities impair the capacity of microglia to ingest and eliminate cellular debris (phagocytosis). Mutations in the TREM-2 and CD33 proteins, which are essential for microglial function, can decrease phagocytosis and raise the risk of AD [39–41].

The development of Parkinson's disease (PD) is influenced by a number of variables. The central nervous system (CNS) is invaded by lymphocytes from the body, and microglial cells are too stimulated as a result [42–44]. Another event is astrocyte gliosis, in which this different type of brain cell turns reactive [45]. Utilising imaging tools, studies have demonstrated early and persistent activation of microglia at sites of neuronal injury. Aggregated alpha-synuclein protein is engulfed by reactive microglia, which activates Nox2 and generates reactive oxygen species (ROS) [46], which aid in the degeneration of dopaminergic neurons in the substantia nigra (SN) [47]. Leucine-rich repeat kinase (LRRK2) is upregulated and other inflammation-related enzymes are activated as a result of alpha-synuclein aggregation. Inflammation in the brain is further exacerbated by the pro-inflammatory chemicals and chemokines released by activated microglia. The development of cognitive impairment in PD has been associated with elevated levels of inflammatory markers such as TNF-α [48, 49]. The inflammatory response causes cognitive dysfunction and oxidative stress, which exacerbates neurodegeneration in Parkinson's disease (PD) [50].

### 2.3.2 Activation of Inflammasomes

In Alzheimer's disease (AD), the abnormal forms of the protein Aβ (oligomeric or fibrillar) can trigger inflammation by activating a signalling pathway called the inflammasome [51]. When AD is still in its early stages, this activation is seen. Clinical research has demonstrated that the monocytes of AD patients emit pro-inflammatory chemicals including IL-18 and IL-1 and include certain proteins such as ASC, NLRP3, and caspase-1 [52]. Through the interaction of NLRP3 and ASC, fibrillar Ab stimulates the release of IL-1β from microglia, the immune cells in the brain. NLRP3 inflammasome is primarily activated by soluble A peptides through CD36 [53]. The NLRP3 inflammasome is controlled by the autophagy-related protein 7 (ATG7), and ATG7 deficiency increases inflammasome activation and IL-1 production in microglia [54]. When they take up Aβ, astrocytes, another kind of brain cell, may similarly create IL-1 through inflammasome signalling. Inflammasome activation in AD is a factor in neuronal damage [55].

Alpha-synuclein (α-syn) protein aggregates in Parkinson's disease (PD) trigger toll-like receptor (TLR) signalling, which in turn helps to assemble the NLRP3 inflammasome and produce IL-1b [56, 57]. The NLRP3 inflammasome is activated when cathepsin B levels rise and mitochondrial reactive oxygen species (ROS) are produced as a result of the transfer of a-syn to microglia, which is made possible by fyn kinase [58, 59]. The inflammasome and pro-inflammatory chemicals can both be stimulated and activated by mitochondrial DNA damage [60]. Studies using PD animal models have revealed that loss of dopaminergic neurons and motor impairments are both caused by NLRP3 inflammasome activation [61]. IL-1b levels in PD patients' blood and cerebrospinal fluid (CSF) have both been reported to be elevated. In transgenic mice models of PD, elevated expression of NLRP3, caspase-1, and IL-1b has also been noted [62, 63]. The inflammatory response in PD is facilitated by the kinase CDK-5, which phosphorylates certain proteins implicated in inflammasome activation [64].

### 2.3.3 Complement System: Involved Activation Mechanisms

A neurodegenerative condition called Alzheimer's disease (AD) is characterised by amyloid-beta plaques and neurofibrillary tangles in the brain [65–67]. Genes producing the complement proteins CR1 and CLU are notably affected by certain genetic variants linked to the risk of late-onset AD, according to genomic research. These complement proteins aid in the development of plaques by aiding in phagocytosis and preventing proteolytic breakdown [68]. Studies using immunohistochemistry have revealed that in the areas of the brain affected by AD, neurofibrillary tangles, amyloid fibrils, and amyloid-beta deposits co-localise with proteins from the classical complement pathway (C1q, C3, and C4). The terminal C5b-9 complement complex (TCC), a marker of the complement system's terminal route that is mostly present in the cortex of AD patients' brains, is linked to the development of neurofibrillary tangles and aggregated amyloid-beta [69]. The buildup of harmful amyloid-beta in AD causes the complement system to become activated, which causes the release of C1q from neurons and the creation of C3 by astrocytes. Synaptic pruning, phagocytosis, and neuroinflammation are caused by this activation, which is mediated by a number of receptors on microglia and astrocytes. Complement activation produces the complement fragment C5a, which functions as a chemotactic factor for glia and can harm neurons or inadvertently activate microglia [70]. The presence of Lewy bodies and melanised neurons in the substantia nigra, which is impacted by PD, causes the classical complement system to become active [71, 72]. The presence of complement activation markers in immunohistochemistry and the stimulation of microglial NADPH oxidase (Nox2) point to the involvement of the complement system in neuroinflammation and the degeneration of dopaminergic neurons [73]. Cognitive and motor impairment are correlated with neuroinflammation in PD, which is characterised by increased C3 and factor H levels and decreased amyloid-beta42 in cerebrospinal fluid [74]. Studies have found negative double staining for a-synuclein and complement system components, which raises the possibility that the complement system may not be responsible for the development of Lewy bodies in the cortex of PD patients [44].

### 2.3.4 Advanced Treatment Approaches and Immunotherapeutics for Age-Related Neurodegenerative Diseases (PD and AD)

The term “immunotherapeutics” refers to therapeutic modalities that use the immune system to target and cure age-related neurodegenerative disorders. These illnesses, like Parkinson’s and Alzheimer’s, are characterised by the buildup of aberrant proteins and the ongoing inflammation of the brain. These treatments aim to target specific mechanisms involved in the progression of these diseases and harness the immune system for therapeutic purposes. The creation of antibodies or vaccinations that selectively target and eliminate the aberrant proteins linked to certain disorders is one method of immunotherapy. Antibodies that target amyloid-beta (Abeta) plaques or tau protein, for instance, have been studied in Alzheimer’s disease. These antibodies may aid in the removal of these aberrant proteins, perhaps delaying the onset of the illness [75]. Modulating the immune response in the brain is another strategy for lowering neurotoxicity and inflammation. Neurodegenerative disorders advance as a result of persistent brain inflammation. Therefore, it has become apparent that targeting the immune system to reduce this inflammatory response is a possible therapeutic approach. Anti-inflammatory medications or the modulation of particular immune cell types engaged in neuroinflammation can accomplish this [76]. Additionally, stem cell-based treatments may be used in immunotherapeutics to encourage tissue regeneration and repair in the sick brain. It is possible to create stem cells to produce neuroprotective agents or to replace injured cells, and they have the capacity to develop into a variety of cell types [77]. It is significant to emphasise that immunotherapeutics are still in the early phases of development for age-related neurodegenerative illnesses, and further study is required to properly understand their safety and efficacy. These strategies show promise in terms of offering new therapeutic choices and maybe halting the advancement of these fatal illnesses [78].

There are already a number of FDA-approved medications used to treat neurodegenerative disease symptoms that target amyloid-beta (Ab) and tau tangles. These medications include rivastigmine, donepezil, galantamine, and AADvac1, a vaccine being tested in phase II clinical trials by Axon Neuroscience [79]. In addition to these strategies, targeting immune cells like macrophages and microglia to control neuroinflammation is gaining popularity. By injecting anti-Ab monoclonal antibodies (mAbs) into the systemic circulation, Ab plaque load can be reduced [80–82]. This strategy has demonstrated potential in lowering Ab plaques. Tyrosine kinase inhibitors are now undergoing phase III clinical trials, and their capacity to prevent mast cell differentiation and degranulation has also been studied. When taken in conjunction with conventional treatment, these inhibitors have demonstrated favourable results. Additionally, it has been discovered that endocannabinoids facilitate the engagement of the CD200 receptor, which inhibits the activation of neuronal microglia. This method might perhaps decrease the spread of the illness and assist in controlling neuroinflammation [83]. Although these methods are being investigated and appear promising, more study is still required to completely understand their efficacy and safety in the treatment of neurodegenerative illnesses.

**Vaccination:** A phase II clinical study for the AN1792 vaccine to prevent Alzheimer's disease (AD) was scheduled to begin in April 2020 by a company by the name of Elan/Wyeth. Preaggregated Aβ1–42 and the adjuvant QS21 are both included in the vaccination [84]. A phase I clinical study for another vaccine called CAD106, created by Novartis Pharmaceuticals, is now underway. It selectively targets Aβ 1–6, a B-cell epitope that is a short fragment of Ab [85]. Phase I studies of this vaccine improved cognitive behaviour in a promising manner, although the findings of the phase II study are still pending. Clinical trials on several vaccines targeting Ab fragments in combination with carrier proteins and adjuvants are also being conducted by Janssen, Pfizer, and Eli Lilly [86]. Monoclonal antibody passive immunisation is a current immunotherapy strategy. Exogenous anti-Ab monoclonal antibodies must be administered without inducing Th1-mediated antibody production [87]. Antibodies like solanezumab, bapineuzumab AAB-001, and bapineuzumab AAB-003 have all been the subject of clinical investigations [87, 88]. In phase III studies, these medications sadly failed to significantly enhance clinical outcomes. Passive immunisation has drawbacks such as the requirement for repeated injections, the danger of haemorrhaging, and the challenge of passing the blood-brain barrier, the high expense, and the possibility of an immunological reaction to the injected antibodies. Immunotherapeutics, such as PPAR-gamma agonists like fenofibrate, can aid with Parkinson's disease (PD) by lowering inflammation, oxidative stress, and raising dopamine levels [88, 89]. Aspirin and statin NSAIDs have also demonstrated anti-inflammatory benefits in PD [89, 90]. In tests, it has been discovered that medications like MCC950 and kaempferol block the NLRP3 inflammasome, which is linked to inflammation [91]. Other immunotherapy strategies for Parkinson's disease (PD) focus on the illness-related protein a-synuclein. In order to reduce the aggregation of a-synuclein and its oligomers, respectively, vaccines like 9E4 and AFF-1 are used [92–94]. However, there remain barriers to the development of immunotherapy, such as difficulties with brain administration directly and the expense and length of time needed for clinical trials to create novel treatments for synucleinopathy. An updated table containing the most recent methods for treating neuroimmune signalling neurodegeneration using sophisticated immunotherapy therapies, together with information about their stages of clinical trials and trial numbers is provided in Table 1.

## 2.4 Impairment of Neuroimmune Signalling in Multiple Sclerosis and Amyotrophic Lateral Sclerosis

The involvement of the complement system in multiple sclerosis (MS) can be better understood by using the model of experimental autoimmune encephalomyelitis (EAE) [95]. The activation of TNF-a and the production of NO, which ultimately lead to demyelination, are caused by the binding of C3d to the CR3 receptor, which

**Table 1** An updated table containing the most recent methods for treating neuroimmune signaling neurodegeneration using sophisticated immunotherapy therapies [169], together with information about their stages of clinical trials and trial numbers, is provided below

| Sr. no. | Disease | Target | Treatment | Approved clinical trials phase | Clinical trial number |
|---|---|---|---|---|---|
| 1 | Alzheimer's | Reduce tau oligomer accumulation | AADvac1103 | Phase 2 | NCT02579252 |
| | | Microglial activation inhibitor | Endocannabinoids107 | – | – |
| 2 | ALS | Suppress glutamate release | Riluzole171 | – | – |
| | | Inhibit nitration of tyrosine residue | Edaravone (MCI-186) | Phase 3 | NCT01492686 |
| | | TDP-43 Proteinopathies | AMX0035 | Phase 2 | NCT03127514 |
| | | Inflammasome | Cyclo(His-pro)178 | – | – |
| 3 | MS | Halt humoral and cell-mediated immunity | IFN-b, glatiramer acetate, and | – | |
| | | | natalizumab181 | – | – |
| | | Sphingosine 1 phosphate modulator | Fingolimod, teriflunomide, and | – | |
| | | | Dimethyl fumarate184 | – | – |
| | | | Siponimod BAF312 (for secondary progressive MS) | Phase 3 | NCT01665144 |
| | | | And ceralifimod183 | | |
| 4 | PD | PPAR-a agonist | Fenofibrate113 | – | – |
| | | Inhibit TNF-a, IL-6, and IL-1b | Statins and aspirin114,115 | – | – |
| | | Reducing the accumulation of a-synuclein oligomers | 9E4 and AFF113 | – | – |
| 5 | Stroke | NLRP3 inflammasome inhibitor | Beta-hydroxy butyrate and MCC950249,244 | – | – |
| | | Treg cell activator | Trichostatin252 | – | – |
| 6 | HD | Semaphorin 4D inhibitor | VX15 | Phase 2 | NCT01632809 |
| | | | Fenofibrate | Phase 2 | NCT03515213 |

causes the microglia to phagocytose myelin [96]. The terminal complex C5b-9 causes demyelination in macrophages, neurons, and oligodendrocyte progenitor cells when it is present in lytic doses [97]. However, oligodendrocyte death may be stopped by a sublytic dosage of C5b-9 by activating the PI3K/Akt pathway [98]. The increase of complement system fragments C1q and Bb in the grey matter of the brain is supported by clinical findings from MS patients. In an EAE model without antibodies, transgenic mice lacking C3 or factor B displayed less severe illness, demonstrating the dualistic function of the complement system in MS [99]. For MS patients, modifying the complement system as opposed to entirely shutting it down may result in better quality of life [100].

In the case of amyotrophic lateral sclerosis (ALS), various components of the complement system, such as C1q and C3, are upregulated [101]. Activation of these components, along with MBL, MASP-1, C4a, and C5a, leads to the destruction of blood-brain barrier integrity [102]. The C5a peptide binds to C5aR and C5L2 receptors, activating MAPK and PKB/Akt pathways [103]. C5a promotes the entry of macrophages into skeletal muscles, contributing to muscle denervation in ALS [104]. Knockout of C5aR1 in an ALS mouse model increased survival, indicating the exacerbating effect of C5aR1 activation in ALS [105]. Another mouse model, TDP-43Q331K, exhibited similar activation of complement C5a in the lumbar region of the spinal cord and tibialis anterior muscle [106]. These findings suggest that C5a could be a potential therapeutic target for ALS [107]. Elevated levels of C1q, C4, C3, and C5b-9 were observed in ALS patients [108]. Activated C1q converts resting microglia into reactive microglia through $Ca^{2+}$ signalling, further stimulating pro-inflammatory markers such as TNF-α and IL-6, which are detrimental to neurons. Therefore, modulating complement system activation holds promise for neuroprotection in ALS [109–111].

### 2.4.1 Activation of the Microglia in ALS

Immune cells called microglia, which are found in the brain, can help or hurt people with ALS (amyotrophic lateral sclerosis). By lowering inflammatory markers, ALS-related microglia provide protection in the rNLS8 mouse model of ALS [112]. However, utilising certain inhibitors to prevent microglial activation did not enhance motor performance in these animals. The M2 phenotype of microglia produces neuroprotective substances like BDNF, whereas the M1 phenotype releases inflammatory cytokines that may be damaging [113]. Microglia can display many activation states. By triggering certain pathways, the chemical fractalkine, which binds to microglial surface receptors, aids in communication between damaged neurons and microglia and so promotes neuroprotection [114].

In order to research ALS, the $SOD1^{G93A}$ transgenic mouse model is frequently employed. According to this theory, astrocytes and activated microglia are responsible for the degeneration of motor neurons [115]. LFA-1, LCA, CR3, CR4, FcγR1, and MHC-II complex are only a few of the immune response-related molecules that are upregulated by reactive microglia [116]. Additionally, they produce superoxide anions, which leads to oxidative stress and deteriorates the situation [117]. Pro-inflammatory chemicals including IL-1, IL-6, and TNF-α are secreted by reactive microglia, and these compounds then attract more inflammatory molecules. Additionally, as the disease progresses, the inflammatory COX-2 protein is raised in astrocytes, microglia, and neurons in the spinal cord [118, 119].

These findings highlight the complex role of microglia in ALS, where they can exhibit both protective and harmful effects through their activation and secretion of various molecules. Understanding these mechanisms may aid in developing potential therapeutic strategies for ALS.

### 2.4.2 Interaction Between T Cells and Microglia in MS

Multiple sclerosis (MS) immune response is still poorly understood. In the central nervous system (CNS), activated microglia, an immune cell subtype, cause oxidative stress and demyelination, which is the loss of the protective sheath that surrounds nerve fibres [120]. About how the immune response in MS is started, there are two primary theories [121]. According to the first theory, the immune response begins peripherally (outside the central nervous system) and then moves inexorably towards the CNS [122]. According to the second theory, it starts in the CNS before moving to the periphery [123].

Particular kinds of T cells, including Th1, Th2, and Th17 cells, which are implicated in the adaptive immune response, can infiltrate more easily when microglia are activated [124]. Pro-inflammatory indicators are released by Th1 cells, anti-inflammatory markers are released by Th2 cells, and particular chemicals are released by Th17 cells [125]. MHC-I and MHC-II, which interact with $CD8^+$ and $CD4^+$ T lymphocytes, are expressed by activated microglia [126]. Demyelination can result from $CD4^+$ T lymphocytes producing antibodies against oligodendrocytes and myelin sheath components. Contrarily, $CD4^+$ T cells can also differentiate into regulatory T cells (iTreg), which dampen the immune system by obstructing other immune cells and secreting neurotrophic substances that promote the health of nerve cells [127, 128]. However, if iTreg activation is compromised, it may aid in the development of autoimmune disorders. These results imply that the immune response in MS involves intricate interactions between diverse immune cells, such as microglia and different kinds of T cells. To completely comprehend these pathways and create efficient MS therapies, more study is required [129].

### 2.4.3 Role of Activated Inflammasomes in MS and ALS

Degeneration of neurons can occur as a result of the NLRP3 inflammasome's dysfunction. Increased levels of TLR-4 and NF-jB in the brain of $SOD1^{G93A}$ mice with ALS imply that activation of the NLRP3 inflammasome results in pyroptosis [130]. The identical mouse model's spinal cord has been shown to have elevated levels of the NLRP3 inflammasome. Caspase-1 is further activated in ALS patients by the NLRP3 inflammasome activation and increased IL-18 levels. The presence of NLRP3 inflammasome in ALS is associated with protein aggregates, damaged mitochondria, and the disease [131].

Elevated levels of IL-1b and caspase-1 in immune cells known as peripheral blood mononuclear cells (PBMC) in a mouse model of multiple sclerosis (MS) point to the involvement of NLRP3 inflammasome [132]. The NLRP3 inflammasome enhances the production of chemotactic proteins and encourages the migration of some immune cells to the central nervous system (CNS). In MS, IL-1b-induced NLRP3 activation contributes to demyelination, whereas NLCR4-stimulated activity protects neurons [133]. Pro-IL-1b and pro-IL-18 are produced as a result of the activation of immune cells, and this process is mediated by the NF-jB pathway. Additionally, IL-1b and IL-18 levels are higher in MS patients, especially during relapses. Specific immune cells ($CD8^+$ mucosal-associated invariant T cells) are stimulated by IL-18 [134], which aids in their invasion of the central nervous system (CNS). Through mechanisms including autophagy, mitophagy, and type 1 interferon response, specific uncommon genetic variations in the NLRP1/3 and CASP1 genes can result in the deactivation of the NLRP3 inflammasome [135]. These findings highlight the involvement of NLRP3 inflammasome in ALS and MS, leading to harmful effects on neurons in both conditions.

### 2.4.4 Advanced Treatment Approach and Immunotherapies for ALS and MS to Modulate the Immune System

The current update on immunotherapeutics for modulating the immune system in ALS and MS includes:

**ALS**

- Riluzole: The only FDA-approved treatment for ALS that can extend life by roughly 2–3 months is this medication. It functions by preventing glutamate from being released at the synaptic cleft [136].
- Edaravone: Its approval by the FDA in May 2017 enables functional recovery by preventing the nitration of a particular amino acid residue in cerebrospinal fluid (CSF) [137].
- COX-2 inhibitors: In animal models and clinical studies, medications including thalidomide, lenalidomide, and cyclosporin have demonstrated promise in reducing inflammatory markers like TNF-a and FasL [138]. Another COX-2 inhibitor, cyclosporin, is now in phase IV of a clinical study (NCT01795872) after being successful in the transgenic mice model [139, 140].
- Glatiramer acetate: An ALS-moderating clinical study employing this medication at a dosage of 40 mg/day was unsuccessful [141].
- Anakinra: An IL-1 antagonist that prevents IL-6 and other pro-inflammatory markers from activating in ALS patients [142].
- PMX205: A C5aR antagonist, which has been shown to lengthen survival and reduce the course of illness [107].
- Cyclo (His-pro): Reduces the production of reactive oxygen species (ROS) and blocks the NLRP3 inflammasome in ALS [143].
- Vaccines: The $hSOD1^{G37R}$ mouse was used to test two vaccinations, tgGDSE2lim and tgG-DSE5b, which led to a late onset and prolonged life span. The lifespan of experimental vaccinations has increased and their late-onset symptoms have been seen in mouse models [144].

**MS**

- NEDA approach: An early MS treatment strategy known as no evidence of disease activity (NEDA) begins with injectable disease-modifying medications and progresses to oral dosages and infusions [145].
- Injectable therapies: Injectable disease-modifying treatments include natalizumab (NTZ), glatiramer acetate (GA), and interferon-bs (IFN-βs) [146].
- Immunomodulating drugs: By concentrating on certain immune cells, mitoxantrone, rituximab, and ocrelizumab are utilised to modify the immune system in MS [147].
- Sphingosine 1 phosphate (S1P) modulator: Fingolimod is an oral medication that alters immune cells by obstructing their blood flow [148].
- Other oral drugs: For relapsing-remitting MS (RRMS), teriflunomide and dimethyl fumarate are authorised, and laquinimod, ponesimod, siponimod, and ceralifimod are under clinical studies [149].
- Combination therapy: Two antibodies, trastuzumab and pertuzumab, can be combined to control complement [150].
- CAR/FoxP3 Treg cells: The experimental autoimmune encephalomyelitis (EAE) model of MS has shown promise in being controlled by intranasal delivery of genetically modified CAR/FoxP3 Treg cells [151].

It is important to note that these therapies are continually evolving, and further research and clinical trials are being conducted to improve the treatment options for ALS and MS patients.

## 2.5 Role of Immune Modulators in Huntington's Disease (HD)

An increase of a three-base-pair (CAG) repeat in exon 1 of the HTT gene results in Huntington's disease (HD), an inherited, dominant, and autosomal neurological condition [152]. The mutant huntingtin protein (mHtt), which has a polyglutamine tract at its N-terminus, is produced as a result of this expansion. In HD, it has been discovered that the presence of mHtt causes the brain immune cells known as microglia to activate the complement system's classical route. mHtt [153] is produced as a result of the gene on chromosome 4p 16.3 being altered.

The striatum, neurons, astrocytes, and myelin sheath in the HD brain demonstrate high deposition of complement components, including C1q, C1r, C3, C4, C9-neoepitope, and iC3b-neoepitope, according to studies utilising in situ hybridisation [154, 155]. It's interesting to note that complement component staining is negative in the early stages of the illness, indicating that early neuronal damage occurs before local complement component production and activation [156]. The caudate nucleus and motor cortex have elevated expression levels of complement subunits C3, C4A, and C4B, according to an analysis of HD brain tissue utilising microarray technology [157].

Animal models have been used to examine the involvement of the complement system in HD, with the R6/2 transgenic mouse model being particularly helpful despite its drawbacks. This model has shed important light on the role of the complement system in the pathogenesis of HD [158].

### 2.5.1 Activation of Microglia

The post-mortem human HD brains have been revealed to have distinct microgliosis [159]. According to research, the striatum's medium spiny GABA neurons (MSNs) are particularly susceptible to the neurotoxicity caused by mHtt. Instead of the soma and other cells, mHtt mostly builds up in nerve terminals and neuronal dendrites, activating the microglia [160]. By increasing inflammatory cytokines, activated and expanded microglia contribute to neurodegeneration [161]. In HD, there are a number of mechanisms that contribute to microglial activation, including NF-κB pathway, Kynurenine pathway, and Cannabinoid pathway.

### 2.5.2 Activation of Inflammasome

The caudate and putamen areas of the brain develop mHtt (mutant huntingtin protein) clumps, which cause the neostriatum to contract. These clusters injure the brain tissue by serving as warning signs [162]. Recent research has shown that even in the absence of any external cues, overexpressed mHtt causes the activation of genes related to inflammation in mouse microglia, a kind of immune cell in the brain [163]. Caspase-1 and Caspase-3 enzymes are active in the R6/2 animal model, exacerbating the degeneration of brain cells [164]. Another study found that by inhibiting the NLRP3 inflammasome signalling pathway, blocking the ATP-sensitive P2 purinergic receptor (P2X7) might lessen the symptoms of Huntington's disease (HD) [165]. Myeloid cells can produce pro-inflammatory cytokines as a result of enhanced expression and activity of certain transcription factors (CCAAT/enhancer-binding protein alpha/beta and Pu.1) [166, 167]. Additionally, the clumping of mHtt causes the complex involved in inflammation, the inflammasome, to activate [168, 169]. Pu.1 and C/EBP activity is stimulated in microglia, although the precise method by which this occurs is still not well known.

### 2.5.3 Advanced Treatment Approach and Immunotherapeutics in HD

Huntington's disease (HD) immunotherapy has promise as a possible treatment. The New York-based biotechnology business Vaccinex is now doing research on an antibody known as VX-15. Due to the fact that the clinical study is being double-blinded, neither the researchers nor the volunteers are aware of who will really be receiving therapy. VX-15 functions by obstructing Semaphorin 4D, a protein that contributes to brain inflammation and degeneration [167, 168].

Antisense oligonucleotide (ASO) therapy is a different treatment modality under investigation and is now in phase I of clinical studies. Short genetic material fragments are used in ASO treatment to target particular disease-related genes. The sole way to provide this treatment, known as lumbar puncture, which entails putting a needle into the lower back to reach the spinal canal, is a drawback [169].

In the final analysis, immunotherapy has potential for treating Huntington's disease, including the use of antibodies like VX-15 and ASO therapy. Clinical trials and ongoing research are examining their efficacy and safety.

## 2.6 Immune System Participation in the Exacerbation of Ischaemic Stroke

Research into and comprehension of the immune system's role in aggravating ischemic stroke are essential. A blood clot or blockage that prevents blood flow to the brain can cause an ischemic stroke, which can have serious neurological effects and cause tissue damage [170]. The immune system is essential for protecting the body from infections and damage, but it can potentially exacerbate ischemic stroke. This section focuses on the inflammatory responseand its impact on stroke progression, exploring the mechanisms through which the immune system intensifies tissue injury during ischemic stroke. New treatment approaches that attempt to control immune responses and reduce the negative consequences of ischemic stroke may be made possible by understanding the immune system's function in this situation.

### 2.6.1 Microglial Responses and the Complement System in Ischemic Strokes

The pathophysiology of ischemic stroke is significantly influenced by the complement system. Following an ischaemia stroke, levels of C3a, C4d, and soluble C5b-9 rise right away, but C5a levels rise later. With accumulation starting soon after a stroke, C1q plays a complicated role as a traditional route starter [171]. However, C1q-deficient animals do not exhibit a protective function, indicating that the presence of C1q in adult mice and newborn mice differs. The alternative pathway has also been implicated, but the lectin pathway, started by MBL, exacerbates stroke damage. Different phenotypes are seen in various areas of the peri-infarct zone, where microglial activation follows diverse patterns. Depending on the stage, microglia have a role in both aggravation and protection in ischemic stroke [172].

### 2.6.2 Activation of Inflammasome Signalling

Inflammasome signalling is involved in atherosclerosis, a major stroke risk factor. The activation of NLRP3 and NLRP1 inflammasomes by foam cells and inflammatory cytokines leads to inflammation and plaque instability. Inflammasome signalling is further activated following ischemic stroke, with DAMPs and PRRs triggering NLRP1 and NLRP3 inflammasomes [173]. This leads to the production of inflammatory cytokines, neuronal cell death, and exacerbation of stroke pathology [174].

### 2.6.3 Immunotherapy for Ischaemic Stroke

Recent research has demonstrated that compounds like b-hydroxybutyrate and MCC950 can prevent the NLRP3 inflammasome from causing macrophages to secrete the pro-inflammatory cytokine IL-1b. Inhibition like this lowers inflammation [175]. Forkhead box transcription factor (FOXP3) is a marker expressed by

Treg cells, which are regarded as neuroinflammation suppressors. It has been discovered that these Treg cells reduce inflammation. It has been discovered that trichostatin A activates FOXP3 and prevents Th1 cells from proliferating, which lowers IFN-c production and increases the activity of Treg cells [176, 177].

## 2.7 Neuroimmune Signalling and Neurodegeneration in HIV: Role of Neurotrophic Factors

HIV-associated neurocognitive disorders (HAND) are neurodegenerative and cognitive illnesses that can result from HIV infection of the neurodegenerative process of HAND. Neurotrophic factors, which are crucial for the survival and functionality of neurons, are important [178].

Neurotrophic factors, such as brain-derived neurotrophic factor (BDNF), nerve growth factor (NGF), and glial cell line-derived neurotrophic factor (GDNF), are dysregulated during HIV infection. Reduced BDNF levels have been linked in studies to cognitive impairment and neurodegeneration in HIV-infected people [179]. Numerous causes, including immunological activation, oxidative stress, inflammation, and HIV viral proteins (including Tat and gp120), can be blamed for the dysregulation of neurotrophic factors. These elements harm and kill neurons by interfering with the signalling pathways that control neurotrophic factor creation, release, and uptake [180, 181].

### 2.7.1 Advanced Treatment Strategies

As a viable therapeutic strategy for treating HIV-related neurodegeneration, targeting neurotrophic factors and associated signalling pathways has come to light. In order to increase the protective benefits of neurotrophic factor, many methods are being investigated [182].

*Pharmacological Interventions*: Histone deacetylase inhibitors and other drugs that boost the production and release of neurotrophic factors are being researched for their potential to increase levels of neurotrophic factors and prevent neurodegeneration.

*Gene Therapy*: In order to improve neuronal survival and function, neurotrophic factors are being delivered and expressed directly in the brain via viral vector-based gene therapy techniques.

*Immunomodulatory Approaches*: To lessen neurotoxicity and maintain neurotrophic factor homeostasis, strategies that target immune activation and inflammation, such as anti-inflammatory medicines and immunosuppressive medications, are being investigated.

*Lifestyle Interventions*: Exercise, mental stimulation, and a healthy lifestyle have all showed promise in boosting levels of neurotrophic factor and neuroprotection in HIV-associated neurodegeneration.

The neurodegenerative process seen in HIV-associated neurocognitive diseases is greatly aided by dysregulation of neurotrophic factors. As prospective therapeutic options for reducing neurodegeneration in HIV, advanced therapy techniques attempt to restore neurotrophic factor balance, minimise neurotoxicity, and boost neuroprotection.

## 2.8 Harnessing Neuroimmune Signalling Pathways for Neuroprotection: Promising Role of Stem Cell Therapy

Stem cell therapy has emerged as a viable strategy to harness neuroimmune signalling pathways for the treatment of diverse neurological illnesses. These pathways are essential for neuroprotection. Mesenchymal stem cells (MSCs) are one type of stem cell treatment used for neuroprotection.

MSCs have demonstrated promise in fostering neuroprotection via a variety of ways. In the first place, MSCs contain immunomodulatory qualities that can control the immune response and lessen inflammation in the central nervous system (CNS). They can modulate the activation and function of various immune cells, such as microglia and T cells, leading to a decrease in pro-inflammatory cytokines and an increase in anti-inflammatory factors. Additionally, MSCs have the ability to release a variety of neurotrophic substances, such as glial cell line-derived neurotrophic factor (GDNF), nerve growth factor (NGF), and brain-derived neurotrophic factor (BDNF). These elements facilitate neural regeneration and repair by encouraging neuronal survival, enhancing neuronal growth and plasticity, and enhancing the development of new blood vessels. Furthermore, through paracrine signalling, MSCs can communicate with local immune cells in the CNS such as microglia, astrocytes, and oligodendrocytes. By modifying microglia activity and polarisation towards an anti-inflammatory phenotype, this interaction can lessen neuroinflammation and foster an environment that is conducive to neuronal survival [183]. MSCs can also develop into neuronal and glial cell lineages, replacing CNS cells that have been lost or injured. Although there is still much to learn about how MSCs differentiate into functioning neurons, it has been shown that they can boost neuronal survival and endogenous neurogenesis. For instance, a research on animal stroke models showed that intravenous treatment of MSCs induced the release of many growth factors and cytokines, which improved neuroprotection and functional recovery. Improved neurological outcomes were the result of the transplanted MSCs' promotion of angiogenesis, suppression of inflammation, and immune response regulation.

Overall, the use of MSCs in stem cell treatment offers considerable promise for neuroprotection via modifying neuroimmune signalling pathways. Stem cells can aid in the maintenance and regeneration of neurons in a variety of neurological illnesses by controlling inflammation, producing neurotrophic factors, encouraging cell replacement, and providing a favourable microenvironment.

## 3 Difficulties of Manipulating the Immune System to Treat Neurodegenerative Disorders

The use of immune system manipulation as a novel therapeutic approach for neurodegenerative disorders, specifically in the context of neuroimmune signalling and immunotherapy, faces several challenges. These challenges include:

*Limited understanding of disease mechanisms*: Our knowledge of the underlying processes of neurodegenerative disorders still has a lot of holes. It is challenging to design efficient immunotherapies that can target certain pathways or elements involved in neuroimmune signalling without a thorough grasp of the disease processes [184].

*Translation from in vivo models to human patients*: Genetic models that resemble family types of neurodegenerative illnesses are often used in preclinical research. These models might not, however, accurately capture the diversity and complexity of human illnesses. Promising findings from animal models must be transferred to human patients, which is a difficult undertaking that calls for careful consideration of species differences and disease-specific characteristics [184].

*Neglect of age-related factors and lifestyle influences*: The effects of age-related variables are rarely well investigated in preclinical models, despite the fact that neurodegenerative illnesses are frequently linked to ageing. Additionally, lifestyle choices like food and exercise can have a big impact on how well the immune system functions and how quickly neurodegenerative diseases proceed. To create treatments that are applicable to the general population, preclinical research must take into account age-related variables and lifestyle impacts [185].

*Influence of the gut microbiota*: It has become clear that the gut microbiota plays a crucial role in affecting both immune responses and brain function. However, more research is still needed to completely understand how the immune system and the gut microbiota interact in neurodegenerative disorders. More investigation is required to determine how the gut microbiota affects neuroimmune signalling and how to target it for therapeutic therapies [169].

*Heterogeneity among individuals and latent CNS infections*: Individuals with neurodegenerative disorders have great heterogeneity. Latent infections of the central nervous system (CNS), such as those linked to Alzheimer's disease (AD), may have an impact on this heterogeneity. For effective immunomodulatory medicines to be developed that may address the many forms of neurodegenerative disorders, it is essential to comprehend the function of latent CNS infections and their effects on immune responses [169].

*Modulating common immunological factors and microglial phenotype*: Certain immunological components are present in all neurodegenerative disorders, and the microglial phenotype is altered. It is challenging to successfully alter these variables and the microglial phenotype in order to obtain therapeutic advantages. It is difficult to create tailored immunomodulatory approaches that can modify these shared variables with selectivity while keeping the immune system in balance [186].

Multidisciplinary research projects, involving partnerships between immunologists, neuroscientists, geneticists, and physicians, are necessary to address these problems. The effective use of immunotherapy in neurodegenerative illnesses depends on further progress in our understanding of disease processes, consideration of age-related variables and lifestyle impacts, investigation of the gut-brain axis, and creation of personalised treatments.

## 4 Summary and Discussion

Understanding the processes of neuroimmune signalling has provided new therapeutic possibilities for the treatment of neurodegenerative illnesses, which play a significant part in this process. The article focuses on conditions including Alzheimer's disease (AD), Parkinson's disease (PD), multiple sclerosis (MS), amyotrophic lateral sclerosis (ALS), Huntington's disease, and ischemic stroke as examples of disorders where neuroimmune signalling is explored in detail and its consequences for neurodegeneration. Microglia, astrocytes, and immune cells are the cellular components of neuroimmune signalling. To control immune responses and preserve the health of the brain cells, these cells interact and communicate through cytokines, chemokines, and other signalling chemicals. Neurodegeneration may result from the dysregulation of this transmission. The central nervous system's resident immune cells, called microglia, play two roles in neurodegenerative illnesses. Chronic activation of microglia can cause the production of pro-inflammatory mediators that are harmful to neurons, even though they also produce neurotrophic factors and direct stem cells to aid in recovery. Complex protein complexes called inflammasomes, which cause inflammation, have been linked to neurodegenerative disorders. Key elements of inflammasomes include pro-caspase-1, apoptosis-associated speck-like protein (ASC), and NOD-like receptors (NLRs). Caspase-1 activation and the release of pro-inflammatory cytokines are both caused by the activation of the inflammasome complex, which also causes neuroinflammation and pyroptosis.

Diseases like AD and PD affect the immune system's functioning. Neurodegeneration is a result of immune cell abnormalities, blood-brain barrier problems, and poor clearance of harmful substances. The immune system is being modulated and disease-specific pathways are being targeted through improvements in therapeutic strategies and immunotherapies. In addition, the essay examines how neuroimmune signalling is compromised in conditions including MS, ALS, Huntington's disease, ischemic stroke, HIV-related neurodegeneration, and how neurotrophic factors and stem cell treatment may be able to use neuroimmune signalling for neuroprotection. In general, effective treatment approaches for neurodegenerative illnesses may be developed by comprehending and targeting neuroimmune signalling networks. A crucial area of study in the realm of neurodegenerative disorders is neuroimmune signalling. Immune responses in the central nervous system (CNS) are tightly controlled by the complex interactions between the immune system and the neurological system. Chronic inflammation, damage to the neurons, and eventually neurodegeneration can result from the dysregulation of this communication. The primary immune cells of the CNS, the microglia, have a role in both neuroprotection and neurotoxicity. Their states of activation, such as the pro-inflammatory M1 phenotype and the anti-inflammatory M2 phenotype, have significant effects on how a disease develops. Potential countermeasures for neurodegenerative processes include focusing on microglial activation and encouraging an anti-inflammatory milieu.

Numerous neurodegenerative illnesses have been linked to inflammasomes, especially the NLRP3 inflammasome. Caspase-1 activation and the production of pro-inflammatory cytokines are both caused by the activation of the inflammasome complex, which also causes neuroinflammation and cell death. Developing therapeutic approaches targeted at controlling inflammation and avoiding neuronal damage is made possible by understanding the inflammasome-mediated pathways. Neurodegenerative illnesses like AD and PD affect the immune system's mechanism. Chronic neuroinflammation and the buildup of hazardous protein aggregates are both caused by immune cell dysfunction, decreased blood-brain barrier integrity, and poor clearance pathways. Targeting disease-specific pathways and enhancing immune responses against pathogenic components are two goals of immunotherapeutic strategies being investigated, including the use of antibodies and vaccines. Additionally, knowledge of neuroimmune signalling in conditions including MS, ALS, Huntington's disease, ischemic stroke, and HIV-related neurodegeneration sheds light on possible treatment targets. For neuroprotection and boosting functional recovery in these diseases, stem cell treatment, harnessing neurotrophic factors, and modulating immunological responses show promise. The investigation of dysregulated neuroimmune signalling in neurodegenerative illnesses brings up new opportunities for creating sophisticated therapeutic strategies and immunotherapies. It could be able to delay or stop the course of these deadly illnesses by focusing on particular neuroimmune system components and regulating immune responses. These encouraging results need to be translated into patient-friendly therapy approaches through more investigation and clinical trials.

## 5 Conclusion

Neuroimmune signalling plays a critical role in maintaining the normal functioning of the central nervous system. Dysregulation of this signalling pathway is now recognised as a critical factor in the development and progression of neurodegenerative diseases. Neurodegenerative disorders such as Alzheimer's disease, Parkinson's disease, and multiple sclerosis are major public health concerns with limited treatment options available. Recent research has highlighted the importance of the interplay between the immune system and the nervous system in the development and progression of these disorders. This has led to the identification of potential therapeutic targets in neuroimmune signalling pathways. This chapter aimed to review the current understanding of neuroimmune signalling pathways and their role in neurodegenerative disorders. We discussed emerging options to modulate these pathways for therapeutic benefit, including immunomodulatory drugs, neurotrophic factors, and stem cell therapy. The potential of targeting neuroimmune signalling pathways is promising, with preclinical and clinical studies showing positive results in improving cognitive function, reducing neuroinflammation, and promoting neuroregeneration. However, further research is needed to fully understand the complexity of these pathways and their specific roles in different neurodegenerative disorders. Overall, targeting neuroimmune signalling pathways represents a promising avenue for the development of novel therapies to treat neurodegenerative disorders. The ongoing research in this field provides hope for improving the lives of individuals affected by these devastating conditions.

## 6 Future Prospects

Understanding the processes of neuroimmune signalling has provided new therapeutic possibilities for the treatment of neurodegenerative illnesses. This field continues to evolve, offering exciting prospects for future research and clinical applications. Some potential future prospects include:

*Advancements in Immunomodulatory Drugs*: Further development and refinement of immunomodulatory drugs that target specific components of neuroimmune signalling pathways could lead to more effective and targeted treatments for neurodegenerative diseases.

*Precision Medicine Approaches*: The development of personalised medicine approaches, taking into account individual genetic, environmental, and immunological factors, could enhance treatment efficacy and improve patient outcomes.

*Combination Therapies*: Investigating the synergistic effects of combining different therapeutic approaches, such as immunomodulatory drugs, neurotrophic factors, and stem cell therapy, may result in more comprehensive and potent treatments for neurodegeneration.

*Biomarker Discovery*: Continued research into identifying reliable biomarkers for neuroinflammation and neurodegeneration could aid in early diagnosis, monitoring disease progression, and assessing treatment response.

*Translational Research*: Bridging the gap between preclinical research and clinical applications is essential for translating promising findings into effective therapies. Translational studies and clinical trials will be crucial in evaluating the safety and efficacy of emerging treatments.

Overall, the future of neuroimmune signalling research holds great potential for developing innovative and effective therapies to combat neurodegenerative diseases. With ongoing advancements in understanding the complex interplay between the immune system and the nervous system, we can expect to witness significant progress in the development of novel treatment options in the coming years.

## References

1. Loftis, J. M., & Janowsky, A. (2014). Neuroimmune basis of methamphetamine toxicity. Neuroimmune signaling in drug actions and addictions. *International Review of Neurobiology, 118*, 165–197. https://doi.org/10.1016/B978-0-12-801284-0.00007-5. ISBN 9780128012840. PMC 4418472. PMID 25175865.
2. Beardsley, P. M., & Hauser, K. F. (2014). Glial modulators as potential treatments of psychostimulant abuse. Emerging targets & therapeutics in the treatment of psychostimulant abuse. *Advances in Pharmacology, 69*, 1–69. https://doi.org/10.1016/B978-0-12-420118-7.00001-9. ISBN 9780124201187. PMC 4103010. PMID 24484974.
3. Rogers, T. J. (2012). The molecular basis for neuroimmune receptor signaling. *Journal of Neuroimmune Pharmacology, 7*(4), 722–724. https://doi.org/10.1007/s11481-012-9398-4. PMC 4011130. PMID 22935971.
4. Gimsa, U., Mitchison, N. A., & Brunner-Weinzierl, M. C. (2013). Immune privilege as an intrinsic CNS property: Astrocytes protect the CNS against T-cell-mediated neuroinflammation. *Mediators of Inflammation, 2013*, 1–11. https://doi.org/10.1155/2013/320519. PMC 3760105. PMID 24023412.
5. Polyzoidis, S., Koletsa, T., Panagiotidou, S., Ashkan, K., & Theoharides, T. C. (2015). Mast cells in meningiomas and brain inflammation. *Journal of Neuroinflammation, 12*(1), 170. https://doi.org/10.1186/s12974-015-0388-3. PMC 4573939. PMID 26377554.
6. Stephan, A. H., Barres, B. A., & Stevens, B. (2012). The complement system: An unexpected role in synaptic pruning during development and disease. *Annual Review of Neuroscience, 35*(1), 369–389. https://doi.org/10.1146/annurev-neuro-061010-113810. PMID 22715882. S2CID 2309037.
7. Sofroniew, M. V., & Vinters, H. V. (2010). Astrocytes: Biology and pathology. *Acta Neuropathologica, 119*(1), 7–35.
8. Ransohoff, R. M., & Perry, V. H. (2009). Microglial physiology: Unique stimuli, specialized responses. *Annual Review of Immunology, 27*, 119–145.
9. Salter, M. W., & Stevens, B. (2017). Microglia emerge as central players in brain disease. *Nature Medicine, 23*(9), 1018–1027.
10. Theoharides, T. C., & Kavalioti, M. (2019). Mast cells and neuroinflammation. *Seminars in Cell & Developmental Biology, 89*, 64–75.
11. Brady, S. T., & Siegel, G. J. (2012). *Basic neurochemistry: Principles of molecular, cellular and medical neurobiology*. Academic Press. ISBN 9780123749475.
12. Talbot, S., Foster, S., & Woolf, C. (2016). Neuroimmune physiology and pathology. *Annual Review of Neuroscience, 34*, 421–447. https://doi.org/10.1146/annurev-immunol-041015-055340. PMID 26907213.
13. Ziv, Y., Ron, N., Butovsky, O., Landa, G., Sudai, E., Greenberg, N., Cohen, H., Kipnis, J., & Schwartz, M. (2006). Immune cells contribute to the maintenance of neurogenesis and spatial learning abilities in adulthood. *Nature Neuroscience, 9*(2), 268–275. https://doi.org/10.1038/nn1629. PMID 16415867. S2CID 205430936.
14. Kipnis, J., Cohen, H., Cardon, M., Ziv, Y., & Schwartz, M. (2004). T cell deficiency leads to cognitive dysfunction: Implications for therapeutic vaccination for schizophrenia and other psychiatric conditions. *Proceedings of the National Academy of Sciences, 101*(21), 8180–8185.
15. McMahon, S., La Russa, F., & Bennett, D. (2015). Crosstalk between the nociceptive and immune systems in host defence and disease. *Nature Reviews Neuroscience, 16*(7), 389–402. https://doi.org/10.1038/nrn3946. PMID 26087680. S2CID 22294761.
16. Heneka, M. T., Carson, M. J., El Khoury, J., Landreth, G. E., Brosseron, F., Feinstein, D. L., et al. (2015). Neuroinflammation in Alzheimer's disease. *The Lancet Neurology, 14*(4), 388–405.
17. Ransohoff, R. M. (2016). How neuroinflammation contributes to neurodegeneration. *Science, 353*(6301), 777–783.

18. Glass, C. K., Saijo, K., Winner, B., Marchetto, M. C., & Gage, F. H. (2010). Mechanisms underlying inflammation in neurodegeneration. *Cell, 140*(6), 918–934.
19. Brites, D., & Vaz, A. R. (2014). Microglia-centered pathogenesis in Alzheimer's disease: Implications for therapy. *Current Alzheimer Research, 11*(6), 608–622.
20. Streit, W. J., & Xue, Q. S. (2013). The brain's aging immune system. *Aging and Disease, 4*(3), 169–183.
21. Colonna, M., & Butovsky, O. (2017). Microglia function in the central nervous system during health and neurodegeneration. *Annual Review of Immunology, 35*, 441–468.
22. Saijo, K., & Glass, C. K. (2011). Microglial cell origin and phenotypes in health and disease. *Nature Reviews Immunology, 11*(11), 775–787.
23. Taylor, S. E., Morganti-Kossmann, M. C., & Lifshitz, J. (2016). Zonulin serves as a marker and potential target in central nervous system injury. *Molecular Diagnosis & Therapy, 20*(5), 437–448.
24. Kaur, C., & Ling, E. A. (2016). The diverse roles of microglia in the neurodegenerative context. *Neuroscience Research, 100*, 1–2.
25. Block, M. L., Zecca, L., & Hong, J. S. (2007). Microglia-mediated neurotoxicity: Uncovering the molecular mechanisms. *Nature Reviews Neuroscience, 8*(1), 57–69.
26. Matuskova, H., Vesela, J., Porizkova, P., Pavlasova, G., Kepak, T., & Trka, J. (2017). Inflammation in CNS disorders: Mechanisms and therapeutic targets. *Mediators of Inflammation, 2017*, 1–11.
27. Kawasaki, T., & Kawai, T. (2014). Toll-like receptor signaling pathways. *Frontiers in Immunology, 5*, 1–8.
28. Patel, A. R., Ritzel, R., McCullough, L. D., & Liu, F. (2013). Microglia and ischemic stroke: A double-edged sword. *International Journal of Physiology, Pathophysiology and Pharmacology, 5*(2), 73–90.
29. Boisvert, M. M., Erikson, G. A., Shokhirev, M. N., & Allen, N. J. (2018). The aging astrocyte transcriptome from multiple regions of the mouse brain. *Cell Reports, 22*(1), 269–285.
30. Swanson, K. V., Deng, M., & Ting, J. P. (2019). The NLRP3 inflammasome: Molecular activation and regulation to therapeutics. *Nature Reviews Immunology, 19*(8), 477–489.
31. Ruscetti, T., & Deverman, B. E. (2020). Enhancing gene delivery to the nervous system by viral vectors. *Frontiers in Cellular Neuroscience, 14*, 606755.
32. Heneka, M. T., Kummer, M. P., & Latz, E. (2014). Innate immune activation in neurodegenerative disease. *Nature Reviews Immunology, 14*(7), 463–477.
33. Kanneganti, T. D., Lamkanfi, M., & Núñez, G. (2007). Intracellular NOD-like receptors in host defense and disease. *Immunity, 27*(4), 549–559.
34. Davis, B. K., Wen, H., & Ting, J. P. (2011). The inflammasome NLRs in immunity, inflammation, and associated diseases. *Annual Review of Immunology, 29*, 707–735.
35. Takeuchi, O., & Akira, S. (2010). Pattern recognition receptors and inflammation. *Cell, 140*(6), 805–820.
36. Liang, Z., Zhao, Y., Ruan, L., Zhu, L., Jin, K., Zhuge, Q., et al. (2017). Impact of aging immune system on neurodegeneration and potential immunotherapies. *Progress in Neurobiology, 157*, 2–8.
37. McCoy, M. K., & Tansey, M. G. (2008). TNF signaling inhibition in the CNS: Implications for normal brain function and neurodegenerative disease. *Journal of Neuroinflammation, 5*(1), 45.
38. Heppner, F. L., Ransohoff, R. M., & Becher, B. (2015). Immune attack: The role of inflammation in Alzheimer disease. *Nature Reviews Neuroscience, 16*(6), 358–372.
39. Yeh, F. L., Hansen, D. V., & Sheng, M. (2017). TREM2, microglia, and neurodegenerative diseases. *Trends in Molecular Medicine, 23*(6), 512–533.
40. Zheng, H., Cheng, B., Li, Y., Li, X., Chen, X., & Zhang, Y. W. (2018). TREM2 in Alzheimer's disease: Microglial survival and energy metabolism. *Frontiers in Aging Neuroscience, 10*, 395.

41. Jonsson, T., Stefansson, H., Steinberg, S., Jonsdottir, I., Jonsson, P. V., Snaedal, J., et al. (2013). Variant of TREM2 associated with the risk of Alzheimer's disease. *The New England Journal of Medicine, 368*(2), 107–116.
42. Jin, S. C., Benitez, B. A., Karch, C. M., Cooper, B., Skorupa, T., Carrell, D., et al. (2014). Coding variants in TREM2 increase risk for Alzheimer's disease. *Human Molecular Genetics, 23*(21), 5838–5846.
43. Song, W., Hooli, B., Mullin, K., Jin, S. C., Cella, M., Ulland, T. K., et al. (2017). Alzheimer's disease-associated TREM2 variants exhibit either decreased or increased ligand-dependent activation. *Alzheimers Dementia, 13*(4), 381–387.
44. Hirsch, E. C., & Hunot, S. (2009). Neuroinflammation in Parkinson's disease: A target for neuroprotection? *Lancet Neurology, 8*(4), 382–397.
45. Ferreira, S. A., & Romero-Ramos, M. (2018). Microglia response during Parkinson's disease: Alphasynuclein intervention. *Frontiers in Cellular Neuroscience, 12*, 2.
46. Zhang, W., Wang, T., Pei, Z., Miller, D. S., Wu, X., Block, M. L., et al. (2005). Aggregated a-synuclein activates microglia: A process leading to disease progression in Parkinson's disease. *The FASEB Journal, 19*(6), 533–542.
47. Daher, J. P., Volpicelli-Daley, L. A., Blackburn, J. P., Moehle, M. S., & West, A. B. (2014). Abrogation of asynuclein–mediated dopaminergic neurodegeneration in LRRK2-deficient rats. *Proceedings of the National Academy of Sciences of the United States of America, 111*(25), 9289–9294.
48. Smith, J. A., Das, A., Ray, S. K., & Banik, N. L. (2012). Role of pro-inflammatory cytokines released from microglia in neurodegenerative diseases. *Brain Research Bulletin, 87*(1), 10–20.
49. Mogi, M., Harada, M., Riederer, P., Narabayashi, H., Fujita, K., & Nagatsu, T. (1994). Tumor necrosis factor-a (TNF-a) increases both in the brain and in the cerebrospinal fluid from parkinsonian patients. *Neuroscience Letters, 165*(1–2), 208–210.
50. Mosley, R. L., Hutter-Saunders, J. A., Stone, D. K., & Gendelman, H. E. (2012). Inflammation and adaptive immunity in Parkinson's disease. *Cold Spring Harbor Perspectives in Medicine, 2*(1), a009381.
51. Liu, Y., Zhou, L., Yang, Y., et al. (2019). NLRP3 inflammasome activation mediates Aβ-induced neuroinflammation and contributes to cognitive impairment in APP/PS1 mice. *Acta Neuropathologica Communications, 7*(1), 113. https://doi.org/10.1186/s40478-019-0769-1
52. Heneka, M. T., McManus, R. M., & Latz, E. (2018). Inflammasome signalling in brain function and neurodegenerative disease. *Nature Reviews Neuroscience, 19*(10), 610–621.
53. Freeman, L., Guo, H., David, C. N., Brickey, W. J., Jha, S., & Ting, J. P. (2017). NLR members NLRC4 and NLRP3 mediate sterile inflammasome activation in microglia and astrocytes. *The Journal of Experimental Medicine, 214*(5), 1351–1370.
54. Saresella, M., La Rosa, F., Piancone, F., Zoppis, M., Marventano, I., Calabrese, E., et al. (2016). The NLRP3 and NLRP1 inflammasomes are activated in Alzheimer's disease. *Molecular Neurodegeneration, 11*(1), 23.
55. Sheedy, F. J., Grebe, A., Rayner, K. J., Kalantari, P., Ramkhelawon, B., Carpenter, S. B., et al. (2013). CD36 coordinates NLRP3 inflammasome activation by facilitating intracellular nucleation of soluble ligands into particulate ligands in sterile inflammation. *Nature Immunology, 14*(8), 812–820.
56. Cho, M. H., Cho, K., Kang, H. J., Jeon, E. Y., Kim, H. S., Kwon, H. J., et al. (2014). Autophagy in microglia degrades extracellular b-amyloid fibrils and regulates the NLRP3 inflammasome. *Autophagy, 10*(10), 1761–1775.
57. Liu, L., & Chan, C. (2014). IPAF inflammasome is involved in interleukin-1b production from astrocytes, induced by palmitate; implications for Alzheimer's disease. *Neurobiology of Aging, 35*(2), 309–321.
58. Lang, Y., Chu, F., Shen, D., Zhang, W., Zheng, C., Zhu, J., et al. (2018). Role of inflammasomes in neuroimmune and neurodegenerative diseases: A systematic review. *Mediators of Inflammation, 2018*, 1549549.

59. Panicker, N., Sarkar, S., Harischandra, D. S., Neal, M., Kam, T. I., Jin, H., et al. (2019). Fyn kinase regulates misfolded a-synuclein uptake and NLRP3 inflammasome activation in microglia. *The Journal of Experimental Medicine, 216*(6), 1411–1430.
60. Bai, H., Yang, B., Yu, W., Xiao, Y., Yu, D., & Zhang, Q. (2018). Cathepsin B links oxidative stress to the activation of NLRP3 inflammasome. *Experimental Cell Research, 362*(1), 180–187.
61. Wang, S., Yuan, Y. H., Chen, N. H., & Wang, H. B. (2019). The mechanisms of NLRP3 inflammasome/pyroptosis activation and their role in Parkinson's disease. *International Immunopharmacology, 67*, 458–464.
62. Lee, E., Hwang, I., Park, S., Hong, S., Hwang, B., Cho, Y., et al. (2019). MPTP-driven NLRP3 inflammasome activation in microglia plays a central role in dopaminergic neurodegeneration. *Cell Death and Differentiation, 26*(2), 213–228.
63. Gordon, R. A., Albornoz, E. C., Christie, D. R., Langley, M. A. B., Kumar, V. S., Mantovani, S., et al. (2018). Inflammasome inhibition prevents asynuclein pathology and dopaminergic neurodegeneration in mice. *Science Translational Medicine, 10*(465), eaah4066.
64. Zhang, P., Shao, X. Y., Qi, G. J., Chen, Q., Bu, L. L., Chen, L. J., et al. (2016). C dk5-D ependent A ctivation of neuronal inflammasomes in Parkinson's disease. *Movement Disorders, 31*(3), 366–376.
65. Alzheimer's Association. (2021). 2021 Alzheimer's disease facts and figures. *Alzheimer's & Dementia, 17*(3), 327–406.
66. Shen, Y., Lue, L. F., Yang, L. B., Roher, A., Kuo, Y. M., Strohmeyer, R., et al. (2001). Complement activation by neurofibrillary tangles in Alzheimer's disease. *Neurobiology of Aging, 22*(6), 955–962.
67. DeMattos, R. B., O'dell, M. A., Parsadanian, M., Taylor, J. W., Harmony, J. A., Bales, K. R., et al. (2002). Clusterin promotes amyloid plaque formation and is critical for neuritic toxicity in a mouse model of Alzheimer's disease. *Proceedings of the National Academy of Sciences of the United States of America, 99*(16), 10843–10848.
68. Nizami, S., Hall-Roberts, H., Warrier, S., Cowley, S. A., & Di Daniel, E. (2019). Microglial inflammation and phagocytosis in Alzheimer's disease: Potential therapeutic targets. *British Journal of Pharmacology, 176*(18), 3515–3532.
69. Fonseca, M. I., McGuire, S. O., Counts, S. E., & Tenner, A. J. (2013). Complement activation fragment C5a receptors, CD88 and C5L2, are associated with neurofibrillary pathology. *Journal of Neuroinflammation, 10*(1), 803.
70. Shi, Q., Chowdhury, S., Ma, R., Le, K. X., Hong, S., Caldarone, B. J., et al. (2017). Complement C3 deficiency protects against neurodegeneration in aged plaque-rich APP/PS1 mice. *Science Translational Medicine, 9*(392), eaaf6295.
71. Yamada, T., McGeer, P. L., & McGeer, E. G. (1992). Lewy bodies in Parkinson's disease are recognized by antibodies to complement proteins. *Acta Neuropathologica, 84*(1), 100–104. 64 Cho, K. (2019). Emerging roles of complement protein C1q in neurodegeneration. *Aging Dis, 10*(3), 652.
72. Depboylu, C., Sch€afer, M. K., Arias-Carrion, O., Oertel, W. H., Weihe, E., & Hoglinger, G. U. (2011). Pos- € sible involvement of complement factor C1q in the clearance of extracellular neuromelanin from the substantia nigra in Parkinson disease. *Journal of Neuropathology and Experimental Neurology, 70*(2), 125–132.
73. Hou, L., Wang, K., Zhang, C., Sun, F., Che, Y., Zhao, X., et al. (2018). Complement receptor 3 mediates NADPH oxidase activation and dopaminergic neurodegeneration through a Src-Erk-dependent pathway. *Redox Biology, 14*, 250–260.
74. Wang, Y., Hancock, A. M., Bradner, J., Chung, K. A., Quinn, J. F., Peskind, E. R., et al. (2011). Complement 3 and factor h in human cerebrospinal fluid in Parkinson's disease, Alzheimer's disease, and multiple-system atrophy. *The American Journal of Pathology, 178*(4), 1509–1516.
75. Kiyota, T., Ingraham, K. L., & Swan, R. J. (2019). Alzheimer's disease therapies: Immunotherapies. *Pharmacology & Therapeutics, 208*, 107477.

76. Vezzani, A., & Viviani, B. (2015). Neuromodulatory properties of inflammatory cytokines and their impact on neuronal excitability. *Neuropharmacology, 96*, 70–82.
77. Tcw, J., & Goate, A. M. (2017). Genetics of β-amyloid precursor protein in Alzheimer's disease. *Cold Spring Harbor Perspectives in Medicine, 7*(2), a024539.
78. Lindström, V., & Gustafsson, G. (2021). Immunotherapy in neurodegenerative diseases: Clinical trials with antiamyloidβ monoclonal antibodies. *Journal of Internal Medicine, 290*(5), 991–1008.
79. Lemere, C. A. (2013). Immunotherapy for Alzheimer's disease: Hoops and hurdles. *Molecular Neurodegeneration, 8*(1), 36.
80. Bard, F., Cannon, C., Barbour, R., Burke, R. L., Games, D., Grajeda, H., et al. (2000). Peripherally administered antibodies against amyloid b-peptide enter the central nervous system and reduce pathology in a mouse model of Alzheimer disease. *Nature Medicine, 6*(8), 916–919.
81. DeMattos, R. B., Bales, K. R., Cummins, D. J., Dodart, J. C., Paul, S. M., & Holtzman, D. M. (2001). Peripheral anti-Ab antibody alters CNS and plasma Ab clearance and decreases brain Ab burden in a mouse model of Alzheimer's disease. *Proceedings of the National Academy of Sciences of the United States of America, 98*(15), 8850–8855.
82. Wisniewski, T., & Goni, F. (2014). Immunotherapy for Alzheimer's disease. *Biochemical Pharmacology, 88*(4), 499–507.
83. Hernangomez, M., Carrillo-Salinas, J. F., Mecha, M., Correa, F., Mestre, L., Lor ıa, F., Feliu, A., Docagne, F., & Guaza, C. (2014). Brain innate immunity in the regulation of neuroinflammation: Therapeutic strategies by modulating CD200-CD200R interaction involve the cannabinoid system. *Current Pharmaceutical Design, 20*(29), 4707.
84. Wisniewski, T., & Frangione, B. (2005). Immunological and anti-chaperone therapeutic approaches for Alzheimer disease. *Brain Pathology, 15*(1), 72–77.
85. Ryan, J. M., & Grundman, M. (2009). Anti-amyloid-b immunotherapy in Alzheimer's disease: ACC-001 clinical trials are ongoing. *Journal of Alzheimer's Disease, 17*(2), 243.
86. Wisniewski, T., & Goni, F. (2015). Immunotherapeutic approaches for Alzheimer's disease. *Neuron, 85*(6), 1162–1176.
87. Doody, R. S., Thomas, R. G., Farlow, M., Iwatsubo, T., Vellas, B., Joffe, S., et al. (2014). Phase 3 trials of solanezumab for mild-to-moderate Alzheimer's disease. *The New England Journal of Medicine, 370*(4), 311–321. 113.
88. Lehmann, J. M., Lenhard, J. M., Oliver, B. B., Ringold, G. M., & Kliewer, S. A. (1997). Peroxisome proliferator-activated receptors a and c are activated by indomethacin and other non-steroidal anti-inflammatory drugs. *The Journal of Biological Chemistry, 272*(6), 3406–3410.
89. Lyman, M., Lloyd, D. G., Ji, X., Vizcaychipi, M. P., & Ma, D. (2014). Neuroinflammation: The role and consequences. *Neuroscience Research, 79*, 1–2.
90. Terblanche, M., Almog, Y., Rosenson, R. S., Smith, T. S., & Hackam, D. G. (2007). Statins and sepsis: Multiple modifications at multiple levels. *The Lancet Infectious Diseases, 7*(5), 358–368.
91. Han, X., Sun, S., Sun, Y., Song, Q., Zhu, J., Song, N., et al. (2019). Small molecule-driven NLRP3 inflammation inhibition via interplay between ubiquitination and autophagy: Implications for Parkinson disease. *Autophagy, 15*(11), 1860–1881.
92. Mo, Y., Xu, E., Wei, R., Le, B., Song, L., Li, D., et al. (2018). Bushen-yizhi formula alleviates neuroinflammation via inhibiting NLRP3 inflammasome activation in a mouse model of Parkinson's disease. *Evidence-Based Complementary and Alternative Medicine, 2018*, 3571604.
93. Lee, J. S., & Lee, S. J. (2016). Mechanism of anti-a-synuclein immunotherapy. *Journal of Movement Disorders, 9*(1), 14.
94. Brundin, P., Dave, K. D., & Kordower, J. H. (2017). Therapeutic approaches to target alpha-synuclein pathology. *Experimental Neurology, 298*, 225–235.

95. Ingram, G., Hakobyan, S., Robertson, N. P., & Morgan, B. P. (2009). Complement in multiple sclerosis: Its role in disease and potential as a biomarker. *Clinical and Experimental Immunology, 155*(2), 128–139.
96. Prineas, J. W., Kwon, E. E., Cho, E. S., Sharer, L. R., Barnett, M. H., Oleszak, E. L., et al. (2001). Immunopathology of secondary-progressive multiple sclerosis. *Annals of Neurology, 50*(5), 646–657.
97. van der Laan, L. J., Ruuls, S. R., Weber, K. S., Lodder, I. J., Dopp, E. A., & Dijkstra, C. D. (1996). Macrophage € phagocytosis of myelin in vitro determined by flow cytometry: Phagocytosis is mediated by CR3 and induces production of tumor necrosis factor-a and nitric oxide. *Journal of Neuroimmunology, 70*(2), 145–152.
98. Tatomir, A., Talpos-Caia, A., Anselmo, F., Kruszewski, A. M., Boodhoo, D., Rus, V., et al. (2017). The complement system as a biomarker of disease activity and response to treatment in multiple sclerosis. *Immunologic Research, 65*(6), 1103–1109.
99. Soane, L., Cho, H. J., Niculescu, F., Rus, H., & Shin, M. L. (2001). C5b–9 terminal complement complex protects oligodendrocytes from death by regulating bad through phosphatidylinositol 3-kinase/Akt pathway. *Journal of Immunology, 167*(4), 2305–2311.
100. Watkins, L. M., Neal, J. W., Loveless, S., Michailidou, I., Ramaglia, V., Rees, M. I., et al. (2016). Complement is activated in progressive multiple sclerosis cortical grey matter lesions. *Journal of Neuroinflammation, 13*(1), 161.
101. Nataf, S., Carroll, S. L., Wetsel, R. A., Szalai, A. J., & Barnum, S. R. (2000). Attenuation of experimental autoimmune demyelination in complement-deficient mice. *Journal of Immunology, 165*(10), 5867–5873.
102. Tran, G. T., Hodgkinson, S. J., Carter, N., Killingsworth, M., Spicer, S. T., & Hall, B. M. (2002). Attenuation of experimental allergic encephalomyelitis in complement component 6-deficient rats is associated with reduced complement C9 deposition, P-selectin expression, and cellular infiltrate in spinal cords. *Journal of Immunology, 168*(9), 4293–4300.
103. Kjaeldgaard, A. L., Pilely, K., Olsen, K. S., Pedersen, S. W., Lauritsen, A. Ø., Møller, K., et al. (2018). Amyotrophic lateral sclerosis: The complement and inflammatory hypothesis. *Molecular Immunology, 102*, 14–25.
104. Woodruff, T. M., Costantini, K. J., Crane, J. W., Atkin, J. D., Monk, P. N., Taylor, S. M., et al. (2008). The complement factor C5a contributes to pathology in a rat model of amyotrophic lateral sclerosis. *Journal of Immunology, 181*(12), 8727–8734.
105. Chen, N. J., Mirtsos, C., Suh, D., Lu, Y. C., Lin, W. J., McKerlie, C., et al. (2007). C5L2 is critical for the biological activities of the anaphylatoxins C5a and C3a. *Nature, 446*(7132), 203–207.
106. Wang, H. A., Lee, J. D., Lee, K. M., Woodruff, T. M., & Noakes, P. G. (2017). Complement C5a–C5aR1 signalling drives skeletal muscle macrophage recruitment in the hSOD1 G93A mouse model of amyotrophic lateral sclerosis. *Skeletal Muscle, 7*(1), 10.
107. Woodruff, T. M., Lee, J. D., & Noakes, P. G. (2014). Role for terminal complement activation in amyotrophic lateral sclerosis disease progression. *Proceedings of the National Academy of Sciences of the United States of America, 111*(1), E3–E4.
108. Lee, J. D., Levin, S. C., Willis, E. F., Li, R., Woodruff, T. M., & Noakes, P. G. (2018). Complement components are upregulated and correlate with disease progression in the TDP-43 Q331K mouse model of amyotrophic lateral sclerosis. *Journal of Neuroinflammation, 15*(1), 1–3.
109. Ferraiuolo, L., Heath, P. R., Holden, H., Kasher, P., Kirby, J., & Shaw, P. J. (2007). Microarray analysis of the cellular pathways involved in the adaptation to and progression of motor neuron injury in the SOD1 G93A mouse model of familial ALS. *The Journal of Neuroscience, 27*(34), 9201–9219.
110. Sta, M., Sylva-Steenland, R. M., Casula, M., de Jong, J. M., Troost, D., Aronica, E., et al. (2011). Innate and adaptive immunity in amyotrophic lateral sclerosis: Evidence of complement activation. *Neurobiology of Disease, 42*(3), 211–220.

111. F€arber, K., Cheung, G., Mitchell, D., Wallis, R., Weihe, E., Schwaeble, W., et al. (2009). C1q, the recognition subcomponent of the classical pathway of complement, drives microglial activation. *Journal of Neuroscience Research, 87*(3), 644–652.
112. Spiller, K. J., Restrepo, C. R., Khan, T., Dominique, M. A., Fang, T. C., Canter, R. G., et al. (2018). Microglia-mediated recovery from ALS-relevant motor neuron degeneration in a mouse model of TDP-43 proteinopathy. *Nature Neuroscience, 21*(3), 329–340. Geloso, M. C., Corvino, V., Marchese, E., Serrano, A., Michetti, F., & D'Ambrosi, N. (2017). The dual role of microglia in ALS: Mechanisms and therapeutic approaches. *Frontiers in Aging Neuroscience, 9*, 242.
113. Harrison, J. K., Jiang, Y., Chen, S., Xia, Y., Maciejewski, D., McNamara, R. K., et al. (1998). Role for neuronally derived fractalkine in mediating interactions between neurons and CX3CR1-expressing microglia. *Proceedings of the National Academy of Sciences of the United States of America, 95*(18), 10896–10901.
114. Boehme, S. A., Lio, F. M., Maciejewski-Lenoir, D., Bacon, K. B., & Conlon, P. J. (2000). The chemokine fractalkine inhibits Fas-mediated cell death of brain microglia. *Journal of Immunology, 165*(1), 397–403.
115. Fischer, L. R., Culver, D. G., Tennant, P., Davis, A. A., Wang, M., Castellano-Sanchez, A., et al. (2004). Amyotrophic lateral sclerosis is a distal axonopathy: Evidence in mice and man. *Experimental Neurology, 185*(2), 232–240.
116. Pehar, M., Cassina, P., Vargas, M. R., Castellanos, R., Viera, L., Beckman, J. S., et al. (2004). Astrocytic production of nerve growth factor in motor neuron apoptosis: Implications for amyotrophic lateral sclerosis. *Journal of Neurochemistry, 89*(2), 464–473.
117. Krenz, N. R., & Weaver, L. C. (2000). Nerve growth factor in glia and inflammatory cells of the injured rat spinal cord. *Journal of Neurochemistry, 74*(2), 730–739.
118. Kawamata, T., Akiyama, H., Yamada, T., & McGeer, P. L. (1992). Immunologic reactions in amyotrophic lateral sclerosis brain and spinal cord tissue. *American Journal of Pathology, 140*(3), 691. 145. Rowland, L. P., Shneider, N. A. (2001). Amyotrophic lateral sclerosis. *The New England Journal of Medicine, 344*(22), 1688–1700.
119. Yasojima, K., Tourtellotte, W. W., McGeer, E. G., & McGeer, P. L. (2001). Marked increase in cyclooxygenase-2 in ALS spinal cord: Implications for therapy. *Neurology, 57*(6), 952–956.
120. Thompson, A. J., Baranzini, S. E., Geurts, J., Hemmer, B., & Ciccarelli, O. (2018). Multiple sclerosis. *Lancet, 391*(10130), 1622–1636.
121. Luo, C., Jian, C., Liao, Y., Huang, Q., Wu, Y., Liu, X., et al. (2017). The role of microglia in multiple sclerosis. *Neuropsychiatric Disease and Treatment, 13*, 1661.
122. Palmer, A. M. (2013). Multiple sclerosis and the blood-central nervous system barrier. *Cardiovascular Psychiatry and Neurology, 2013*, 530356.
123. Hemmer, B., Kerschensteiner, M., & Korn, T. (2015). Role of the innate and adaptive immune responses in the course of multiple sclerosis. *Lancet Neurology, 14*(4), 406–419.
124. Kasper, L. H., & Shoemaker, J. (2010). Multiple sclerosis immunology: The healthy immune system vs the MS immune system. *Neurology, 74*(Supplement 1), S2–S8.
125. Strachan-Whaley, M., Rivest, S., & Yong, V. W. (2014). Interactions between microglia and T cells in multiple sclerosis pathobiology. *Journal of Interferon & Cytokine Research, 34*(8), 615–622.
126. Henderson, A. P., Barnett, M. H., Parratt, J. D., & Prineas, J. W. (2009). Multiple sclerosis: Distribution of inflammatory cells in newly forming lesions. *Annals of Neurology, 66*(6), 739–753.
127. Khattar, M., Chen, W., & Stepkowski, S. M. (2009). Expanding and converting regulatory T cells: A horizon for immunotherapy. *Archivum Immunologiae et Therapiae Experimentalis (Warsz), 57*(3), 199–204.
128. Benner, E. J., Banerjee, R., Reynolds, A. D., Sherman, S., Pisarev, V. M., Tsiperson, V., et al. (2008). Nitrated a–synuclein immunity accelerates degeneration of nigral dopaminergic neurons. *PLoS One, 3*(1), e1376.
129. Costantino, C. M., Baecher-Allan, C. M., & Hafler, D. A. (2008). Human regulatory T cells and autoimmunity. *European Journal of Immunology, 38*(4), 921–924.

130. Gugliandolo, A., Giacoppo, S., Bramanti, P., & Mazzon, E. (2018). NLRP3 inflammasome activation in a transgenic amyotrophic lateral sclerosis model. *Inflammation, 41*(1), 93–103.
131. Johann, S., Heitzer, M., Kanagaratnam, M., Goswami, A., Rizo, T., Weis, J., et al. (2015). NLRP3 inflammasome is expressed by astrocytes in the SOD1 mouse model of ALS and in human sporadic ALS patients. *Glia, 63*(12), 2260–2273.
132. Kadhim, H., Deltenre, P., Martin, J. J., & Sebire, G. (2016). In-situ expression of Interleukin-18 and associated mediators in the human brain of sALS patients: Hypothesis for a role for immune-inflammatory mechanisms. *Medical Hypotheses, 86*, 14–17.
133. Debye, B., Schmulling, L., Zhou, L., Rune, G., Beyer, C., & Johann, S. (2018). Neurodegeneration and € NLRP3 inflammasome expression in the anterior thalamus of SOD1 (G93A) ALS mice. *Brain Pathology, 28*(1), 14–27.
134. Meissner, F., Molawi, K., & Zychlinsky, A. (2010). Mutant superoxide dismutase 1-induced IL-1b accelerates ALS pathogenesis. *Proceedings of the National Academy of Sciences of the United States of America, 107*(29), 13046–13050.
135. Deora, V., Lee, J. D., Albornoz, E. A., McAlary, L., Jagaraj, C. J., Robertson, A. A., et al. (2020). The microglial NLRP3 inflammasome is activated by amyotrophic lateral sclerosis proteins. *Glia, 68*(2), 407–421.
136. Dharmadasa, T., & Kiernan, M. C. (2018). Riluzole, disease stage and survival in ALS. *Lancet Neurology, 17*(5), 385–386.
137. Yoshino, H. (2019). Edaravone for the treatment of amyotrophic lateral sclerosis. *Expert Review of Neurotherapeutics, 19*(3), 185–193.
138. Kiaei, M., Petri, S., Kipiani, K., Gardian, G., Choi, D. K., Chen, J., et al. (2006). Thalidomide and lenalidomide extend survival in a transgenic mouse model of amyotrophic lateral sclerosis. *The Journal of Neuroscience, 26*(9), 2467–2473.
139. Karlsson, J., Fong, K. S., Hansson, M. J., Elmar, E., Csiszar, K., & Keep, M. F. (2004). Life span extension and reduced neuronal death after weekly intraventricular cyclosporin injections in the G93A transgenic mouse model of amyotrophic lateral sclerosis. *Journal of Neurosurgery, 101*(1), 128–137.
140. Appel, S. H., Stewart, S. S., Appel, V., Harati, Y., Mietlowski, W., Weiss, W., et al. (1988). A doubleblind study of the effectiveness of cyclosporine in amyotrophic lateral sclerosis. *Archives of Neurology, 45*(4), 381–386.
141. Meininger, V., Drory, V. E., Leigh, P. N., Ludolph, A., Robberecht, W., & Silani, V. (2009). Glatiramer acetate has no impact on disease progression in ALS at 40 mg/day: A double-blind, randomized, multicentre, placebo-controlled trial. *Amyotrophic Lateral Sclerosis, 10*(5–6), 378–383.
142. Maier, A., Deigendesch, N., Muller, K., Weishaupt, J. H., Krannich, A., R € ohle, R., et al. (2015). Inter- € leukin-1 antagonist anakinra in amyotrophic lateral sclerosis—A pilot study. *PLoS One, 10*(10), e0139684.
143. Grottelli, S., Mezzasoma, L., Scarpelli, P., Cacciatore, I., Cellini, B., & Bellezza, I. (2019). Cyclo (HisPro) inhibits NLRP3 inflammasome cascade in ALS microglial cells. *Molecular and Cellular Neurosciences, 94*, 23–31.
144. Zhao, B., Marciniuk, K., Gibbs, E., Yousefi, M., Napper, S., & Cashman, N. R. (2019). Therapeutic vaccines for amyotrophic lateral sclerosis directed against disease specific epitopes of superoxide dismutase 1. *Vaccine, 37*(35), 4920–4927.
145. Smith, A. L., Cohen, J. A., & Hua, L. H. (2017). Therapeutic targets for multiple sclerosis: Current treatment goals and future directions. *Neurotherapeutics, 14*(4), 952–960.
146. Goodin, D. S. (2008). Disease-modifying therapy in multiple sclerosis: Update and clinical implications. *Neurology, 71*(24 suppl 3), S8–S13.
147. Edan, G., Miller, D., Clanet, M., Confavreux, C., Lyon-Caen, O., Lubetzki, C., et al. (1997). Therapeutic effect of mitoxantrone combined with methylprednisolone in multiple sclerosis: A randomised multicentre study of active disease using MRI and clinical criteria. *Journal of Neurology, Neurosurgery, and Psychiatry, 62*(2), 112–118.
148. TaSKapilioGLu, O. (2018). Recent advances in the treatment for multiple sclerosis; current € new drugs specific for multiple sclerosis. *Archives of Neuropsychiatry, 55*(Suppl 1), S15.

149. Eckstein, C., & Bhatti, M. T. (2016). Currently approved and emerging oral therapies in multiple sclerosis: An update for the ophthalmologist. *Survey of Ophthalmology, 61*(3), 318.
150. Miller, A. E., Wolinsky, J. S., Kappos, L., Comi, G., Freedman, M. S., Olsson, T. P., et al. (2014). Oral teriflunomide for patients with a first clinical episode suggestive of multiple sclerosis (TOPIC): A randomised, double-blind, placebo-controlled, phase 3 trial. *Lancet Neurology, 13*(10), 977–986.
151. Mills, E. A., Ogrodnik, M. A., Plave, A., & Mao-Draayer, Y. (2018). Emerging understanding of the mechanism of action for dimethyl fumarate in the treatment of multiple sclerosis. *Frontiers in Neurology, 9*, 5.
152. Walker, F. O. (2007). Huntington's disease. *Lancet, 369*(9557), 218–228.
153. Nayak, A., Ansar, R., Verma, S. K., Bonifati, D. M., & Kishore, U. (2011). Huntington's disease: An immune perspective. *Neurology Research International, 2011*, 563784.
154. MacDonald, M. E., Ambrose, C. M., Duyao, M. P., Myers, R. H., Lin, C., Srinidhi, L., et al. (1993). A novel gene containing a trinucleotide repeat that is expanded and unstable on Huntington's disease chromosomes. *Cell, 72*(6), 971–983.
155. Rawlins, M. D., Wexler, N. S., Wexler, A. R., Tabrizi, S. J., Douglas, I., Evans, S. J., et al. (2016). The prevalence of Huntington's disease. *Neuroepidemiology, 46*(2), 144–153.
156. Singhrao, S. K., Neal, J. W., Morgan, B. P., & Gasque, P. (1999). Increased complement biosynthesis by microglia and complement activation on neurons in Huntington's disease. *Experimental Neurology, 159*(2), 362–376.
157. Hodges, A., Strand, A. D., Aragaki, A. K., Kuhn, A., Sengstag, T., Hughes, G., et al. (2006). Regional and cellular gene expression changes in human Huntington's disease brain. *Human Molecular Genetics, 15*(6), 965–977.
158. Larkin, P. B., & Muchowski, P. J. (2012). Genetic deficiency of complement component 3 does not alter disease progression in a mouse model of Huntington's disease. *Journal of Huntington's Disease, 1*(1), 107–118.
159. Zhao, T., Hong, Y., Li, S., & Li, X. J. (2016). Compartment-dependent degradation of mutant huntingtin accounts for its preferential accumulation in neuronal processes. *The Journal of Neuroscience, 36*(32), 8317–8328.
160. Wang, N., Gray, M., Lu, X. H., Cantle, J. P., Holley, S. M., Greiner, E., et al. (2014). Neuronal targets for reducing mutant huntingtin expression to ameliorate disease in a mouse model of Huntington's disease. *Nature Medicine, 20*(5), 536–541.
161. Ochaba, J., Monteys, A. M., O'Rourke, J. G., Reidling, J. C., Steffan, J. S., Davidson, B. L., et al. (2016). PIAS1 regulates mutant huntingtin accumulation and Huntington's disease-associated phenotypes in vivo. *Neuron, 90*(3), 507–520.
162. Ellrichmann, G., Reick, C., Saft, C., & Linker, R. A. (2013). The role of the immune system in Huntington's disease. *Clinical & Developmental Immunology, 2013*, 1–11.
163. Mellman, I., & Steinman, R. M. (2001). Dendritic cells: Specialized and regulated antigen processing machines. *Cell, 106*(3), 255–258.
164. Chen, M., Ona, V. O., Li, M., Ferrante, R. J., Fink, K. B., Zhu, S., et al. (2000). Minocycline inhibits caspase-1 and caspase-3 expression and delays mortality in a transgenic mouse model of Huntington disease. *Nature Medicine, 6*(7), 797–801.
165. Sperlagh, B., & Illes, P. (2014). P2X7 receptor: An emerging target in central nervous system diseases. *Trends in Pharmacological Sciences, 35*(10), 537–547.
166. Labzin, L. I., Heneka, M. T., & Latz, E. (2018). Innate immunity and neurodegeneration. *Annual Review of Medicine, 69*, 437–449.
167. Rodrigues, F. B., & Wild, E. J. (2018). Huntington's disease clinical trials corner: August 2018. *Journal of Huntington's Disease, 7*(3), 279–286.
168. Southwell, A. L., Franciosi, S., Villanueva, E. B., Xie, Y., Winter, L. A., Veeraraghavan, J., et al. (2015). Antisemaphorin 4D immunotherapy ameliorates neuropathology and some cognitive impairment in the YAC128 mouse model of Huntington disease. *Neurobiology of Disease, 76*, 46–56.
169. Baidya, F., Bohra, M., Datta, A., Sarmah, D., Shah, B., Jagtap, P., Raut, S., Sarkar, A., Singh, U., Kalia, K., Borah, A., Wang, X., Dave, K. R., Yavagal, D. R., & Bhattacharya, P. (2021).

Neuroimmune crosstalk and evolving pharmacotherapies in neurodegenerative diseases. *Immunology, 162*(2), 160–178. https://doi.org/10.1111/imm.13264
170. Iadecola, C., & Anrather, J. (2011). The immunology of stroke: From mechanisms to translation. *Nature Medicine, 17*(7), 796–808.
171. Niemela, M., Tulamo, R., Hernesniemi, J., Scozzafava, J., Findlay, J. M., Macdonald, R. L., & Dempsey, R. J. (2006). Alterations in plasma complement levels after human ischemic stroke-comments. *Neurosurgery, 59*(1), 32–33.
172. Sch€afer, M. K., Schwaeble, W. J., Post, C., Salvati, P., Calabresi, M., Sim, R. B., et al. (2000). Complement C1q is dramatically up-regulated in brain microglia in response to transient global cerebral ischemia. *Journal of Immunology, 164*(10), 5446–5452.
173. Duewell, P., Kono, H., Rayner, K. J., Sirois, C. M., Vladimer, G., Bauernfeind, F. G., et al. (2010). NLRP3 inflammasomes are required for atherogenesis and activated by cholesterol crystals. *Nature, 464*(7293), 1357–1361.
174. Hua, K. F., Yang, S. M., Kao, T. Y., Chang, J. M., Chen, H. L., Tsai, Y. J., et al. (2013). Osthole mitigates progressive IgA nephropathy by inhibiting reactive oxygen species generation and NFjB/NLRP3 pathway. *PLoS One, 8*(10), e77794.
175. Saver, J. L., Goyal, M., Bonafe, A., Diener, H. C., Levy, E. I., Pereira, V. M., et al. (2015). Stent-retriever thrombectomy after intravenous t-PA vs. t-PA alone in stroke. *The New England Journal of Medicine, 372*(24), 2285–2295.
176. Hernandez, A. L., Kitz, A., Wu, C., Lowther, D. E., Rodriguez, D. M., Vudattu, N., et al. (2015). Sodium chloride inhibits the suppressive function of FOXP3+ regulatory T cells. *Journal of Clinical Investigation, 125*(11), 4212–4222.
177. Okoye, I. S., Coomes, S. M., Pelly, V. S., Czieso, S., Papayannopoulos, V., Tolmachova, T., et al. (2014). MicroRNA-containing T-regulatory-cell-derived exosomes suppress pathogenic T helper 1 cells. *Immunity, 41*(1), 89–103.
178. Antinori, A., Arendt, G., Becker, J. T., Brew, B. J., Byrd, D. A., Cherner, M., Clifford, D. B., Cinque, P., Epstein, L. G., Goodkin, K., Gisslen, M., Grant, I., Heaton, R. K., Joseph, J., Marder, K., Marra, C. M., McArthur, J. C., Nunn, M., Price, R. W., et al. (2007). Updated research nosology for HIV-associated neurocognitive disorders. *Neurology, 69*(18), 1789–1799. https://doi.org/10.1212/01.WNL.0000287431.88658.8b
179. Kaul, M., Garden, G. A., & Lipton, S. A. (2001). Pathways to neuronal injury and apoptosis in HIV-associated dementia. *Nature, 410*(6831), 988–994. https://doi.org/10.1038/35073667
180. Gelman, B. B., & Lisinicchia, J. G. (2013). Neuropathology of HIV infection. In M. Gendelman, H. Friedman, & S. S. Grant (Eds.), *Encyclopedia of AIDS* (pp. 1–13). Springer. https://doi.org/10.1007/978-1-4419-1004-8_327
181. Nath, A., & Clements, J. E. (2011). Eradication of HIV from the brain: Reasons for pause. *AIDS, 25*(5), 577–580. https://doi.org/10.1097/QAD.0b013e3283437b56
182. Kuchibhotla, K. V., & Lattarulo, C. R. (2020). Neurotrophic factors and Alzheimer's disease. In S. A. Lipton & D. J. Yahr (Eds.), *Neurodegeneration: The molecular pathology of dementia and movement disorders* (2nd ed., pp. 289–305). Wiley. https://doi.org/10.1002/9781119487938.ch20
183. Chen, G., Zhang, Z., Wang, L., et al. (2013). Neural stem cell-like cells derived from autologous bone mesenchymal stem cells for the treatment of patients with cerebral palsy. *Journal of Translational Medicine, 11*, 21.
184. Readhead, B., Haure-Mirande, J. V., Funk, C. C., Richards, M. A., Shannon, P., Haroutunian, V., et al. (2018). Multiscale analysis of independent Alzheimer's cohorts finds disruption of molecular, genetic, and clinical networks by human herpesvirus. *Neuron, 99*(1), 64–82.
185. Baruch, K., Rosenzweig, N., Kertser, A., Deczkowska, A., Sharif, A. M., Spinrad, A., et al. (2015). Breaking immune tolerance by targeting Foxp3+ regulatory T cells mitigates Alzheimer's disease pathology. *Nature Communications, 6*(1), 1–2.
186. Keren-Shaul, H., Spinrad, A., Weiner, A., Matcovitch-Natan, O., Dvir-Szternfeld, R., Ulland, T. K., et al. (2017). A unique microglia type associated with restricting development of Alzheimer's disease. *Cell, 169*(7), 1276–1290.

# Nanotheranostic Targeting Neuroimmune Signaling: The Mind Knockout

Jayant Kumar, Raghuraj Singh, Krishna Jadhav, Rahul Kumar Verma, and Anoop Kumar

**Abstract**

Nanotheranostic represents a cutting-edge arena of therapeutic medicine that combines the diagnostic and therapeutic agent into a single nanotechnology platform, transforming the landscape of precision medicine targeting neuroimmune signaling. Nanotheranostic targeting the neuroimmune signaling pathways includes expanding and applying nanomaterials or nanoparticles (NPs) to target and modulate the neuroimmune signaling pathways precisely. The final goal is to enhance the diagnosis and treatment of neuroimmune signaling related neurodegenerative diseases like PD, AD, and autoimmune disorders. This chapter provides knowledge of nanotheranostic targeting neuroimmune signaling pathways and explaining its multifaceted therapeutic applications, recent improvements, and future prospects of nanotheranostic platform. We explore the various types of miscellaneous diagnostic techniques combined within nanotheranostic platform, encompassing imaging and sensing techniques. The combination of diagnostic and therapeutic platforms within the nanotheranostic offers enormous potential for the initial detection of CNS disorders, treatment strategies of personalized medicine, and improved potential therapeutic outcomes.

J. Kumar
Department of Pharmaceutical Sciences Hemwati Nandan Bahuguna Garhwal University (A Central University), Srinagar, Uttarakhand, India

R. Singh · K. Jadhav · R. K. Verma
Pharmaceutical Nanotechnology Lab, Institute of Nano Science and Technology (INST), Sector-81, Mohali, Punjab, India

A. Kumar (✉)
Department of Pharmacology, Delhi Pharmaceutical Sciences and Research University (DPSRU), New Delhi, India
e-mail: anoopdpsru@dpsru.edu.in



R. Shukla et al. (eds.), *Neuroendocrine-Immune Axis*,
https://doi.org/10.1007/978-981-95-5088-3_10

**Keywords**

Nanotheranostic · Neuroimmune signaling · Nanoparticles (NPs) · Nanotechnology

## Abbreviations

| | |
|---|---|
| AuNPs | Gold nanoparticles |
| BBB | Blood-brain barrier |
| CNS | Central nervous system |
| CNTs | Carbon nanotubes |
| CT | Computed tomography |
| DOX | Doxorubicin |
| MRI | Magnetic resonance imaging |
| NIR | Near infrared region |
| NPs | Nanoparticles |
| PEG | Polyethylene glycol |
| PET | Positron emission tomography |
| PLA | Poly (lactic acid) |
| PLGA | Poly(lactide-coglycolide) |
| PNS | Peripheral nervous system |
| PVA | Polyvinyl alcohol |
| SPECT | Single-photon emission computed tomography |

## 1 Introduction

Neuroimmune signaling encompasses the intricate and dynamic interactions between the nervous and immune systems within the body, serving as a pivotal process in regulating immune responses, maintaining homeostasis, and orchestrating inflammation [1]. Within the brain, neuroimmune signaling assumes a critical role in governing diverse physiological phenomena, including neuroinflammation, synaptic plasticity, and neurodegeneration [2]. Notably, the dysregulation of neuroimmune signaling has emerged as a key factor in the pathogenesis of numerous neurological disorders, encompassing Alzheimer's disease, Parkinson's disease, multiple sclerosis, and autism spectrum disorders [2]. Consequently, neuroimmune signaling encompasses a wide range of processes, including both pro-inflammatory and anti-inflammatory mechanisms, which can impact neurotransmission at various sites within the brain. This emerging area of research presents a novel frontier in our comprehension of normal and pathological brain function and the intricate interplay between systemic and central nervous system (CNS) disorders [3].

Nonetheless, diagnosing inflammation and neurodegenerative disorders presents significant challenges due to the formidable barrier posed by the blood-brain barrier

(BBB), impeding the entry of many diagnostic and therapeutic agents into the brain [4]. In response to this hurdle, the emerging field of nanotheranostic has emerged as a promising solution, seamlessly combining therapeutic and diagnostic capabilities to advance the development of personalized medicine for a diverse array of diseases. Nanotheranostic, by synergistically amalgamating therapeutic and diagnostic functions, holds tremendous potential to revolutionize the treatment landscape of neurological disorders, particularly those that involve intricate neuroimmune signaling pathways [5]. By leveraging nanotechnology, this approach enables the design and fabrication of nanoparticles that can precisely target specific tissues or cells within the body [6]. The integration of molecular imaging techniques with nanotechnology not only empowers researchers to visualize and interrogate biological processes at the cellular and molecular levels but also facilitates the development of highly targeted therapies tailored to the unique characteristics of neurological disorders [7]. Moreover, the seamless integration of molecular imaging techniques with nanotheranostic empowers investigators to visualize and comprehensively assess the distribution and pharmacokinetics of nanoparticles within the body [8]. Such real-time insights about the efficacy and safety of nanotheranostic therapies contribute to optimizing treatment protocols and further refining therapeutic interventions. In pursuit of targeting neuroimmune signaling within the brain, an array of nanoparticle formulations has been developed, encompassing liposomes, polymeric nanoparticles, dendrimers, and gold nanoparticles [9]. These nanoparticles can be adeptly loaded with imaging agents, including magnetic resonance imaging (MRI) contrast agents, positron emission tomography (PET) tracers, or fluorescent dyes, alongside drugs that are specifically tailored to target distinct components of the neuroimmune signaling cascade [10, 11].

Further customization of the nanoparticle surface is possible through the attachment of ligands that exhibit specific affinity toward cells implicated in neuroinflammatory and neurodegenerative processes, such as microglia and astrocytes [12]. Another avenue of exploration lies in harnessing the potential of nanoscale exosomes, which are naturally occurring extracellular vesicles that can be sourced from diverse cell types. Through engineering, exosomes can be harnessed as carriers of therapeutic drugs and imaging agents, with the added advantage of their inherent capability to traverse the blood-brain barrier and accumulate within targeted brain regions [13, 14]. These innovative platforms hold tremendous promise by enabling precise visualization and targeted intervention at the level of specific components within the intricate neuroimmune signaling pathway. Encouragingly, nanotheranostic have already exhibited remarkable efficacy in preclinical models of neuroinflammation and neurodegeneration, paving the way for their translation into clinical applications [9]. The future holds great promise for nanotheranostic, heralding a new era of personalized medicine that combines diagnostics and therapeutics to address the intricate complexities of neuroinflammatory and neurodegenerative diseases.

## 2 Insights of Molecular Mechanisms Underlying Neuroimmune Signaling

Neuroimmune signaling involves a complex interplay of molecular mechanisms facilitating communication and coordination between the nervous and immune systems. These mechanisms provide insights into the intricate pathways and interactions that underlie the regulation of immune responses, inflammation, and neuroinflammatory processes [15]. Neuroimmune signaling encompasses a complex interplay of glial cells, neurons, immune cells, cytokines, chemokines, toll like receptors, neurotransmitters, and neuropeptides. These intricate molecular mechanisms orchestrate the communication and coordination between the nervous and immune systems, regulating immune responses, inflammation, and neuroinflammatory processes [16]. This bidirectional cross-talk plays a critical role in maintaining homeostasis, regulating immune responses, and modulating various physiological processes in both the central nervous system (CNS) and the peripheral nervous system (PNS). Understanding the molecular mechanisms underlying neuroimmune signaling is essential for unraveling the pathogenesis of various neurological disorders and developing targeted therapeutic interventions. One key insight is the involvement of glial cells, which include microglia, astrocytes, and oligodendrocytes, in neuroimmune signaling [17].

When the immune system is activated, whether in response to illness, injury, or experimental administration of immune triggers like lipopolysaccharides (LPS), a complex cascade of cytokines and immune molecules is induced. This immune activation leads to a unique state in the brain and affects various aspects of neural function and behavior. Cytokines such as interleukin-1β (IL-1β), interleukin-6 (IL-6), and tumor necrosis factor-α (TNF-α) are key players in the neuroimmune response. Elevated levels of IL-1β disrupt memory and synaptic plasticity, while IL-6 acts as a brake on plasticity [18]. These cytokines, along with other immune signaling pathways like complement signaling, contribute to the acute neuroimmune activation state. During infection, both neurons and microglia can release complement proteins that activate microglia, contributing to the degradation of synaptic and impairments in spatial memory [19]. Immune signaling molecules are therefore crucial for the outcomes of acute neuroimmune activation.

Microglia, the resident immune cells of the central nervous system (CNS), play a crucial role in detecting and responding to immune challenges within the brain. They act as immune sentinels, constantly surveilling the microenvironment and rapidly responding to pathogens or damage [20]. Astrocytes, on the other hand, support neuronal functions and are involved in immune modulation by releasing cytokines and chemokines. Oligodendrocytes, responsible for myelination of neuronal axons, have also been implicated in neuroimmune signaling, suggesting their involvement in regulating immune responses within the CNS [21, 22].

Microglia and astrocytes undergo morphological and transcriptional changes during neuroimmune activation. Microglia transition from a divided, surveying state to an ameboid state, extending phagocytic processes toward the source of activation [23]. Similarly, astrocytes exhibit distinct reactive transcriptional profiles in response to immune stimulation [24]. Activated microglia are a primary source of proinflammatory cytokine and chemokine generation during immune events, altering neuronal-glia communication [20]. Inhibition of microglial activity has been shown to prevent immune challenge-induced memory impairments, highlighting the importance of these neuroimmune cells in cognitive deficits [25].

The activity of specific immune signaling molecules is influenced by the adjacent immune environment and the nature of the immune stimulus. For example, interferon-γ (IFN-γ) can either enhance or downregulate inflammatory responses in the brain, depending on the context [26]. The function of IL-1 inside the brain varies between neurodegenerative and neuroinflammatory conditions. Differences in immune cell activation states and transcriptional profiles, shaped by the nature of the immune stimulus, lead to specific cytokine signaling pathways and gene expression patterns [27]. Understanding the broader context of cytokine and immune cell interactions is crucial when considering the separate roles of inflammatory cytokines in modulating neural processes [28]. Chronic inflammation represents a special subset of neuroimmune activation characterized by tenacious activation of inflammatory cytokines and glial cells. In conditions such as sepsis or traumatic brain injury, microglial activation can persist for extended periods, contributing to long-term cognitive deficits [29, 30].

Another important molecular mechanism is the release of cytokines and chemokines by immune cells and glial cells. Cytokines are small signaling molecules that mediate immune responses and play a crucial role in neuroinflammation. They can have both pro-inflammatory and anti-inflammatory effects, depending on the context and the specific cytokine involved. Chemokines, on the other hand, regulate immune cell migration and recruitment to sites of inflammation or injury. The interaction between cytokines and chemokines contributes to the orchestration of immune responses within the CNS [31].

Toll-like receptors (TLRs) are a family of pattern recognition receptors that play a crucial role in neuroimmune signaling. TLRs detect pathogen-associated molecular patterns (PAMPs) and damage-associated molecular patterns (DAMPs), initiating immune responses. Activation of TLRs in glial cells triggers the release of cytokines, chemokines, and other immune molecules, contributing to neuroinflammation. TLRs are also expressed in neurons, where they participate in neuroprotective and neuroinflammatory processes [32, 33]. Neurotransmitters and neuropeptides are additional components involved in neuroimmune signaling. Neurotransmitters, such as glutamate, gamma-aminobutyric acid (GABA), and dopamine, play crucial roles in neuronal communication but can also modulate immune responses. Neuropeptides, including substance P and neuropeptide Y, have been implicated in regulating inflammation and immune cell functions within the CNS. The interaction between neurotransmitters, neuropeptides, and immune cells further highlights the intricate cross-talk between the nervous and immune systems [34, 35].

Furthermore, the involvement of molecular pathways, such as the nuclear factor-kappa B (NF-κB) pathway and the inflammasome pathway, provides insights into the regulation of immune responses and neuroinflammation. Activation of the NF-κB pathway leads to the transcriptional upregulation of pro-inflammatory cytokines and chemokines, promoting neuroinflammatory processes. The inflammasome pathway, on the other hand, is responsible for the activation of caspase-1 and the release of pro-inflammatory cytokines, such as interleukin-1β (IL-1β) and IL-18 [36, 37]. Overall, the molecular mechanisms underlying neuroimmune signaling offer valuable insights into the complex interactions between the nervous and immune systems. Understanding these mechanisms is critical for deciphering the pathogenesis of neurological disorders characterized by dysregulated immune responses and inflammation. Moreover, these insights pave the way for the development of targeted therapeutic strategies aimed at modulating neuroimmune signaling to restore immune homeostasis and mitigate neuroinflammatory processes in the context of various neurological conditions.

## 3 Nanotheranostic

Nanotechnology is a well-known, rapidly developing platform with a fascinating interest in nanomedicine's current era. The word "nano" designates one billionth of a meter. In pharmaceuticals, "nanotechnology" is typically applied to mention the fabrication of nanocarriers (NCs) with dimensions from 10 to 1000 nm for the delivery of therapeutic drugs. In the USA, the National Institute of Health (NIH) defined nanomedicine as the application of nanotechnology platforms in the diagnosis and treatment with concurrent monitoring of living systems [38]. In 2002, the PharmaNetics president and CEO, John Funkhouser, was attributed by researchers for announcing the term "theranostic", which combines diagnostics and therapeutics for personalized nanomedicine [39]. It is a highly developing and promising field of advanced medicine with significant therapeutic potential to treat many diseases [40]. Theranostics refers to an inclusive effort that incorporates both diagnostics and therapeutics in a single platform [41] (Fig. 1).

With the development of nanotechnologies-based nanomaterials in health care, the concept of nanotheranostic was then announced. Nanotheranostic is a highly progressive form of theranostic that includes a "nanotechnology platform" for advanced diagnosis and drug therapy of different disorders and diseases with low prognoses [42]. Nanotheranostic has established the development of highly advanced diagnosis and treatment platforms for brain disorders at the initial stages. It can offer non-invasive targeting, imagery, and therapy at particular disease-specific sites without disturbing neighboring normal cells [43].

The main aim of nanotheranostic is to diagnose pathological conditions and treat diseases or disorders at their earliest stage of progress. It is a new-angled and innovative therapeutic class of nanomaterials that can analyze, prevent, and cure diseases at sub-cellular and cellular levels by advancing nanobiotechnology [44]. In the last few years, nanotheranostics has also developed as a highly promising therapeutic tool in smart nanomedicine for diagnosing and treating different fatal neurological disorders. In brain nanotheranostic, diagnostic and therapeutic molecular entities are encapsulated inside the matrix of a single nanotechnology platform (NPs), which can be further industrialized as a potential clinical nano-formulation for effectively targeting numerous brain disorders [45].

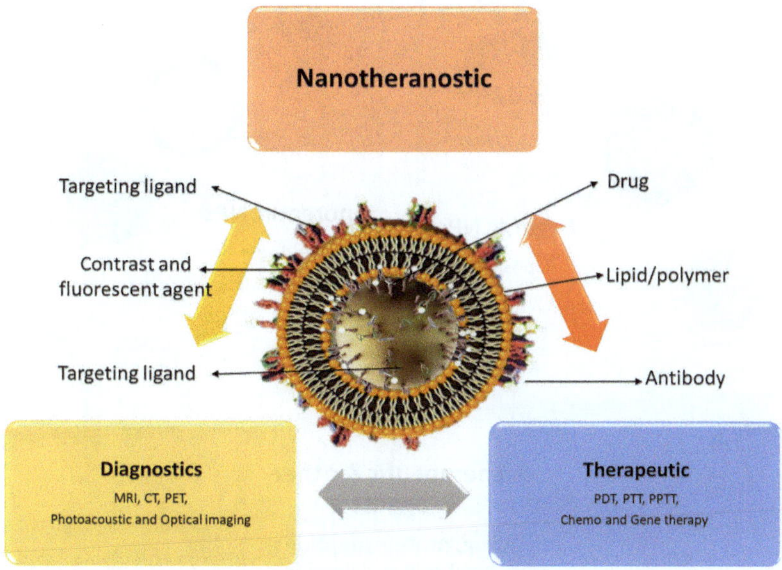


**Fig. 1** Multifunctional nanotheranostic modalities: Nanotheranostic encapsulating drug and diagnostic agent, and functionalized with targeting ligand to achieve site-specific delivery

NPs can be generally classified according to their structural composition, such as organic, inorganic, metal, and carbon NPs [46, 47]. Generally, theranostic NPs can be developed by different chemical concepts: targeting moiety for site-specific targeting, a therapeutic drug for effective delivery of disease management, diagnostic agents for imaging, and polymer coating for colloidal stability and bioconjugation [48]. Nanotheranostic contains the various types of new generations of smart NPs, such as micelles, polymer-drug conjugates, liposomes, lipid NPs, dendrimers, metal and inorganic NPs, and carbon nanotubes (CNTs) to fabricate souped-up nanomedicine. The important advantages of NPs used in nanotheranostics are their small size and large effective surface area to volume ratio, as well as stimulating electrical and photosensitive properties. These characteristics allow both diagnostic and therapeutic molecules to be encapsulated, adsorbed, noncovalently, and covalently attached at the surface of NPs, forming NP-drug/imaging/targeting complexes. The developed complex can subsequently indicator (diagnostics) and deliver the drug to the targeted sites for effective therapy of a particular disease [49].

## 3.1 Nanotheranostic Platform for Neuroimmune Signaling

Nanotheranostic is a novel combination of modern diagnostic and therapeutic agents in a single device and potentially plays a role in neuroimmune signaling. NPs are the main structural components of the nanotheranostic platform, which includes gold-NPs, iron oxide-NPs, silver-NPs, magnetic-NPs, mesoporous silica NPs, lipid-NPs, polymeric-NPs, and quantum dots (Fig. 2).

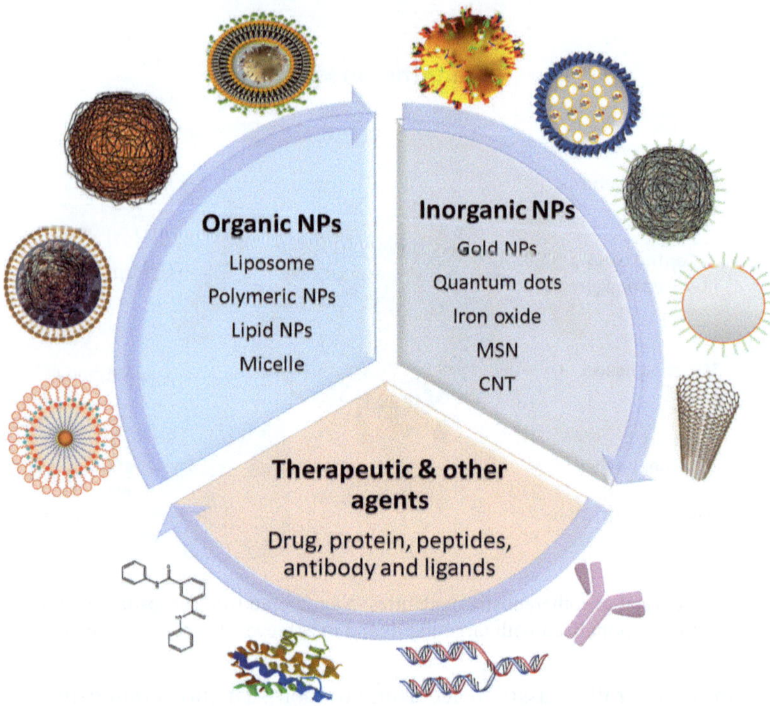


**Fig. 2** Different types of NPs explored for nanotheranostic targeting neuroimmune signaling

### 3.1.1 Organic NPs

#### Lipid NPs

Several lipid NPs, such as solid lipid NPs, nanostructured lipid carriers (NLCs), and liposomes, have been applied as carriers for nanotheranostic platforms in managing brain disorders. Solid lipid NPs are a progressive form of nanomaterial with potential therapeutic applications in targeted brain nanotheranostic. They are made up by using biocompatible solid lipids as the core material of the hydrophobic part enclosed by a phospholipid monolayer [50]. This structural arrangement enhances their stability and biocompatibility in a biological atmosphere, making them appropriate as a nanocarrier system [51]. They can encapsulate both hydrophilic and hydrophobic molecules, specifically showing high benefits in peptides and protein delivery. It is also advised as a highly flexible nano-platform for targeted brain therapeutic and imaging purposes. From the previous few decades, numerous brain-targeting solid lipid NPs have been studied for various CNS disorders [52].

Liposomes are widely used nanoplatforms with an established track record in drug delivery for delivering the therapeutic drug, genes, and imaging agents. There are several examples of liposome-based nanotheranostic systems with a diversity of characteristics that combine imaging and therapeutic proficiencies. Several nanotheranostic platforms have also shown considerable potential for hybrid liposome-QD systems [53]. Saesoo et al. developed the SPIONs containing liposomes, and surface functionalization was done using rituximab for CNS lymphoma diagnosis and therapy. The developed nanotheranostic not only improves CNS delivery of therapeutic drug, but also offers effective diagnostic and disease monitoring proficiencies [54].

**Polymeric NPs**

Polymeric NPs (PNPs) are described with several advantages in brain nanotheranostic due to their high drug encapsulation capacity, therefore, avoiding them from drug biotransformation and fast elimination, and in the targeted drug delivery transversely the BBB without changing the biological barrier features [55]. PNPs are nano-sized range colloidal particles in which the therapeutic drug is either entrapped, encapsulated, or adsorbed onto the polymeric matrix. Several conventional polymers are used for the development of NPs, which include gelatin, albumin, chitosan, alginate, poly(lactic acid-co-glycolic acid) (PLGA), polylactic acid (PLA), polyglycolide, and poly-glutamic acid [56]. The physicochemical characteristics of the different polymers, such as molecular weight, crystallinity, lipophilicity, hydrophilicity, and polydispersity index (PDI), are found to control dissolution and kinetics of drug delivery. The NPs can be developed using different methods such as solvent evaporation, nanoprecipitation, ultrasonication, solvent diffusion, emulsification, and salting out. Typically, biocompatible, biodegradable, less toxic polymers and clinically viable, e.g., PLA and PLGA, are generally chosen in the development of nanotheranostic and their co-polymers which are voluntarily removed after metabolism [57]. Chitosan is the most commonly used natural polymer to develop nanotheranostic for CNS diseases due to its high biocompatible, biodegradable, and nontoxic nature [58]. Similarly, synthetic polymers, such as PLGA and PLA, are also most frequently employed for the development of nanotheranostic for CNS disorders owing to their biodegradability, biocompatible, and low immunogenicity. PNPs are the multipurpose and extensively used NPs for delivering therapeutic drugs to the brain due to their high encapsulation efficiency and can be surface engineered with appropriate targeting ligands and /or cell-penetrating peptides (CPP) for successful brain drug delivery [59]. Presently, these NPs are utilized as effective nanotheranostic for diagnosing and treating different CNS disorders by considering their exclusive conjugation capability and encapsulating diagnostic and therapeutic agents on their structural matrix [5].

**Micelles**

Micelles are amphiphilic structures with both hydrophobic (inner core) and hydrophilic (outer shell). Micelles have unique advantages such as both kinetic as well as thermodynamic stability, smaller dimension, and higher payload [60]. Numerous scientific studies have shown to develop targeting micelles for diagnosis and therapy of several brain disorders. The developed micelles that can load a variety of diagnostics agents and therapeutic drugs to deliver drugs at specifically targeted sites are used preclinically for both diagnostic and therapeutic purposes [61]. Various scientific studies have been conducted on developing targeted micelles for diagnosing and treating brain disorders. Micelles are not only biocompatible; they are also highly flexible in terms of development and functionalization. Surface functionalization of micelles with targeting ligands allows precise recognition and site-specific binding to disease-related biomarkers. This targeted approach allows the precise accumulation of diagnostic agents at the targeted site, improving the sensitivity and selectivity of diagnostic modalities [61]. Micelles have increased the significant attention in the field of nanotheranostic for their potential applications in the diagnosis and therapy of different CNS disorders. In present-day, micelles are effectively being used in animal and human studies [60, 62]. Singh et al. developed curcumin-loaded self-traceable micelles that can cross the BBB after administration systemically. It significantly improved the therapeutic potential against brain cancer at low doses, holding high therapeutic potential for brain cancer nanotheranostic [59]. Sun et al. developed novel gold and SPIO-loaded micelles (GSMs). They functionalized with modified polyethylene-glycol-polycaprolactone (PEG-PCL) copolymer as both diagnostic and therapeutic platforms in managing glioblastoma. As a diagnostic device for glioblastoma, GSMs were used as diagnostic agents for both MRI and CT of glioblastoma [63].

### 3.1.2 Inorganic NPs

**Gold NPs**

Gold (Au)-NPs synthesized from Au cores are innovative delivery systems that demonstrate unique features for nanotheranostic platforms. Au-NPs are biocompatible and typically synthesized by chemical modification of hydrogen tetrachloroaurate. The synthesized Au-NPs have different shapes, such as spherical, rods, cubes, hexagonal, and cages. The comfort of modifying Au-NPs into numerous shapes and sizes and the conjugation with different functionalities simplify the scientists' digging into the ultimate applications of Au-NPs as nanotheranostics, specifically for the diagnosis as well as for treatment of neurological disorders. The spheroidal Au-NPs with a size of 10 nm demonstrate an exclusive ultraviolet (UV) absorption at the wavelength 520 nm, and differences in size correspond to the blue or red shifts. The Au nanorods show the absorption shifts to the near-infrared radiation (NIR) at a range from 690–900 nm. These characteristic photosensitive features allow Au-NPs to be used as diverse nanotheranostic for clinical applications [58]. The innovative characteristics of Au-NPs include diagnostic properties, monodispersity, surface plasmon absorption, high surface-volume ratio, tunable size, high ability to bind with biological molecules through Au–Sulfur bonds, low systemic toxicity, ease of production, and light-scattering properties [64]. Au-NPs have been explored in various diagnostic applications that comprise photoacoustic, surface-enhanced Raman spectroscopy (SERS), and computed tomography (CT) scan. It can be synthesized with therapeutic drugs and specific targeting moieties as a sophisticated nanotheranostic which clearly identify the receptor at specific target site for active therapeutic targeting [58, 65].

Li et al. developed innovative peptide-conjugated Au multi-functional Au-nanorods for the diagnostic and therapeutic purpose of AD. They shared the high NIR absorption feature of Au-nanorods with two Aβ-important inhibitors, polyoxometalates (POMs) and Aβ-targeted peptide inhibitors for active Aβ aggregation inhibition and similarly amyloid deposits dissociation by NIR irradiation. Moreover, Au-nanorods' size and shape-dependent photosensitive properties were used as highly effective investigative probes for finding Aβ-aggregates. Therefore, this scientific study demonstrates a multi-functional nanotheranostic containing several components such as a reporter, inhibitors, and targeting ligand in a single nano-system for treating AD [63].

Melancon and co-workers fabricated multi-functional Au-based nanoshells with optical and magnetic properties, which were additionally functionalized with targeting ligand and explored in brain and neck cancer [66]. An innovative concept, using Au-NPs with radiation therapy, has significantly encouraged long-standing survival compared to radiation therapy alone. It showed high therapeutic potential to advance glioma treatment [66]. In recent scientific research, matrix metalloproteinase-2 (MMP-2) responsive gold-gelatin NPs were developed, allowing the pH-triggered drug release of encapsulated drug with an RGD-directing peptide and octarginine for specific glioma targeting. The in-vivo study showed that targeting of glioma with gold nanotheranostic was realized well with co-localization inside the neovessels [67].

Sun et al. developed a novel gold and super-paramagnetic iron oxide (SPIO) encapsulated micelles surface functionalized with polyethylene glycol-polycaprolactone (PEG-PCL) co-polymer for both diagnostic and therapeutic nanomaterial in the treatment of glioblastoma multiforme (GBM). To analyze the potential of developed nanotheranostic, they examined the radio-sensitizing efficiency by quantifying DNA impairment attentions in glioblastoma cell lines. They found that developed micelles administration in combination with radiation therapy led to a two-fold rise in the density of dsDNA disruptions. As an investigative tool for GBM, developed Au-micelles were used as contrasting agents for MRI and CT of GBM tumors applied in a mouse model [63].

**Magnetic NPs**

Magnetic nanoparticles (MNPs) are potential carriers for targeted delivery at neuronal sites with the extra benefit of MRI traceability. Several scientists have shown earlier that these carriers can be entered into the different neuronal sites of the brain with the help of an applied magnetic field. Due to the high magnetic response, the core of iron oxide effectively facilitates targeted brain delivery [68]. Iron oxide NPs usually contain a magnetic core and are coated with polymeric layers such as dextran. Easy and convenient methods such as co-precipitation and thermal decomposition are used to develop the iron oxide NPs. Additionally, the surface modification with various targeting ligands, including QSHYRHISPAQV (QSH) and Calreticulin (CRT) peptides, inorganic molecules, and stabilizers is employed, which facilitate the use of Iron oxide NPs in theranostic applications [69].

Ruan et al. developed a novel Curcumin (Cur)-superparamagnetic iron oxide (SPIO) nanotheranostic and modified it with Calreticulin (CRT) and QSHYRHISPAQV (QSH) peptides. The developed nanotheranostic more competently passes through BBB and effectively binds to the target β-amyloid plaques, which can be detected through MRI. It also induces neurogenesis and neuroprotection and reduces the burden of β-amyloid plaque by effectively regulating the neuroinflammation by inhibiting the NLR family Pyrin-Domain containing-3 (NLRP-3) inflammasome and reversing cognitive deficits of Alzheimer's disease [70]. Chartok et al. developed a nanotheranostic platform and determined the suitability of developed magnetic nanotheranostic for both non-invasive MRI screening and magnetically enhanced accumulation in the brain. They determined that the accumulation of magnetic nanotheranostic in gliosarcomas could be considerably improved with magnetic targeting and MRI effectively measured this. Hence, this nanotheranostic demonstrated a promising approach for the targeted delivery therapeutic [71] (Fig. 1).

**Carbon Nanotubes**
Carbon nanotubes (CNTs) are cylindrical structures made up of graphene sheets. CNTs are the allotropes of carbon with poor biocompatibility and gradual biodegradation. They have distinct mechanical and electronic qualities appropriate for the theranostic implication in brain disorders. CNTs have emerged as promising nanocarriers capable of enhancing chemotherapeutic agents' targeted delivery and therapeutic efficacy in treating brain cancer, thereby presenting a compelling avenue for future clinical translation [72]. CNTs produce mortal heat upon NIR-irradiation. The inherent property of NIR-light absorption has been used to destroy tumor cells in vitro. In contrast, the NIR-photoluminescence property of CNT has been used for in vitro cellular probing and imaging [73]. Costa et al. developed functionalized CNT that can be used as drug carriers in nanotheranostic applications, including brain delivery of BBB impervious agents [74] (Table 1).

## 3.2 Nanotheranostic as Both Diagnostic and Therapeutic Tools

Treatment of different types of CNS diseases can be highly effective if diagnosed at an initial stage of disease development. However, as a matter of apprehension, current diagnostic methods are typically incompetent in identifying CNS disorders at their early stages. Consequently, nanotheranostic, including NPs, encapsulating diagnostic and therapeutic agents, can be game-altering in personalized medicine. The most frequently used diagnostic materials in nanotheranostic are quantum dots or different fluorescent materials such as dyes for optical imaging, radioactive molecules for imaging, iron oxides for MRI, and heavy metal ions for CT [44]. Besides the diagnostic and therapeutic agents, NPs are the main structural components of a nanotheranostic system. NPs enhance biodistribution and foster pharmacokinetics of the loaded diagnostic and therapeutic agent at the target site. It increases the drug's therapeutic efficacy and decreases systemic toxicity by reducing non-specific tissue distribution [80]. These NPs have been distinguished as surface functionalized with targeting ligands, biomarkers, or other imaging moieties for achieving site-specific drug delivery and real-time diagnostic ability. Nanotheranostic proposed an inclusive range of diagnostic techniques, including magnetic MRI, CT,

**Table 1** List of nanotheranostic modalities used for the treatment of neurological disorders.

| Nanotheranostic modalities | Targeting ligand/ mechanism | Neurological disease | Diagnostic and therapeutic agents | Results | Ref. |
|---|---|---|---|---|---|
| Organic NPs | | | | | |
| Liposome | Anti-CD20 | CNS lymphoma | SPIONs Rituximab | Novel targeted nanotheranostic showing potent anti-lymphoma activity | [75] |
| Liposome | Endocytosis | PD | QD, Apomorphine | Improved brain targeting with great fluorescence intensity | [76] |
| Stealth Immuno-liposome | Anti-HSP72 | Cerebral ischemia | Rhodamine/ gadolinium, Citicoline | Novel targeted nanotheranostic platform | [77] |
| Polymeric NPs | Endocytosis | Malignant glioma | Temozolamide | Traced by iron-oxide used as MRI contrast agent and efficiently reduced glioma growth | [78] |
| Polyfluorene chitosan NPs | Transcytosis | AD | Polyfluorene | Inhibits or modulates amyloid aggregates and sense by their unique optical properties | [79] |
| Polymeric NPs | F3 peptide | Brain tumor | Photofrin, iron oxide | Improved therapeutic efficiency | [80] |
| Inorganic NPs | | | | | |
| Iron oxide NPs | G23 peptide/ receptor-mediated endocytosis | Brain tumor | Doxorubicin | MRI contrast agent for in vitro and in vivo diagnosis and significant tumor reduction | [81] |
| SPIONs-NPs | Endocytosis | AD | SPIONs | Real-time monitoring and ameliorates neurological change | [82] |
| Iron oxide magnetic NPs | Penetration | AD | Iron oxide, Rutin | Aβ aggregates detected by MRI and provide controlled drug release | [83] |
| Gold NPs | Transferrin peptide | Brain tumor | Au and pro-drug (pc −4) | Diagnosis and treatment | [84] |

positron emission tomography (PET), and optical imaging. Through the combination of diagnostic agents, such as quantum dots, iron-oxide, or gold nanoparticles, nanotheranostic allows for the non-invasive imaging of disease biomarkers and tissue morphology at the molecular level [60]. This smart diagnostic ability aids disease detection in an initial state, accurate and precise site-specific drug delivery. The multifunctionality of nanotheranostic also allows the theranostic response, wherein the therapeutic drug efficacy can be guided in real-time by using the encapsulated diagnostic agents. This response loop facilitates the optimization of treatment, evaluation of therapeutic efficacy and dose adjustment, leading to personalized medication. Moreover, nanotheranostic grip potential for minimizing drug resistance and significantly enhancing drug delivery to the BBB [85]. Nanotheranostics offer several characteristic opportunities for targeted brain therapy in addition to their promise for diagnostics. Therapeutic drugs, diagnostic agents, peptides, and genes can all be encapsulated or conjugated onto a signal nanoplatform as the therapeutic components to complete the aim of nanotheranostic in the management of CNS disorder. The efficiency of these therapeutic drugs can be amplified while off-target side effects are significantly reduced by sensibly controlling their release at the target site. Additionally, smart nanotheranostic can be developed to respond to internal stimuli such as enzymes, pH and outside stimuli like light, magnetic fields, enabling localized therapy and on-demand drug delivery [86].

## 3.3 Nanotheranostic as Personalized Medicines

"Personalized nanomedicine" defines the use of nanosized materials to elaborate optimized dose regimens of treatment that are provided to each patient. Personalized medicine can be defined as giving "the right drug to the right patient at the right time". Pharmacoproteomic, pharmacogenomic, and numerous other -omic techniques have been established recently to produce a patient-specific medicine [87, 88]. These methods enable a thorough molecular and genetic profile of every patient, helping to pinpoint the molecular indicators that would influence the progression of the illness and how well it would respond to the therapeutic treatments. As a result, personalized medicine relies not only on the study of genetic polymorphisms and biomarkers but also on the creation of methods for diagnosing illnesses and predicting how treatment would work. Nanotheranostic medicines, which combine therapeutic and imaging agents in a single nanocarrier, have high potential to help provide a personalized therapeutic approach to treating severe brain diseases and/or disorders [89]. Nanotheranostic, which was designed for monitoring the progression of a disease non-invasively while it is being treated, will, in other words, promote the personalization of clinical therapies to reflect the unique characteristics of the disease in each patient. The unique diagnostic and therapeutic potential of the smart nano-theranostics facilitates patient screening based on non-invasive imaging, while achieving personalized therapeutic goals by controlling the therapeutic efficacy of the system and predicting patient response to nanomedicine treatments. Although these achievements are encouraging, they are still in the minority and require continued efforts to improve the sensitivity and accuracy of diagnosis and improve therapeutic efficacy for numerous neurological disorders [90].

## 4 Nanotheranostic Targeting Neuroimmune Signaling

Neuroimmune signaling targeting requires crossing the BBB, a highly discriminating biological barrier that limits the entry of foreign materials into the brain. Overcoming the BBB postures is an essential challenge for NPs-based nanotheranostic treatments of CNS disorders. Neuroimmune signaling is a complex relationship between the nervous system and the immune system, which plays a vital role in brain disorders [91]. Nanotheranostic targeting neuroimmune signaling includes utilizing nanoscale materials and developing nanotherapeutics to target and modulate the neuroimmune signaling pathways precisely. These nanotherapeutics are designed to have both diagnostic and therapeutic capabilities, allowing real-time monitoring of disease development and targeted drug delivery [92] (Fig. 3).

The surface of NPs can be functionalized by several active and passive targeting ligands to improve the site-specific diagnostic and therapeutic efficacy. Active neuronal targeting of these theragnostic NPs can be achieved by surface engineering with different targeting ligands such as aptamers, peptides, and antibodies. Ligand-functionalized NPs precisely bind to the specific receptors which are overexpressed

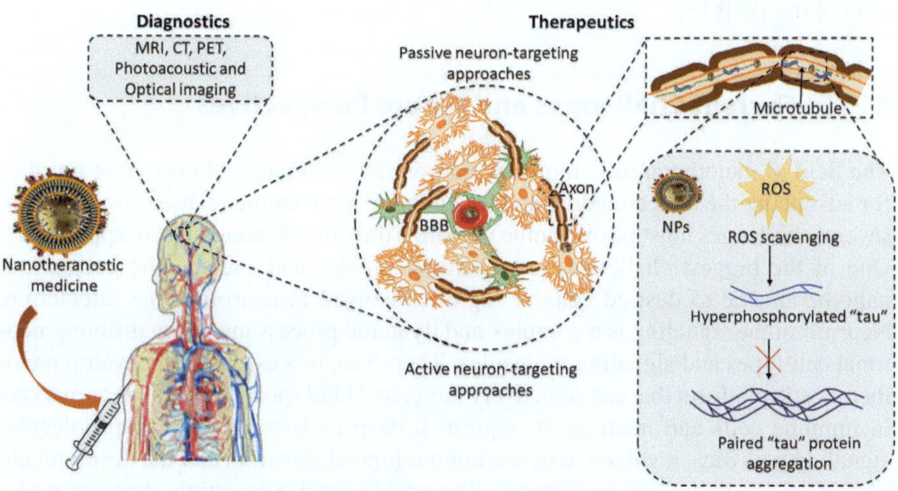


**Fig. 3** Diagnostic and therapeutic approaches for nanotheranostic targeting neuroimmune signaling in neurological disorders. The decoration of NPs with neuron-specific targeting ligands (active and passive targeting) enhances the precise binding to the specific receptors which are overexpressed on the brain endothelial cell surface and allow to cross the BBB in that way reaching to the neurons. The multifunctional nanotheranostic injected precisely targets hyperphosphorylated tau. These nanotheranostics can scavenge ROS to prevent hyperphosphorylation of tau in neurons, and release therapeutic. This leads to the protection neuron from ROS-arbitrated cell death

on the brain endothelial cell surface and allow to cross the BBB [93]. Active targeting is based on the surface engineering of NPs with ligands articulated in neurons. In addition to this, some other studies used dual-targeting approach based on the functionalization of both BBB-penetrating and neuron-targeting ligands on the exterior of NPs. Tet-1 is a peptide having 12-amino acid in their structure which binds precisely to neurons. Therefore, it has been used as a neuron-specific targeting ligand in neuroimmune signaling [94].

Similarly, for passive targeting, the NPs surface can be decorated with hydrophilic polymers such as PEG, PVA, and poly[N-(2-hydroxypropyl) methacrylamide and others. Intriguingly, these polymers improve systemic circulation time and simultaneously prevent recognition by monocytes and macrophages. Thus, decreasing the risk of rapid clearance and opsonization of NPs. These decorated NPs surface functionalized with targeting ligand, which hold the potential for diagnosis and treatment of illness enacted by a single nanoformulation of biodegradable and biocompatible NPs [95]. Presently, several nanotheranostic platforms such as lipidic NPs, polymer-based NPs, inorganic NPs, and others are being revealed for the site-specific diagnosis and management of several neurological disorders [96].

Several scientists recently innovated a smart nanoplatform with a design for more effective drug delivery strategies to treat different brain disorders that do not cause brain damage. Through computer modeling, surface-engineered NPs, such as liposomes, polymeric NPs, nanobubbles, nanosomes, dendrimers, and polymeric micelles, perform remarkably effectively by avoiding the BBB. The delivery of NPs to the CNS is also impeded by the CNS and the extracellular and intracellular barriers of the BBB [5].

## 5 Current Challenges and Future Perspectives

The field of nanotheranostic targeting neuroimmune signaling holds great promise for advancing the understanding and treatment of neuroimmune diseases. However, several challenges must be overcome to realize the full potential of these approaches. One of the biggest challenges is to achieve precise and site-specific delivery of nanotheranostic to desired cells or regions involved in neuroimmune interaction. Neuroimmune signaling is a complex and dynamic process involving different neuronal cell types and signaling molecules. Therefore, it is essential to develop nanotheranostic platform that can selectively target and bind specific receptors or markers in immune cells and neurons. It requires a deep understanding of the molecular signaling pathways involved in neuroimmunological signaling and the development of surface modifications and ligands that enable precise targeting. Another major challenge is crossing the BBB to deliver nanotheranostic to the brain. The BBB is a highly selective barrier that limits the entry of diagnostic and therapeutic agents into the brain. To enable effective neurotheranostics, operate nanotheranostic systems with BBB-targeting ligands, receptor-mediated transcytosis, or perturbation techniques such as focused ultrasound or transient BBB opening. Crossing the BBB barrier facilitates the delivery of nanotheranostic to their targets in the brain, enabling effective real-time monitoring and therapy of neuroimmune diseases. Imaging is essential in neurotherapy because it enables high-resolution, real-time imaging of neuroimmune interactions.

The development of improved imaging techniques and probes that can specifically target neuroimmune markers and signaling molecules is of great importance. Integrating multiple imaging modalities such as MRI, PET, or optical imaging in a nanotheranostic system provides comprehensive and complementary information about neuroimmunological signaling. Future research should focus on developing new imaging probes and techniques that can accurately and sensitively detect neuroimmune processes in vivo, facilitating accurate diagnosis and therapeutic monitoring. Ensuring the efficacy and safety of nanotheranostic approaches is also a significant challenge. Optimization of therapeutic payload, drug release kinetics, and dosage regimens is necessary to achieve desired therapeutic outcomes while minimizing unwanted side effects. In addition, the long-term potential of immunogenicity, biocompatibility, and clearance mechanisms of nanotheranostic must be thoroughly investigated to ensure their safety in clinical application. Understanding the interactions between nanotheranostic and the immune system is critical to minimizing unwanted immune reactions and maximizing therapeutic efficacy. The future prospects of nanotheranostic targeting of neuroimmune signaling are exciting and have great potential. An individual medical approach is the key to patient-specific diagnosis and treatment. Integrating nanotheranostic with new technologies such as genomics, proteomics, and bioinformatics can identify specific biomarkers as guides for individualized treatment strategies. This individualized approach leads to more effective and targeted treatment, ultimately improving patient outcomes and quality of life. In addition, advances in nanotechnology, such as the development of multi-functional nanocarriers, enable the simultaneous administration of therapeutic agents and imaging probes. These nanotheranostic systems can monitor the treatment response in real-time, allowing the optimization of individual treatment. Combining nanotheranostic with other emerging fields, such as immunotherapy, provides synergistic approaches to regulate neuroimmune signaling and improve treatment outcomes.

## 6 Conclusion

Nanotheranostic platform has emerged as a developing approach in health care by integrating both diagnostic and therapeutic capabilities into a single NP. The convergence of sensing technologies, imaging techniques, gene therapy, phototherapy, and targeted drug delivery within nanotheranostic holds remarkable promise for precision medication and enhanced patient outcomes. While challenges persist, constant improvements and scientific research efforts transport us faster to understand the full potential of nanotheranostic platforms in clinical practice.

## References


1. Tchessalova, D., Posillico, C. K., & Tronson, N. C. (2018). Neuroimmune activation drives multiple brain states. *Frontiers in Systems Neuroscience, 12*, 39.
2. Morrow, A. L., & Price, L. H. (2019). Neuroimmune signaling in neuropsychiatric disease. In *Psychopharmacology* (pp. 2855–2855). Springer.
3. Tian, L., et al. (2012). Neuroimmune crosstalk in the central nervous system and its significance for neurological diseases. *Journal of Neuroinflammation, 9*(1), 1–10.
4. Ahmad, F., et al. (2022). Smart nanoformulations for brain cancer Theranostics: Challenges and promises. *Cancers, 14*(21), 5389.
5. Sharma, M., et al. (2019). Nanotheranostics, a future remedy of neurological disorders. *Expert Opinion on Drug Delivery, 16*(2), 113–128.
6. Mitchell, M. J., et al. (2021). Engineering precision nanoparticles for drug delivery. *Nature Reviews Drug Discovery, 20*(2), 101–124.
7. Woźniak, M., et al. (2022). Molecular imaging and nanotechnology—Emerging tools in diagnostics and therapy. *International Journal of Molecular Sciences, 23*(5), 2658.
8. Mushtaq, S., et al. (2021). Recent progress in technetium-99m-labeled nanoparticles for molecular imaging and cancer therapy. *Nanomaterials, 11*(11), 3022.
9. Kumar, R., et al. (2021). Review of nanotheranostics for molecular mechanisms underlying psychiatric disorders and commensurate nanotherapeutics for neuropsychiatry: The mind knockout. *Nano, 5*(3), 288.
10. Refaat, A., et al. (2022). In vivo fluorescence imaging: Success in preclinical imaging paves the way for clinical applications. *Journal of Nanobiotechnology, 20*(1), 1–22.
11. Estelrich, J., Sánchez-Martín, M. J., & Busquets, M. A. (2015). Nanoparticles in magnetic resonance imaging: From simple to dual contrast agents. *International Journal of Nanomedicine, 10*, 1727.
12. Teleanu, D. M., et al. (2019). Nanomaterials for drug delivery to the central nervous system. *Nanomaterials, 9*(3), 371.
13. Xiao, L., Hareendran, S., & Loh, Y. P. (2021). Function of exosomes in neurological disorders and brain tumors. *Extracellular Vesicles Circulating Nucleic Acids, 2*, 55.
14. Benameur, T., Panaro, M. A., & Porro, C. (2022). Exosomes and their cargo as a new avenue for brain and treatment of CNS-related diseases. *The Open Neurology Journal, 16*(1).
15. Dantzer, R. (2018). Neuroimmune interactions: From the brain to the immune system and vice versa. *Physiological Reviews, 98*(1), 477–504.
16. Klein Wolterink, R. G., et al. (2022). Neuroimmune interactions in peripheral organs. *Annual Review of Neuroscience, 45*, 339–360.
17. Liberman, A. C., et al. (2019). Neuroimmune and inflammatory signals in complex disorders of the central nervous system. *Neuroimmunomodulation, 25*(5–6), 246–270.
18. Donzis, E. J., & Tronson, N. C. (2014). Modulation of learning and memory by cytokines: Signaling mechanisms and long term consequences. *Neurobiology of Learning Memory, 115*, 68–77.
19. Jacob, A., et al. (2007). The role of the complement cascade in endotoxin-induced septic encephalopathy. *Laboratory Investigation, 87*(12), 1186–1194.
20. Colonna, M., & Butovsky, O. (2017). Microglia function in the central nervous system during health and neurodegeneration. *Annual Review of Immunology, 35*, 441–468.
21. Nutma, E., et al. (2020). Astrocyte and oligodendrocyte cross-talk in the central nervous system. *Cells, 9*(3), 600.
22. Domingues, H. S., et al. (2016). Oligodendrocyte, astrocyte, and microglia crosstalk in myelin development, damage, and repair. *Frontiers in Cell Developmental Biology, 4*, 71.
23. Stence, N., Waite, M., & Dailey, M. E. (2001). Dynamics of microglial activation: A confocal time-lapse analysis in hippocampal slices. *Glia, 33*(3), 256–266.
24. Zamanian, J. L., et al. (2012). Genomic analysis of reactive astrogliosis. *Journal of Neuroscience, 32*(18), 6391–6410.

25. Paolicelli, R. C., Bisht, K., & Tremblay, M.-È. (2014). Fractalkine regulation of microglial physiology and consequences on the brain and behavior. *Frontiers in Cellular Neuroscience, 8*, 129.
26. Heremans, H., et al. (1989). The role of cytokines in various animal models of inflammation. *Lymphokine Research, 8*(3), 329–333.
27. Becher, B., Spath, S., & Goverman, J. (2017). Cytokine networks in neuroinflammation. *Nature Reviews Immunology, 17*(1), 49–59.
28. Chovatiya, R., & Medzhitov, R. (2014). Stress, inflammation, and defense of homeostasis. *Molecular Cell, 54*(2), 281–288.
29. Olivieri, R., et al. (2018). The additive effect of aging on sepsis-induced cognitive impairment and neuroinflammation. *Journal of Neuroimmunology, 314*, 1–7.
30. Singer, B. H., et al. (2016). Cecal ligation and puncture results in long-term central nervous system myeloid inflammation. *PLoS One, 11*(2), e0149136.
31. Ramesh, G., MacLean, A. G., & Philipp, M. T. (2013). Cytokines and chemokines at the crossroads of neuroinflammation, neurodegeneration, and neuropathic pain. *Mediators of Inflammation, 2013*, **1**.
32. Kumar, V. (2019). Toll-like receptors in the pathogenesis of neuroinflammation. *Journal of Neuroimmunology, 332*, 16–30.
33. Li, L., et al. (2021). Role of astroglial toll-like receptors (TLRs) in central nervous system infections, injury and neurodegenerative diseases. *Brain, Behavior, Immunity, 91*, 740–755.
34. Teleanu, R. I., et al. (2022). Neurotransmitters—Key factors in neurological and neurodegenerative disorders of the central nervous system. *International Journal of Molecular Sciences, 23*(11), 5954.
35. Hodo, T. W., et al. (2020). Critical neurotransmitters in the neuroimmune network. *Frontiers in Immunology, 11*, 1869.
36. Yu, H., et al. (2020). Targeting NF-κB pathway for the therapy of diseases: Mechanism and clinical study. *Signal Transduction Targeted Therapy, 5*(1), 209.
37. Guo, H., Callaway, J. B., & Ting, J. P. (2015). Inflammasomes: Mechanism of action, role in disease, and therapeutics. *Nature Medicine, 21*(7), 677–687.
38. Markman, J. L., et al. (2013). Nanomedicine therapeutic approaches to overcome cancer drug resistance. *Advanced Drug Delivery Reviews, 65*(13–14), 1866–1879.
39. Jagtap, P., et al. (2017). Nanotheranostic approaches for management of bloodstream bacterial infections. *Nanomedicine: Nanotechnology, Biology and Medicine, 13*(1), 329–341.
40. Sumer, B., & Gao, J. (2008). Theranostic nanomedicine for cancer. *Nanomedicine, 2*, 137–140.
41. Caldorera-Moore, M. E., Liechty, W. B., & Peppas, N. A. (2011). Responsive theranostic systems: Integration of diagnostic imaging agents and responsive controlled release drug delivery carriers. *Accounts of Chemical Research, 44*(10), 1061–1070.
42. Wong, X. Y., et al. (2020). Nanomaterials for nanotheranostics: Tuning their properties according to disease needs. *ACS Nano, 14*(3), 2585–2627.
43. Shukla, R., et al. (2020). Nanophytomedicine market: Global opportunity analysis and industry forecast. In *Nanophytomedicine: Concept to clinic* (pp. 19–31).
44. Janib, S. M., Moses, A. S., & MacKay, J. A. (2010). Imaging and drug delivery using theranostic nanoparticles. *Advanced Drug Delivery Reviews, 62*(11), 1052–1063.
45. Jadhav, K., et al. (2022). Neurodegenerative disorders due to inhalation of various small particles. In *Nanomedical drug delivery for neurodegenerative diseases* (pp. 41–54). Elsevier.
46. Lim, E.-K., et al. (2015). Nanomaterials for theranostics: Recent advances and future challenges. *Chemical Reviews, 115*(1), 327–394.
47. Singh, R., et al. (2023). New generation smart drug delivery systems for rheumatoid arthritis. *Current Pharmaceutical Design, 29*, 984.
48. Patel, K. D., Singh, R. K., & Kim, H.-W. (2019). Carbon-based nanomaterials as an emerging platform for theranostics. *Materials Horizons, 6*(3), 434–469.
49. Kojima, R., Aubel, D., & Fussenegger, M. (2015). Novel theranostic agents for next-generation personalized medicine: Small molecules, nanoparticles, and engineered mammalian cells. *Current Opinion in Chemical Biology, 28*, 29–38.

50. Ahmad, J., et al. (2020). Nanostructured lipid carriers (NLCs): Nose-to-brain delivery and theranostic application. *Current Drug Metabolism, 21*(14), 1136–1143.
51. Hersh, A. M., Alomari, S., & Tyler, B. M. (2022). Crossing the blood-brain barrier: Advances in nanoparticle technology for drug delivery in neuro-oncology. *International Journal of Molecular Sciences, 23*(8), 4153.
52. Kaur, I. P., et al. (2008). Potential of solid lipid nanoparticles in brain targeting. *Journal of Controlled Release, 127*(2), 97–109.
53. Anwar, A., Siddiqui, R., & Khan, N. A. (2018). Importance of theranostics in rare brain-eating amoebae infections. *ACS Chemical Neuroscience, 10*(1), 6–12.
54. Saesoo, S., et al. (2018). Characterization of liposome-containing SPIONs conjugated with anti-CD20 developed as a novel theranostic agent for central nervous system lymphoma. *Colloids Surfaces B: Biointerfaces, 161*, 497–507.
55. Van Vlerken, L. E., & Amiji, M. M. (2006). Multi-functional polymeric nanoparticles for tumour-targeted drug delivery. *Expert Opinion on Drug Delivery, 3*(2), 205–216.
56. Koo, Y.-E. L., et al. (2006). Brain cancer diagnosis and therapy with nanoplatforms. *Advanced Drug Delivery Reviews, 58*(14), 1556–1577.
57. Masserini, M. (2013). Nanoparticles for brain drug delivery. *International Scholarly Research Notices, 2013*.
58. Xie, J., Lee, S., & Chen, X. (2010). Nanoparticle-based theranostic agents. *Advanced Drug Delivery Reviews, 62*(11), 1064–1079.
59. Singh, A., et al. (2016). Multifunctional photonics nanoparticles for crossing the blood–brain barrier and effecting optically trackable brain theranostics. *Advanced Functional Materials, 26*(39), 7057–7066.
60. Muthu, M. S., Mei, L., & Feng, S.-S. J. N. (2014). Nanotheranostics: Advanced nanomedicine for the integration of diagnosis and therapy. *Nanomedicine, 9*(9), 1277–1280.
61. Viswanadh, M. K., et al. (2018). Nanotheranostics: Emerging strategies for early diagnosis and therapy of brain cancer. *Nano, 2*(1), 70.
62. Meyers, J. D., et al. (2013). Nanoparticles for imaging and treating brain cancer. *Nanomedicine, 8*(1), 123–143.
63. Sun, L., et al. (2016). Theranostic application of mixed gold and superparamagnetic iron oxide nanoparticle micelles in glioblastoma multiforme. *Journal of Biomedical Nanotechnology, 12*(2), 347–356.
64. Fan, Z., et al. (2014). Theranostic nanomedicine for cancer detection and treatment. *Journal of Food and Drug Analysis, 22*(1), 3–17.
65. Norden, A. D., Drappatz, J., & Wen, P. Y. (2008). Novel anti-angiogenic therapies for malignant gliomas. *The Lancet Neurology, 7*(12), 1152–1160.
66. Melancon, M. P., et al. (2011). Targeted multifunctional gold-based nanoshells for magnetic resonance-guided laser ablation of head and neck cancer. *Biomaterials, 32*(30), 7600–7608.
67. Ruan, S., He, Q., & Gao, H. (2015). Matrix metalloproteinase triggered size-shrinkable gelatin-gold fabricated nanoparticles for tumor microenvironment sensitive penetration and diagnosis of glioma. *Nanoscale, 7*(21), 9487–9496.
68. Pankhurst, Q. A., et al. (2003). Applications of magnetic nanoparticles in biomedicine. *Journal of Physics D: Applied Physics, 36*(13), R167.
69. Frimpong, R. A., & Hilt, J. Z. (2010). Magnetic nanoparticles in biomedicine: Synthesis, functionalization and applications. *Nanomedicine, 5*(9), 1401–1414.
70. Ruan, Y., et al. (2022). Highly sensitive Curcumin-conjugated nanotheranostic platform for detecting amyloid-beta plaques by magnetic resonance imaging and reversing cognitive deficits of Alzheimer's disease via NLRP3-inhibition. *Journal of Nanobiotechnology, 20*(1), 322.
71. Chertok, B., et al. (2008). Iron oxide nanoparticles as a drug delivery vehicle for MRI monitored magnetic targeting of brain tumors. *Biomaterials, 29*(4), 487–496.
72. Doane, T. L., & Burda, C. (2012). The unique role of nanoparticles in nanomedicine: Imaging, drug delivery and therapy. *Chemical Society Reviews, 41*(7), 2885–2911.
73. Ren, J., et al. (2012). The targeted delivery of anticancer drugs to brain glioma by PEGylated oxidized multi-walled carbon nanotubes modified with angiopep-2. *Biomaterials, 33*(11), 3324–3333.

74. Costa, P. M., et al. (2018). Functionalised carbon nanotubes enhance brain delivery of amyloid-targeting Pittsburgh compound B (PiB)-derived ligands. *Nano, 2*(2), 168.
75. Shazeeb, M. S., et al. (2014). Liposome-encapsulated superoxide dismutase mimetic: Theranostic potential of an MR detectable and neuroprotective agent. *Contrast Media, 9*(3), 221–228.
76. Wen, C.-J., et al. (2012). Theranostic liposomes loaded with quantum dots and apomorphine for brain targeting and bioimaging. *International Journal of Nanomedicine*, 1599–1611.
77. Agulla, J., et al. (2014). In vivo theranostics at the peri-infarct region in cerebral ischemia. *Theranostics, 4*(1), 90.
78. Bernal, G. M., et al. (2014). Convection-enhanced delivery and in vivo imaging of polymeric nanoparticles for the treatment of malignant glioma. *Nanomedicine: Nanotechnology, Biology, 10*(1), 149–157.
79. Roy Chowdhury, S., et al. (2018). Remarkably efficient blood–brain barrier crossing polyfluorene–chitosan nanoparticle selectively tweaks amyloid oligomer in cerebrospinal fluid and Aβ1–40. *ACS Omega, 3*(7), 8059–8066.
80. Bhojani, M. S., et al. (2010). Targeted imaging and therapy of brain cancer using theranostic nanoparticles. *Molecular Pharmaceutics, 7*(6), 1921–1929.
81. Su, C.-H., et al. (2016). Evaluation of blood–brain barrier-stealth nanocomposites for in situ glioblastoma theranostics applications. *Nanoscale, 8*(15), 7866–7870.
82. Zhang, R., et al. (2016). Traceable nanoparticle delivery of small interfering RNA and retinoic acid with temporally release ability to control neural stem cell differentiation for Alzheimer's disease therapy. *Advanced Materials, 28*(30), 6345–6352.
83. Hu, B., et al. (2015). Nanotheranostics: Congo red/Rutin-MNPs with enhanced magnetic resonance imaging and H2O2-responsive therapy of Alzheimer's disease in APPswe/PS1dE9 transgenic mice. *Advanced Materials, 27*(37), 5499–5505.
84. Dixit, S., et al. (2015). Transferrin receptor-targeted theranostic gold nanoparticles for photosensitizer delivery in brain tumors. *Nanoscale, 7*(5), 1782–1790.
85. Bhowmik, A., Khan, R., & Ghosh, M. K. (2015). Blood brain barrier: A challenge for effectual therapy of brain tumors. *BioMed Research International, 2015*.
86. Wang, Y., et al. (2019). Stimuli-responsive nanotheranostics based on lanthanide-doped upconversion nanoparticles for cancer imaging and therapy: Current advances and future challenges. *Nano Today, 25*, 38–67.
87. Vizirianakis, I. S. (2011). Nanomedicine and personalized medicine toward the application of pharmacotyping in clinical practice to improve drug-delivery outcomes. *Nanomedicine: Nanotechnology, Biology Medicine, 7*(1), 11–17.
88. Vizirianakis, I. S. (2002). Pharmaceutical education in the wake of genomic technologies for drug development and personalized medicine. *European Journal of Pharmaceutical Sciences, 15*(3), 243–250.
89. Jain, K. (2009). Role of nanobiotechnology in the development of personalized medicine. *Nanomedicine, 4*, 249.
90. Mura, S., & Couvreur, P. (2012). Nanotheranostics for personalized medicine. *Advanced Drug Delivery Reviews, 64*(13), 1394–1416.
91. Hajal, C., et al. (2018). In vitro models of molecular and nano-particle transport across the blood-brain barrier. *Biomicrofluidics, 12*(4), 042213.
92. Ahn, S. I., et al. (2020). Microengineered human blood–brain barrier platform for understanding nanoparticle transport mechanisms. *Nature Communications, 11*(1), 175.
93. Xu, S., et al. (2013). Targeting receptor-mediated endocytotic pathways with nanoparticles: Rationale and advances. *Advanced Drug Delivery Reviews, 65*(1), 121–138.
94. Garcia-Chica, J., et al. (2020). An overview of nanomedicines for neuron targeting. *Nanomedicine, 15*(16), 1617–1636.
95. Kevadiya, B. D., et al. (2019). Neurotheranostics as personalized medicines. *Advanced Drug Delivery Reviews, 148*, 252–289.
96. Lee, S. W. L., et al. (2020). Modeling nanocarrier transport across a 3D in vitro human blood-brain–barrier microvasculature. *Advanced Healthcare Materials, 9*(7), 1901486.

# Mitochondrial Health and Critical Neurotransmitters in Neurodegeneration

Nisha Sharma, Manav Jain, and Umashanker Navik

**Abstract**

The primary regulators of neuronal proliferation, differentiation, and survival are neurotransmitters (NT). Numerous physical, psychological, and neurodegenerative diseases are brought on by the impairment of brain physiological functions due to aberrant NT levels. Neurodegenerative symptoms have been linked to deficits in several NTs. While cholinergic and glutamatergic impairments have been connected to cognitive decline in Alzheimer's, monoamine neurotransmission changes have been intimately linked to neuropsychiatric symptoms. Parkinson's disease (PD) results mainly from the death of dopaminergic neurons, majorly in the substantia nigra pars compacta, which results in dopamine insufficiency manifested in movement dysfunction. Huntington's disease, manifested via cognitive and motor changes, is primarily described by altered levels of dopamine, glutamate, adenosine, and acetylcholine. Amyotrophic lateral sclerosis (ALS) covers a variety of symptoms, such as severe metabolic dysregulation, muscular atrophy, paralysis, and motor neuron degeneration, where increased glutamate levels and accompanying neurotoxicity are suggested mechanisms. Modulating neurotransmitter transporters (NTT), which regulate brain NT homeostasis, can be a possible therapeutic approach in certain neurodegenerative illnesses. Also, the studies indicate a nexus between anomalous neuronal metabolism and neurodegeneration. The high energy requirements of the brain make it particularly vulnerable to mitochondrial damage. Most neurodegenerative disorders exhibit aberrant mitochondrial morphology. For cellular energy needs, mito-

N. Sharma · M. Jain
Department of Pharmacology, Postgraduate Institute of Medical Education and Research (PGIMER), Chandigarh, India

U. Navik (✉)
Department of Pharmacology, Central University of Punjab, Bathinda, India

 
R. Shukla et al. (eds.), *Neuroendocrine-Immune Axis*,
https://doi.org/10.1007/978-981-95-5088-3_11

chondria synthesize ATP, and reactive oxygen species (ROS) are toxic byproducts produced during impaired oxidative phosphorylation in mitochondria. The buildup of intracellular ROS can damage mtDNA and nuclear DNA, possibly leading to many age-related disease states. In addition to its evident role in energy homeostasis, mitochondrial dysfunction also interferes with calcium homeostasis, which can have severe consequences in neuropathologies. For instance, in Alzheimer's disease, defects in calcium homeostasis can result in intracellular calcium accumulation, glutamate-induced excitotoxicity, and neuronal loss. In this chapter, we have discussed the role of NTs and mitochondria in the pathogenesis of common neurodegenerative disorders like Parkinson's disease, Alzheimer's disease, amyotrophic lateral sclerosis, and Huntington's disease.

**Keywords**

Neurotransmitters · Neurodegeneration · Mitochondrial health · Therapeutic strategy

## 1 Introduction

Mitochondrial health and critical neurotransmitters play crucial roles in neurodegeneration, a complex and debilitating category of disorders that affect the nervous system. The mitochondria, often referred to as the "powerhouses of the cell," are essential for producing adenosine triphosphate (ATP), the prime energy source consumed by the cells. In neurons, which are highly energy-demanding cells, maintaining optimal mitochondrial function is essential for their survival and proper functioning. Neurotransmitters, on the other hand, are chemical messengers that facilitate communication between neurons. They are crucial for controlling synaptic plasticity, neuronal proliferation, differentiation, and survival, as well as learning, memory, and behavior [1]. The levels of NT, however, can potentially precipitate or contribute to neuronal dysfunction and degeneration when they are altered or disrupted by numerous causes, including genetic mutations, environmental pollutants, aging, or stress. Cognitive, motor, and sensory impairments result from neurodegenerative diseases (ND), a set of disorders characterized by the loss of neuronal populations and their connections throughout time. Since there are currently no effective therapies or cures for the majority of ND, it poses a significant issue for society and the healthcare system. NT systems and their transporters (NTT), which control the synthesis, release, reuptake, and degradation of NT in the synaptic cleft, are linked to many NDs. For instance, the most prevalent form of dementia, Alzheimer's disease (AD), is characterized by deficiencies in cholinergic and glutamatergic transmission as well as changes in monoamine neurotransmission [2]. The second most prevalent ND, Parkinson's disease (PD), is caused by the selective death of dopaminergic neurons in the substantia nigra pars compacta (SNpc), which lowers levels of the neurotransmitter and impairs motor skills [3]. The numerous NT systems affected by Huntington's disease (HD), an autosomal dominant ND brought on by an enlarged CAG repeat positioned in the huntingtin gene, include dopamine, glutamate, adenosine, and acetylcholine [4]. Excitotoxicity and elevated glutamate levels are linked to the fatal neurodegenerative disease (ND) known as amyotrophic lateral sclerosis (ALS) [5]. Modulating the activity or expression of NT and NTT may provide a viable treatment approach for ND, given the significance of NT and NTT in preserving neuronal function and integrity. Several pharmacological drugs that specifically target NT receptors or NTT have been developed to treat ND. For instance, acetylcholinesterase inhibitors (AChEIs) and N-methyl-D-aspartate receptor (NMDAR) antagonists are frequently used in AD patients to address cognitive problems. The major treatment for PD motor symptoms is levodopa (L-DOPA), a dopamine precursor that may pass the blood-brain barrier (BBB).

In recent years, researchers have increasingly recognized the intricate interplay between mitochondrial health and critical neurotransmitters in neurodegenerative disorders. Dysfunction of mitochondria can lead to increased oxidative stress, impaired energy production, and the accumulation of toxic byproducts, all of which can contribute to neuronal damage and cell death [6, 7]. Additionally, disrupted neurotransmitter signaling can hamper neuronal communication and can impair essential processes required for proper brain function. By studying this intricate relationship, the researchers hope to identify potential therapeutic targets and develop novel interventions that could slow down or even halt the progression of neurodegenerative diseases [8, 9]. Approaches aimed at improving mitochondrial function, reducing oxidative stress, and restoring neurotransmitter balance hold promise. Additionally, modulating the neurotransmitter system through pharmacological agents or deep-brain stimulation techniques may provide symptomatic relief and slow down disease progression.

The goal of this book chapter is to give a summary of what is now known about how NT and NTT function in the pathogenesis and treatment of ND. The primary NT systems and their transporters that are engaged in ND are reviewed first. After that, we will talk about how NT dysregulation can aggravate or cause neuronal injury. The current pharmaceutical therapies for ND that target NT receptors or NTT are outlined.

Furthermore, we will explore the significance of mitochondrial health and significant neurotransmitters in the context of neurodegeneration. We will delve into the impact of mitochondrial dysfunction on neurotransmitter regulation and how imbalances in critical neurotransmitters can influence mitochondrial function. Furthermore, we will discuss the potential implications of these interactions for developing new therapeutic strategies to combat neurodegenerative diseases. By shedding light on this convoluted relationship, we aim to contribute to the growing body of knowledge surrounding neurodegeneration and inspire further research in this critical field.

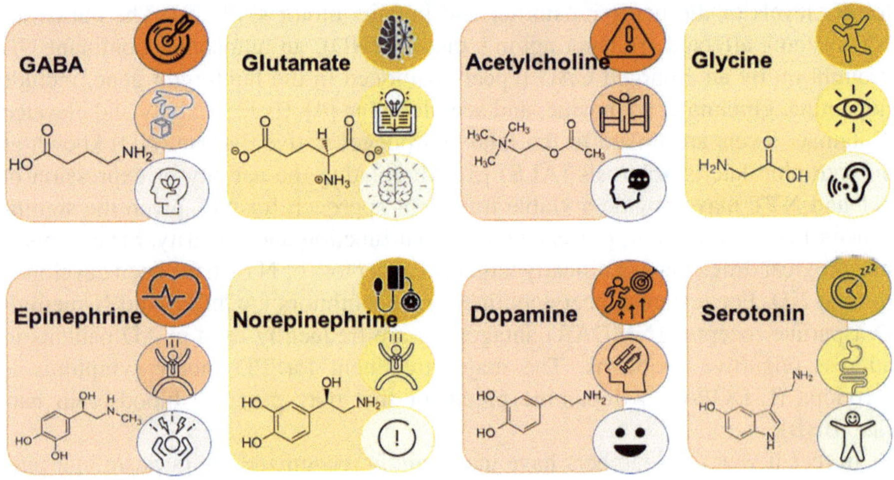


**Fig. 1** Major neurotransmitters involved in the proper functioning of the brain, behavior, and cognition

## 2 Neurotransmitters: Basic Information

Neurotransmitters (NTs) are molecules that intensify, conduct, and transform impulses in cells, and they are indispensable for the proper operation of the brain, behavior, and cognition (Fig. 1). Since 1921, more than 200 NTs have been found; however, the precise number is still unknown because more biomolecules have been discovered with neuroactivity [10]. It must elicit a certain behavior in the postsynaptic neuron, have the same impact on exogenous injection, be formed and released by a similar neuron to be designated as an NT, as well as have a particular mechanism for terminating its action [11]. The effects of various NT types on brain health have been investigated; canonical NTs are generally accepted, but noncanonical NTs are still debatable [12].

### 2.1 Canonical Neurotransmitters

Chemically speaking, canonical neurotransmitters (NTs) are comprised of amines, amino acids, and certain other compounds. Conditional to their actions, they may be central or peripheral, positive or negative [12]. They can be divided into central and peripheral varieties and are produced in various parts of the brain.

#### 2.1.1 Amino Acids

The central nervous system (CNS) depends heavily on amino acids as chemical messengers. Glutamate, glycine, and other amino acids, as well as ϒ-aminobutyric acid (GABA), are essential for normal brain function and play a role in the development of diseases like epilepsy, stroke, and dementia [13].

A precursor of GABA and an excitatory NT in the CNS, glutamate is created from glutamine. N-methyl-d-aspartate (NMDA) and alpha-amino-3-hydroxy-5-methyl-4-isoxazolepropionic acid (AMPA) receptors are activated as a result of its release from presynaptic neurons [14]. Neurologic disorders like Parkinson's disease (PD) may be impacted by an excess of glutamate that causes excessive $Ca^{2+}$ influx in the postsynaptic neuron and causes severe neuronal firing and excitotoxicity. Through release and absorption mechanisms, astrocytes are crucial for maintaining healthy extracellular levels of glutamate. At the tripartite glutamatergic synapse, several mechanisms control glutamate homeostasis [15]. The primary inhibitory neurotransmitter (NT) is GABA, which is either generated by commensal bacteria from the gut microbiota or synthesized by glutamate decarboxylase. Nevertheless, research has revealed that GABA first acts as an excitatory NT, producing depolarization rather than hyperpolarization in a number of nervous system locations. The excessive excitability of neurons is caused by low GABA levels [13]. Another major inhibitory NT is glycine, which co-activates NMDA receptors with glutamate. It participates in auditory, sensory processing, cardiovascular and respiratory functions, voluntary motor control, and exhibits excitatory activity during early development [13]. A new NT made from L-serine through serine racemase is called D-serine, which is released by glial cells. It is found in the corpus striatum, amygdala, anterior olfactory nuclei, olfactory tubercule, rostral cerebral cortex, and hippocampus. It is most concentrated in the protoplasm of the astrocytes that ensheath synapses in the grey matter [13]. Another NT in the CNS, called L-aspartate, has generated debate because of its diverse localizations [13].

### 2.1.2 Amines

Neurotransmitters called monoamines play a crucial role in behavior, motivation, and emotional and motor control. These chemicals are produced by presynaptic neurons and interact with specific receptors on the postsynaptic membrane. Monoamine oxidase or catechol-O-methyltransferase continues to break down the excess monoamines that are still present in the synaptic cleft, which causes severe neurological conditions like Alzheimer's disease, Parkinson's disease, and schizophrenia [16]. Dopamine, also known as 4-(2-aminoethyl)-1,2-benzenediol, is an essential amine NT that is important for both motivational and motor activity. It is made and released by dopaminergic neurons, especially those in the ventral tegmental region and substantia nigra pars compacta. Dopamine also acts as a precursor to other catecholamines, including epinephrine and norepinephrine, and is vital in maintaining homeostasis. Schizophrenia, PD, and HD are a few psychiatric conditions that can be brought on by dopamine dysregulation [16]. Another amine NT implicated in controlling various biological processes is serotonin or 5-hydroxytryptamine. It affects brain function, behavior, vasoconstriction, respiration, sleep and waking states, and gastrointestinal secretion. Through the action of the tryptophan hydroxylase enzyme, enterochromaffin cells in the stomach synthesize the majority of serotonin [16]. Monoamine molecules epinephrine and norepinephrine serve as both NTs and hormones. The locus coeruleus contains norepinephrine neurons that project to several parts of the brain. Epinephrine has been shown to affect the fight-or-flight reaction by means of elevated blood sugar levels, vasodilatation, and pupil dilation [16]. Histamine serves as a signaling molecule and an NT in several physiological processes. Histaminergic neurons in the hypothalamic tuberomammillary nucleus, which branch out to various parts of the brain, produce and release it. Histamine activity has been related in studies to conditions like Alzheimer's disease and schizophrenia [16].

#### 2.1.3 Other Molecules

Molecules called neurotransmitters (NTs) are essential for many biological processes, including cognitive and motor functioning. The neurotransmitter acetylcholine, which is found in the peripheral nervous system, is a well-researched NT and is in charge of causing muscle contraction in the neuromuscular system. Additionally, it has a role in consciousness, voluntary movement control, learning, memory, and attention [11]. Adenosine triphosphate (ATP), soluble gases including carbon monoxide, nitric oxide, and hydrogen sulfide, and a number of neuropeptides such somatostatin, endorphins, vasopressin, neurotensin, substance P, and neuropeptide Y are examples of non-neuronal substances (NTs) [16]. The currency of the cell, ATP, is needed for vital functions such as DNA/RNA synthesis, active transport, signaling, and muscle contraction [17]. A well-researched neurotransmitter (NT), nitric oxide (NO), is a key signaling molecule and a synaptic plasticity regulator [18]. As a feedback inhibitory mechanism, it has been seen to affect D-serine biosynthesis and may diffuse to cells that produce D-serine. Widespread NT Neuropeptide Y expression in the nervous system raises worries about the emergence of neurological disorders, including epilepsy [19].

### 2.2 Noncanonical Neurotransmitters

Small extracellular vesicles (EVs) with a bilipid coating known as exosomes have drawn interest as noncanonical neurotransmitters because of their function in controlling intercellular communication, growth and development, inflammation, antigen presentation, and cancer [20]. Exosomes resemble synaptic vesicles in terms of morphology, but they are released by their parent cells into extracellular regions [12]. They alter pre- and postsynaptic signaling, regulate the release of neurotransmitters, maintain synapses, and promote or inhibit neuronal growth and death, regeneration of axons, and NT production and recycling. Synaptotagmin-4 (Syt4), Wnt, and AMPAR are a few of the cargos found in exosomes that control synaptic plasticity. Due to their release from presynaptic neurons in response to changes in action potential, neuronal exosomes are also suggested to function as neurotransmitters. Inositol 1,4,5-triphosphate (IP3) is produced when G protein-coupled receptors (GPCRs) are activated by neuropeptides and other ligands. Exosomes can cause $Ca^{2+}$ inflow through calcium channels and IP3 receptors to cause $Ca^{2+}$ release from the endoplasmic reticulum [12, 21]. Exosomes have been observed to play a significant role in the progression of neurodegenerative diseases like Alzheimer's disease (AD) and Parkinson's disease (PD), as well as in neuronal communication, nerve regeneration, synaptic function, and immune response [22]. Another group of substances referred to as NTs, steroids, have properties similar to those of neurotransmitters, including initiating intracellular signaling cascades, stimulating intracellular calcium release, gating membrane channels, and activating phosphoinositide 3-kinase-AKT, MAPK, and Src pathways [23]. The discovery of D-aspartic acid in the nervous tissues of animals and human brains suggests that it has a role in developing the nervous system and neurological activity. This little molecule is gaining ground as a possible NT [24].

## 3 Alteration in Neurotransmitters in Neurodegenerative Diseases

Numerous disorders are caused by differences in the levels of various NTs, as was briefly discussed in the section above [25, 26]. NT disorders can be considered as the class of neurometabolic diseases that affect amino acids, monoamines, cholinergic transmission, purines, and other molecules (such as ion channels, neuropeptides, and other neuromodulators) [27]. These defects are mostly inborn and extremely uncommon illnesses. But NT imbalances can also gradually arise as a result of faulty neural receptors, abnormal intracellular signaling, and release, or other synaptic issues [28]. The purpose of the following subsections is to highlight the relationship between changes in neurotransmitters and the most common CNS neurodegenerative illnesses (Fig. 2).

### 3.1 Neurodegenerative Diseases and Their Associated Neurotransmitters

**Alzheimer's Disease**

Alzheimer's disease (AD) is a widespread neurodegenerative condition with a high mortality rate that primarily affects neocortical regions and is described by episodic memory loss, substantial behavioral abnormalities, and behavioral deterioration. Although the exact cause of AD is still unknown, amyloid beta plaques and hyperphosphorylated tau protein clumps are thought to be the primary culprits. Other events have been proposed as potential explanations for neuronal and synaptic degeneration in AD, including alterations in synaptic proteins, NT levels, mitochondrial dysfunction, altered redox signaling, $Ca^{2+}$ deregulation, metabolic stresses, and inflammation [6].

The "amyloid cascade hypothesis," which contends that increased production of amyloidogenic Aβ-42 peptides causes AD by reducing synapses and causing degeneration of neuronal cells, is one of several theories that have been put up to explain the occurrence of AD [29]. The "calcium hypothesis," which contends that an imbalance in cellular calcium homeostasis is the primary cause of neurodegeneration in AD, is another alternative [26]. Peptides decrease spontaneous synaptic activity by forming calcium influx through L-type voltage-sensitive $Ca^{2+}$ channels

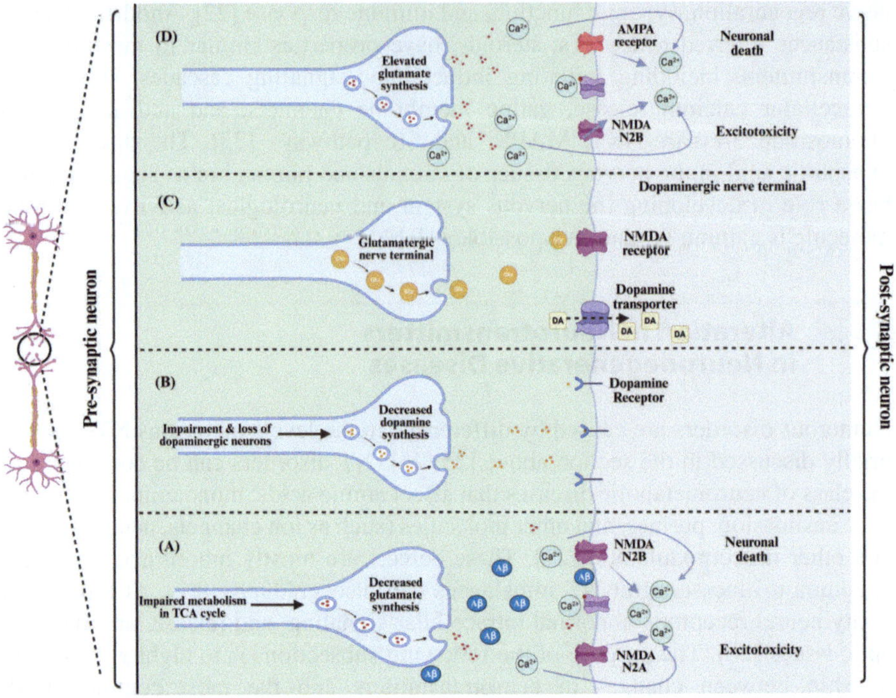


**Fig. 2** Relationship between changes in major neurotransmitters and the most common CNS neurodegenerative illnesses: (**a**) Alzheimer's disease, (**b**) Parkinson's disease, (**c**) Huntington's disease, and (**d**) ALS. (Created with BioRender.com)

(VGCCs) [30]. Cognitive decline, synaptic plasticity impairments, epileptiform symptoms, and neuropsychiatric symptoms are only a few of the neurodegenerative signs of AD that have been allied to deficiencies in different neurotransmitters (NTs) [31]. Amyloid precursor protein and tau phosphorylation metabolism alterations are regulated and encouraged by an aberrant cholinergic system, which results in neurotoxicity, neuroinflammation, and neuronal death [32]. Memory loss results from reduced levels of acetylcholine, which can control proper cholinergic signal transduction in AD patients [31].

It has been demonstrated that restoring dopaminergic neurotransmission can reverse pathologies and cognitive abnormalities in AD patients who have dopaminergic deficiencies. Since dopaminergic impairment is intimately linked to brain reward, it may also be responsible for behavioral abnormalities in AD patients [31]. Last but not least, severe serotonergic denervation and alteration have been observed in AD patients and have been proposed as contributing to AD pathogenesis. The serotonergic system can be a prospective target for managing AD symptoms because it has been demonstrated that restoring serotonergic function can alter behavioral and cognitive symptoms [31].

**Parkinson's Disease**

Lewis bodies and striatal dopamine deficit are symptoms of Parkinson's disease (PD), a continuous progressive neurodegenerative condition marked by the death of dopaminergic neurons, particularly in the substantia nigra pars compacta. Movement abnormalities, hyposomnia, dysautonomia, sleep, and psychiatric/cognitive disorders are some of the symptoms of PD. The striatum's dysregulation of glutamate homeostasis is a newly recognized important component of PD pathogenesis [32]. Astrocytic glutamate excitotoxicity is brought on by inflammation, which changes the expression of glutamate transporters and receptors. By activating extra-synaptic NMDA receptors, the buildup of α-synuclein increases the $Ca^{2+}$ depolarization-dependent release of presynaptic glutamate [7, 26]. Glutamate release is upregulated when AMPA receptors are activated by elevated glutamate levels. The glutamate vesicles that α-synuclein releases cause mGluR5 overexcitation [33].

**Huntington's Disease**

A genetic neurodegenerative brain illness called Huntington's disease impairs movement and cognitive function. Polyglutamine expansions and pathogenic aggregation in striatal and cortical motor and prefrontal areas are the results of the CAG trinucleotide repeat expansion in the first exon of the huntingtin gene, which is the source of the disease [34]. An imbalance in the activity equilibrium of direct and indirect channels is the root cause of abnormal bodily motions. Due to their dense dopaminergic innervation and dopamine receptor populations, the caudate and putamen experience the most severe neurodegeneration. In the initial stages of the disease, augmented thalamocortical glutamatergic transmission results in hyperkinetic movements, but in the later stages, marked by compromised direct and indirect routes, hypokinesia sets in [35]. Dopamine and glutamate signaling pathways may work in concert to increase toxicity by activating the D1 receptor, according to studies [36].

**Amyotrophic Lateral Sclerosis**

A multisystem ailment with clinical, genetic, and neuropathological variability, amyotrophic lateral sclerosis (ALS) was once solely thought to be a motor neuron disease. Multiple symptoms of ALS include the loss of motor neurons, muscle atrophy, paralysis, and severe metabolic dysregulation [37]. Mitochondrial dysfunction, elevated oxidative stress, weakened homeostasis, proteostatic deficits, prominent glutamate excitotoxicity, disturbed RNA metabolism, high AMPA receptors in motor neurons, low $Ca^{2+}$ buffering capacity, neuroinflammation, and neurotrophin depletion are some of the mechanisms underlying ALS. Survival motor neuron protein copy number variation, elongator complex protein 3 variants, and insertion or deletion mutations in the neurofilament heavy polypeptide (NEFH) gene are among the genetic risk factors for ALS. SOD1 mutations are present in 2% of disease cases [38]. It has been attempted to develop therapies based on reducing the concentration of NT in order to treat ALS patients with accumulated glutamate levels and neurotoxicity [7]. The only disease-modifying medication approved in most European nations is riluzole, a medication with anti-glutamatergic properties. It has been shown to extend patient survival by 3–6 months but comes with adverse effects, including nausea, diarrhea, lethargy, dizziness, and liver issues [37].

## 4 Targeting Neurotransmitter Pathways as a Therapeutic Strategy

Drugs that alter the amount or activity of neurotransmitters at the synaptic cleft are the most often used pharmacological strategy to modify neurotransmitter pathways. These medications can affect neurotransmitter production, transport, release, reuptake, or degradation, or they can function as agonists or antagonists at their receptors. For example, acetylcholinesterase inhibitors (AChEIs) are used to treat Alzheimer's disease by increasing its availability and cholinergic transmission [8]. Monoamine oxidase inhibitors (MAOIs) are being used in the treatment of Parkinson's disease by inhibiting the enzyme that degrades dopamine, norepinephrine, and serotonin, increasing their availability and enhancing dopaminergic, noradrenergic, and serotonergic transmission [8]. Glutamate receptor antagonists are used to treat amyotrophic lateral sclerosis (ALS) by blocking excessive activation of glutamate receptors, reducing excitotoxicity and neuronal death [8]. However, these drugs have limitations such as low efficacy, short duration of action, adverse effects, drug interactions, and poor bioavailability. Additionally, they do not address the underlying cause of neurodegeneration and may lose effectiveness as the disease progresses and more neurons die. Therefore, there is a need for more effective and safe pharmacological interventions that target multiple neurotransmitter pathways simultaneously and modify the disease course.

Neurotransmitter transporters (NTTs) are essential for fine-tuning brain NT homeostasis and are attractive targets for treating neurological and neuropsychiatric diseases [39]. NTTs, such as glutamate-aspartate transporter (GLAST) and glutamate transporter-1 (GLT-1) and their human homologs, are mainly present in astrocytes and are accountable for maintaining optimal extracellular glutamic levels, checking accumulation in the synaptic cleft, and thereby eluding excitotoxicity [40, 41]. Pharmacological approaches, such as β-lactam antibiotics, growth factors, estrogen receptor modulators, and histone deacetylase inhibitors, can improve the expression level of these NTTs [40]. Otherwise, mGluR agonists can target glutaminergic signaling and reduce glutamate levels in the synaptic cleft [41]. Also, many agents, including antidepressants exert their actions through monoamine transporters (MATs) for neuropsychiatric and neurodegenerative illnesses [42]. It has been reported that caffeine is an effective NT modulator in the mesocorticolimbic brain areas, exhibiting the potential for alleviating the severity of many neurological disorders through the regulation of dopaminergic signaling. As a promising treatment for Parkinson's disease, caffeine has been demonstrated to be protective against the loss of dopaminergic neurons [43]. Consuming prebiotic chitooligosaccharides has been shown to significantly reduce acetylcholinesterase activity, making it an advantageous substance for preventing or treating Alzheimer's disease

[44]. Drugs that act on specific receptors can mimic NTs and modify neurotransmission in pathological conditions. For example, heroin which is structurally similar to the brain's natural opioids, stimulates many more receptors and, therefore, can lead to the stimulation of opioid receptor activity [8].

Novel therapeutic strategies are being developed to target neurotransmitter pathways in neurodegenerative diseases, aiming to restore or replace lost or damaged neurons or their components or modulate endogenous mechanisms that regulate neurotransmitter homeostasis. These strategies include gene therapy, stem cell therapy, nanotechnology, and microbiota modulation. For instance, tyrosine hydroxylase (TH), aromatic L-amino acid decarboxylase (AADC), or glial cell line-derived neurotrophic factor (GDNF) genes have been delivered via gene therapy using adeno-associated virus (AAV) vectors to the striatum of PD patients to improve dopaminergic transmission and restore DA synthesis [45]. In order to increase cholinergic transmission and restore ACh synthesis, gene therapy has also been utilized to deliver genes encoding for choline acetyltransferase (ChAT), vesicular ACh transporter (VAChT), or nerve growth factor (NGF) to the basal forebrain of AD patients [46]. Similarly, stem cell therapy has been utilized to create dopaminergic neurons and transplant them into the striatum of PD patients to improve dopaminergic transmission and restore DA production [47]. Nanotechnology can overcome some of the limitations of conventional pharmacological approaches, such as poor bioavailability, low specificity, short half-life, and adverse effects. For example, AChEIs, MAOIs, or glutamate receptor antagonists have been delivered via liposomes, polymeric nanoparticles, or magnetic nanoparticles through the blood-brain barrier (BBB) and to the affected brain regions in neurodegenerative diseases to increase their efficacy and lower their toxicity [48]. Microbiota modulation involves manipulating the gut microbiota to influence the production and metabolism of neurotransmitters in the brain. The gut microbiota can modulate neurotransmitter pathways by producing neurotransmitters or their precursors, such as GABA, glutamate, DA, NE, and 5-HT, or by influencing the expression and activity of enzymes, transporters, and receptors involved in neurotransmitter pathways [32].

In conclusion, targeting neurotransmitter pathways is a promising therapeutic strategy for neurodegenerative diseases. However, more research is needed to optimize existing pharmacological approaches and develop novel therapeutic strategies that can target multiple neurotransmitter pathways simultaneously and modify the disease course.

## 5 Mitochondria: Biogenesis, Functioning, and Regulation

Mitochondria are dynamic organelles in mammalian cells that exhibit a key role in energy production, metabolism, and various cellular processes [49, 50]. Mitochondrial biogenesis is the process through which fresh mitochondria are formed from the existing mitochondria within a cell. It involves the replication of mitochondrial DNA (mtDNA) and synthesizing proteins necessary for mitochondrial function. The biogenesis of mitochondria is tightly regulated and can be influenced by various factors such as energy demands, cellular signaling, and environmental cues [51, 52].

Mitochondria possess their genetic material known as mitochondrial DNA (mtDNA), a circular double-stranded DNA molecule [53]. It encodes a limited number of genes, typically 37 in human mtDNA. These genes include 22 transfer RNA (tRNAs) and two ribosomal RNAs (rRNAs) responsible for the mitochondrial translational machinery. Additionally, there are 13 genes encoding structural proteins essential for the functioning of the oxidative phosphorylation (OXPHOS) system [53, 54]. These proteins comprise seven subunits of NADH: ubiquinone oxidoreductase (complex I), one subunit of ubiquinone: cytochrome c oxidoreductase (complex III), three subunits of cytochrome c oxidase (complex IV), and two subunits of the $H^+$ ATP synthase (complex V). Other OXPHOS subunits and factors involved in mtDNA metabolism are nuclear-encoded and subsequently imported into mitochondria through translocases in inner and outer mitochondrial membranes [55]. Notably, mtDNA exhibits unique characteristics, such as maternal inheritance, high copy numbers per cell, minimal intergenic spacing, overlapping genes, absence of introns, and a distinct genetic code [56]. Recent studies propose the existence of nucleoid structures, composed of multiple mtDNA molecules and associated proteins, likely linked to the inner mitochondrial membrane, which serves as functional units for separation and maintenance [57–59].

Regulation of the mitochondrial network involves more than just fusion, fission, and transport. It encompasses broader behaviors like cristae remodeling, mitophagy, and endoplasmic reticulum-mitochondrial contact sites [60]. The dynamic environment of the mitochondrial complex enables it to respond and adapt to cellular stress and fluctuating metabolic demands [61, 62]. Mitochondrial biogenesis is modulated in order to adjust the number of mitochondria in a cell according to its energy requirements, which are influenced by factors such as growth, cell division, oxidative stress, and hormones [63]. The process of biogenesis involves the coordinated functioning of genes expressed from both nuclear and mitochondrial DNA (Fig. 3). Several factors contribute to this process, including members of the PPARγ coactivator-1 (PGC-1) family, Mitochondrial Transcription Factor A (TFAM), sirtuins, Nuclear Respiratory Factors 1 and 2 (NRF1 and NRF2), and AMP-activated protein kinase (AMPK) [63, 64].

The primary regulator of mitochondrial biogenesis, PGC-1, interacts with NRF1 directly to promote the transcription of the genes [65]. Increased expression of PGC-1α in cerebellar neurons leads to a 30% increase in mitochondrial density and protects neurons against degeneration induced by mutant synuclein A53T-α or a mutant Huntingtin gene (Htt) [66]. In fact, the impaired function of PGC-1α is a common characteristic of several neurodegenerative diseases [67]. Among other factors are sirtuins, a class of protein deacetylases known as class III deacetylases. Studies have shown that upregulation of SIRT1 expression can improve various pathological conditions, including neurodegenerative illnesses [68–70]. In conditions of energy deficiency caused by injuries, SIRT1 upregulation exerts neuroprotective effects by boosting mitochondrial biogenesis and activating the turnover of impaired mitochondria [70, 71]. In contrast to SIRT1, primarily located in the

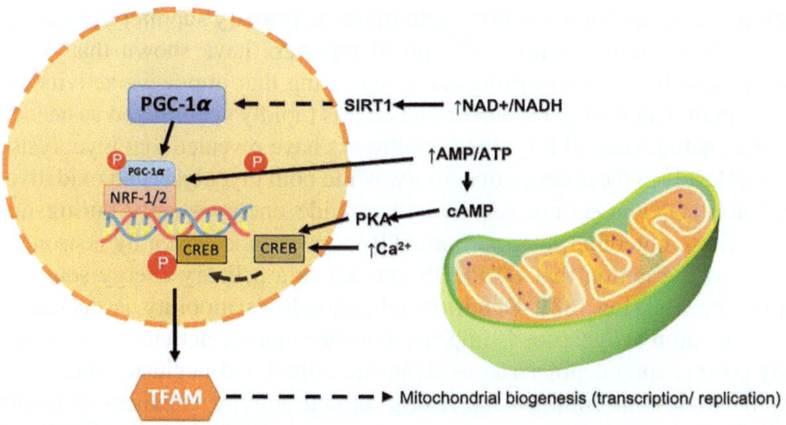


**Fig. 3** Representation of the process of mitochondrial biogenesis involving the coordinated functioning of genes expressed from both nuclear and mitochondrial DNA

nucleus, SIRT3 is localized to mitochondria, where it has a prime role in controlling oxidative stress via activation of superoxide dismutase-2 (SOD2) [72]. SIRT3 also deacetylates SOD2, enhancing its ability to scavenge reactive oxygen species (ROS) [73].

AMP-activated protein kinase (AMPK) is a crucial regulator of cellular metabolism in mammalians. It is stimulated by a rise in the ratio of AMP to ATP, leading to better glucose transport, fatty acid oxidation, and other metabolic processes [74, 75]. AMPK also plays a role in regulating mitochondrial function. On the one hand, it stimulates mitochondrial biogenesis by increasing the transcription of genes controlled by PGC-1α. While AMPK triggers the removal of damaged mitochondria by a process called mitophagy, mediated by Unc-51-like autophagy activating kinase (ULK1). Eventually, the damaged mitochondria are effectively replaced with new, functional ones attributable to these dual AMPK-controlled mechanisms [76].

Neurons, which are polarized, excitable cells with complex morphology, have high energy requirements [77]. Mitochondria play a vital role in neuronal function and are directed to areas with increased energy needs. Interestingly, there are distinct morphological differences between mitochondria in the axonal and dendritic compartments. Dendritic mitochondria occupy a larger volume than axonal mitochondria, which are smaller and more widely spaced out [78]. Axonal mitochondria are more dynamic than dendritic mitochondria in terms of mobility and response to neural activity and metabolism [78]. Neurons need significant molecules of ATP to support various energy-intensive processes. Despite constituting only about 2% of body mass, the human brain consumes approximately 20% of the body's oxygen, highlighting the substantial energy consumption [79]. Conserving resting membrane potential, reinstating ionic balance after depolarization, and synaptic vesicle recycling all require substantial ATP. Moreover, the intricate structural network of neurons presents a difficult task of energy supply to distal sites [79, 80].

Traditionally, glycolysis was thought to be the primary supplier of neuronal ATP [81]. However, studies using ATP optical reporters have shown that ATP levels remain relatively stable in resting cells, indicating that impulsive activity may not inflict a significant energy burden or that ATP is rapidly synthesized as needed [82]. Experiments blocking ATP synthesis pathways have revealed that glycolysis maintains ATP levels in the absence of activity, while both glycolysis and oxidative phosphorylation (OXPHOS) are required to provide energy supply during neuronal activity. These outcomes confirm that ATP production is driven by neuronal action and support the notion that OXPHOS can act as a primary energy source during synaptic transmission [83]. Dendritic mitochondria's mobility is decreased, and their localization at synapses is improved as the synaptic activity is increased [78]. During postsynaptic depolarization, dendritic mitochondria cluster and extend into the spinal bases. In contrast, increasing axonal activity enhances mitochondrial localization at synapses with a shortening in mitochondrial length [84]. The number of presynapses with mitochondria reduces in response to the suppression of synaptic activity, however, reduction in axonal activity increases mitochondrial mobility. These observations highlight the profound effects of synaptic activity on mitochondrial behavior and localization.

Mitochondria also play a crucial role in buffering intracellular calcium ($Ca^{2+}$) at synapses. Voltage-gated $Ca^{2+}$ channels open during an action potential, leading to $Ca^{2+}$ influx into the presynaptic terminal and neurotransmitter release [85]. The electron transport chain's activity establishes a mitochondrial membrane potential (MMP), which favors the sequestration of $Ca^{2+}$ within the mitochondrial matrix. Mitochondria possess a significant capacity to stock and buffer intracellular $Ca^{2+}$, thereby modulating neurotransmission. Presynaptic mitochondria have been shown to regulate presynaptic $Ca^{2+}$ levels, influencing synaptic strength and neurotransmitter release [86].

In summary, neurotransmission and neuronal plasticity are directly influenced by mitochondrial dynamics at both pre-and post-synaptic sides of the synapse.

## 6 Dysfunctional Mitochondria: A Hub of Oxidative Stress

Mitochondria are well-known sources of reactive oxygen species (ROS) production in mammalian cells, both in normal and pathological conditions [87]. The generation of mitochondrial ROS (mtROS) was first discovered in the 1970s, and it is estimated that 1%–2% of cellular oxygen consumption is used for ROS production [88]. mtROS are formed as byproducts of bioenergetic metabolism during OXPHOS, where one-electron transfers produce superoxide ($O_2\cdot$), which is converted to hydrogen peroxide ($H_2O_2$) by superoxide dismutase enzymes [89].

Multiple sites within mitochondria contribute to mtROS production, including the electron transport chain (ETC) positioned on the inner mitochondrial membrane [90]. Among them, complexes I and III of the ETC are major sites of ROS generation [90]. Additionally, other mitochondrial proteins such as dehydrogenase of α-phosphate dehydrogenase, acyl-CoA dehydrogenase, and adrenodoxin reductase-adrenodoxin-cytochrome P450scc system are involved in mtROS production [91]. The regulation of mtROS production involves various factors, including mitochondrial membrane potential, metabolic state, and oxygen levels.

Uncoupling proteins (UCPs) present in the inner mitochondrial membrane decrease ROS production by inducing mild-uncoupling events, which decreases ROS production from the ETC [92, 93]. Sirtuins are also involved in modulating ROS levels and have antioxidant and redox signaling functions [94]. SOD enzymes, including SOD1 and SOD2, convert $O_2\cdot$ to $H_2O_2$, and other antioxidant systems, such as catalase and glutathione peroxidase, neutralize $H_2O_2$ [95]. Overall, the levels of mtROS are influenced by the interplay of ROS generation and antioxidant systems within mitochondria.

## 7 Interplay between Mitochondrial Abnormalities and Neurodegeneration Process

The interplay between mitochondrial abnormalities and the neurodegeneration process is a complex but crucial aspect of several neurodegenerative diseases. Mitochondrial dysfunction can act as both a cause and consequence of neurodegeneration, creating a vicious cycle that ultimately contributes to disease progression. Mitochondrial dysfunction has been associated with the pathogenesis of illnesses such as AD, PD, HD, ALS, and others.

### 7.1 Mitochondrial Regulation of Neurogenesis

Neurons in the central nervous system have high energy requirements, with the bulk of ATP being utilized for maintaining neuronal excitability and synaptic function. Mitochondria, through oxidative phosphorylation, play a crucial role in meeting these energy demands during neurogenesis, the process of forming new neurons from neural precursor cells (NPCs). NPCs can be found in neonatal, embryonic brains, and even adult brains as the brain matures, although the extent of neurogenesis in adult humans is still debated [96].

During neurogenesis, there is a metabolic shift from glycolysis in undifferentiated stem cells to a reliance on mitochondrial OXPHOS in mature neurons [97]. As neural cells differentiate, their mitochondria multiply and take on an extended shape to promote OXPHOS [97]. These changes in energy metabolism and mitochondrial morphology are not just by-products of neuronal differentiation but are instrumental in initiating neurogenesis. NPCs with fused mitochondria maintain stemness, while those with fragmented mitochondria possess the ability to develop into neurons [98]. This causative link observed between neurogenesis and mitochondrial dynamics provides novel insights for possible therapeutic goals for the treatment of neurodegenerative disorders associated with mitochondrial dysfunction.

Mitochondrial activity is crucial for adult neurogenesis, as proliferating immediate progenitor cells become reliant on OXPHOS, and disturbance of mitochondrial

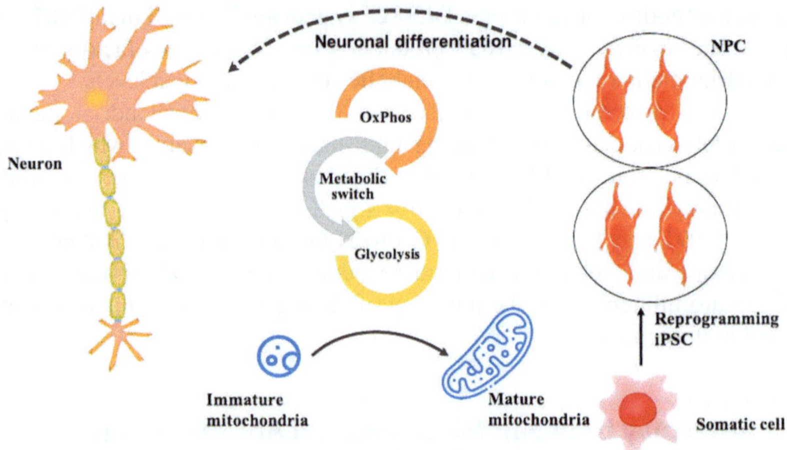


**Fig. 4** Representation of the role of mitochondria in mammalian neurogenesis

function ultimately leads to premature aging phenotypes [99]. Redox regulation and mitochondrial function impact the balance between stemness and neuronal commitment of NPCs. Impairing mitochondrial function in NPCs can cause calcium imbalance, delayed cortical neurogenesis, and disruption of neuronal generation from embryonic stem cells [97, 100]. Furthermore, proper mitochondrial functioning is essential not only for the survival of postmitotic neurons but also for maintaining self-renewing capacity in NPCs. Proteins such as MCPH1 and ARHGAP11B, known to be essential for neurogenesis and neocortex expansion, act within the mitochondria of NPCs [101]. Therefore, mitochondrial function plays a pivotal role in mammalian neurogenesis and the growth of the neocortex (Fig. 4).

## 7.2 Age-Associated Changes in Neuronal Metabolism

The function of neuronal mitochondria deteriorates as individuals age, as evidenced by smaller mitochondrial size and increased brain damage in patients with age-related disorders [102]. This decline in mitochondrial function during aging is commonly characterized by reduced respiratory capacity, altered activity of mitochondrial enzymes, disrupted mitochondrial proteostasis, and enhanced generation of reactive oxygen species (ROS) [103]. Changes in mitochondrial morphology, such as progressive fragmentation, have also been observed in aging cells, particularly in neurons and muscle cells. Maintaining proper mitochondrial dynamics in the nervous system has been shown to delay aging [102, 103].

The worsening in neuronal mitochondrial function with age can have implications for the connectivity and plasticity of neurons, potentially contributing to the onset of neurological diseases. In mice and rats, neuronal mitochondria synthesize melatonin, an endogenous free radical scavenger that decreases with aging, leading to the activation of neuronal inflammatory responses [104]. Age-related loss or damage of mitochondria also results in reduced ATP production, which further accelerates cerebral blood flow decline and mild cognitive impairment [103]. In aging mice, overexpression of the cytochrome c oxidase subunit Va (COX5A) gene supports sustaining mitochondrial function and marginally enhances hippocampus synaptic plasticity and spatial recognition memory [104].

## 7.3 Mitostasis Dysfunction in Neuropathy

The association of mitochondria with neurodegenerative diseases is extensively documented, making it challenging to distinguish between incidental damage to mitochondria and their actual role in the disease process. The numerous functions that mitochondria perform in the disrupted nervous systems can be organized using the idea of mitostatic neurological disorders. These disorders are characterized by specific disruptions in mitostasis, such as altered mitochondrial biosynthesis, impaired fusion/fission dynamics, or dysfunctional clearance mechanisms [105]. Conditions solely characterized by bioenergetics failure, depolarization, or changed mitochondrial shape are excluded from this definition [105].

Genetic evidence strongly supports mitostatic origin in certain diseases. For example, disruptions in mitochondrial fusion and fission genes lead to neuropathies affecting optic and peripheral nerves, highlighting the significance of mitochondrial dynamics in neurons [105]. Numerous rare mutations affecting mitostatic pathways have been discovered, including those in genes such as DRP1, mitofusin 2, Opa1, and TRAK1 [106]. Parkinson's disease is the most commonly identified mitostatic disease, with mutations in PINK1 and Parkin impacting mitochondrial clearance [107, 108]. However, the interplay between mitochondrial movement and clearance in Parkinson's disease is complex, and other factors, such as LRRK2 mutations, can also contribute to mitochondrial dysfunction [109].

Mitochondrial pathology is also prominent in diseases like Alzheimer's, ALS, hypoxia, stroke, and conditions involving mutations in motor proteins or microtubule-associated proteins [110]. Mitochondrial damage is increasingly recognized as a probable factor in axonal injury in multiple sclerosis, where altered mitochondrial transport and density occur due to inflammation and demyelination [111]. It is complicated and yet not fully known how mitochondria function in multiple sclerosis and how neuroinflammatory demyelination affects mitochondrial dynamics.

Recent progress has revealed the influence of mitochondria on axonal growth during neurodegeneration. Mitochondria present in the developing axon tip are crucial for persistent axon growth after axotomy, leading to the upregulation of mitochondrial export into the axonal stump [112]. These findings highlight the importance of mitostatic processes in neuronal survival and present opportunities for investigating the effect of neuronal mitochondria on the pathophysiology of disease using genetic screens.

In summary, neurons employ diverse mechanisms to regulate mitostasis, reflecting its critical role in their survival. Enhancing mitostasis may hold promise for improving mitochondrial health and neuronal survival in various pathological conditions. Cell biologists still have a lot to discover to fully comprehend mitochondrial dynamics in neurons.

## 8 Mitochondrial-Targeted Therapies: Implication in Neurodegenerative Diseases

Given the importance of mitochondrial dysfunction in neurodegenerative diseases, researchers have explored various mitochondrial-targeted therapies as potential treatments. These therapies aim to improve mitochondrial function, reduce oxidative stress, and enhance cellular energy production.

### 8.1 Employing Mitochondrial-Directed Antioxidants Against Neurodegeneration

ROS significantly contributes to initiating the cascade of mitochondrial dysfunction. And so, it is important to identify effective antioxidant systems and therapies to address oxidative stress-induced neuropathologies. Mitochondrial-targeted antioxidants (MTAs) have demonstrated the ability to intrude into the intramitochondrial cascade triggered by oxidative stress. MTAs can accumulate in neurons and supporting cells (microglia and oligodendrocytes), effectively reducing ROS levels, decreasing inflammation, promoting myelin formation by oligodendrocytes, and increasing ATP production in neurons [113]. In Alzheimer's disease, elevated levels of ROS activate neurons to synthesize excess lipids, which can be taken on by ROS, leading to the production of harmful lipid peroxides [114]. Several MTAs have shown beneficial effects by reinstating neuronal functions in Alzheimer's models by reducing ROS levels, restoring energy supply, relieving cellular stress, and improving the clearance of amyloid-β aggregates [114].

An emerging approach is the conjugation of antioxidant agents with appropriate carriers to develop targeted treatments with minimal off-target effects. Lipophilic cations, liposomes, and peptides have received significant attention as carriers to enhance permeation into mitochondria and reestablish their activity [115]. Triphenylphosphonium (TPP) has been extensively exploited as a carrier for delivering antioxidant molecules to dysfunctional mitochondria, particularly hydrophobic ones that have difficulty crossing cellular membranes to reach the inner mitochondrial matrix [116]. Due to its positive charge, which is drawn to the mitochondrial transmembrane potential, TPP and other lipophilic carriers enable targeted delivery to mitochondria. Once inside the cell, the TPP moiety enters the mitochondrial matrix and confines in the inner mitochondrial membrane, having the hydrophobic functional group placed in between lipid tails and the polar TPP head facing the matrix [117].

Over the years, various compounds, including MitoQ, SkQ1, MitoVitE, and MitoTEMPO, have been conjugated with TPP in order to assist penetration in the mitochondrial compartment [116]. These compounds have been explored and utilized in cellular and pre-clinical models to study neurodegenerative diseases [116].

MitoQ, a conjugate of TPP with a modified ubiquinone, has been extensively studied as an MTA [118]. MitoQ is available as a dietary supplement due to its tolerability and safety when administered orally. Its antioxidant action involves scavenging superoxide, peroxyl, and peroxynitrite ROS in mitochondria [119]. SkQ1, another MTA composed of TPP conjugated with plastiquinone, has demonstrated similar antioxidant activity to MitoQ [120]. SkQ1 has shown protective effects in animal models of neurodegenerative diseases and has been explored in Parkinson's and Alzheimer's disease models [120, 121]. MitoVitE, a mitochondrially targeted form of Vitamin E, has been shown to counter lipid peroxidation and protect mitochondria as well as cells from oxidative damage. MitoTEMPO, composed of TPP and TEMPO (2,2,6,6-Tetramethylpiperidin-1-yl)oxyl), is another MTA with antioxidant properties. It has been investigated in pre-clinical models of neurodegenerative diseases, particularly Parkinson's and Alzheimer's disease, and has demonstrated the ability to alleviate oxidative stress and display protective effects against ROS-induced damage [122, 123].

## 8.2 Modulating Mitophagy as a Therapeutic Approach

In Parkinson's disease, the imbalance between the generation of new mitochondria and the removal of damaged ones via mitophagy plays a crucial role in neurodegeneration [124]. Mutations in genes involved in mitophagy contribute to the aggregation of dysfunctional mitochondria, which impairs neuronal function. Mitophagy alterations have been reported in Alzheimer's disease, where amyloid-β and tau aggregates disturb the axonal transport, and exacerbate ROS production and oxidative stress [125]. Mutant HTT impairs mitochondrial dynamics and transport as well as mitophagy in Huntington's disease, which results in an accumulation of damaged mitochondria and ROS [126, 127]. Targeting the mitophagy aids the endogenous antioxidant system in managing ROS levels in accumulated defective mitochondria, facilitating the recovery and restoration of damaged mitochondria.

## 8.3 Mitochondrial Proteins as Viable Therapeutic Targets

Recent research has raised questions regarding whether controlling mitochondrial dynamics can protect against mitochondrial dysfunction. A growing body of evidence suggests that inhibiting the fission process mediated by DRP1 (dynamic-related protein 1) has neuroprotective effects in various disease models [128]. The use of an inhibitor called mdivi-1, which targets DRP1, has shown neuroprotective properties in models of ischemic injury [129]. In the context of Alzheimer's disease (AD), mdivi-1 has been reported to reverse the synaptic transmission impairment caused by amyloid-β and improve cognitive function while also reducing neuroinflammation and oxidative stress [130]. Inhibiting the connection between Fis1 and DRP1 with the small molecule inhibitor P110 restored mitochondrial function, relieved aberrant mitochondrial shape, and decreased ROS production in patient-derived fibroblasts with an ALS (amyotrophic lateral sclerosis) mutation in SOD1 [131]. In motor neuron-like cells, treatment with P110 led to reduced expression of mitophagy proteins with reduced markers of apoptosis [131]. Interestingly, P110 also improved behavioral outcomes and mitochondrial abnormalities in an ALS mouse model, resulting in an extended lifespan [132]. Importantly, disrupting the Fis1-DRP1 interaction with P110 had no adverse effects in healthy cells [132]. However, preventing the MFF-DRP1 interaction by a compound called P259 had a deleterious impact on cognitive and motor functions in Huntington's disease (HD) mouse model, leading to decreased survival rates [133]. Interestingly, this compound had no noticeable effect on control mice, suggesting that inhibiting DRP1 or its association with specific receptors could be detrimental in certain pathological conditions. Generally, the consequences of the Fis1-DRP1 and MFF-DRP1 interactions appear to differ under physiological conditions and in disease states.

Pieces of evidence also exist that state that diseased conditions alter the post-translational modifications (PTMs) of proteins in the mitochondria. For example, the phosphorylation of DRP1 at S616 (pDRP1S616) is upregulated in HD, and inhibiting the MAPK-pDRP1S616 pathway has been found to correct mitochondrial fragmentation, restore mitochondrial membrane potential, and reduce ROS levels [134]. S-nitrosylation of DRP1 is also increased in HD, resulting in hyperactivation of DRP1 through an elevation in pDRP1S616 levels [134]. Similar protective effects have been reported with a cell-based model of Alzheimer's disease, where expression of DRP1C644A decreases cell death [135]. In neurons, amyloid-β triggers a CDK5-dependent stimulation in pDRP1S616 levels. Moreover, amyloid-β induces an elevation in $Ca^{2+}$ levels, triggering the CaMKII-pDRP1S616 pathway. Inhibiting this pathway prevents DRP1 recruitment and reduces amyloid-β-induced neurotoxicity and apoptosis [135]. Collectively, these researches reveal that while DRP1-mediated fission is associated with plasticity and overactivation of this process can progress to neurodegeneration.

Furthermore, calcineurin-mediated dephosphorylation of pDRP1S637 has been observed in the recruitment of DRP1 to the mitochondrial outer membrane (MOM) and neuronal cell death in various models of ischemia [136]. Inhibiting the interaction between calcineurin and DRP1 protects against ischemia-induced neuronal cell death. Additionally, MEK/ERK-mediated phosphorylation of Mfn1 at T562 has been associated with apoptosis, and expressing a mutant form of Mfn1 that cannot be phosphorylated at T562 provides protection against neuronal cell death [137].

It's important to note that while mitochondrial-targeted therapies show promise in preclinical studies and some early-stage clinical trials, their efficacy and safety in treating neurodegenerative diseases in humans still require further investigation. The complexity of these diseases and the multifactorial nature of mitochondrial dysfunction make their treatment challenging, and a comprehensive understanding of disease mechanisms is essential for developing effective therapies.

## 9 Summary

Neurotransmitters (NTs) are crucial for neuronal proliferation, differentiation, and survival. Aberrant NT levels can lead to various physical, psychiatric, and neurodegenerative illnesses. Alzheimer's, Parkinson's, Huntington's, and Amyotrophic lateral sclerosis are all linked to deficits in NTs. Modulating neurotransmitter transporters (NTT) can be a therapeutic approach for certain neurodegenerative illnesses. Mitochondrial dysfunction, particularly in energy homeostasis, can lead to age-related disease states and damage mtDNA and nuclear DNA. Defects in calcium homeostasis can result in intracellular calcium accumulation, glutamate-induced excitotoxicity, and neuronal loss. Understanding the role of NTs and mitochondria in neurodegenerative disorders is essential for effective treatment.

## References


1. Sheffler, Z. M., Reddy, V., & Pillarisetty, L. S. (2019). *Physiology, neurotransmitters*. StatPearls Publishing.
2. Tiwari, S., Atluri, V., Kaushik, A., Yndart, A., & Nair, M. (2019). Alzheimer's disease: Pathogenesis, diagnostics, and therapeutics. *International Journal of Nanomedicine, 14*, 5541–5554.
3. Poewe, W., Seppi, K., Tanner, C. M., Halliday, G. M., Brundin, P., Volkmann, J., Schrag, A. E., & Lang, A. E. (2017). Parkinson disease. *Nature Reviews Disease Primers, 3*(1), 1–21.
4. Ross, C. A., & Tabrizi, S. J. (2011). Huntington's disease: From molecular pathogenesis to clinical treatment. *The Lancet Neurology, 10*(1), 83–98.
5. Kiernan, M. C., Vucic, S., Cheah, B. C., Turner, M. R., Eisen, A., Hardiman, O., Burrell, J. R., & Zoing, M. C. (2011). Amyotrophic lateral sclerosis. *The Lancet, 377*(9769), 942–955.
6. Trigo, D., Avelar, C., Fernandes, M., Sá, J., & da Cruz e Silva, O. (2022). Mitochondria, energy, and metabolism in neuronal health and disease. *FEBS Letters, 596*(9), 1095–1110.
7. Devine, M. J., & Kittler, J. T. (2018). Mitochondria at the neuronal presynapse in health and disease. *Nature Reviews Neuroscience, 19*(2), 63–80.
8. Briston, T., & Hicks, A. R. (2018). Mitochondrial dysfunction and neurodegenerative proteinopathies: Mechanisms and prospects for therapeutic intervention. *Biochemical Society Transactions, 46*(4), 829–842.
9. Mangrulkar, S. V., Wankhede, N. L., Kale, M. B., Upaganlawar, A. B., Taksande, B. G., Umekar, M. J., Anwer, M. K., Dailah, H. G., Mohan, S., & Behl, T. (2023). Mitochondrial dysfunction as a signaling target for therapeutic intervention in major neurodegenerative disease. *Neurotoxicity Research, 41*(6), 1–22.
10. Tavakolian-Ardakani, Z., Hosu, O., Cristea, C., Mazloum-Ardakani, M., & Marrazza, G. (2019). Latest trends in electrochemical sensors for neurotransmitters: A review. *Sensors, 19*(9), 2037.
11. Niyonambaza, S. D., Kumar, P., Xing, P., Mathault, J., De Koninck, P., Boisselier, E., Boukadoum, M., & Miled, A. (2019). A review of neurotransmitters sensing methods for neuro-engineering research. *Applied Sciences, 9*(21), 4719.
12. Xia, X., Wang, Y., Qin, Y., Zhao, S., & Zheng, J. C. (2022). Exosome: a novel neurotransmission modulator or non-canonical neurotransmitter? *Ageing Research Reviews, 74*, 101558.
13. Dalangin, R., Kim, A., & Campbell, R. E. (2020). The role of amino acids in neurotransmission and fluorescent tools for their detection. *International Journal of Molecular Sciences, 21*(17), 6197.
14. Bak, L. K., Schousboe, A., & Waagepetersen, H. S. (2006). The glutamate/GABA-glutamine cycle: Aspects of transport, neurotransmitter homeostasis and ammonia transfer. *Journal of Neurochemistry, 98*(3), 641–653.
15. Satarker, S., Bojja, S. L., Gurram, P. C., Mudgal, J., Arora, D., & Nampoothiri, M. (2022). Astrocytic glutamatergic transmission and its implications in neurodegenerative disorders. *Cells, 11*(7), 1139.

16. Teleanu, R. I., Niculescu, A. G., Roza, E., Vladâcenco, O., Grumezescu, A. M., & Teleanu, D. M. (2022). Neurotransmitters—Key factors in neurological and neurodegenerative disorders of the central nervous system. *International Journal of Molecular Sciences, 23*(11), 5954.
17. Dunn, J., & Grider, M. H. (2020). *Physiology, adenosine triphosphate*. StatPearls Publishing.
18. Mustafa, A. K., Gadalla, M. M., & Snyder, S. H. (2009). Signaling by gasotransmitters. *Science Signaling, 2*(68), re2.
19. Yi, M., Li, H., Wu, Z., Yan, J., Liu, Q., Ou, C., & Chen, M. (2018). A promising therapeutic target for metabolic diseases: Neuropeptide Y receptors in humans. *Cellular Physiology and Biochemistry, 45*(1), 88–107.
20. You, Y., & Ikezu, T. (2019). Emerging roles of extracellular vesicles in neurodegenerative disorders. *Neurobiology of Disease, 130*, 104512.
21. Verweij, F. J., Bebelman, M. P., Jimenez, C. R., Garcia-Vallejo, J. J., Janssen, H., Neefjes, J., Knol, J. C., de Goeij-De, H. R., Piersma, S. R., Baglio, S. R., & Verhage, M. (2018). Quantifying exosome secretion from single cells reveals a modulatory role for GPCR signaling. *Journal of Cell Biology, 217*(3), 1129–1142.
22. Saeedi, S., Israel, S., Nagy, C., & Turecki, G. (2019). The emerging role of exosomes in mental disorders. *Translational Psychiatry, 9*(1), 122.
23. Rudolph, L. M., Cornil, C. A., Mittelman-Smith, M. A., Rainville, J. R., Remage-Healey, L., Sinchak, K., & Micevych, P. E. (2016). Actions of steroids: New neurotransmitters. *Journal of Neuroscience, 36*(45), 11449–11458.
24. D'Aniello, S., Somorjai, I., Garcia-Fernàndez, J., Topo, E., & D'Aniello, A. (2011). D-aspartic acid is a novel endogenous neurotransmitter. *The FASEB Journal, 25*(3), 1014–1027.
25. Guzman-Martinez, L., Maccioni, R. B., Andrade, V., Navarrete, L. P., Pastor, M. G., & Ramos-Escobar, N. (2019). Neuroinflammation as a common feature of neurodegenerative disorders. *Frontiers in Pharmacology, 10*, 1008.
26. Pchitskaya, E., Popugaeva, E., & Bezprozvanny, I. (2018). Calcium signaling and molecular mechanisms underlying neurodegenerative diseases. *Cell Calcium, 70*, 87–94.
27. Brennenstuhl, H., Jung-Klawitter, S., Assmann, B., & Opladen, T. (2019). Inherited disorders of neurotransmitters: Classification and practical approaches for diagnosis and treatment. *Neuropediatrics, 50*(01), 002–014.
28. Mastrangelo, M. (2021). Epilepsy in inherited neurotransmitter disorders: Spotlights on pathophysiology and clinical management. *Metabolic Brain Disease, 36*(1), 29–43.
29. Agarwal, M., Alam, M. R., Haider, M. K., Malik, M. Z., & Kim, D. K. (2020). Alzheimer's disease: An overview of major hypotheses and therapeutic options in nanotechnology. *Nanomaterials, 11*(1), 59.
30. Ueda, K., Shinohara, S., Yagami, T., Asakura, K., & Kawasaki, K. (1997). Amyloid β protein potentiates Ca2+ influx through L-type voltage-sensitive Ca2+ channels: A possible involvement of free radicals. *Journal of Neurochemistry, 68*(1), 265–271.
31. Ju, Y., & Tam, K. Y. (2022). Pathological mechanisms and therapeutic strategies for Alzheimer's disease. *Neural Regeneration Research, 17*(3), 543.
32. Chen, Y., Xu, J., & Chen, Y. (2021). Regulation of neurotransmitters by the gut microbiota and effects on cognition in neurological disorders. *Nutrients, 13*(6), 2099.
33. Iovino, L., Tremblay, M. E., & Civiero, L. (2020). Glutamate-induced excitotoxicity in Parkinson's disease: The role of glial cells. *Journal of Pharmacological Sciences, 144*(3), 151–164.
34. Jimenez-Sanchez, M., Licitra, F., Underwood, B. R., & Rubinsztein, D. C. (2017). Huntington's disease: Mechanisms of pathogenesis and therapeutic strategies. *Cold Spring Harbor Perspectives in Medicine, 7*(7), a024240.
35. Jamwal, S., & Kumar, P. (2019). Insight into the emerging role of striatal neurotransmitters in the pathophysiology of Parkinson's disease and Huntington's disease: A review. *Current Neuropharmacology, 17*(2), 165–175.
36. Klein, M. O., Battagello, D. S., Cardoso, A. R., Hauser, D. N., Bittencourt, J. C., & Correa, R. G. (2019). Dopamine: Functions, signaling, and association with neurological diseases. *Cellular and Molecular Neurobiology, 39*(1), 31–59.

37. Masrori, P., & Van Damme, P. (2020). Amyotrophic lateral sclerosis: A clinical review. *European Journal of Neurology, 27*(10), 1918–1929.
38. Morgan, S., & Orrell, R. W. (2016). Pathogenesis of amyotrophic lateral sclerosis. *British Medical Bulletin, 119*(1), 87–97.
39. Niello, M., Gradisch, R., Loland, C. J., Stockner, T., & Sitte, H. H. (2020). Allosteric modulation of neurotransmitter transporters as a therapeutic strategy. *Trends in Pharmacological Sciences, 41*(7), 446–463.
40. Pajarillo, E., Rizor, A., Lee, J., Aschner, M., & Lee, E. (2019). The role of astrocytic glutamate transporters GLT-1 and GLAST in neurological disorders: Potential targets for neurotherapeutics. *Neuropharmacology, 161*, 107559.
41. Underhill, S. M., Ingram, S. L., Ahmari, S. E., Veenstra-VanderWeele, J., & Amara, S. G. (2019). Neuronal excitatory amino acid transporter EAAT3: Emerging functions in health and disease. *Neurochemistry International, 123*, 69–76.
42. Cheng, M. H., & Bahar, I. (2019). Monoamine transporters: Structure, intrinsic dynamics and allosteric regulation. *Nature Structural & Molecular Biology, 26*(7), 545–556.
43. Alasmari, F. (2020). Caffeine induces neurobehavioral effects through modulating neurotransmitters. *Saudi Pharmaceutical Journal, 28*(4), 445–451.
44. Yoon, N. Y., Ngo, D. N., & Kim, S. K. (2009). Acetylcholinesterase inhibitory activity of novel chitooligosaccharide derivatives. *Carbohydrate Polymers, 78*(4), 869–872.
45. Axelsen, T. M., & Woldbye, D. P. (2018). Gene therapy for Parkinson's disease, an update. *Journal of Parkinson's Disease, 8*(2), 195–215.
46. Lennon, M. J., Rigney, G., Raymont, V., & Sachdev, P. (2021). Genetic therapies for Alzheimer's disease: A scoping review. *Journal of Alzheimer's Disease, 84*(2), 491–504.
47. Sivandzade, F., & Cucullo, L. (2021). Regenerative stem cell therapy for neurodegenerative diseases: An overview. *International Journal of Molecular Sciences, 22*(4), 2153.
48. Jagaran, K., & Singh, M. (2021). Nanomedicine for neurodegenerative disorders: Focus on Alzheimer's and Parkinson's diseases. *International Journal of Molecular Sciences, 22*(16), 9082.
49. Duchen, M. R. (2000). Mitochondria and calcium: From cell signalling to cell death. *The Journal of Physiology, 529*(1), 57–68.
50. Susin, S. A., Lorenzo, H. K., Zamzami, N., Marzo, I., Snow, B. E., Brothers, G. M., Mangion, J., Jacotot, E., Costantini, P., Loeffler, M., & Larochette, N. (1999). Molecular characterization of mitochondrial apoptosis-inducing factor. *Nature, 397*(6718), 441–446.
51. Ng, M. Y., Wai, T., & Simonsen, A. (2021). Quality control of the mitochondrion. *Developmental Cell, 56*(7), 881–905.
52. Hock, M. B., & Kralli, A. (2009). Transcriptional control of mitochondrial biogenesis and function. *Annual Review of Physiology, 71*, 177–203.
53. Taanman, J. W. (1999). The mitochondrial genome: structure, transcription, translation and replication. *Biochimica et Biophysica Acta - Bioenergetics, 1410*(2), 103–123.
54. Chomyn, A., Mariottini, P., Cleeter, M. W., Ragan, C. I., Matsuno-Yagi, A., Hatefi, Y., Doolittle, R. F., & Attardi, G. (1985). Six unidentified reading frames of human mitochondrial DNA encode components of the respiratory-chain NADH dehydrogenase. *Nature, 314*(6012), 592–597.
55. Tang, J. X., Thompson, K., Taylor, R. W., & Oláhová, M. (2020). Mitochondrial OXPHOS biogenesis: Co-regulation of protein synthesis, import, and assembly pathways. *International Journal of Molecular Sciences, 21*(11), 3820.
56. Zardoya, R. (2020). Recent advances in understanding mitochondrial genome diversity. *F1000Research, 9*, 9.
57. Lee, S. R., & Han, J. (2017). Mitochondrial nucleoid: Shield and switch of the mitochondrial genome. *Oxidative Medicine and Cellular Longevity, 2017*, 8060949.
58. Gilkerson, R., Bravo, L., Garcia, I., Gaytan, N., Herrera, A., Maldonado, A., & Quintanilla, B. (2013). The mitochondrial nucleoid: Integrating mitochondrial DNA into cellular homeostasis. *Cold Spring Harbor Perspectives in Biology, 5*(5), a011080.

59. Wang, Y., & Bogenhagen, D. F. (2006). Human mitochondrial DNA nucleoids are linked to protein-folding machinery and metabolic enzymes at the mitochondrial inner membrane. *Journal of Biological Chemistry, 281*(35), 25791–25802.
60. Yu, R., Lendahl, U., Nistér, M., & Zhao, J. (2020). Regulation of mammalian mitochondrial dynamics: Opportunities and challenges. *Frontiers in Endocrinology, 11*, 374.
61. Picard, M., McEwen, B. S., Epel, E. S., & Sandi, C. (2018). An energetic view of stress: Focus on mitochondria. *Frontiers in Neuroendocrinology, 49*, 72–85.
62. Ferree, A., & Shirihai, O. (2012). Mitochondrial dynamics: The intersection of form and function. In *Mitochondrial oxidative phosphorylation: Nuclear-encoded genes, enzyme regulation, and pathophysiology* (pp. 13–40). Springer.
63. Jornayvaz, F. R., & Shulman, G. I. (2010). Regulation of mitochondrial biogenesis. *Essays in Biochemistry, 47*, 69–84.
64. Scarpulla, R. C., Vega, R. B., & Kelly, D. P. (2012). Transcriptional integration of mitochondrial biogenesis. *Trends in Endocrinology & Metabolism, 23*(9), 459–466.
65. Liang, H., & Ward, W. F. (2006). PGC-1α: A key regulator of energy metabolism. *Advances in Physiology Education, 30*, 145.
66. Wareski, P., Vaarmann, A., Choubey, V., Safiulina, D., Liiv, J., Kuum, M., & Kaasik, A. (2009). PGC-1α and PGC-1β regulate mitochondrial density in neurons. *Journal of Biological Chemistry, 284*(32), 21379–21385.
67. Wang, Y., Xu, E., Musich, P. R., & Lin, F. (2019). Mitochondrial dysfunction in neurodegenerative diseases and the potential countermeasure. *CNS Neuroscience & Therapeutics, 25*(7), 816–824.
68. Donmez, G. (2013). Sirtuins as possible targets in neurodegenerative diseases. *Current Drug Targets, 14*(6), 644–647.
69. Wang, R., Wu, Y., Liu, R., Liu, M., Li, Q., Ba, Y., & Huang, H. (2022). Deciphering therapeutic options for neurodegenerative diseases: Insights from SIRT1. *Journal of Molecular Medicine, 100*(4), 537–553.
70. Donmez, G. (2012). The effects of SIRT1 on Alzheimer's disease models. *International Journal of Alzheimer's Disease, 2012*, 509529.
71. Liu, H., Xu, S., Wang, C., Deng, Y., Xu, B., Yang, T., Sun, J., & Liu, W. (2022). The beneficial role of SIRT1 in preventive or therapeutic options of neurodegenerative diseases. *Neuroscience, 504*, 79–92.
72. Salvatori, I., Valle, C., Ferri, A., & Carrì, M. T. (2017). SIRT3 and mitochondrial metabolism in neurodegenerative diseases. *Neurochemistry International, 109*, 184–192.
73. Mishra, Y., & Kaundal, R. K. (2023). Role of SIRT3 in mitochondrial biology and its therapeutic implications in neurodegenerative disorders. *Drug Discovery Today, 28*, 103583.
74. Trefts, E., & Shaw, R. J. (2021). AMPK: Restoring metabolic homeostasis over space and time. *Molecular Cell, 81*(18), 3677–3690.
75. Garcia, D., & Shaw, R. J. (2017). AMPK: Mechanisms of cellular energy sensing and restoration of metabolic balance. *Molecular Cell, 66*(6), 789–800.
76. Reznick, R. M., & Shulman, G. I. (2006). The role of AMP-activated protein kinase in mitochondrial biogenesis. *The Journal of Physiology, 574*(1), 33–39.
77. Marek, K. W., & Davis, G. W. (2003). Controlling the active properties of excitable cells. *Current Opinion in Neurobiology, 13*(5), 607–611.
78. Seager, R., Lee, L., Henley, J. M., & Wilkinson, K. A. (2020). Mechanisms and roles of mitochondrial localisation and dynamics in neuronal function. *Neuronal Signaling, 4*(2), NS20200008.
79. Roy, C. S., & Sherrington, C. S. (1890). On the regulation of the blood-supply of the brain. *The Journal of Physiology, 11*(1–2), 85.
80. Bordone, M. P., Salman, M. M., Titus, H. E., Amini, E., Andersen, J. V., Chakraborti, B., Diuba, A. V., Dubouskaya, T. G., Ehrke, E., Espindola de Freitas, A., & Braga de Freitas, G. (2019). The energetic brain—A review from students to students. *Journal of Neurochemistry, 151*(2), 139–165.

81. Lujan, B., Kushmerick, C., Banerjee, T. D., Dagda, R. K., & Renden, R. (2016). Glycolysis selectively shapes the presynaptic action potential waveform. *Journal of Neurophysiology, 116*(6), 2523–2540.
82. Rangaraju, V., Calloway, N., & Ryan, T. A. (2014). Activity-driven local ATP synthesis is required for synaptic function. *Cell, 156*(4), 825–835.
83. Hall, C. N., Klein-Flügge, M. C., Howarth, C., & Attwell, D. (2012). Oxidative phosphorylation, not glycolysis, powers presynaptic and postsynaptic mechanisms underlying brain information processing. *Journal of Neuroscience, 32*(26), 8940–8951.
84. Sheng, Z. H., & Cai, Q. (2012). Mitochondrial transport in neurons: Impact on synaptic homeostasis and neurodegeneration. *Nature Reviews Neuroscience, 13*(2), 77–93.
85. Südhof, T. C. (2012). Calcium control of neurotransmitter release. *Cold Spring Harbor Perspectives in Biology, 4*(1), a011353.
86. Dolphin, A. C., & Lee, A. (2020). Presynaptic calcium channels: Specialized control of synaptic neurotransmitter release. *Nature Reviews Neuroscience, 21*(4), 213–229.
87. Kowaltowski, A. J., de Souza-Pinto, N. C., Castilho, R. F., & Vercesi, A. E. (2009). Mitochondria and reactive oxygen species. *Free Radical Biology and Medicine, 47*(4), 333–343.
88. Boveris, A., & Chance, B. (1973). The mitochondrial generation of hydrogen peroxide. General properties and effect of hyperbaric oxygen. *Biochemical Journal, 134*(3), 707–716.
89. Boveris, A. (1977). Mitochondrial production of superoxide radical and hydrogen peroxide. In *Tissue hypoxia and ischemia* (pp. 67–82). Springer.
90. Murphy, M. P. (2009). How mitochondria produce reactive oxygen species. *Biochemical Journal, 417*(1), 1–3.
91. Derouet-Hümbert, E., Roemer, K., & Bureik, M. (2005). Adrenodoxin (Adx) and CYP11A1 (P450scc) induce apoptosis by the generation of reactive oxygen species in mitochondria. *Biological Chemistry, 386*(5), 453–461.
92. Mailloux, R. J., & Harper, M. E. (2011). Uncoupling proteins and the control of mitochondrial reactive oxygen species production. *Free Radical Biology and Medicine, 51*(6), 1106–1115.
93. Casteilla, L., Rigoulet, M., & Pénicaud, L. (2001). Mitochondrial ROS metabolism: Modulation by uncoupling proteins. *IUBMB Life, 52*(3–5), 181–188.
94. Lombard, D. B., Tishkoff, D. X., & Bao, J. (2011). Mitochondrial sirtuins in the regulation of mitochondrial activity and metabolic adaptation. In *Histone deacetylases: The biology and clinical implication* (pp. 163–188). Springer.
95. Holley, A. K., Bakthavatchalu, V., Velez-Roman, J. M., & St. Clair, D. K. (2011). Manganese superoxide dismutase: Guardian of the powerhouse. *International Journal of Molecular Sciences, 12*(10), 7114–7162.
96. Fisher, L. J. (1997). Neural precursor cells: Applications for the study and repair of the central nervous system. *Neurobiology of Disease, 4*(1), 1–22.
97. Coelho, P., Fão, L., Mota, S., & Rego, A. C. (2022). Mitochondrial function and dynamics in neural stem cells and neurogenesis: Implications for neurodegenerative diseases. *Ageing Research Reviews, 80*, 101667.
98. Brunetti, D., Dykstra, W., Le, S., Zink, A., & Prigione, A. (2021). Mitochondria in neurogenesis: Implications for mitochondrial diseases. *Stem Cells, 39*(10), 1289–1297.
99. Arrázola, M. S., Andraini, T., Szelechowski, M., Mouledous, L., Arnauné-Pelloquin, L., Davezac, N., Belenguer, P., Rampon, C., & Miquel, M. C. (2019). Mitochondria in developmental and adult neurogenesis. *Neurotoxicity Research, 36*(2), 257–267.
100. Khacho, M., Clark, A., Svoboda, D. S., Azzi, J., MacLaurin, J. G., Meghaizel, C., Sesaki, H., Lagace, D. C., Germain, M., Harper, M. E., & Park, D. S. (2016). Mitochondrial dynamics impacts stem cell identity and fate decisions by regulating a nuclear transcriptional program. *Cell Stem Cell, 19*(2), 232–247.
101. Iwata, R., & Vanderhaeghen, P. (2021). Regulatory roles of mitochondria and metabolism in neurogenesis. *Current Opinion in Neurobiology, 69*, 231–240.

102. Gao, X., Yu, X., Zhang, C., Wang, Y., Sun, Y., Sun, H., Zhang, H., Shi, Y., & He, X. (2022). Telomeres and mitochondrial metabolism: Implications for cellular senescence and age-related diseases. *Stem Cell Reviews and Reports, 18*(7), 2315–2327.
103. Srivastava, S. (2017). The mitochondrial basis of aging and age-related disorders. *Genes, 8*(12), 398.
104. Jauhari, A., Baranov, S. V., Suofu, Y., Kim, J., Singh, T., Yablonska, S., Li, F., Wang, X., Oberly, P., Minnigh, M. B., & Poloyac, S. M. (2020). Melatonin inhibits cytosolic mitochondrial DNA–induced neuroinflammatory signaling in accelerated aging and neurodegeneration. *The Journal of Clinical Investigation, 130*(6), 3124–3136.
105. Misgeld, T., & Schwarz, T. L. (2017). Mitostasis in neurons: Maintaining mitochondria in an extended cellular architecture. *Neuron, 96*(3), 651–666.
106. Waterham, H. R., Koster, J., van Roermund, C. W., Mooyer, P. A., Wanders, R. J., & Leonard, J. V. (2007). A lethal defect of mitochondrial and peroxisomal fission. *New England Journal of Medicine, 356*(17), 1736–1741.
107. Pickrell, A. M., Huang, C. H., Kennedy, S. R., Ordureau, A., Sideris, D. P., Hoekstra, J. G., Harper, J. W., & Youle, R. J. (2015). Endogenous Parkin preserves dopaminergic substantia nigral neurons following mitochondrial DNA mutagenic stress. *Neuron, 87*(2), 371–381.
108. Pickrell, A. M., & Youle, R. J. (2015). The roles of PINK1, parkin, and mitochondrial fidelity in Parkinson's disease. *Neuron, 85*(2), 257–273.
109. Hsieh, C. H., Shaltouki, A., Gonzalez, A. E., da Cruz, A. B., Burbulla, L. F., Lawrence, E. S., Schüle, B., Krainc, D., Palmer, T. D., & Wang, X. (2016). Functional impairment in miro degradation and mitophagy is a shared feature in familial and sporadic Parkinson's disease. *Cell Stem Cell, 19*(6), 709–724.
110. Schon, E. A., & Przedborski, S. (2011). Mitochondria: The next (neurode) generation. *Neuron, 70*(6), 1033–1053.
111. Barcelos, I. P., Troxell, R. M., & Graves, J. S. (2019). Mitochondrial dysfunction and multiple sclerosis. *Biology, 8*(2), 37.
112. Han, S. M., Baig, H. S., & Hammarlund, M. (2016). Mitochondria localize to injured axons to support regeneration. *Neuron, 92*(6), 1308–1323.
113. Fields, M., Marcuzzi, A., Gonelli, A., Celeghini, C., Maximova, N., & Rimondi, E. (2023). Mitochondria-targeted antioxidants, an innovative class of antioxidant compounds for neurodegenerative diseases: Perspectives and limitations. *International Journal of Molecular Sciences, 24*(4), 3739.
114. Khan, S., Barve, K. H., & Kumar, M. S. (2020). Recent advancements in pathogenesis, diagnostics and treatment of Alzheimer's disease. *Current Neuropharmacology, 18*(11), 1106–1125.
115. Oyewole, A. O., & Birch-Machin, M. A. (2015). Mitochondria-targeted antioxidants. *The FASEB Journal, 29*(12), 4766–4771.
116. Jiang, Q., Yin, J., Chen, J., Ma, X., Wu, M., Liu, G., Yao, K., Tan, B., & Yin, Y. (2020). Mitochondria-targeted antioxidants: A step towards disease treatment. *Oxidative Medicine and Cellular Longevity, 2020*, 8837893.
117. Apostolova, N., & Victor, V. M. (2015). Molecular strategies for targeting antioxidants to mitochondria: Therapeutic implications. *Antioxidants & Redox Signaling, 22*(8), 686–729.
118. Oliver, D. M., & Reddy, P. H. (2019). Small molecules as therapeutic drugs for Alzheimer's disease. *Molecular and Cellular Neuroscience, 96*, 47–62.
119. Tauskela, J. S. (2007). MitoQ—A mitochondria-targeted antioxidant. *IDrugs: The Investigational Drugs Journal, 10*(6), 399–412.
120. Kolosova, N. G., Tyumentsev, M. A., Muraleva, N. A., Kiseleva, E., Vitovtov, A. O., & Stefanova, N. A. (2017). Antioxidant SkQ1 alleviates signs of Alzheimer's disease-like pathology in old OXYS rats by reversing mitochondrial deterioration. *Current Alzheimer Research, 14*(12), 1283–1292.
121. Pavshintsev, V. V., Podshivalova, L. S., Frolova, O. Y., Belopolskaya, M. V., Averina, O. A., Kushnir, E. A., Marmiy, N. V., & Lovat, M. L. (2017). Effects of mitochondrial antioxi-

dant SkQ1 on biochemical and behavioral parameters in a Parkinsonism model in mice. *Biochemistry (Moscow), 82*, 1513–1520.
122. Hu, H., & Li, M. (2016). Mitochondria-targeted antioxidant mitotempo protects mitochondrial function against amyloid beta toxicity in primary cultured mouse neurons. *Biochemical and Biophysical Research Communications, 478*(1), 174–180.
123. Zhelev, Z., Bakalova, R., Aoki, I., Lazarova, D., & Saga, T. (2013). Imaging of superoxide generation in the dopaminergic area of the brain in Parkinson's disease, using mito-TEMPO. *ACS Chemical Neuroscience, 4*(11), 1439–1445.
124. Wang, X. L., Feng, S. T., Wang, Y. T., Yuan, Y. H., Li, Z. P., Chen, N. H., Wang, Z. Z., & Zhang, Y. (2022). Mitophagy, a form of selective autophagy, plays an essential role in mitochondrial dynamics of Parkinson's disease. *Cellular and Molecular Neurobiology, 42*(5), 1–9.
125. Kerr, J. S., Adriaanse, B. A., Greig, N. H., Mattson, M. P., Cader, M. Z., Bohr, V. A., & Fang, E. F. (2017). Mitophagy and Alzheimer's disease: Cellular and molecular mechanisms. *Trends in Neurosciences, 40*(3), 151–166.
126. Šonský, I., Vodička, P., Kepková, K. V., & Hansikova, H. (2021). Mitophagy in Huntington's disease. *Neurochemistry International, 149*, 105147.
127. Franco-Iborra, S., Plaza-Zabala, A., Montpeyo, M., Sebastian, D., Vila, M., & Martinez-Vicente, M. (2021). Mutant HTT (huntingtin) impairs mitophagy in a cellular model of Huntington disease. *Autophagy, 17*(3), 672–689.
128. Yang, D., Ying, J., Wang, X., Zhao, T., Yoon, S., Fang, Y., Zheng, Q., Liu, X., Yu, W., & Hua, F. (2021). Mitochondrial dynamics: A key role in neurodegeneration and a potential target for neurodegenerative disease. *Frontiers in Neuroscience, 15*, 654785.
129. Liu, J. M., Yi, Z., Liu, S. Z., Chang, J. H., Dang, X. B., Li, Q. Y., & Zhang, Y. L. (2015). The mitochondrial division inhibitor mdivi-1 attenuates spinal cord ischemia–reperfusion injury both in vitro and in vivo: Involvement of BK channels. *Brain Research, 1619*, 155–165.
130. Wang, W., Yin, J., Ma, X., Zhao, F., Siedlak, S. L., Wang, Z., Torres, S., Fujioka, H., Xu, Y., Perry, G., & Zhu, X. (2017). Inhibition of mitochondrial fragmentation protects against Alzheimer's disease in rodent model. *Human Molecular Genetics, 26*(21), 4118–4131.
131. Shi, W., Tan, C., Liu, C., & Chen, D. (2023). Mitochondrial fission mediated by Drp1-Fis1 pathway and neurodegenerative diseases. *Reviews in the Neurosciences, 34*(3), 275–294.
132. Joshi, A. U., Saw, N. L., Vogel, H., Cunnigham, A. D., Shamloo, M., & Mochly-Rosen, D. (2018). Inhibition of Drp1/Fis1 interaction slows progression of amyotrophic lateral sclerosis. *EMBO Molecular Medicine, 10*(3), e8166.
133. Kornfeld, O. S., Qvit, N., Shamloo, M., & Mochly-Rosen, D. (2018). A novel pharmacological modulator abrogates physiological mitochondrial fission though specifically inhibiting the Mff-Drp1 protein-protein interaction. *The FASEB Journal, 32*, 543–521.
134. Roe, A. J., & Qi, X. (2018). Drp1 phosphorylation by MAPK1 causes mitochondrial dysfunction in cell culture model of Huntington's disease. *Biochemical and Biophysical Research Communications, 496*(2), 706–711.
135. Manczak, M., Calkins, M. J., & Reddy, P. H. (2011). Impaired mitochondrial dynamics and abnormal interaction of amyloid beta with mitochondrial protein Drp1 in neurons from patients with Alzheimer's disease: Implications for neuronal damage. *Human Molecular Genetics, 20*(13), 2495–2509.
136. Slupe, A. M., Merrill, R. A., Flippo, K. H., Lobas, M. A., Houtman, J. C., & Strack, S. (2013). A calcineurin docking motif (LXVP) in dynamin-related protein 1 contributes to mitochondrial fragmentation and ischemic neuronal injury. *Journal of Biological Chemistry, 288*(17), 12353–12365.
137. Pyakurel, A., Savoia, C., Hess, D., & Scorrano, L. (2015). Extracellular regulated kinase phosphorylates mitofusin 1 to control mitochondrial morphology and apoptosis. *Molecular Cell, 58*(2), 244–254.

# Evolving Pharmacotherapies Involving Neuroimmune Crosstalk

Ranika Maurya, Kaifi Ali, Hemant Soni, Aishwarya Mhatre, and Saba Naqvi

**Abstract**

Neurodegenerative diseases exhibit a gradual onset and present a challenge due to the scarcity of specific biomarkers. While factors such as infection, trauma, and genetic mutation contribute to its development, the intricate interplay between the immune system and the central nervous system (CNS) of the body is widely implicated in causing damage of neurons in neurodegenerative disorders. The immune system assumes a distinctive role in both disease development/progression and cellular equilibrium of the body. It initiates cellular as well as humoral responses, which facilitates repairing of tissue, rejuvenation of cells, and the elimination of cellular waste. However, prolonged and aberrant activation of immune system can have detrimental effects on healthy neurons. The secretion of pro-inflammatory substances by key innate immune components such as microglia, compliment system, and inflammasomes can intensify cytotoxicity. Moreover, these inflammatory agents expedite the instigation of microglia, leading to a gradual decline in neuronal populations. The intricate interplay between the nervous and immune systems influences various aspects of CNS function, including immune surveillance, inflammation, neurogenesis, synaptic plasticity, and neurotransmission. This book chapter explores the significance of neuroimmune crosstalk in the context of neurological diseases, shedding light on the underlying mechanisms and potential therapeutic interventions.

**Keywords**

Neuroimmune crosstalk · Immunotherapy · Neurodegeneration · Complement system

R. Maurya · K. Ali · H. Soni · A. Mhatre · S. Naqvi (✉)
Department of Pharmacology and Toxicology, National Institute of Pharmaceutical Education and Research (NIPER)-Raebareli, Lucknow, Uttar Pradesh, India
e-mail: saba.naqvi@niperraebareli.edu.in

 
R. Shukla et al. (eds.), *Neuroendocrine-Immune Axis*,
https://doi.org/10.1007/978-981-95-5088-3_12

# 1 Introduction

The reciprocal interplay between nervous system and immune system, scientifically known as neuroimmune crosstalk, plays a critical role in preserving homeostasis within the central nervous system (CNS) of the body. Neuroimmune crosstalk refers to the bidirectional communication between immune cells, glial cells, and neurons, which collectively regulate the inflammatory response and immune surveillance in the CNS. However, dysregulation of this reciprocal relationship between nervous system and immune system is associated in the development of various neurological disorders such as Parkinson's disease (PD), Alzheimer's disease (AD), Multiple sclerosis (MS), etc, and also responsible for immunological processes within the CNS [1].

However, brain has its own immune system but it differs from peripheral immune system which creates challenges for therapeutic intervention in various neurological disorders [1, 2]. For developing the therapeutic approaches in neurodegenerative disorders immunological processes have important role since they are involved in the neurodevelopment and neuropathology of CNS disorders [3]. Together, neurons, glia, and peripheral immune cells actively control immunological processes that have impact on brain functions [4]. Neuroimmune crosstalk functions as a complex system of biochemical signals composed of neuropeptides, neurotransmitters, neurohormones, chemokines, cytokines, and growth factors produced by the immune system and nervous system, in order to maintain systemic homeostasis. Sensor cells in any tissue detect noxious stimuli, whether endogenous or exogenous to alert the brain [5].

The two systems transmit information across both short and long distances. Axons, dendrites, and synapses are the primary components of the intercellular connections of nervous system to communicate across these distances. The immune system primarily relies on immune cells that migrate in an organized and coordinated manner as well as by systemically acting soluble substances such as cytokines, antibodies, and chemokines for long-range communication. Short-range communication within the immune system occurs through either direct cell-to-cell interaction at specialized sites known as immunological synapses or through the action of locally acting soluble substances. In addition to these parallel intercellular communications, there are various factors in the immune system, like cytokines, and in the neurological system, such as neurotrophic factors, that support cell growth and differentiation [6].

In reaction to toxins, stress, and injury, the immune system stimulates the healing of damaged tissues, the synthesis of neurotrophic factors, and remyelination, maintaining the brain's integrity in terms of both structure and function. Neuroinflammation can be incited during neurodegeneration by a variety of mechanisms, including the activation of microglia, T cells, oligodendrocytes, astrocytes, inflammasomes, and the complement system, all of which constitute the brain's unique immune system [7, 8] but the complex nature of balancing between neuroprotection and neurotoxicity leads one to hypothesize that continuous or excessive activation of glial cells results in more severe and long-lasting damage of neurons, which ultimately can cause neuroinflammation and neurodegeneration [9].

Microglia and inflammasomes are produced as a defensive mechanism of the host to combat infections and unhealthy neurons [10, 11]. The abnormal activation of microglia, complement system, and inflammasomes can, however, result from the accumulation of amyloid beta (Aβ), alpha synuclein, superoxide dismutase, mutant huntingtin, and other proteins [12, 13]. These proteinopathies exhibit organelle ramification and synaptic dysfunction of nervous system [14]. Immune and central neurological systems are both impacted by neurodegenerative disorders [9]. In this chapter, we will explore about the neuroimmune crosstalk, its significance and challenges encountered in treating disorders involving neuroimmune crosstalk and the significance of understanding these challenges for effective therapeutic interventions.

## 2 Key Molecular Players Involved in Neuroimmune Interactions

Neuroimmune interactions in the central nervous system involve a multifaceted interplay of molecular players that regulate communication between immune cells, glial cells, and neurons. Understanding the roles of these key molecular players is crucial for unravelling the mechanisms underlying neuroimmune crosstalk and its significance for neurological diseases [15]. Here, we discuss the major molecular players involved in neuroimmune interactions:

1. *Microglia*: Microglia, which are the occupant immune cells of the central nervous system (CNS), have a crucial function in facilitating communication between the immune system and the nervous system. In response to immune signals, microglia undergo activation, transitioning from a resting state to an activated state. Activated microglia release a variety of immune mediators, including cytokines (such as IL-1β, IL-6, and TNF-α), chemokines, reactive oxygen species, and nitric oxide. These immune mediators can modulate neuronal activity, synaptic transmission, and inflammation within the CNS [1].
2. *Astrocytes*: Astrocytes are another essential glial cell type involved in neuroimmune interactions. Activated astrocytes release cytokines (such as IL-1β and IL-6), chemokines, and growth factors that can influence immune cell behaviour and neuronal function. Astrocytes also contribute to the formation of the blood-brain barrier (BBB), which regulates the passage of immune cells and molecules into the CNS [16].
3. *Neurons*: Neurons actively participate in neuroimmune crosstalk through the release of neurotransmitters and neuropeptides. Neurotransmitters, such as glutamate, gamma-aminobutyric acid (GABA), dopamine, and serotonin, can modulate immune cell function and behaviour. Neuropeptides, including substance-P, neuropeptide-Y, and vasoactive intestinal peptide (VIP), can regulate immune cell recruitment, activation, and cytokine production. Furthermore, immune mediators can also influence neuronal activity, synaptic transmission, and plasticity [2].

4. *Cytokines and chemokines*: Cytokines and chemokines are crucial molecular players in neuroimmune interactions. Cytokines, including interleukins, tumour necrosis factors, and interferons, are small proteins released by immune cells, glial cells, and neurons. They act as signalling molecules that regulate immune responses and communication between cells. Chemokines are a subclass of cytokines that specifically guide immune cell migration and trafficking within the CNS. Cytokines and chemokines can impact neuronal function, synaptic transmission, and neuroinflammatory processes [17].
5. *Toll-like Receptors* (*TLRs*): TLRs are pattern recognition receptors expressed on various cell types, including microglia, astrocytes, and neurons. TLRs recognize pathogen-associated molecular patterns (PAMPs) and damage-associated molecular patterns (DAMPs) and play a significant role in initiating and modulating immune responses within the CNS. TLRs activation induces the intracellular signalling cascades, leading to the formation and release of pro-inflammatory cytokines, chemokines, and other immune mediators. TLR signalling can be both protective and detrimental, depending on the context and duration of activation [18].
6. *Neurotransmitter receptors*: Neurotransmitter receptors expressed on immune cells, glial cells, and neurons are involved in neuroimmune interactions. For instance, immune cells express neurotransmitter receptors such as glutamate receptors and GABA receptors, enabling them to respond to neurotransmitter signalling and modulate their function. Conversely, neurotransmitter receptors on neurons can be influenced by immune mediators, leading to alterations in neuronal activity and synaptic plasticity [17].

Understanding the roles of these key molecular players provides insights into the intricate communication between the nervous and immune systems in the CNS. Dysregulation of neuroimmune crosstalk involving these molecular players has been involved in various neurological diseases, including neuroinflammatory conditions, neurodegenerative disorders, and psychiatric disorders. Targeting these molecular players holds promise for developing therapeutic interventions aimed at modulating neuroimmune crosstalk and ameliorating the pathogenesis of neurological diseases [17].

## 3 Molecular Mechanism Involved in Neuroimmune Crosstalk

### 3.1 Complement System

The liver is primarily involved in the component synthesis for the complement system, a component associated with body's innate immunity, which is circulated throughout the body and gets activated [19, 20] as they can't pass through the blood-brain barrier (BBB), glial cells and neurons locally manufacture them [21]. The complement system includes over 30 fluid phases that work together to eliminate disease-causing microbes and external cells [22]. The relationship between innate immunity and adaptive immunity, clearance of immune product waste, and protection against microbial infections are the three complement system activation pathways known as classical (CP), lectin (LP), and alternative (AP) [23, 24].

## 3.2 Complement System in CNS

By battling infections and removing waste materials, the complement system aids in preserving the brain homeostasis [25].

The classical way of the complement system begins with C1q, which binds to the Fc region (fragment crystallizable region) of antibodies [26]. Microglia primarily serve as the main source of C1q, triggering the activation of C1 by stimulating C1r and cleaving C1s [26, 27], C4 and activated C1 interact to form C4a and C4b. C2a and C2b are activated when C4b binds to the pathogen's surface. Together, C2a and C4b combine to form C3 convertase (C4b2a and C3bBb) [28].

The pathogen's carbohydrates are bound by the mannan-binding lectin (MBL) pathway, which then triggers MBL-associated serine proteases (MASPs). C3 convertase is created when MASP-2 divides C4 and C2 [29].

The alternative pathway, which is the second complement system pathway, is a separate pathway of antibody responsible for identifying self from non-self. Activation of C3 ($H_2O$), which results from the hydrolysis of native C3, combines with factor-B (serine protease, which prompts B cells to activate inflammatory responses), and develops into C3 convertase [30]. Properdin is recognized as a recognition factor for alternate pathway and has the ability to immediately activate and amplify it [31]. The C3 convertase then divides into C3a and C3b in all three routes. In order to create C5 convertase, C3b connects to C3 convertase, which then causes activation of the membrane attack complex (MAC) that leads to phagocytosis [32].

## 3.3 Role of Microglia

Microglia have the capacity to regenerate themselves and are responsible for triggering immune responses in the CNS [33]. In response to numerous neurological conditions, it undergoes proliferation and structural modifications transforming into a reactive state, known as microgliosis [34]. However, these activated microglia produce neurotrophic factors and direct stem cells to the site of injury [35]. However, continuous microglia activation causes the release of pro-inflammatory mediators including IL-1β, TNF-α, hydrogen peroxide, hydroxyl radicals ($OH^-$), and superoxide anions that are harmful to neurons [36]. Toll-like receptor (TLR) and activated microglia stimulate the adaptor protein (MyD88), which encourages the autophosphorylation of IL-1 receptor-associated kinase (IRAK) and causes the construction of pro-inflammatory cytokines. Additionally, it facilitates the release of reactive oxygen species, leads to neuroinflammation and synaptic loss in diseases like Parkinson's disease, and Alzheimer's disease which is associated with the slow onset of dementia, and is also a result of microgliosis [37]. It has been noted that 24 h after stroke, M1 microglia emits proinflammatory cytokines such as IL-6, IL-1β, and TNF-α, which leads to secondary damage a few days later. In contrast, M2 microglia are anti-inflammatory in nature and are said to have elevated level and increased number in the core region at the same time. Pro-inflammatory M1 phenotype is activated by TLR or IFN-γ modification, whereas regulatory factors like TGF-β and IL-4, IL-10, and IL-13 stimulate anti-inflammatory M2 phenotype [36].

## 3.4 Inflammation-Mediated Inflammasome

The inflammation-mediated response is controlled by a complex protein known as inflammasome (oligomeric protein complex) which has three chief components such as nucleotide-binding oligomerization domain-like receptors, or NOD-like receptors (also known as nucleotide-binding leucine-rich repeat receptors), apoptosis-associated speck-like protein (ASC), and procaspase-1 [38, 39]. NOD-like receptors represent a group of innate immune receptors. They were initially discovered for their ability to detect and react to intracellular pathogens, including bacterial cell wall components, toxic metabolites, uric acids, damaged membranes, and various endogenous by-products resulting from injury/damage of tissues.

Nucleotide-binding domain and leucine-rich repeat containing proteins, also known as NLRs, constitute a group of cytosolic pattern recognition receptors (PRRs) that share a consistent tripartite domain arrangement. This arrangement comprises an N-terminal domain referred to as pyrin domain (PYD) or caspase recruitment domain (CARD), a central nucleotide-binding and oligomerization domain known as NACHT or NOD, which facilitates the oligomerization and activation of NLRs, and a variable number of C-terminal leucine-rich repeats (LRRs) domain responsible for recognizing specific ligands [39]. The NOD-like receptor, pyrin domain-containing 1 and 3 inflammasome (NLRP1 and NLRP3) serves as a sensing site, with an adapter protein (apoptosis-associated speck-like protein) containing a caspase-activating recruitment domain (ASC), and a cysteine protease procaspase-1 that acts as the effector site, which is activated either through the MAPK or NF-ĶB pathway [40]. Neurodegenerative disorders such as Alzheimer's disease, Parkinson's disease, multiple sclerosis, and amyotrophic lateral sclerosis, NLRP3 plays a crucial role in causing inflammation and injury in the central nervous system (CNS), while both NLRP1 and NLRP3 are significantly involved in the pathogenesis of ischemic stroke [41]. The NLRP3 inflammasome is present in microglia and astrocytes. Upon activation, the NLRP3 protein interacts with ASC, which in turn interacts with the CARD domain through PYD domain and causes the mobilization of procaspase-1. This mobilization results in the formation of NLRP3-ASC-procaspase-1 complex, also known as the NLRP3 inflammasome [41, 42]. It causes the formation of active inflammatory cytokines (IL-1β, IL-18) from inactive proinflammatory cytokines (pro IL-1β, pro IL-18) and also activates caspase-1. It eventually causes neuroinflammation and pyroptosis. Gasdermin-D (GSDMD) causes pyroptosis which causes pore formation in the plasma membrane [43]. Pattern recognition receptors (PRRs) are expressed by microglia whose elevated expression causes activation of inflammasome signalling. They inform and affect the adaptive immunity via TLRs. PRRs possess the nucleotide-binding domain and oligomerization domain, leucine-rich repeat, and pyrin-containing domain and are represented by NLRP-1, NLRP-3 NLRC-4, and (AIM2) [44].

## 4 Immune Machinery Destruction in Neurodegenerative Diseases

### 4.1 Complement System Activation Pathways in Alzheimer's and Parkinson's Diseases

Alzheimer's disease (AD) is a neurodegenerative tauopathy, affecting more than 6% of individuals above 65 years, causing dementia. It is characterized by the formation of neurofibrillary tangles comprising hyperphosphorylated tau and the deposition of insoluble amyloid-beta (Aβ) plaques. Genome-wide studies have linked single-nucleotide polymorphisms (SNPs) in complement protein genes, namely complement receptor 1 (CR1) and clusterin (CLU), to late-onset AD risk [45]. CR1, a transmembrane protein, enhances phagocytosis of opsonized particles through interactions with C3b, C4b, and C1q. CLU prevents proteolytic degradation of Aβ oligomers, promoting plaque formation. Immunohistochemistry studies have shown the co-localization of classical complement pathway proteins (C1q, C3, and C4) with amyloid fibrils, neurofibrillary tangles, and Aβ deposits, primarily in the hippocampus, temporal cortex, and amygdala. The terminal complement complex (TCC) marker, C5b-9, is mainly present in the cortex of AD brains, allied with the formation of aggregated Aβ and neurofibrillary tangles. Additionally, the predominant expression of C5aR1 and C5L2 receptors on dystrophic neurites within Aβ plaques and neurofibrillary tangles suggests the involvement of complement activation in AD pathology. The interaction between C5a and its receptor, C5aR1, may have detrimental effects on neurons or indirectly activate microglia. Conversely, the activated C5L2 receptor exhibits anti-inflammatory properties [45].

Accumulation of toxic Aβ leads to complement system activation through the secretion of C1q by neurons and the construction of C3 by astrocytes. C1q binds to Aβ plaques and interacts with C1q receptor (C1qR) on microglia, initiating phagocytosis and synaptic pruning. Soluble Aβ oligomers and fibrils activate microglia through CD14, CD36, CD47, α6β1 integrins, TLRs, and RAGE receptors, inducing phagocytosis. Binding of Aβ oligomers to astrocytic TLR4 or CD36 receptors activates NF-κB signalling, triggering the production of inflammatory cytokines and chemokines [46].

In Parkinson's disease (PD), the activation of the classical complement system is triggered by the presence of Lewy bodies and melanized neurons in the substantia nigra (SN). Immunohistochemistry studies have shown positive staining for early complement activation markers (C3b, C4d, and C7) and late activation marker C9 in these depositions. Microglia-derived C1q and C5a have also exhibited positive staining in the SN. Aforementioned involvement of C1q in the removal of aggregated neuromelanin in the SN. MAC activation in Lewy bodies and oligodendrocytes leads to phagocytosis in the SN in PD. The stimulation of microglial NADPH oxidase (Nox2) through complement receptor CR3 indicates the role of the complement system in neuroinflammation-mediated dopaminergic neuron loss [46].

### 4.2 Inference of Microglia in Alzheimer's and Parkinson's Disease

In AD, microglia recognize the structure of Aβ oligomers as damage-associated molecular patterns (DAMPs), leading to their morphological alteration into a "reactive" or "primed" state. Microglial activation in AD has both degenerative and protective effects. On one hand, activated microglia produce pro-inflammatory cytokines (TNF-α and IL-1β, IL-6) that induce apoptosis in cholinergic neurons. On the other hand, they release proteolytic enzymes such as insulin-degrading enzyme (IDE), MMPs, and neprilysin, which reduce Aβ plaques, providing neuroprotection. However, a mutation in immune-related proteins, TREM-2 and CD33, can inhibit microglial phagocytic function, impairing Aβ clearance and increasing the risk of AD. TREM-2, expressed in macrophages, microglia, and dendritic cells, plays a vital role in cell survival, phagocytosis, proliferation, and inflammatory cytokine production. Studies have linked TREM-2 mutations (e.g., R47H, D87N, T96K, and R62H) to increased AD risk due to functional loss of the protein [47].

In PD, disease progression is attributed to lymphocyte infiltration into the CNS, excessive activation of microglial cells, and astrogliosis. Reactive microglia phagocytose aggregated α-synuclein and stimulate Nox2, generating reactive oxygen species (ROS) that contribute to dopaminergic neurodegeneration in the SN. Aggregation of α-synuclein upregulates leucine-rich repeat kinase (LRRK2) and activates iNOS and CD68 through macroglia and invading peripheral cells. Activated microglia release pro-inflammatory cytokines (IL-12, IL-23, IL-1β, EGF, and TNF-α) and chemokines (CCL2, CCL5, and CXCL10) in the striatum, CSF, and peripheral blood. Elevated TNF-α levels in serum are associated with exacerbated cognitive decline, indicating the role of inflammatory markers in cognitive function. Additionally, these cytokines trigger the generation of ROS through iNOS and COX-2, leading to oxidative stress and exacerbating neurodegeneration in PD [47].

## 5 Disease Treatment Involving Neuroimmune Crosstalk

### 5.1 Immunotherapy for Parkinson's Disease

Immunotherapeutic approaches for Parkinson's disease (PD) aim to alleviate inflammation and oxidative stress while also increasing dopamine levels. One such therapy involves using fenofibrate, a PPAR-γ agonist, which helps to lower inflammation and oxidative stress while boosting dopamine levels [48, 49]. Aspirin, on the other hand, acts as an irreversible COX-1 inhibitor, reducing inflammation and oxidative stress by increasing lipoxin levels [50]. Statins are another treatment option that reduces inflammation by inhibiting IL-1b, IL-6, and TNF-α [51]. In a study conducted by Gordon et al., the daily oral dose of MCC950 at a dose of 20 mg/kg was found to prevent the activation of caspase-1 in the striatum by inhibiting the NLRP3 inflammasome [52].

Additionally, according to Han et al., kaempferol has shown promise in reducing NLRP3 inflammasome development by promoting autophagy in microglia, ultimately leading to decreased NLRP3 protein expression [53]. These approaches offer potential avenues for immunotherapeutic interventions in Parkinson's disease. PMX-205, a C5a antagonist, can prevent dopaminergic neurodegeneration. Anti-α-synuclein immunotherapy is another effective approach that prevents the spread of extracellular α-synuclein to neighbouring nerve cell [54]. Vaccines like 9E4 (a humanized mouse monoclonal antibody) and AFF-1 target calpain-cleaved α-synuclein and α-synuclein oligomers, reducing their accumulation [55]. However, direct delivery of α-synuclein antibodies to the brain, which is crucial for maintaining α-synuclein homeostasis, presents a challenge for immunotherapy development [56] (Fig. 1).

## 5.2 Immunotherapy for Alzheimer's Disease (AD)

FDA medications for symptomatic conduct targeting Aβ and Tau tangles include donepezil, galantamine, rivastigmine, and AADvac1 [36]. There is a growing interest in directing immune cells like microglia and macrophages to control neuroinflammation. Injections of anti-Aβ monoclonal antibodies (mAbs) in the systemic circulation have shown promise in reducing Aβ plaque burden [57, 58]. Tyrosine kinase inhibitors, currently in phase III trials, can suppress mast cell differentiation and degranulation, offering potential benefits as an adjunct to conventional therapy (NCT01872598). Endocannabinoids can decrease microglial activation by increasing CD200 receptor binding. AD pathogenesis is strongly linked to Aβ, leading to neurodegeneration and impaired synapses, which contribute to cognitive dysfunction [59, 60]. Thus, strategies targeting Aβ could effectively AD progression. Current anti-Aβ treatments primarily aim to reduce Aβ production, inhibit aggregation, and enhance clearance.

### 5.2.1 Immunotherapies Based on Aβ

**Active Immunotherapy**

*AN1792*

AN1792, the first anti-Aβ vaccination to undergo clinical testing, utilized synthetic full-length Aβ-42 with QS-21 adjuvant. In phase IIa trial, 19.7% of AN1792-treated patients showed significant anti-AN1792 IgG titres, making them antibody responders. Positive outcomes, such as reduced Aβ deposition and

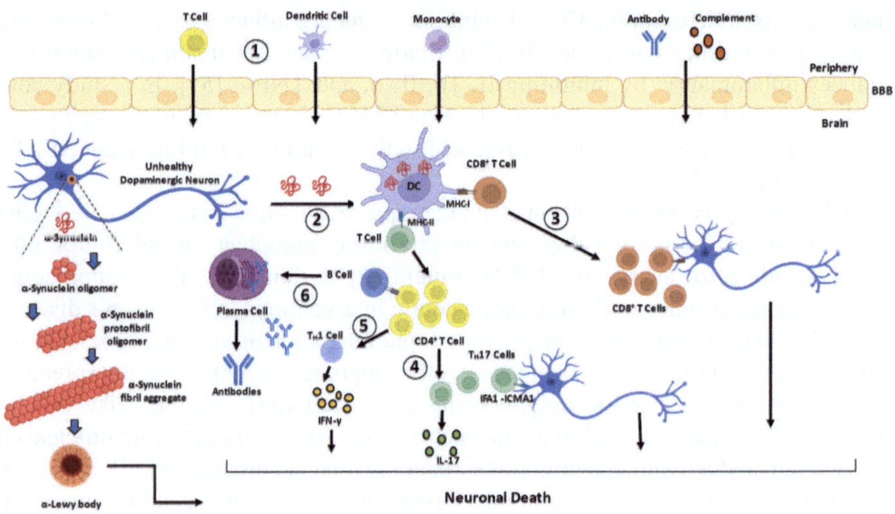


**Fig. 1** Mechanisms of autoimmunity in Parkinson's disease. *BBB* blood-brain barrier, *DC* dendritic cell, *IFN-γ* interferon gamma, *Th cell* T helper cell

favourable effects on NTB score and cerebrospinal fluid (CSF) tau levels, were observed in these antibody responders. However, 6% of treated patients experienced T cell-mediated meningoencephalitis, leading to the discontinuation of the investigation [61].

*Amilomotide*

Amilomotide (CAD106) vaccine contains N-terminal Aβ1–6 B-cell epitope and creates anti-Aβ antibody without eliciting an Aβ-specific T-cell response [38]. Phase I trial (NCT00411580) demonstrated favourable safety and sufficient antibody response [62]. In subsequent phase IIa trials (NCT00733863, NCT00795418, NCT00956410, NCT01023685) and phase IIb trial (NCT01097096), CAD106 showed promising results with a good balance between antibody response and tolerability. However, as it entered the phase II/III trial (NCT02565511), unexpected variations in cognitive performance, brain capacity, and body weight were observed, leading to an early end of the study.

*UB-311*

UB-311 is a vaccine comprising two synthetic Aβ1–14-targeting peptides as B-cell epitopes, each linked to a distinct helper T-cell peptide epitope (UBITh®). This design utilizes a Th2-influenced delivery strategy to enhance immunogenicity while minimizing T-cell inflammatory response [63]. In a phase II trial (NCT02551809), UB-311 demonstrated an impressive 100% response rate, showing high on-target immunogenicity and the probability to advance cognitive performance in patients with early to mild-stage Alzheimer's disease [63].

## Passive Immunotherapy

*Aducanumab*

Aducanumab (BIIB037), human IgG1 monoclonal antibody, adopts an extended conformation and binds to the N terminus of Aβ [64]. It targets various Aβ totals, counting insoluble fibrils and soluble oligomers. In the phase Ib randomized study PRIME (NCT01677572), aducanumab-treated patients exhibited significantly lower amyloid PET standard uptake value ratio (SUVRr) composite scores, particularly those receiving 10 mg/kg of aducanumab at 54 weeks. Patients with prodromal or moderate AD experienced a dose- and time-dependent reduction in brain amyloid levels. Aducanumab administration also led to delayed MMSE scores, indicating positive effects on cognition and clinical outcomes [65]. However, a JAMA article from March 2021 reported dissent among three FDA Peripheral and Central Nervous System Drugs Advisory Committee members regarding the use of aducanumab to treat AD [66]. Despite this disagreement, the U.S. FDA employed the accelerated approval process and officially approved aducanumab in June 2021 for AD treatment, based on the clearance of amyloid plaque from the brain.

*Donanemab*

Donanemab (LY3002813) is humanized monoclonal IgG1 antibody. It demonstrated a significant reduction in the integrated Alzheimer's disease rating score in individuals with early-stage AD, indicating less cognitive and functional loss. However, results for secondary endpoints were inconsistent. Notably, donanemab-treated patients showed a significant decrease in amyloid plaque at 76 weeks based on 18F-florbetapir PET findings, with 54.7% of participants having an amyloid-negative status at 52 weeks [67]. Moreover, larger reductions in amyloid plaque induced by donanemab therapy at 24 weeks were directly associated with less cognitive deterioration and slower tau advancement. Additionally, donanemab treatment led to a sharp decline in plasma P-tau217, a biomarker for AD pathology, within 12 weeks. It is important to note that the incidence of ARIA-E was higher in the donanemab group compared to the placebo group (26.7% vs. 0.8%), necessitating more extensive and extended studies to evaluate the safety and effectiveness of donanemab in AD [67].

*Crenezumab*

Crenezumab (RG7412) is humanized IgG1 monoclonal antibody. It targets various forms of Amyloid β, including monomers and aggregates [68] with greater affinity for oligomers [50]. Two phase III trials, CREAD (NCT02670083) and CREAD2 (NCT03114657), halted prematurely due to the unlikely chance to enhance CDR-SB scores. The phase III CREAD OLE (NCT03491150) study was also terminated based on interim analysis [68].

*Gantenerumab*

Gantenerumab (RO4909832), a human IgG1 monoclonal antibody, possesses high affinity for aggregated Aβ and facilitates Aβ clearance through Fc receptor-mediated phagocytosis. In February 2020, a phase II trial (DIAN-TU, NCT04623242) on patients with hereditary AD did not meet its primary endpoint. Subsequently, a phase II trial named DIAN-TU-001 (NCT01760005) was conducted on individuals with mutations associated with early-onset AD. In this trial, gantenerumab treatment led to significant reductions in A-plaque levels, CSF total tau, and phospho-tau181, along with inhibition of neurofilament light chain growth. However, there were no observed positive effects on cognitive tests. Notably, 19.2% of the individuals experienced oedema associated with amyloid-related imaging abnormalities [69].

*Solanezumab*

Solanezumab (LY2062430), a humanized monoclonal antibody, targets the mid-domain of the Aβ-peptide (Aβ-13-28) to enhance Aβ clearance [70]. Unfortunately, in two completed phase III clinical trials, EXPEDITION 1 (NCT00905372) and EXPEDITION 2 (NCT00904683), solanezumab did not demonstrate effectiveness in slowing cognitive decline or improving functional capacity in individuals with mild-to-moderate AD [71].

### 5.2.2 Immunotherapies Based on Tau Protein

Neurofibrillary tangles are a crucial feature of AD and consist of improperly phosphorylated tau (p-tau) protein. Tau normally functions as a cytoplasmic protein, aiding in the stabilization of microtubules by binding to tubulin during its polymerization. However, in AD, hyperphosphorylated tau loses its ability to effectively bind microtubules, resulting in the formation of neurofibrillary tangles and aggregates [61].

#### Active Immunotherapy

*AADvac1*

Axon Neuroscience's AADvac1 is a first-generation active immunotherapy vaccine designed to target a 12-amino-acid sequence in the microtubule-binding region of tau protein. Has a favourable safety profile, with a very low occurrence of adverse events in both the placebo and immunized groups [72]. Promising results from a phase I trial (NCT02031198) indicated that patients with mild-to-moderate AD have experienced reduced brain atrophy and less cognitive deterioration with the treatment [73].

*ACI-35*

ACI-35 targets abnormal forms of hyperphosphorylated tau. The vaccine contains 16 copies of a synthetic tau, specifically recognizes problematic tau phosphorylation sites S396 and S404. Currently, ACI-35 is undergoing phase I/II clinical trial in Finland to evaluate its safety and potency to treat mild to moderate Alzheimer's (NCT04445831) [74].

**Passive Immunotherapy**

*Semorinemab*

Semorinemab (RO705705), a humanized anti-tau monoclonal antibody, is designed to target all six human tau isoforms extracellularly. It belongs to the immunoglobulin G4 isotype, offering potential protection to neurons. While AC Immune SA has disclosed its safety profile, clinical trials have not demonstrated significant evidence of the drug's efficacy in treating Alzheimer's disease [75].

*BIIB076 and Gosuranemab*

Biogen has developed two monoclonal antibodies, Gosuranemab (BIIB092) and BIIB076 (NI-105). BIIB076 (IgG1) has completed phase I trial (NCT03056729), and now it is in phase 2 trial. On the other hand, Gosuranemab (humanized monoclonal antibody) targets the tau present in the interstitial fluid (ISF) and cerebrospinal fluid (CSF). In a phase-2 trial conducted in advanced supranuclear palsy (NCT03068468), gosuranemab group showed a significant 98% reduction of unbound N-terminal tau in CSF, while the placebo group had an 11% increase. However, despite this N-terminal tau neutralization, it did not lead to clinical effectiveness in the trial [76].

*Tilavonemab*

Tilavonemab (ABBV-8E12), an antibody that targets the N-terminal of tau, recognizes aggregated extracellular form of pathogenic tau. Developed by C2N Diagnostics and AbbVie, a phase I trial (NCT02880956) has verified its safety [74, 77]. However, the phase-2 trial assessing the effectiveness and safety of tilavonemab in 453 patients with early Alzheimer's disease (AD) did not produce the predictable results, leading to the discontinuation of tilavonemab use in AD treatment (NCT02880956) [74]. Furthermore, a separate phase-2 trial evaluating the enduring safety and tolerability of tilavonemab in 364 individuals with early AD was completed on September 30, 2021. However, the final hearsays of this trial are not yet available (NCT03712787) [74].

*Bepranemab*

Bepranemab (UCB0107) is a humanized monoclonal IgG4 antibody that targets the mid-region of tau (amino acids 235–250). Mid-region antibodies show more promise than N-terminal-targeting anti-tau antibodies in preventing the cell-to-cell spread of pathogenic and aggregated tau [78]. Currently, bepranemab is undergoing enrolment in a phase II study to assess its effectiveness, safety, and tolerability in treating moderate cases of Alzheimer's disease (NCT04867616) [78]. This trial is expected to include 450 participants and is projected to be completed in November 2025.

*Zagotenemab*

MCI-1, a mouse monoclonal antibody developed by Peter Davies against an early pathogenic conformation of tau, is the source of the humanized anti-tau antibody zagotenemab (LY3303560). Two phase I trials of zagotenemab have been finished so far (NCT02754830, NCT02754830) [74], on AD patients and healthy volunteers. Phase II testing began in August 2019 and lasted until August 2021, enrolling 285 participants who had suffered from memory loss for at least 6 months. Lilly declared that zagotenemab's development was stopped in October 2021 because the trial had failed to meet its primary aim [74].

## 5.3 Immunotherapy for Multiple Sclerosis (MS)

The success of treatments targeting CD20 in MS patients demonstrated the significant role of B cells in the development of MS. Despite limited penetration through the BBB, anti-CD20 monoclonal antibodies like rituximab [83] and ocrelizumab [84] effectively eliminate CD20+ B cells in the blood, CSF, [79], and CNS perivascular space [80]. Memory ($CD27^{+}$) and naive ($CD27^{-}$) B cells are efficiently depleted following anti-CD20 monoclonal antibody therapy [81, 82]. Subsequently, reconstitution occurs primarily favouring predominance of naive ($CD27^{-}$) B cell phenotype.

In MS patients, memory B cells tend to yield higher levels of proinflammatory cytokines like granulocyte, macrophage, IL-6, colony-stimulating factor, and TNF-Alpha, while showing reduced production of anti-inflammatory cytokines like IL-10 [85]. IL-6 produced by B cells enhances the differentiation of proinflammatory T H17 cells [86], and granulocyte-macrophage colony-stimulating factor stimulates myeloid cells to produce proinflammatory cytokines such as IL-6 and IL-12 [87, 88]. After the termination of anti-CD20 therapy, reconstitution of B cells exhibits decreased proinflammatory cytokines and an increase in IL-10, aligning with a higher proportion of naive B cells.

## 5.4 Amyotrophic Lateral Sclerosis (ALS)

In ALS, there is an observed overexpression of numerous components of the complement system, such as C1q and C3 [89]. This primes the disruption of the (BBB) blood-brain barrier integrity, involving activation of constituents like MASP-1, MBL, C3, C4a, and C5a [89]. Notably, C5a peptide plays a critical role as it interacts with the C5aR receptor on neurons and glia, and also binds to the C5a-like receptor 2 (C5L2), which activates the PKB/Akt and MAPK pathways [90, 91]. Thus, C5a emerges as a potential treatment target for ALS. Elements of the complement system, astrocytes and microglia, are found in adjacent to the BBB supporting CNS homeostasis and protection. They release neurotrophic factors like BDNF and neurotrophin-5 to support neurogenesis, facilitate waste removal, and reduce inflammation, ultimately providing neuroprotection [36].

**Immunotherapy for Modulating the Immune System in ALS**

Riluzole, an FDA-approved medication, is known to extend life by approximately 2–3 months through the suppression of glutamate release at the synaptic cleft [92]. The suppression of glutamate release at synaptic cleft is the mechanism of action [92]. In May 2017, edaravone also received FDA approval as the second medication. It functions by inhibiting the nitration of a tyrosine residue in the cerebrospinal fluid, leading to functional recovery [93]. In the SOD-1G93A mouse model, COX-2 inhibitors, thalidomide, and its derivative lenalidomide demonstrated a reduction in the synthesis of inflammatory markers such as TNF-alpha and FasL [94]. Cyclosporin, another COX-2 inhibitor, is currently in phase IV of a clinical trial (NCT01795872) after showing promise in a transgenic mice model [95–97]. Anakinra, an inhibitor of IL-1, has shown promising results in preventing pro-inflammatory markers in ALS [98].

## 6 Challenges to Treat Neurodegenerative Disorder Via Neuroimmune Crosstalk

Disorders involving neuroimmune crosstalk play a significant part in the development and advancement of various neurological diseases. The intricate interplay between the central nervous system and the immune system has crucial implications for pathogenesis and treatment of these disorders. However, addressing the challenges associated with treating such disorders is a complex task that requires a comprehensive consideration of the underlying mechanisms to develop innovative therapeutic approaches. This section aims to discuss the challenges faced in treating disorders involving neuroimmune crosstalk [47, 99].

(a) *Limited understanding of neuroimmune interactions*: One of the major challenges in treating disorders involving neuroimmune crosstalk is the limited understanding of the complex interactions between the nervous system and the immune system. The intricate signalling pathways and involvement of various immune cells in the CNS change the dynamic of immune responses during disease progression presenting significant hurdles in emerging targeted therapies. Further research is required to explicate the specific mechanisms involved in neuroimmune crosstalk to identify potential therapeutic targets [99, 100].

(b) *Blood-brain barrier (BBB) permeability*: The BBB acts as a protective barrier, regulating the passage of molecules and immune cells into the CNS. However, in certain neurological disorders, the integrity of the BBB may be compromised, leading to increased permeability and infiltration of immune cells. This poses a challenge in designing therapies that can effectively target and modulate the neuroimmune interactions while considering the limitations imposed by the BBB. Developing strategies to enhance drug delivery across the BBB or selectively modulate immune cell trafficking remains a significant obstacle in the treatment of these disorders [100].

(c) *Heterogeneity of neuroimmune responses*: Neuroimmune responses vary across different neurological diseases and even within the same disease. The heterogeneity of these responses poses challenges in identifying common therapeutic targets and developing generalized treatment approaches. Tailoring treatments to the specific neuroimmune profiles of individual patients requires a deeper understanding of disease subtypes, biomarkers, and personalized medicine approaches. Precision medicine and molecular profiling techniques hold promise for addressing this challenge, but further research and clinical validation are needed [101].

(d) *Timing and duration of neuroimmune interventions*: Neuroimmune responses are dynamic and can vary depending on the stage of disease progression. Effective treatment strategies must consider the optimal timing and duration of neuroimmune interventions. In some cases, early intervention to modulate excessive immune responses may be beneficial, while in others, targeting later stages of neuroinflammation or immune dysregulation might be more effective. Identifying critical time windows for intervention and developing therapeutics with the appropriate pharmacokinetic profiles are key challenges in optimizing treatment outcomes [102].

(e) *Limited biomarkers and diagnostic tools*: Currently, there is a lack of reliable biomarkers and diagnostic tools for assessing the neuroimmune status in patients with neurological disorders. This hampers early detection, accurate diagnosis, and monitoring of disease progression. Identifying specific biomarkers and developing non-invasive diagnostic techniques will aid in stratifying patients and selecting appropriate treatment strategies [103].

(f) *Potential side effects and immune dysregulation*: Manipulating neuroimmune crosstalk for therapeutic purposes carries the risk of immune dysregulation and potential side effects. The delicate balance between immune activation and suppression must be carefully considered to avoid exacerbating disease pathology or compromising the overall immune defence mechanisms. Developing therapies that selectively target disease-associated immune responses while preserving the immune system's overall functionality is a significant challenge in treating these disorders [104].

## 7 Evolving Pharmacotherapies Involving Neuroimmune Crosstalk in CNS-Related Diseases

### 7.1 Immunomodulatory Therapies

Immunomodulatory therapies represent a class of drugs designed to regulate the immune system's response, particularly in the context of neuroinflammation. These therapies aim to dampen excessive immune activation without compromising overall immune function [105].

(a) *Monoclonal antibodies*: Monoclonal antibodies (mAbs) are engineered to target specific inflammatory molecules involved in neuroinflammation. For example, mAbs against tumour necrosis factor-alpha (TNF-alpha) have shown promise in reducing inflammation in conditions like multiple sclerosis. Additionally, antibodies targeting beta-amyloid in Alzheimer's disease aim to clear amyloid plaques and modulate microglial responses [106].

E.g., Natalizumab is a monoclonal antibody that targets alpha-4 integrin and is approved for the treatment of relapsing forms of multiple sclerosis (MS). It prevents immune cells, particularly lymphocytes, from crossing the blood-brain barrier, reducing neuroinflammation and relapses in MS patients.

Ocrelizumab: Ocrelizumab targets CD20 on B cells and is approved for the treatment of relapsing-remitting MS and primary progressive MS. It depletes B cells, which are involved in the immune response in MS, leading to reduced inflammation and neuroprotection [106].

(b) *Small molecule inhibitors*: Small molecule inhibitors interfere with specific signalling pathways involved in the immune response. Agents targeting Janus kinases (JAK) or sphingosine-1-phosphate receptors (S1P) have shown potential in multiple sclerosis by inhibiting pro-inflammatory cytokine signalling and modulating lymphocyte migration, respectively.

E.g., Fingolimod is a sphingosine-1-phosphate receptor modulator approved for relapsing forms of multiple sclerosis. It sequesters lymphocytes in lymph nodes, preventing them from entering the CNS and causing inflammation [107].

*Tofacitinib*: Tofacitinib is a Janus kinase (JAK) inhibitor used to treat rheumatoid arthritis, and it is being investigated for its potential application in CNS-related diseases. JAK inhibitors have shown anti-inflammatory effects by blocking cytokine signalling pathways [108].

*Baricitinib*: Baricitinib is another JAK inhibitor with potential immunomodulatory effects that may be beneficial in CNS-related diseases [108].

(c) *Cytokine blockers*: Several cytokines play key roles in the immune response and neuroinflammation. Blocking these cytokines, such as interleukin-6 (IL-6) or interleukin-17 (IL-17), has been investigated as a therapeutic strategy in CNS-related diseases [109].

## 7.2 Anti-inflammatory Drugs

Traditional anti-inflammatory drugs, such as corticosteroids and non-steroidal anti-inflammatory drugs (NSAIDs), have been used to manage neuroinflammatory conditions. However, novel formulations and delivery methods are being explored to enhance their effectiveness and reduce adverse effects [95].

(a) *Liposomal formulations*

Liposomal formulation encapsulates the drug within liposomes, e.g., prednisolone. It has shown promise in reducing inflammation in experimental models of CNS-related diseases, and clinical trials are ongoing to evaluate its efficacy [95].

(b) *Targeted drug delivery*

Nanoparticle-based drug delivery: Nanoparticles, such as liposomes or polymeric nanoparticles, can be engineered to carry anti-inflammatory drugs and target specific cells or tissues within the CNS. These approaches aim to enhance drug delivery and minimize off-target effects [95].

## 7.3 Nanotechnology

The area of medicine currently uses nanomaterials for a variety of purposes, including bioimaging, medication delivery, and biosensors. Quantum dots, nanotubes, nanoparticles, and nanofibrils are some of these uses. Nanoparticles are developed to treat neurodegenerative diseases in the recently emerging field of neuro-nanomedicine. In order to easily pass the BBB and reach the intended target spot in the cells, some nanoparticles are designed in this manner. The main problem with treating neurodegenerative diseases is that some medications can't cross the blood-brain barrier, which reduces their effectiveness; here, nanoparticles play a major role [110].

(a) *Gold nanoparticles*: Gold nanoparticles combined with exosome-derived membranes effectively deliver drugs to particular targets in the brain, according to M. Sanati et al. Additionally, investigations using bioluminescence imaging demonstrated that exosomes coated in gold nanoparticles easily enter cells that are intended for the brain [111]. Targeted exosome delivery is made possible by the alteration of the surface with gold nanoparticles, and this successful and cutting-edge method of delivery has been demonstrated. Using beta-lactalbumin as a potential model for amyloid formation, the effects of gold nanoparticles on this procedure were investigated. The results supported the possibility that AUNPs could block amyloid fibrils since they absorb more NPs into proteins and don't cause structural alterations. As a result, it may be a useful treatment for amyloid disorders [112].
(b) *Magnetic nanomaterials* (*MNPs*) have recently gained importance in the medical industry due to their magnetic force response potential, which affects cell sorting and targeted medication administration. According to Materón, Miyazaki et al. 2021, magnetic targeting aims to reduce the toxicity of the nanoparticles and prevent side effects from higher doses of medication by loading potential pharmaceuticals into magnetic nanoparticles so that magnetic field gradients can effectively localize the drug at the target site [113].
(c) *Cerium oxide nanoparticles*: Nanoceria is another name for cerium nanoparticles. The study demonstrated that ser616 hyperphosphorylation, which is highly linked to neurodegenerative illnesses, especially Alzheimer's disease, and brain protective properties of nanoceria against mitochondrial fragmentation caused by amyloid beta protein [114].

(d) *Graphene quantum dots (GQDs)*: The building blocks of graphene quantum dots (GQDs) are one or more layers of graphene with a thickness of less than 100 nm. GQDs are also utilized to prevent A from aggregating in Alzheimer's disease. The His13-Lys16 (HHQK) portion of the β-amyloid peptide, which is made up of numerous amino acids and various sections, is essential for the production and oligomerization of fibrils. It is also recognized as a crucial element of the glycosaminoglycan (GAG) binding site. Tramiprosate, a GAGSs mimic, made up the construction, and GQDs displayed a decrease in A-sheet aggregation. In a second study, the same design also has a synergistic impact and protects the PC12 cells from cytotoxicity [115].

## 7.4 Stem Cell Therapies

Stem cell-based therapies hold promise in CNS-related diseases by promoting tissue repair and modulating the neuroimmune response. Mesenchymal stem cells (MSCs) are particularly promising due to their immunomodulatory properties [116].

(a) *MSC transplantation*: MSC transplantation has shown beneficial effects in preclinical studies and early clinical trials for CNS-related diseases. MSCs can modulate the immune response, reduce inflammation, and promote tissue repair.
(b) *Exosome therapy*: Exosomes are small vesicles released by MSCs that carry bioactive molecules involved in cell-to-cell communication. Exosome therapy represents a non-cellular approach to harnessing the immunomodulatory properties of MSCs.

## 7.5 Gene Therapies

Gene therapies offer a unique approach to target specific immune-related genes and pathways implicated in CNS diseases. By modulating gene expression, these therapies have the potential to alter the neuroimmune response and provide long-term benefits [117].

(a) *Gene editing*: Gene editing technologies like CRISPR-Cas9 can be used to modify immune cells, introducing beneficial genetic changes that enhance the immune response against disease targets while reducing harmful effects [118, 119].

Clustered Regularly Interspaced Short Palindromic Repeats (CRISPR)-based Gene Editing for Neurological Disorders is a revolutionary gene editing technology that allows for precise and targeted modifications of the genome. In the context of CNS-related diseases, CRISPR-based gene editing holds significant promise for correcting genetic mutations or modulating immune-related genes involved in neuroinflammation [119].

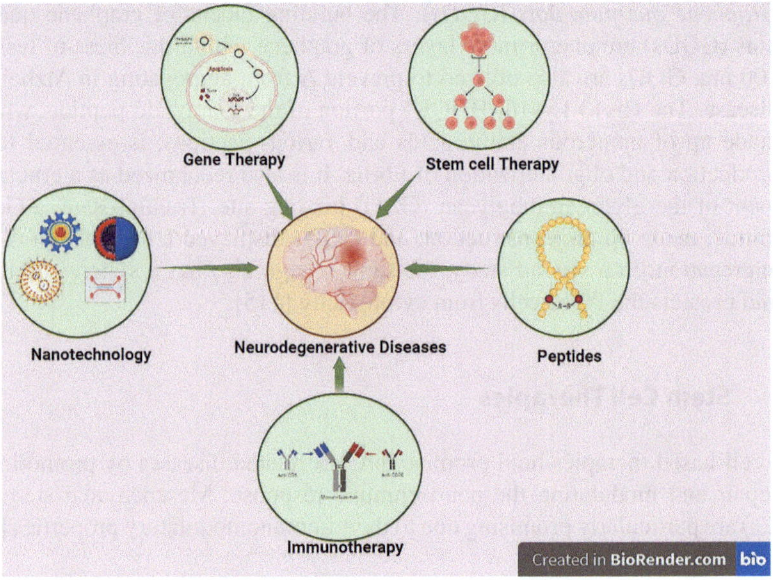


**Fig. 2** Evolving pharmacotherapies for neurodegenerative diseases

For example, in neurodegenerative diseases caused by specific genetic mutations, CRISPR-Cas9 can be used to correct the mutation directly within the affected cells, potentially halting disease progression. Moreover, CRISPR-Cas9 has the potential to modulate immune-related genes to dampen neuroinflammation, thereby providing therapeutic benefits in various CNS-related disorders [119].

(b) *RNA interference*: RNA interference (RNAi) can be employed to silence specific genes involved in neuroinflammation, reducing the production of pro-inflammatory cytokines and chemokines [119, 120] (Fig. 2).

## 8 Conclusion

This chapter provides an in-depth analysis of the intricate communication among the nervous and immune systems in the context of neurodegenerative diseases. Treating disorders involving neuroimmune crosstalk poses several challenges due to the complexity of neuroimmune interactions. The involvement of innate immune proteins and the existence of adaptive immune cells within the brain environment are key factors contributing to neurodegeneration. The dysregulation of various immune components, including T cells microglia, the complement system, and inflammasome signalling, has a crucial role in the degeneration of neurons. Additionally, the interplay between genetic factors, gender, and environmental influences further contributes to disease progression.

By addressing these challenges and exploring alternative therapeutic approaches alongside the current treatment options, it becomes possible to effectively reduce neuroinflammation and toxicity, thereby preventing the initiation and worsening of neurodegenerative conditions such as Amyotrophic lateral sclerosis, Alzheimer's disease, Parkinson's disease, multiple sclerosis, Huntington's disease, and ischemic stroke.

As the field continues to evolve, continued research, a multidisciplinary approach, including collaborations between neuroscientists, immunologists, clinicians, and drug developers, will pave the way for transformative breakthroughs, ultimately bringing new hope to patients and their families worldwide.

## References


1. Carson, M. J., et al. (2006). CNS immune privilege: Hiding in plain sight. *Immunological Reviews, 213*, 48–65.
2. Tian, L., Rauvala, H., & Gahmberg, C. G. (2009). Neuronal regulation of immune responses in the central nervous system. *Trends in Immunology, 30*(2), 91–99.
3. Bailey, S. L., et al. (2006). Innate and adaptive immune responses of the central nervous system. *Critical Reviews in Immunology, 26*(2), 149–188.
4. Tian, L., et al. (2012). Neuroimmune crosstalk in the central nervous system and its significance for neurological diseases. *Journal of Neuroinflammation, 9*, 155.
5. Cacabelos Ramon, T. C., Lucia, F.-N., & Gjumrakch, A. (2016). Neuroimmune crosstalk in CNS disorders: The histamine connection. *Current Pharmaceutical Design, 22*(7), 819–848.
6. Kerschensteiner, M., et al. (2003). Neurotrophic cross-talk between the nervous and immune systems: Implications for neurological diseases. *Annals of Neurology, 53*(3), 292–304.
7. Lucas, S. M., Rothwell, N. J., & Gibson, R. M. (2006). The role of inflammation in CNS injury and disease. *British Journal of Pharmacology, 147*(Suppl 1), S232–S240.
8. Block, M. L., & Hong, J. S. (2005). Microglia and inflammation-mediated neurodegeneration: Multiple triggers with a common mechanism. *Progress in Neurobiology, 76*(2), 77–98.
9. Meisel, C., & Meisel, A. (2011). Suppressing immunosuppression after stroke. *The New England Journal of Medicine, 365*(22), 2134–2136.
10. Bennett, M. L., & Bennett, F. C. (2020). The influence of environment and origin on brain resident macrophages and implications for therapy. *Nature Neuroscience, 23*(2), 157–166.
11. Krakauer, T. (2019). Inflammasomes, autophagy, and cell death: The trinity of innate host defense against intracellular bacteria. *Mediators of Inflammation, 2019*, 2471215.
12. Hickman, S., et al. (2018). Microglia in neurodegeneration. *Nature Neuroscience, 21*(10), 1359–1369.
13. Voet, S., et al. (2019). Inflammasomes in neuroinflammatory and neurodegenerative diseases. *EMBO Molecular Medicine, 11*(6), e10248.
14. Ghavami, S., et al. (2014). Autophagy and apoptosis dysfunction in neurodegenerative disorders. *Progress in Neurobiology, 112*, 24–49.
15. Madeira, J. M., Schindler, S. M., & Klegeris, A. (2015). A new look at auranofin, dextromethorphan and rosiglitazone for reduction of glia-mediated inflammation in neurodegenerative diseases. *Neural Regeneration Research, 10*(3), 391–393.
16. Galea, I., Bechmann, I., & Perry, V. H. (2007). What is immune privilege (not)? *Trends in Immunology, 28*(1), 12–18.
17. Bailey, S. L., et al. (2007). CNS myeloid DCs presenting endogenous myelin peptides 'preferentially' polarize CD4+ T(H)-17 cells in relapsing EAE. *Nature Immunology, 8*(2), 172–180.

18. Godbout, J. P., et al. (2005). Exaggerated neuroinflammation and sickness behavior in aged mice following activation of the peripheral innate immune system. *The FASEB Journal, 19*(10), 1329–1331.
19. Bonifati, D. M., & Kishore, U. (2007). Role of complement in neurodegeneration and neuroinflammation. *Molecular Immunology, 44*(5), 999–1010.
20. Orsini, F., et al. (2014). Versatility of the complement system in neuroinflammation, neurodegeneration and brain homeostasis. *Frontiers in Cellular Neuroscience, 8*, 380.
21. Veerhuis, R., Nielsen, H. M., & Tenner, A. J. (2011). Complement in the brain. *Molecular Immunology, 48*(14), 1592–1603.
22. Wagner, E., & Frank, M. M. (2010). Therapeutic potential of complement modulation. *Nature Reviews. Drug Discovery, 9*(1), 43–56.
23. Carpanini, S. M., Torvell, M., & Morgan, B. P. (2019). Therapeutic inhibition of the complement system in diseases of the central nervous system. *Frontiers in Immunology, 10*, 362.
24. Walport, M. J. (2001). Complement. First of two parts. *The New England Journal of Medicine, 344*(14), 1058–1066.
25. Morgan, B. P. (2018). Complement in the pathogenesis of Alzheimer's disease. *Seminars in Immunopathology, 40*(1), 113–124.
26. Ehrnthaller, C., et al. (2011). New insights of an old defense system: Structure, function, and clinical relevance of the complement system. *Molecular Medicine, 17*(3–4), 317–329.
27. Veerhuis, R., et al. (1999). Cytokines associated with amyloid plaques in Alzheimer's disease brain stimulate human glial and neuronal cell cultures to secrete early complement proteins, but not C1-inhibitor. *Experimental Neurology, 160*(1), 289–299.
28. Garcia, B. L., et al. (2016). Novel evasion mechanisms of the classical Complement pathway. *Journal of Immunology, 197*(6), 2051–2060.
29. Dahl, M. R., et al. (2001). MASP-3 and its association with distinct complexes of the mannan-binding lectin complement activation pathway. *Immunity, 15*(1), 127–135.
30. Müller-Eberhard, H. J. (1988). Molecular organization and function of the complement system. *Annual Review of Biochemistry, 57*, 321–347.
31. Harboe, M., & Mollnes, T. E. (2008). The alternative complement pathway revisited. *Journal of Cellular and Molecular Medicine, 12*(4), 1074–1084.
32. Alexander, J. J., et al. (2008). The complement cascade: Yin-Yang in neuroinflammation–neuro-protection and -degeneration. *Journal of Neurochemistry, 107*(5), 1169–1187.
33. Song, W. M., & Colonna, M. (2018). The microglial response to neurodegenerative disease. *Advances in Immunology, 139*, 1–50.
34. McGeer, P. L., et al. (1988). Reactive microglia are positive for HLA-DR in the substantia nigra of Parkinson's and Alzheimer's disease brains. *Neurology, 38*(8), 1285–1291.
35. Walter, L., & Neumann, H. (2009). Role of microglia in neuronal degeneration and regeneration. *Seminars in Immunopathology, 31*(4), 513–525.
36. Baidya, F., et al. (2021). Neuroimmune crosstalk and evolving pharmacotherapies in neurodegenerative diseases. *Immunology, 162*(2), 160–178.
37. Subhramanyam, C. S., et al. (2019). Microglia-mediated neuroinflammation in neurodegenerative diseases. *Seminars in Cell & Developmental Biology, 94*, 112–120.
38. Yang, Y., et al. (2019). Recent advances in the mechanisms of NLRP3 inflammasome activation and its inhibitors. *Cell Death & Disease, 10*(2), 128.
39. Chen, L., et al. (2021). NOD-like receptors in autoimmune diseases. *Acta Pharmacologica Sinica, 42*(11), 1742–1756.
40. Agostini, L., et al. (2004). NALP3 forms an IL-1beta-processing inflammasome with increased activity in Muckle-Wells autoinflammatory disorder. *Immunity, 20*(3), 319–325.
41. Song, L., et al. (2017). NLRP3 inflammasome in neurological diseases, from functions to therapies. *Frontiers in Cellular Neuroscience, 11*, 63.
42. Schroder, K., Zhou, R., & Tschopp, J. (2010). The NLRP3 inflammasome: A sensor for metabolic danger? *Science, 327*(5963), 296–300.
43. van de Veerdonk, F. L., et al. (2011). Inflammasome activation and IL-1β and IL-18 processing during infection. *Trends in Immunology, 32*(3), 110–116.

44. Taylor, R. A., & Sansing, L. H. (2013). Microglial responses after ischemic stroke and intracerebral hemorrhage. *Clinical & Developmental Immunology, 2013*, 746068.
45. McGeer, E. G., Klegeris, A., & McGeer, P. L. (2005). Inflammation, the complement system and the diseases of aging. *Neurobiology of Aging, 26*(Suppl 1), 94–97.
46. Benveniste, E. N., Nguyen, V. T., & O'Keefe, G. M. (2001). Immunological aspects of microglia: Relevance to Alzheimer's disease. *Neurochemistry International, 39*(5–6), 381–391.
47. Mohammad, H. M. F., et al. (2023). Protective effects of evening primrose oil on behavioral activities, nigral microglia and histopathological changes in a rat model of rotenone-induced parkinsonism. *Journal of Chemical Neuroanatomy, 127*, 102206.
48. Tan, E. K., et al. (2020). Parkinson disease and the immune system—Associations, mechanisms and therapeutics. *Nature Reviews. Neurology, 16*(6), 303–318.
49. Lehmann, J. M., et al. (1997). Peroxisome proliferator-activated receptors alpha and gamma are activated by indomethacin and other non-steroidal anti-inflammatory drugs. *The Journal of Biological Chemistry, 272*(6), 3406–3410.
50. Lyman, M., et al. (2014). Neuroinflammation: The role and consequences. *Neuroscience Research, 79*, 1–12.
51. Terblanche, M., et al. (2007). Statins and sepsis: Multiple modifications at multiple levels. *The Lancet Infectious Diseases, 7*(5), 358–368.
52. Gordon, R., et al. (2018). Inflammasome inhibition prevents alpha-synuclein pathology and dopaminergic neurodegeneration in mice. *Science Translational Medicine, 10*(465), eaah4066.
53. Han, X., et al. (2019). Small molecule-driven NLRP3 inflammation inhibition via interplay between ubiquitination and autophagy: Implications for Parkinson disease. *Autophagy, 15*(11), 1860–1881.
54. Lee, J. S., & Lee, S. J. (2016). Mechanism of anti-alpha-synuclein immunotherapy. *Journal of Movement Disorders, 9*(1), 14–19.
55. Liang, Z., et al. (2017). Impact of aging immune system on neurodegeneration and potential immunotherapies. *Progress in Neurobiology, 157*, 2–28.
56. Brundin, P., Dave, K. D., & Kordower, J. H. (2017). Therapeutic approaches to target alpha-synuclein pathology. *Experimental Neurology, 298*(Pt B), 225–235.
57. Bard, F., et al. (2000). Peripherally administered antibodies against amyloid beta-peptide enter the central nervous system and reduce pathology in a mouse model of Alzheimer disease. *Nature Medicine, 6*(8), 916–919.
58. Wisniewski, T., & Goni, F. (2014). Immunotherapy for Alzheimer's disease. *Biochemical Pharmacology, 88*(4), 499–507.
59. Hardy, J., & Selkoe, D. J. (2002). The amyloid hypothesis of Alzheimer's disease: Progress and problems on the road to therapeutics. *Science, 297*(5580), 353–356.
60. Selkoe, D. J., & Hardy, J. (2016). The amyloid hypothesis of Alzheimer's disease at 25 years. *EMBO Molecular Medicine, 8*(6), 595–608.
61. Ait-Bouziad, N., et al. (2020). Phosphorylation of the overlooked tyrosine 310 regulates the structure, aggregation, and microtubule- and lipid-binding properties of tau. *The Journal of Biological Chemistry, 295*(23), 7905–7922.
62. Winblad, B., et al. (2012). Safety, tolerability, and antibody response of active Abeta immunotherapy with CAD106 in patients with Alzheimer's disease: Randomised, double-blind, placebo-controlled, first-in-human study. *Lancet Neurology, 11*(7), 597–604.
63. Wang, C. Y., et al. (2017). UB-311, a novel UBITh((R)) amyloid beta peptide vaccine for mild Alzheimer's disease. *Alzheimer's & Dementia: Translational Research & Clinical Interventions, 3*(2), 262–272.
64. Arndt, J. W., et al. (2018). Structural and kinetic basis for the selectivity of aducanumab for aggregated forms of amyloid-beta. *Scientific Reports, 8*(1), 6412.
65. Sevigny, J., et al. (2016). The antibody aducanumab reduces Abeta plaques in Alzheimer's disease. *Nature, 537*(7618), 50–56.

66. Alexander, G. C., Emerson, S., & Kesselheim, A. S. (2021). Evaluation of aducanumab for Alzheimer disease: Scientific evidence and regulatory review involving efficacy, safety, and futility. *JAMA, 325*(17), 1717–1718.
67. Mintun, M. A., et al. (2021). Donanemab in early Alzheimer's disease. *The New England Journal of Medicine, 384*(18), 1691–1704.
68. Adolfsson, O., et al. (2012). An effector-reduced anti-beta-amyloid (Abeta) antibody with unique abeta binding properties promotes neuroprotection and glial engulfment of Abeta. *The Journal of Neuroscience, 32*(28), 9677–9689.
69. Salloway, S., et al. (2021). A trial of gantenerumab or solanezumab in dominantly inherited Alzheimer's disease. *Nature Medicine, 27*(7), 1187–1196.
70. Honig, L. S., et al. (2018). Trial of Solanezumab for mild dementia due to Alzheimer's disease. *The New England Journal of Medicine, 378*(4), 321–330.
71. Doody, R. S., et al. (2014). Phase 3 trials of solanezumab for mild-to-moderate Alzheimer's disease. *The New England Journal of Medicine, 370*(4), 311–321.
72. Novak, P., et al. (2018). FUNDAMANT: An interventional 72-week phase 1 follow-up study of AADvac1, an active immunotherapy against tau protein pathology in Alzheimer's disease. *Alzheimer's Research & Therapy, 10*(1), 108.
73. Bittar, A., Bhatt, N., & Kayed, R. (2020). Advances and considerations in AD tau-targeted immunotherapy. *Neurobiology of Disease, 134*, 104707.
74. Song, C., et al. (2022). Immunotherapy for Alzheimer's disease: Targeting beta-amyloid and beyond. *Translational Neurodegeneration, 11*(1), 18.
75. Ayalon, G., et al. (2021). Antibody semorinemab reduces tau pathology in a transgenic mouse model and engages tau in patients with Alzheimer's disease. *Science Translational Medicine, 13*(593), eabb2639.
76. Dam, T., et al. (2021). Safety and efficacy of anti-tau monoclonal antibody gosuranemab in progressive supranuclear palsy: A phase 2, randomized, placebo-controlled trial. *Nature Medicine, 27*(8), 1451–1457.
77. West, T., et al. (2017). Preclinical and clinical development of ABBV-8E12, a humanized anti-tau antibody, for treatment of Alzheimer's disease and other tauopathies. *The Journal of Prevention of Alzheimer's Disease, 4*(4), 236–241.
78. Albert, M., et al. (2019). Prevention of tau seeding and propagation by immunotherapy with a central tau epitope antibody. *Brain, 142*(6), 1736–1750.
79. Cross, A. H., et al. (2006). Rituximab reduces B cells and T cells in cerebrospinal fluid of multiple sclerosis patients. *Journal of Neuroimmunology, 180*(1–2), 63–70.
80. Martin Mdel, P., et al. (2009). Depletion of B lymphocytes from cerebral perivascular spaces by rituximab. *Archives of Neurology, 66*(8), 1016–1020.
81. Roll, P., et al. (2006). Regeneration of B cell subsets after transient B cell depletion using anti-CD20 antibodies in rheumatoid arthritis. *Arthritis and Rheumatism, 54*(8), 2377–2386.
82. Palanichamy, A., et al. (2014). Rituximab efficiently depletes increased CD20-expressing T cells in multiple sclerosis patients. *Journal of Immunology, 193*(2), 580–586.
83. Hauser, S. L., et al. (2008). B-cell depletion with rituximab in relapsing-remitting multiple sclerosis. *The New England Journal of Medicine, 358*(7), 676–688.
84. Hauser, S. L., et al. (2017). Ocrelizumab versus interferon beta-1a in relapsing multiple sclerosis. *The New England Journal of Medicine, 376*(3), 221–234.
85. Guerrier, T., et al. (2018). Proinflammatory B-cell profile in the early phases of MS predicts an active disease. *Neurology: Neuroimmunology and Neuroinflammation, 5*(2), e431.
86. Barr, T. A., et al. (2012). B cell depletion therapy ameliorates autoimmune disease through ablation of IL-6-producing B cells. *The Journal of Experimental Medicine, 209*(5), 1001–1010.
87. Rojas, O. L., et al. (2019). Recirculating intestinal IgA-producing cells regulate Neuroinflammation via IL-10. *Cell, 176*(3), 610 e18–624 e18.
88. Li, R., et al. (2015). Proinflammatory GM-CSF-producing B cells in multiple sclerosis and B cell depletion therapy. *Science Translational Medicine, 7*(310), 310ra166.

89. Kjaeldgaard, A. L., et al. (2018). Amyotrophic lateral sclerosis: The complement and inflammatory hypothesis. *Molecular Immunology, 102*, 14–25.
90. Woodruff, T. M., et al. (2008). The complement factor C5a contributes to pathology in a rat model of amyotrophic lateral sclerosis. *Journal of Immunology, 181*(12), 8727–8734.
91. Chen, N. J., et al. (2007). C5L2 is critical for the biological activities of the anaphylatoxins C5a and C3a. *Nature, 446*(7132), 203–207.
92. Dharmadasa, T., & Kiernan, M. C. (2018). Riluzole, disease stage and survival in ALS. *Lancet Neurology, 17*(5), 385–386.
93. Yoshino, H. (2019). Edaravone for the treatment of amyotrophic lateral sclerosis. *Expert Review of Neurotherapeutics, 19*(3), 185–193.
94. Kiaei, M., et al. (2006). Thalidomide and lenalidomide extend survival in a transgenic mouse model of amyotrophic lateral sclerosis. *The Journal of Neuroscience, 26*(9), 2467–2473.
95. Karlsson, J., et al. (2004). Life span extension and reduced neuronal death after weekly intraventricular cyclosporin injections in the G93A transgenic mouse model of amyotrophic lateral sclerosis. *Journal of Neurosurgery, 101*(1), 128–137.
96. Appel, S. H., et al. (1988). A double-blind study of the effectiveness of cyclosporine in amyotrophic lateral sclerosis. *Archives of Neurology, 45*(4), 381–386.
97. Meininger, V., et al. (2009). Glatiramer acetate has no impact on disease progression in ALS at 40 mg/day: A double- blind, randomized, multicentre, placebo-controlled trial. *Amyotrophic Lateral Sclerosis, 10*(5–6), 378–383.
98. Maier, A., et al. (2015). Interleukin-1 antagonist anakinra in amyotrophic lateral sclerosis—A pilot study. *PLoS One, 10*(10), e0139684.
99. Liu, Y., et al. (2023). Neuroimmune mechanisms underlying Alzheimer's disease: Insights into central and peripheral immune cell crosstalk. *Ageing Research Reviews, 84*, 101831.
100. Delpech, J. C., et al. (2019). Neuroimmune crosstalk through extracellular vesicles in health and disease. *Trends in Neurosciences, 42*(5), 361–372.
101. Sanford, L. D., et al. (2023). Modeling integrated stress, sleep, fear and neuroimmune responses: Relevance for understanding trauma and stress-related disorders. *Neurobiology of Stress, 23*, 100517.
102. Kondaparthi, P., Flora, S. J. S., & Naqvi, S. (2019). Selenium nanoparticles: An insight on its pro-oxidant and antioxidant properties. *Frontiers in Nanoscience and Nanotechnology, 6*, 1–5.
103. Davis, K. D., et al. (2020). Discovery and validation of biomarkers to aid the development of safe and effective pain therapeutics: Challenges and opportunities. *Nature Reviews Neurology, 16*(7), 381–400.
104. Mancuso, S., et al. (2022). Effects of B-cell lymphoma on the immune system and immune recovery after treatment: The paradigm of targeted therapy. *International Journal of Molecular Sciences, 23*(6), 3368.
105. Taheri, F., et al. (2022). The role of efferocytosis in neuro-degenerative diseases. *Neurological Sciences, 43*(3), 1593–1603.
106. Salloway, S., et al. (2009). A phase 2 multiple ascending dose trial of bapineuzumab in mild to moderate Alzheimer disease. *Neurology, 73*(24), 2061–2070.
107. Mandal, S. M., et al. (2014). Challenges and future prospects of antibiotic therapy: From peptides to phages utilization. *Frontiers in Pharmacology, 5*, 105.
108. Brummer, T., et al. (2021). Treatment approaches to patients with multiple sclerosis and coexisting autoimmune disorders. *Therapeutic Advances in Neurological Disorders, 14*, 17562864211035542.
109. Kunz, M., & Ibrahim, S. M. (2009). Cytokines and cytokine profiles in human autoimmune diseases and animal models of autoimmunity. *Mediators of Inflammation, 2009*, 979258.
110. Naqvi, S., Panghal, A., & Flora, S. J. S. (2020). Nanotechnology: A promising approach for delivery of neuroprotective drugs. *Frontiers in Neuroscience, 14*, 494. https://doi.org/10.3389/fnins.2020.00494
111. Sanati, M., et al. (2019). Impact of gold nanoparticles on amyloid beta-induced Alzheimer's disease in a rat animal model: Involvement of STIM proteins. *ACS Chemical Neuroscience, 10*(5), 2299–2309.

112. Singh, A., Ujjwal, R. R., Naqvi, S., Verma, R. K., Tiwari, S., Kesharwani, P., & Shukla, R. (2022). Formulation development of tocopherol polyethylene glycol nanoengineered polyamidoamine dendrimer for neuroprotection and treatment of Alzheimer disease. *Journal of Drug Targeting, 30*(7), 777–791. https://doi.org/10.1080/1061186X.2022.2063297
113. Materón, E. M., et al. (2021). Magnetic nanoparticles in biomedical applications: A review. *Applied Surface Science Advances, 6*, 100163.
114. Dowding, J. M., et al. (2014). Cerium oxide nanoparticles protect against Aβ-induced mitochondrial fragmentation and neuronal cell death. *Cell Death & Differentiation, 21*(10), 1622–1632.
115. Pichla, M., Bartosz, G., & Sadowska-Bartosz, I. (2020). The antiaggregative and antiamyloidogenic properties of nanoparticles: A promising tool for the treatment and diagnostics of neurodegenerative diseases. *Oxidative Medicine and Cellular Longevity, 2020*, 3534570.
116. Vissers, C., Ming, G. L., & Song, H. (2019). Nanoparticle technology and stem cell therapy team up against neurodegenerative disorders. *Advanced Drug Delivery Reviews, 148*, 239–251.
117. Sudhakar, V., & Richardson, R. M. (2019). Gene therapy for neurodegenerative diseases. *Neurotherapeutics, 16*(1), 166–175.
118. Zhu, X., et al. (2021). Gene therapy for neurodegenerative disease: Clinical potential and directions. *Frontiers in Molecular Neuroscience, 14*, 618171.
119. Sun, J., & Roy, S. (2021). Gene-based therapies for neurodegenerative diseases. *Nature Neuroscience, 24*(3), 297–311.
120. Paulson, H. (2006). RNA interference as potential therapy for neurodegenerative disease: applications to inclusion-body myositis? *Neurology, 66*(2 Suppl 1), S114–S117.

# Drug Repositioning and Repurposing for Neurodegenerative Diseases

Kisan Pawar, Somidi Srinu, and Abha Sharma

**Abstract**

Neurodegenerative diseases are characterized by degeneration of neurons, which is becoming more prevalent day by day, and the number of patients suffering from the neurodegenerative diseases also increases globally. The currently available treatment is only for the management of symptoms or the control of progression of disease. The new therapies come from natural products, synthesis, and existing drug; the therapies that come from existing drugs are drug repurposing and repositioning. Drug repurposing is a strategy of finding novel properties from well-known medications. This is more advantageous than other sources because it is reducing the time as well as cost and it is an alternative approach for the rapid development and identification of drugs for the diagnosis and treatment of rare and complex diseases. The cost required for the development of existing drugs is less than the development of new drug. There are several drug repurposing studies reported for finding out the new role of the existing drugs for the treatment of neurodegenerative diseases including Parkinson's disease, Alzheimer's disease, amyotrophic lateral sclerosis, Huntington's disease, and multiple sclerosis.

**Keywords**

Drug repurposing · Neurodegenerative diseases · Pharmacokinetic · Pharmacodynamic · Parkinson's disease · Alzheimer's disease · Amyotrophic lateral sclerosis · Multiple sclerosis · Huntington's disease

K. Pawar · S. Srinu · A. Sharma (✉)
Department of Medicinal Chemistry, National Institute of Pharmaceutical Education and Research, Raebareli, India
e-mail: abha.sharma@niperraebareli.edu.in

 
R. Shukla et al. (eds.), *Neuroendocrine-Immune Axis*,
https://doi.org/10.1007/978-981-95-5088-3_13

## Abbreviation

| | |
|---|---|
| AD | Alzheimer's Disease |
| PD | Parkinson's Disease |
| HD | Huntington's Disease |
| ALS | Amyotrophic Lateral Sclerosis |
| NMDA | N-Methyl-D-Aspartate |
| HIV | Human Immunodeficiency Virus |
| Aβ | Amyloid Beta |
| APP | Amyloid Precursor Protein |
| ROS | Reactive Oxygen Species |
| DNA | Deoxyribonucleic Acid |
| CAG | Cytosine Adenine Guanine |
| CNS | Central Nervous System |
| SOD | Superoxide Dismutase |
| FDA | Food and Drug Administration |
| P-gp | P-glycoprotein |
| TB | Tuberculosis |
| AT | Angiotensin |
| eIF2α-P | Eukaryotic Initiation Factor 2 |
| cGMP | Cyclic Guanosine Monophosphate |
| GSK | Glycogen Synthase Kinase 3 |
| ADAM10 | ADAM Metallopeptidase Domain 10 |
| JAK2 | Janus Activated Kinase 2 |
| SIRT | Sirtuin 1 |
| JNK1 | C-Jun N-terminal Kinases |
| HIF1 | Hypoxia-Inducible Factor 1 |
| BACE1 | Amyloid Precursor Protein (APP) Cleaving Enzyme-1 |
| PPARα | Peroxisome Proliferator-Activated Receptor Alpha |
| GABA-A Receptor | γ-Aminobutyric Acid-A Receptor |
| GLP-1 receptors | Glucagon-Like Peptide-1 Receptor |
| SNCA gene | Alpha-Synuclein Gene |
| CLEAR network | Coordinated Lysosomal Expression and Regulation network |
| GBA | Glucocerebrosidase |
| PKA | Protein Kinase |
| ASIC1 | Acid-Sensing Ion Channel 1 |

## 1 Introduction

Neurodegenerative diseases, such as Alzheimer's diseases (AD) and Parkinson's diseases (PD), are characterized by progressive neuronal damage in the brain and develop various symptoms related to the neuronal system [1]. AD is caused due to accumulation of phosphorylated tau and amyloid *β* protein in neuron leading to cognitive dysfunction and dementia [2]. On the other hand, PD is caused due to loss of dopaminergic neurons in the substantia nigra leading to motor dysfunction including bradykinesia, muscular rigidity, tremor, and postural impairment [3]. Both diseases have same pathophysiology like oxidative stress [4], neuroinflammation [5], etc. The available treatment for neurodegenerative diseases is limited, and mainly symptomatic treatment is available [6]. The drug approved for AD treatment includes acetylcholinesterase inhibitors, e.g., Rivastigmine, Galantamine, Donepezil, and N-methyl-D-aspartate (NMDA) receptor antagonist, e.g., Memantine [2] and the recently approved monoclonal antibody for AD treatment is aducanumab in the USA [7]. On the other hand, drugs approved for PD treatment are dopamine precursor, i.e., Levodopa, monoamine oxidase inhibitors, e.g., Rasagiline, dopamine receptor agonist, e.g., Pramipexole, Ropinirole [3]. Therefore, it is crucial to create novel treatments for these diseases. Starting with an existing drug is the most successful way to discover a new drug. Drug repurposing is the strategy to find out new uses of already approved drug, the pharmacokinetic and pharmacodynamic profile of approved drug is well known so it is easy to identify the new uses of these drugs [8].

## 2 Neurodegenerative Diseases

Neurodegeneration refers to the progressive and irreversible loss of structure and function of neurons, which are the cells responsible for transmitting information in the brain and nervous system. It is characterized by the gradual deterioration of neural pathways and the death of neurons, leading to a decline in cognitive function [2], movement, and other neurological processes. Neurodegenerative diseases are a group of disorders that involve neurodegeneration as their primary characteristic [9]. Some well-known examples of neurodegenerative diseases include AD, PD, Huntington's disease (HD), and amyotrophic lateral sclerosis (ALS) [10]. These conditions share common features, such as the accumulation of abnormal proteins in the brain and the loss of specific populations of neurons [11]. The causes of neurodegeneration are multifactorial and often involve a combination of genetic, environmental, and lifestyle factors [1]. In some cases, specific gene mutations can lead to the development of certain neurodegenerative diseases. Environmental factors, such as exposure to toxins or head injuries, may also contribute to the onset or progression of neurodegeneration. Age is another significant risk factor, as many neurodegenerative diseases are more prevalent in older individuals. Neurodegenerative diseases typically manifest with a wide range of symptoms, depending on the specific condition and the areas of the brain affected. Common symptoms include memory loss, cognitive decline, impaired motor function, changes in behavior and mood, and difficulties with speech and language [12].

### 2.1 Alzheimer's Disease

AD is a complicated neurodegenerative condition that predominantly impacts the brain, resulting in gradual deterioration of cognitive function, memory loss, and alterations in behavior and personality. While the exact cause of AD is not fully understood, several key pathological mechanisms have been identified [2, 13].

### 2.1.1 Epidemiology

Worldwide, around 50 million are AD patients and this number is double every 5 years and will reach 152 billion in 2050. AD is making the sixth-leading cause of death in the United States and the fifth leading cause of death in elderly people aged 65. From 2000 to 2019 the deaths from stroke, HIV, and heart disease decreased, whereas the deaths from AD increased; the death trajectory from AD was increased in 2020 by the COVID-19 pandemic [13].

### 2.1.2 Pathophysiology

(a) Amyloid Beta (A$\beta$) Plaques: One hallmark feature of AD is the accumulation of abnormal protein deposits called amyloid beta plaques in the brain. The amyloid precursor protein (APP), which is processed by enzymes, yields amyloid beta. In AD, there is an imbalance between the production and clearance of A$\beta$, leading to its accumulation and subsequent formation of plaques. These plaques disrupt neuronal communication and trigger inflammatory responses, contributing to neurodegeneration [2].

(b) Neurofibrillary Tangles: The development of neurofibrillary tangles inside neurons is another hallmark of AD. Upon aggregation of a protein tau it forms tangles, which normally stabilize the internal structure of neurons. In Alzheimer's, tau becomes abnormally phosphorylated, leading to its misfolding and subsequent accumulation into tangles. Tau tangles disrupt the movement of essential nutrients and other molecules in neurons, impairing their normal function and contributing to cell death [14, 15].

(c) Neurotransmitter Imbalance: AD is associated with disruptions in the levels of various neurotransmitters, which are chemical messengers involved in communication between neurons. The neurotransmitter acetylcholine is particularly affected in AD, with reduced levels in the brain. Acetylcholine is crucial for learning, memory, and attention. The depletion of acetylcholine and other neurotransmitters contributes to the cognitive decline observed in AD [14].

(d) Oxidative Stress and Inflammation: AD is anticipated to be associated with oxidative stress, which is characterized by an in equilibrium between the generation of reactive oxygen species (ROS) and an individual's antioxidant mechanisms. ROS can damage cellular structures, including proteins and DNA, leading to neuronal dysfunction and death. Additionally, the accumulation of amyloid beta plaques triggers an inflammatory response, involving the activation of immune cells and the release of inflammatory molecules. Chronic inflammation exacerbates neurodegeneration and contributes to the progression of the disease [4].

(e) Neuronal Loss and Synaptic Dysfunction: AD leads to the loss of synapses, the connections between neurons, and the subsequent death of nerve cells. Synaptic dysfunction disrupts the communication between neurons, impairing cognitive processes such as memory and thinking. Over time, the widespread loss of neu-

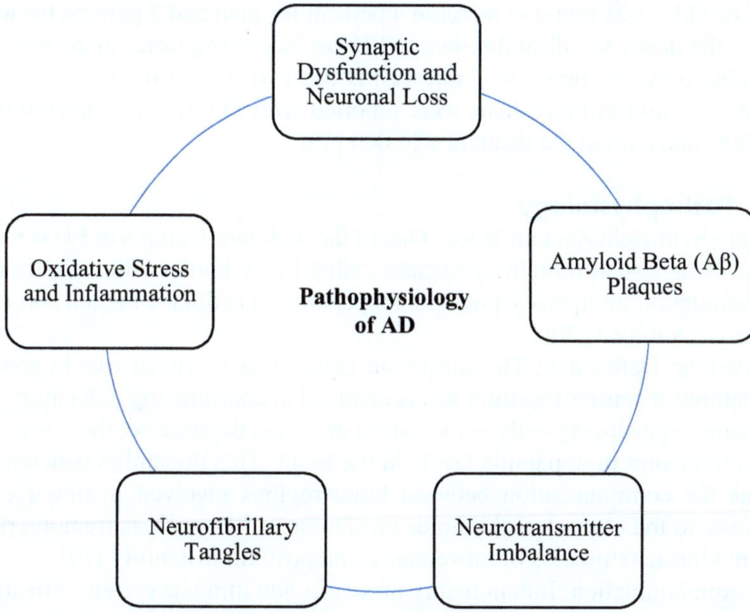


**Fig. 1** Pathophysiology of AD

rons, particularly in brain regions crucial for memory and cognition, further exacerbates the cognitive decline seen in AD [16].

It's important to note that while these pathological mechanisms are well-established in AD (Fig. 1), the precise sequence of events and their interplay is still a topic of ongoing research. Additionally, other factors, such as genetic predisposition, environmental influences, and lifestyle factors, may also accountable to develop AD [14].

## 2.2 Parkinson Disease (PD)

PD is a neurodegenerative disorder characterized by the progressive loss of dopamine-producing cells in a region of the brain called the substantia nigra. This loss of dopamine leads to a variety of motor and non-motor symptoms. The exact cause of PD is not fully understood, but several key pathophysiological processes have been identified [3, 17, 18].

### 2.2.1 Epidemiology

Prevalence rate of PD disease increased day by day, as estimated data show 35 cases in 100,000 are reported every year. In the meta analysis study of the prevalence of PD cases in North American populations of ages 45-54 years, was found to increase from less than 1 percent for both women and men. The prevalence of PD cases in aged 85 or older was found to increase 4 percent for men and 2 percent for women. Globally, the death and disability due to PD are increasing faster than other neurological disorders, the prevalence rate of PD has two-fold in the past 25 years. In 2019, the 8.5 million individuals were reported with PD. It is an increase of 81% since 2000, and caused the death of 329,000 [19].

### 2.2.2 Pathophysiology

(a) Alpha-Synuclein Accumulation: One of the hallmark features of PD is the presence of abnormal protein aggregates called Lewy bodies within neurons. The accumulation of alpha-synuclein disrupts normal cellular function and impairs neuronal survival [20].
(b) Dopamine Deficiency: The substantia nigra plays a crucial role in producing dopamine, a neurotransmitter that is involved in coordinating movement. As the dopamine-producing cells in the substantia nigra degenerate, there is a significant reduction in dopamine levels in the brain. This dopamine deficiency disrupts the communication between brain regions involved in motor control, leading to the characteristic motor symptoms of PD, such as tremors, rigidity, bradykinesia (slowness of movement), and postural instability [19].
(c) Neuroinflammation: Inflammatory processes and immune system activation are thought to play a role in the pathophysiology of PD. Activation of microglia, the immune cells of the central nervous system, occurs in response to the accumulation of alpha-synuclein and the resulting neuronal damage. Chronic inflammation contributes to the degeneration of dopamine-producing cells and exacerbates neurodegeneration in PD [3].
(d) Disrupted protein handling and clearance: The normal clearance and degradation of proteins within cells is essential for maintaining cellular health. In PD, there are impairments in the mechanisms responsible for clearing misfolded or aggregated proteins, including alpha-synuclein. This leads to the accumulation of toxic protein aggregates, such as Lewy bodies, contributing to neuronal dysfunction and cell death.

It's important to note that PD is a complex disorder with various contributing factors (Fig. 2), and the interplay between these pathophysiological processes is still an area of active research. Genetic factors, environmental toxins, and other mechanisms may also influence the development and progression of PD [17].

## 2.3 Huntington's Disease (HD)

HD is a hereditary neurodegenerative disorder caused by a mutation in the huntingtin gene. This mutation leads to the production of an abnormal form of the huntingtin protein, which ultimately results in the progressive degeneration of neurons in certain areas of the brain. The pathophysiology of HD involves several key mechanisms [20].

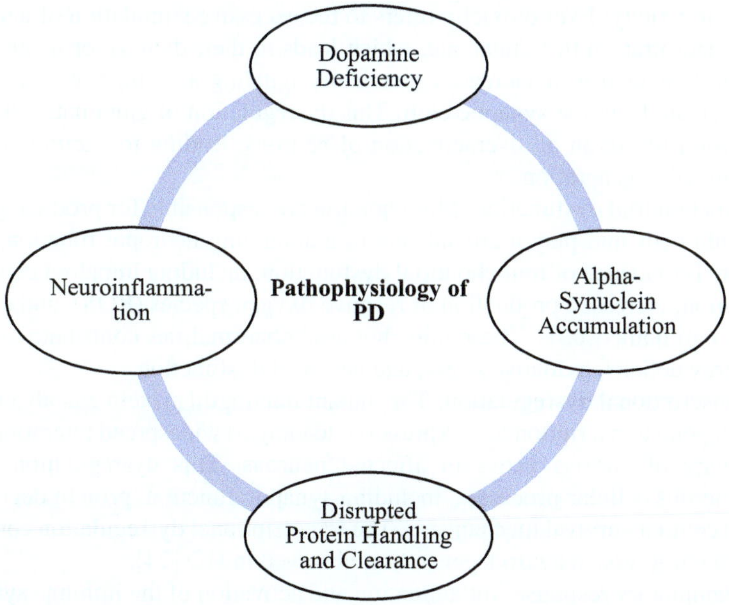


**Fig. 2** Pathophysiology of PD

### 2.3.1 Epidemiology

Worldwide, prevalence rate of HD is 5–10 patients per 100,000 persons, the occurrence rate in men and women is similar. The occurrence rate of HD patients is higher in Western European average of around seven per 100,000 and one per million in Asian and African populations. The 2013 epidemiological study of HD in the UK found that the occurrence rate is 12.3 per 100,000 [21].

### 2.3.2 Pathophysiology

(a) CAG Repeat Expansion: The genetic mutation responsible for HD involves an abnormal expansion of a specific segment of DNA called CAG trinucleotide repeats within the huntingtin gene. Normally, this CAG repeat segment is repeated between 10 and 35 times. However, in individuals with HD, the CAG repeat segment is abnormally expanded, with more than 35 repeats. The length of the CAG repeat is directly correlated with the age of onset and severity of the disease [22].

(b) Mutant huntingtin protein aggregation: The abnormal huntingtin protein produced as a result of the CAG repeat expansion has a tendency to misfold and aggregate. These protein aggregates, known as huntingtin aggregates or inclusion bodies, accumulate within neurons and disrupt their normal function. The precise mechanisms underlying the toxicity of mutant huntingtin aggregates are still being investigated, but it is believed that their accumulation contributes to cellular dysfunction and neuronal death [23].

(c) Excitotoxicity: Excitotoxicity refers to the excessive stimulation of neurons by the neurotransmitter glutamate, which leads to their damage or death. In HD, there is evidence of increased glutamate signaling and impaired clearance of glutamate from the synaptic cleft. This dysregulation of glutamate neurotransmission results in an overactivation of neurons, leading to excitotoxicity and neuronal degeneration.
(d) Mitochondrial dysfunction: Mitochondria are responsible for producing energy within cells and play a crucial role in maintaining neuronal function. In HD, there is evidence of mitochondrial dysfunction, including impaired energy production, increased production of reactive oxygen species (ROS), and disrupted calcium homeostasis. These mitochondrial abnormalities contribute to cellular energy deficits, oxidative stress, and neuronal dysfunction.
(e) Transcriptional dysregulation: The mutant huntingtin protein can also interfere with gene transcription and expression, leading to widespread alterations in the activity of various genes in affected neurons. This dysregulation disrupts numerous cellular processes, including synaptic function, protein degradation, and cellular survival mechanisms. The transcriptional dysregulation contributes to the progressive neurodegeneration observed in HD [24].
(f) Inflammatory response: Inflammation and activation of the immune system are observed in the brains of individuals with HD. The presence of mutant huntingtin protein and neuronal damage triggers an immune response, leading to the activation of microglia, the immune cells of the central nervous system. Upon stimulation of microglia facilitates the release of pro-inflammatory molecules, exacerbating neuroinflammation and contributing to the progression of the disease.

These pathophysiological processes in HD (Fig. 3) collectively result in the progressive loss of neurons, particularly in regions of the brain involved in motor control, cognition, and emotional regulation. The characteristic symptoms of HD, including motor abnormalities, cognitive decline, and psychiatric disturbances, arise as a consequence of the progressive degeneration of these specific brain regions [25].

## 2.4 Multiple Sclerosis (MS)

MS is a chronic autoimmune disease characterized by inflammation, demyelination, and damage to the central nervous system (CNS), it involves spinal cord and brain. The exact cause of MS is not fully understood, but it is believed to involve a combination of genetic, environmental, and immunological factors. The pathophysiology of MS involves several key mechanisms (Fig. 4) [26].

### 2.4.1 Epidemiology

Around 400,000 cases in the United States and 2.5 million cases worldwide have MS. MS is threefold more common in females than in males, with the age of

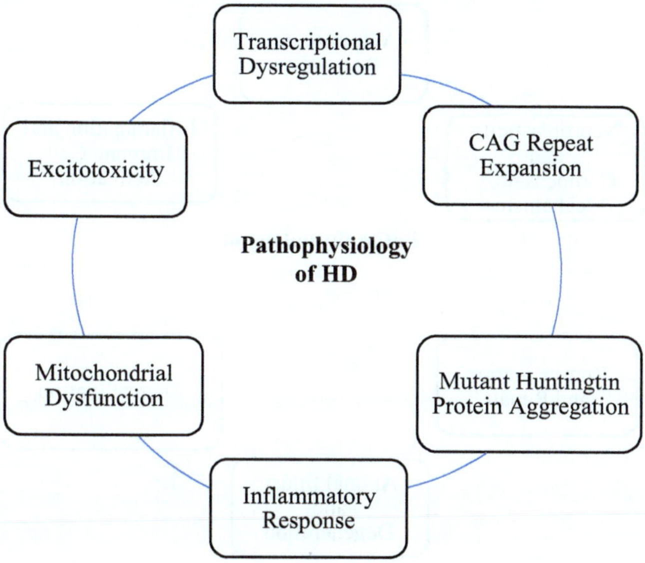


**Fig. 3** Pathophysiology of HD

occurrence of MS between 20 and 40 years. Prevalence of 1 in 1000 is observed in the European population, and lower prevalence in the Asian and African population, the recent studies have noted high prevalence in the African-American population same like the European population. The various studies have noted that the population migrates from the area of high prevalence, also high risk of causing MS [27].

### 2.4.2 Pathophysiology

(a) Autoimmune Response: In MS, the immune system mistakenly targets components of the CNS as if they were foreign invaders. Immune cells, particularly T cells, become activated and cross the blood-brain barrier, entering the CNS. Once inside, these immune cells launch an inflammatory attack on myelin, the protective covering that surrounds nerve fibers. The immune response leads to the activation of other immune cells, such as macrophages and B cells, which further contribute to the inflammation and damage.

(b) Inflammation and Immune Cell Activation: The inflammatory response in MS involves the release of pro-inflammatory molecules, such as cytokines and chemokines, by immune cells. These molecules attract more immune cells to the affected areas, resulting in an accumulation of immune cells around blood vessels and the formation of inflammatory lesions. The inflammation disrupts the normal functioning of the CNS, leading to damage to myelin and, in later stages, to axons (nerve fibers).

(c) Demyelination: Demyelination refers to the loss or damage of the myelin sheath that insulates nerve fibers. In MS, the immune system's attack on myelin leads to the breakdown of myelin, leaving nerve fibers exposed and vulnerable.

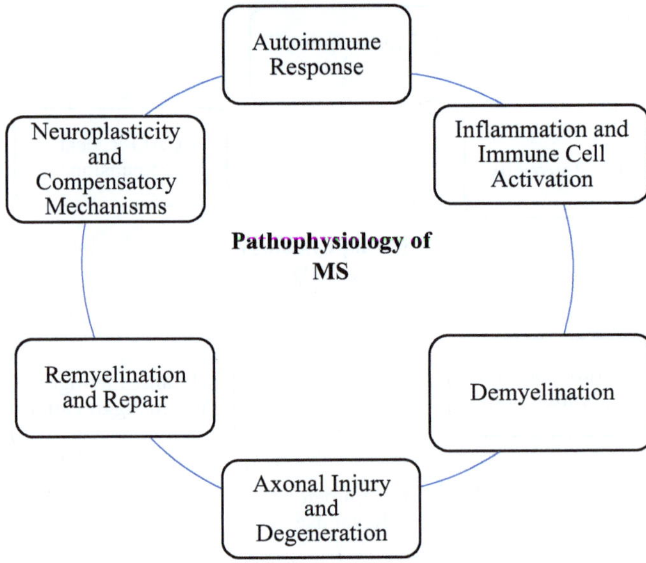


**Fig. 4** Pathophysiology of MS

Demyelination disrupts the transmission of nerve impulses, causing a wide range of neurological symptoms, depending on the location and extent of the demyelinated areas.

(d) Neuroplasticity and compensatory mechanisms: The brain has a remarkable capacity for adaptation and compensation, known as neuroplasticity. In response to the damage caused by MS, the brain can reorganize and reroute neural connections to partially compensate for lost function. These compensatory mechanisms can help to mitigate the impact of the disease and enable individuals with MS to adapt and regain some level of function [28]. The interplay of these pathophysiological processes in MS results in the formation of multiple areas of inflammation, demyelination, and axonal damage throughout the CNS. This leads to a wide range of symptoms, including fatigue, muscle weakness, impaired coordination, sensory disturbances, cognitive changes, and visual impairments. The course of MS can vary among individuals, with periods of relapse and remission (relapsing-remitting MS) or a more progressive and gradual accumulation of disability (primary progressive MS or secondary progressive MS) [29].

## 2.5 Amyotrophic Lateral Sclerosis (ALS)

ALS, also known as Lou Gehrig's disease, is a progressive neurodegenerative disorder that affects the nerve cells responsible for controlling voluntary muscle

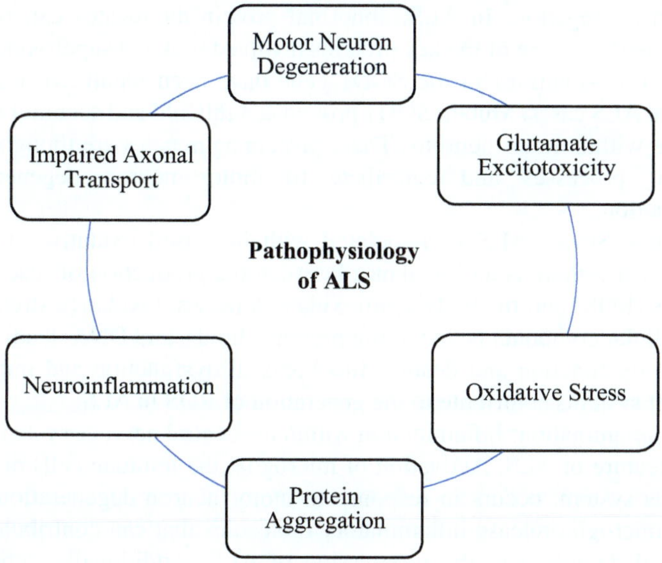


**Fig. 5** Pathophysiology of ALS

movement. The exact cause of ALS is not fully understood, but multiple pathological mechanisms contribute to its pathophysiology (Fig. 5) [30].

### 2.5.1 Epidemiology

The incident rate of ALS is between the range of 0.6 and 3.8 per 100,000-person. ALS incident rate in Scotland and Stockholm is 3.8 per 100,000 persons in 2014, China ALS incident rate is reported 1.2 in 2011 to 2015 and 0.8 in 2010 to 2015 per 100,000 persons, according to the US national ALS registry reported prevalence of 5.0 per 100,000 per in 2014 [31].

### 2.5.2 Pathophysiology

(a) Motor neuron degeneration: ALS primarily affects motor neurons, which are the nerve cells responsible for transmitting signals from the brain and spinal cord to the muscles, enabling voluntary movement. In ALS, motor neurons in the brain's motor cortex, brainstem, and spinal cord gradually degenerate and die. The loss of motor neurons disrupts the communication between the brain and muscles, leading to muscle weakness, atrophy, and ultimately paralysis.

(b) Glutamate excitotoxicity: Glutamate is a neurotransmitter that plays a vital role in transmitting signals between nerve cells. In ALS, there is evidence of increased levels of glutamate in the spaces around motor neurons. Excessive glutamate can overstimulate motor neurons, leading to a phenomenon known as excitotoxicity. Excitotoxicity causes calcium ions inward movement into motor nerves, triggering a cascade of toxic events, including mitochondrial dysfunction, oxidative stress, and ultimately neuronal death.

(c) Protein aggregation: In ALS, abnormal protein aggregates can form within motor neurons. One of the key proteins involved is called superoxide dismutase 1 (SOD1). Mutations in the SOD1 gene have been identified in a subset of familial ALS cases. Mutant SOD1 protein can misfold and accumulate, forming clumps within motor neurons. These protein aggregates are thought to disrupt cellular processes and contribute to motor neuron degeneration and dysfunction.
(d) Oxidative Stress: ALS is associated with increased oxidative stress, which occurs when there is an imbalance between the production of reactive oxygen species (ROS) and the body's antioxidant defenses. Oxidative stress can damage cellular components, including proteins, lipids, and DNA, leading to motor neuron dysfunction and death. Mitochondrial dysfunction and impaired antioxidant systems contribute to the generation of ROS in ALS.
(e) Neuroinflammation: Inflammation within the central nervous system is a prominent feature of ALS. Activation of microglia, the immune cells of the central nervous system, occurs in response to motor neuron degeneration. The activated microglia release inflammatory molecules that can contribute to further neuronal damage and the progression of ALS. Additionally, other immune cells, such as T cells and B cells, are also implicated in the inflammatory response in ALS.
(f) Impaired Axonal Transport: Axonal transport is the process by which molecules, including proteins, organelles, and other essential components, are transported along the length of nerve fibers (axons). In ALS, disruptions in axonal transport have been observed, leading to impaired delivery of essential materials to motor neurons and compromised cellular function. Axonal transport defects contribute to the degeneration and death of motor neurons in ALS.

The combination of these pathophysiological processes in ALS results in the progressive loss of motor neurons, leading to muscle weakness, muscle atrophy, and ultimately paralysis. The disease can also affect respiratory muscles, leading to respiratory failure, which is a common cause of death in ALS.

## 3 Approaches for the Treatment of Neurodegenerative Diseases

Neurodegenerative disorders, such as AD, PD, HD, and ALS, pose significant challenges to healthcare systems worldwide. These conditions are characterized by the progressive degeneration and dysfunction of neurons, leading to debilitating symptoms and, in many cases, irreversible damage. Developing effective treatments for neurodegenerative disorders has proven to be an immense scientific and clinical undertaking. One promising avenue for discovering new therapies is drug repurposing. Traditionally, drug discovery has focused on identifying novel compounds specifically designed to target the underlying mechanisms of a particular disease. However, this process can be time-consuming, costly, and prone to failure during clinical trials. Drug repurposing, on the other hand, involves exploring the therapeutic potential of existing drugs that were originally developed for different indications. By repurposing drugs, researchers can leverage existing knowledge about their safety profiles and mechanisms of action, potentially accelerating the path to clinical use. There are two primary approaches to drug repurposing: synthetic and traditional. Synthetic approaches involve the design and synthesis of novel compounds inspired by existing drugs or known disease mechanisms. These compounds are chemically modified to enhance their efficacy, safety, or specificity for a particular neurodegenerative disorder. Synthetic drug repurposing holds the promise of developing targeted treatments with improved pharmacokinetic properties and reduced side effects. In contrast, traditional drug repurposing involves investigating the therapeutic potential of existing drugs that have already been approved for other indications. This approach capitalizes on the vast libraries of drugs already available, potentially by passing lengthy preclinical and early clinical development stages. Traditional drug repurposing typically relies on various sources of evidence, including *in vitro* studies, animal models, and retrospective analyses of patient data. By repurposing existing drugs, researchers can quickly move them into clinical trials for new indications, thereby shortening the time and costs associated with drug development [32–34]. Both synthetic and traditional approaches to drug repurposing have their advantages and challenges. Synthetic approaches offer the possibility of designing highly specific compounds tailored to target the underlying disease mechanisms of neurodegenerative disorders. However, the process of designing and synthesizing new compounds can be time-consuming and resource-intensive. Additionally, ensuring the safety and efficacy of synthetic repurposed drugs requires rigorous preclinical and clinical testing. Traditional drug repurposing, on the other hand, benefits from the wealth of information available on already approved drugs. By leveraging existing knowledge about drug safety and mechanisms of action, traditional repurposing allows for faster translation to clinical use. However, identifying the most suitable candidates for repurposing from the vast pool of existing drugs can be challenging. Moreover, the success of traditional repurposing relies heavily on the availability of relevant preclinical models and the availability of comprehensive patient data. Drug repurposing, whether through synthetic or traditional approaches, offers an exciting opportunity to expedite the discovery and development of effective treatments for neurodegenerative disorders. By repurposing existing drugs, researchers can potentially bypass the lengthy and costly process of developing new compounds from scratch. While challenges remain, the application of drug repurposing holds immense promise in addressing the urgent need for therapies to combat neurodegenerative diseases and improve the lives of millions of affected individuals worldwide [8].

## 4 Drug Repurposing and Repositioning

A growing trend in drug development is the repurposing of existing drugs, commonly referred to as drug repositioning or drug reprofiling. Drug repurposing and repositioning in neurodegeneration refers to the process of identifying and utilizing existing drugs for the treatment of neurological disorders and diseases. Neurodegenerative diseases, such as AD, PD, HD, and ALS, are characterized by the progressive degeneration of the structure and function of the nervous system. Developing new drugs from scratch can be a lengthy, expensive, and risky process. However, drug repurposing offers a potential solution by leveraging drugs that have already been approved for other indications or are in various stages of clinical development. By repurposing existing drugs, researchers can bypass the time-consuming early stages of drug discovery, such as toxicity testing and optimization, and focus on investigating their potential efficacy in treating neurodegenerative diseases. The concept of repurposing drugs stems from the growing understanding that many diseases share common underlying molecular pathways and mechanisms. While a drug may have been originally developed for a specific condition, it may also exhibit beneficial effects in other diseases due to shared targets or biological pathways involved in both conditions. This approach offers the possibility of accelerating the availability of effective treatments for neurodegenerative diseases. The process of drug repurposing and repositioning involves several steps. First, researchers analyze the molecular pathways and targets associated with the neurodegenerative disease of interest. They then identify drugs that are known to interact with these pathways or targets, either through computational methods or experimental screening. Once potential drug candidates are identified, they undergo preclinical and clinical evaluation to assess their safety and efficacy in treating disease. One advantage of drug repurposing is that repurposed drugs have often already undergone rigorous safety testing in humans, which can expedite their development process. Additionally, repurposing existing drugs can potentially reduce costs associated with drug development and provide access to treatments more quickly. Drug repurposing and repositioning in neurodegeneration hold promise for discovering new therapeutic options for patients with neurodegenerative diseases. By leveraging existing knowledge and approved drugs, researchers can accelerate the drug discovery process, potentially leading to improved treatment outcomes and quality of life for individuals affected by these debilitating conditions [8, 32, 33]. The World Health Organization has forecast that neurodegenerative illnesses that affect body functions will become the second leading cause of death in the next 20 years. There are three main places where new therapies can come from: synthesis, natural products, and changes to current drugs. This last option, called "drug repurposing," is the best because the drug's metabolic and pharmacodynamic traits are already known, and this method doesn't require as much money as the traditional way of making new drugs. Several studies have looked at the possibility of using old drugs to treat the most important neurological diseases, such as AD, PD, HD, MS, and ALS. *In silico* screening of chemical libraries in novel targets is one of the most often employed medication repositioning methodologies [33]. The drug thalidomide is a great example of how drugs can be used for different purposes. It was first used as an over-the-counter medicine to treat morning sickness in pregnant women, but it was quickly taken off the market after reports of teratogenicity and dysmyelia. But in 1998, the Food and Drug Administration (FDA) gave approval for thalidomide to be used to treat the skin symptoms of erythema nodosum leprosum. Because it blocks blood vessel growth, thalidomide was approved in 2006 to treat myeloma [35]. This shows how important it can be to find new uses for old drugs. However, several more instances of traditional medications being used in novel ways have been documented [36, 37]. Neurodegenerative diseases are diseases that get worse with age and have very different pathophysiology. The causes and processes of these diseases are not well understood, so there is no treatment for them. The frequency of chronic illnesses that affect memory, cognition, and mobility is rising together with the aging of the world population. Since there is a growing need of treatment for neurodegenerative disease and drug repurposing holds a lot of potential, it makes sense to test current drugs on these diseases. This study looks at drugs that have been changed to treat AD, PD, HD, MS, and ALS, which are the most studied neurological diseases [10, 38].

## 5 Repurposed Drugs

### 5.1 Drug Repurposing for Alzheimer's Disease

AD is a prominent neurodegenerative disorder that accounts for approximately 80% of dementia cases in the elderly population. The clinical manifestations of this condition entail a gradual deterioration of cognitive abilities, including memory impairment, learning difficulties, and a decline in both behavioral and functional capacities. The etiology of AD remains incompletely comprehended; however, Researchers hypothesized that continuous amyloid plaques accumulation in the synapses of the brain is implicated, culminating in neuronal and synaptic degeneration. At present, a cure for AD remains elusive, and pharmacological interventions are primarily aimed at managing cognitive symptoms and other associated manifestations. These interventions are most effective when initiated during the early stages of the disease [2, 11, 14]. Curiously, galantamine, one of the medications used to treat AD, was used for a different reason. Galantamine, an alkaloid found in Galanthus sp., is capable of enhancing nerve impulse transmission, making it a promising treatment option for neuromuscular disorders following anesthesia. This alkaloid also inhibits muscle acetylcholinesterase, which makes it a promising natural product to cure neuromuscular disorders. Galantamine is especially intriguing since it possesses a tertiary ammonium base, which makes the drug to invade the BBB, then it inhibits the acetylcholinesterase in the brain. Studying therapeutic activity of galantamine was started in 1980 and it is used to treat Alzheimer's in 2000, galantamine has been studied for its therapeutic effects in the treatment of AD. It is still one of the most popular medications for postponing the onset of severe symptoms in AD patients [39, 40]. Researchers conducted studies to know the anti-Alzheimer's activity of anti-tumor drugs. This hypothesis is supported by the observation that dementia and cancer might overlap signaling pathways, including mitochondrial malfunction, stress due to free radicals, impaired metabolic activities of cells, and the emergence of proteins formed by abnormal folding. Chemotherapy for breast cancer patients has been shown to reduce their likelihood of getting AD [41]. Nitrosourea carmustine is a brain cancer treatment alkylating agent. Hydrophobicity of the small molecule makes it to cross BBB easily. Carmustine showed an impact decrease in the production of amyloid aggregates in cells with more quantity of the protein precursor at a non-toxic dosage [42, 43]. A retinoid X receptor antagonist known as bexarotene, which is currently used in the treatment of skin cancer, has the capability to restore the damaged neurons, improving cognitive functioning, and lowering the amyloid plaques in the mice that overexpress similar AD genes [44]. Tamibarotene, clinically used in the management of type of blood cancer in Japan, acts through the retinoic acid receptor. It has the ability to affect a number of mechanisms involved in the pathophysiology of Alzheimer's disease, including lowering the levels of cortical acetylcholine and decreasing the formation of pro-inflammatory cytokines and chemokines by the cells of brain. It can also improve behavior in mice with accelerated senescence [45]. It has been proposed that imatinib, approved for the clinical management of bone marrow cancer and tumors, acts by inhibiting the tyrosine kinase enzyme, may be helpful to cure AD by the lowering of amyloid aggregates and promoting the neuroprotection. Imatinib has low capability to cross BBB, P-glycoprotein (P-gp) quickly effluxes it [46]. It has also been investigated as a possible therapy for AD, paclitaxel, an antimitotic drug already licensed for treating a variety of cancers. It is especially effective in treating neurodisorders, which occur due to tau protein, it is found in large amounts in CNS cells, and it is responsible for stabilizing microtubules. This protein becomes more fibrillating when it is phosphorylated because it loses its capacity to attach to microtubules. The phosphorylation is decreased by paclitaxel. Although it is also prone to P-gp and has low BBB penetration, paclitaxel has the same issue as imatinib. This makes it less effective [47, 48]. Thalidomide, an anticancer medication that has been shown to reduce angiogenesis, replication of endothelial cells, and damage of BBB, is another anticancer medication that may have anti-AD characteristics. By suppression of TNF, it was also able to lessen the death of hippocampus neurons [49, 50]. Researchers also studied the antimicrobial drugs to know the anti-Alzheimer's activity. The macrolide antibiotics erythromycin and azithromycin have both shown suppression of the protein responsible for the production of amyloid aggregates, resulting in a reduction of amyloid aggregates in the cerebral part of the brain. Tetracyclines are found to be accelerators for the disintegration of produced fibrils and increase the resistance of amyloid to trypsin digestion, in addition to reducing the synthesis of amyloid. Additionally, they reduced oxidative stress, pointing to a variety of mechanisms of action [51]. In this regard, doxycycline has shown promise, both alone and in combination with rifampicin [52, 53]. Rifampicin, the most often given antibiotic for Mycobacterium infections, has been demonstrated to reduce amyloid aggregates, depends on dose, most likely owing to reduced synthesis, improved amyloid aggregates elimination [54]. An antibiotic used to treat leprosy, dapsone gained notoriety in the 1990s when leprosy patients who had undergone dapsone treatment had a lower incidence of dementia. Dapsone may be able to reduce senile plaques, although the evidence for this claim is ambiguous. This led to the theory that this phenomenon may operate as a barrier to amyloid formation. Studies revealed comparable rates of AD in leprosy and TB patients, despite variations in the proportion of patients who had received medication treatments in the two groups [15, 55]. Acyclovir, penciclovir, and foscarnet are antiviral medications in AD cell models that have been shown to reduce tau protein hyperphosphorylation, amyloid aggregates, suggesting they may be effective treatments for AD [56]. It was found that amphotericin B, an antifungal drug, shows the amyloid aggregate reduction activity. Amphotericin B is toxic; thus, it is not a good choice for the treatment of AD, according to more recent research, which did not come to the same conclusions [57]. In transgenic mice, the antifungal and antiparasitic medication clioquinol exhibits the reduction of amyloid plaques [58]. Valproic acid, an antiepileptic medication, has been proposed as a neuroprotective treatment for AD because of transgenic mice studies showing improved memory impairments and decreased amyloid-plaque development. Although the postulated pathology pathways are complicated, it may be accomplished by reducing the generation of cytokines, enhancing microglial phagocytosis of amyloid [59, 60]. Valsartan, a commonly used antihypertensive acts through angiotensin receptors. Use of this class of medications for AD is justified by the fact that prolonged unfavorable stress, the main environmental factor contributing to the development and occurrence of Alzheimer's, may result in increased angiotensin II in the brain, which acts through subtype AT1, AT2 receptors. Additionally, it has been proposed that angiotensin II elevations are related to the development of amyloid, and that inhibiting AT1 using angiotensin receptor blockers may help postpone the decrease of cognitive function. In addition to this mode of action, valsartan also reduces acetylcholine release and suppresses vasoconstriction, inflammation, and mitochondrial dysfunction. Valsartan medication has been associated with decreased levels of amyloid and decline in dementia. Medication having higher capability to cross BBB, additional research is necessary before the usage of the drug in the treatment of Alzheimer's [61, 62]. Calcium channel blockers are medications that are approved for treating high blood pressure and angina pectoris. Dihydropyridine derivatives, like nilvadipine, have shown reduction of amyloid aggregates production, polymerization, and abnormal accumulation in the in vitro studies; they also improve the cell survivability and decrease the neurotoxic effect. They can also get through the blood-brain barrier well and increase brain blood flow through their vasodilatory effects [63, 64]. Since type 2 diabetes has been established as an AD risk factor, antidiabetics have also been repurposed for the condition. Research suggests that AD patients' brains have become less sensitive to insulin signaling. Insulin exhibits neuroprotective activity; it also controls the tau protein hyperphosphorylation and improves memory and cognitive functioning. Insulin may also trigger functionalities of the stem cells [65]. This suggests that drugs that promote insulin secretion could be useful in treating Alzheimer's disease. Glucagon-like peptide 1, a hormone, increases the production of insulin; its derivatives could be useful for treating Alzheimer's disease pathological pathways like hyperphosphorylation of tau protein, amyloid aggregation, neuronal function impairment, and cell death [66]. Zileuton, a 5-lipoxygenase-blocking antiasthma medication, was hypothesized to have advantageous uses in Alzheimer's disease. The results from the observation that 5-lipoxygenase is upregulated in Alzheimer's patients make it an interesting target within this context. In fact, investigations using the mouse compound zileuton showed a decrease in the production of amyloid [67]. Trazodone, an antidepressant medication, has demonstrated promising results in inhibiting the signaling pathway of the PERK/eIF2α-P branch of the unfolded protein response. This pathway is responsible for regulating the synthesis of protein cells and is excessively activated in individuals with Alzheimer's disease. The efficacy of Trazodone in reversing eIF2α-P translational attenuation has been demonstrated through *in vitro* and *in vivo* studies. Moreover, it exhibited neuroprotective effects, facilitated memory restoration, and hindered neurodegeneration as reported [68]. Sildenafil and tadalafil, two pharmaceutical agents utilized in the clinical management of erectile dysfunction, function as inhibitors of phosphodiesterase-5. This enzyme serves as a regulator of cyclic guanosine monophosphate (cGMP), a signaling molecule that has been implicated in the pathogenesis of AD. The cyclic guanosine monophosphate (cGMP) molecule is implicated in the regulation of cognitive functioning, and neuroplasticity, research data from a study, suggests a negative impact. However, cGMP may also play a beneficial role in the modulation of memory impairments that arise from A$\beta$. The efficacy of Sildenafil in the inhibition of neuroinflammation and reduction of A$\beta$ in aged mouse models has been demonstrated [69]. Tadalafil exhibited cognitive improvement and neuroprotective properties. It was also able to cross the blood-brain barrier in concentrations that were more potent in inhibiting phosphodiesterase-5 [70]. The hormone melatonin, which is produced by the pineal gland, was initially utilized to enhance sleep quality and eliminate free radicals. Melatonin has the potential to reduce amyloid aggregate generation by both receptor-dependent and receptor-independent pathways. The inhibition of GSK-3$\beta$ is crucial in promoting the polymerization of amyloid aggregates and the formation of A$\beta$. Melatonin has been found to inhibit GSK-3$\beta$ through the melatonin receptor MTI/2-mediated PLC/PKC and PI3K/Akt pathways. Furthermore, it has been observed that the activation of melatonin receptor MTI/2 can lead to the suppression of ADAM10 through ERK1/CREB and ERK1/Oct1 signaling pathways. Concurrently, it is plausible that melatonin may exert direct regulatory effects on associated proteins such as JAK2, SIRT1, JNK1, and HIF1, thereby suppressing BACE1 or promoting ADAM10 via pathways that are independent of receptors [71]. Rhus verniciflua flavonoid sulfuretin prevents A$\beta$ aggregation through numerous mechanisms. Promoter region at nucleotides 444 to 300 is responsible for Sulfuretin's effect on ADAM10 transcription through the RXR/PPAR$\alpha$ signaling pathway. Sulfuretin enhances the formation of immature APP while decreasing the amount of beta-site APP-cleaving enzyme-1 (BACE1), the enzyme responsible for cleaving APP to create APPsb [72].

## 5.2 Drug Repurposing for Parkinson's Disease

Worldwide, PD impacts people and is the second most prevalent neurodegenerative disease after AD. Though other neurotransmitter systems also seem to be impacted, the motor symptoms of PD are most well-known and are believed to be predominantly caused by the death of dopaminergic neurons in the substantia nigra. As of today, it is understood that PD includes a number of non-motor symptoms, including cognitive decline, sleep issues, and depression. Even though there is no cure for PD, deep brain stimulation and dopamine replacement are effective management methods that keep the condition under control [3, 17, 18]. Amantadine, a medication used to treat PD, was also repurposed, much like AD. Amantadine was really first created to treat influenza, and only then was it focused on the clinical management of Parkinson's disease acting through glutamate receptors as a weak antagonist, boosting dopamine and its reuptake inhibition [5]. An agent approved for treatment of chronic myeloid leukemia is nilotinib. It is a tyrosine kinase A$\beta$1 inhibitor. Ab1 phosphorylation is caused by the $\alpha$-synuclein overexpression. Nilotinib enhances the breakdown of $\alpha$-synuclein because it prevents A$\beta$ 1 phosphorylation [73, 74]. Researchers have also looked at the anti-PD properties of the antibiotic doxycycline, which was previously covered as a possible anti-AD option. Doxycycline concentration variations really allow for the distinction between antibacterial and anti-inflammatory action. Studies have shown that smaller concentrations than those used to make antibiotics do not alter the susceptibility of bacteria, but instead demonstrate anti-inflammatory action, which is linked to its neuroprotective properties. Apart from the antioxidant property of doxycycline, it has the ability to convert the oligomeric species of synuclein into the soluble and non-toxic species promotes neuroprotection. The normal monomeric forms of synuclein are retained, but it was found that doxycycline specifically binds to the synuclein oligomers [75]. Zonisamide, an antiepileptic drug having sulfonamide as a basic nucleus, exhibits diverse pathological actions, rendering it suitable for administration in various pathological conditions. The mechanisms of action encompass the obstruction of sodium and calcium channels, regulation of the GABA-A receptor, suppression of carbonic anhydrase, and hindrance of glutamate release. Research conducted on rats has demonstrated a rise in dopamine levels within the striatum region upon administration of therapeutic dosages. Conversely, administration of elevated dosages resulted in a reduction of intracellular dopamine levels. Regarding PD, it exhibited favorable efficacy in addressing motor, non-motor symptoms; however, the precise pathological action remains uncertain [76]. Zonisamide exhibits monoamine oxidase-B inhibitory activity. The enzyme, primarily found in astrocytes, is accountable for the breakdown of hormone dopamine in both neural and glial cells, resulting in the production of free radicals that may significantly contribute to the development of PD. The stabilization of dopamine quantity in the synapses and the enhancement of dopamine's effect can be achieved through its inhibition, as reported in previous studies. Selegiline, an antiparkinsonian medication, is an instance of a monoamine oxidase-B inhibitor. Its administration has been linked to increased astrocyte activation following striatal injury, as per studies [77]. Methylphenidate, a stimulant of CNS, functions through obstructing presynaptic dopamine transporter, noradrenaline transporter. This results in the inhibition of the reuptake of dopamine and noradrenaline in the striatum and prefrontal cortex. This therapeutic intervention has been employed for the management of attention-deficit hyperactivity disorder. Several research investigations have demonstrated the efficacy of this medication in mitigating gait disturbances associated with PD, along with non-motor symptoms [78]. Glucagon-like peptide-1 analog, Exenatide, is utilized in treating type 2 diabetes, similar to liraglutide which was previously discussed. The efficacy of this intervention has been investigated in the context of PD, revealing its potential for conferring neuroprotection and promoting advantageous neuroplastic alterations that may impede or forestall the advancement of the condition. The substance has the ability to traverse the blood-brain barrier and elicits its neuroprotective effects by stimulating GLP-1 receptors. Exenatide has been reported to have positive effects in the treatment of AD. The drug Liraglutide, which has been previously discussed, is presently in phase II clinical trial stage and is anticipated to yield results in 2024 [79, 80]. The anti-Parkinson's disease potential of $\beta_2$-adrenoreceptor agonists, which are a class of antiasthma drugs, has been subject to investigation. Recent research has established a correlation between the $\beta_2$-adrenoreceptor and the regulation of the $\alpha$-synuclein gene SNCA. It has been demonstrated that the activation of $\beta_2$-adrenoreceptors exhibits neuroprotective effects. Among the tested pharmaceutical agents, three anti-asthmatic drugs exhibited the most favorable outcomes. Notably, salbutamol demonstrated the ability to traverse the blood-brain barrier and is presently sanctioned for therapeutic use. The conducted study demonstrated that the three drugs exhibited response in terms of lowering the SNCA-mRNA and $\alpha$-synuclein [81]. The clinical utilization of Ambroxol hydrochloride dates back to more than four decades ago, primarily for the treatment of respiratory ailments. Presently, it continues to be extensively employed for this intended purpose. The substance in question was recognized as a viable chaperone for enhancing the performance of glucosylceramidase. This, in turn, has the potential to bolster lysosomal activity by stimulating the CLEAR network, ultimately leading to the elimination of $\alpha$-synuclein. The Phase I clinical trial evaluating the efficacy of ambroxol in treating PD associated with glucocerebrosidase (GBA) has concluded. The results indicate that ambroxol is capable of traversing the blood-brain barrier and augmenting GBA activity. A clinical trial in Phase II is currently in progress [82]. The natural alkaloid, Lycorine, has been authorized for anti-amoeba therapy and has been found to activate cAMP-dependent protein kinase (PKA) potentially enhancing the degradation of $\alpha$-synuclein, thereby improving the proteasome's function [83]. Squalamine, an organic compound with anticancer and antiviral properties, was initially identified in the Squalus acanthias species of dogfish shark in 1993. Squalamine possesses a significant number of positive charges, which enables it to selectively bind with lipid membranes. Oligomeric aggregates interact due to this binding and ultimately synuclein aggregates are formed [84].

## 5.3 Drug Repurposing for Huntington's Disease (HD)

HD is a genetic disorder inherited in an autosomal dominant pattern, and is recognized as the prevailing monogenic neurological disease in developed nations. The condition is distinguished by spontaneous choreatic motions, behavioral and psychological dysfunctions, and cognitive decline. The manifestation of this phenomenon is attributed to a genetic mutation that gives rise to an altered variant of the versatile protein huntingtin. This variant is responsible for inducing toxicity, which in turn results in impaired neuronal function and eventual demise. HD typically presents in adulthood and exhibits progressive symptomatology that ultimately culminates in mortality within a few years. The absence of a known cure for this ailment necessitates the adoption of a symptom management approach as the sole recourse [20, 21]. The compound Tetrabenazine was initially created with the objective of designing uncomplicated substances that exhibit antipsychotic activity similar to reserpine. It functions as a potent, monoamine reuptake inhibitor in presynaptic neurons and also acts as a mild D2 receptor antagonist in postsynaptic neurons. The efficacy of this compound in antipsychotic studies yielded inconclusive results, prompting its redirection toward the treatment of conditions characterized by anomalous, involuntary hyperkinetic movements, such as HD. In addition, it has been observed that tetrabenazine is a safer alternative for treating HD compared to dopamine receptor blockers, as there is no documented evidence of it causing dyskinetic symptoms [85]. In light of this, alternative pharmaceuticals possessing dopamine antagonistic properties have undergone examination for their potential therapeutic application in the management of HD. The subject under consideration pertains to tiapride, which functions as an antipsychotic by acting as dopamine D2 receptor blocker. Huntington's chorea was treated with selegiline commonly in Europe [86]. Clozapine is a pharmacological agent belonging to the class of neuroleptic drugs, which is commonly employed for the management of schizophrenia. The compound exhibits a notable inclination toward binding with D1 and D4 receptors, while demonstrating minimal blocking effects on the D2 receptors. The drug was proposed to treat the symptoms of chorea; it showed less extrapyramidal adverse effects. However, the outcomes of clinical trials were inconsistent, as reported in reference [87]. Olanzapine, a commonly prescribed antipsychotic medication, is utilized for the management of both motor and behavioral manifestations of HD. The pharmaceutical compound in question exhibits a notable inclination toward serotoninergic receptors, while concurrently acting as an antagonist toward dopamine D2 receptors. The aforementioned treatment is deemed to be secure and exhibits a high level of tolerance. It is advisable to prescribe this treatment in cases where symptoms such as irritability, sleep disturbances, weight loss, and chorea are present [88, 89]. Risperidone is an antipsychotic medication that is commonly prescribed for individuals with schizophrenia and bipolar disorder. Its mechanism of action involves acting as an antagonist to D2 receptors and an agonist to serotonin. As a result, it has been found to be effective in treating HD chorea. According to a study [90], there were positive outcomes observed in terms of mitigating motor decline and psychiatric symptoms. An atypical antipsychotic drug, quetiapine, exhibits notable affinity toward both dopamine and serotonin receptors. While there is limited documentation of the administration of quetiapine for addressing HD symptoms, the available reports underscore its efficacy in managing chorea, particularly when concomitant psychiatric symptoms are present [91]. The pharmacological agent known as Memantine, which is derived from adamantane, is utilized in the therapeutic management of AD. This compound functions as a non-competitive N-methyl-D-aspartate (NMDA) inhibitor. Overstimulation of the NMDA receptor results in a significant influx of calcium ions into the cell, ultimately culminating in cellular death. Hence, it has been observed that memantine has the potential to inhibit the influx of calcium in neuronal cells, thereby mitigating the risk of cerebral cell death. The efficacy of Memantine in treating HD was investigated, and it revealed that the drug has the ability to reduce the excitotoxicity of glutamate on neurons [92].

## 5.4 Drug Repurposing for Multiple Sclerosis

MS is a chronic autoimmune disorder that affects the central nervous system. This is a persistent and inflammatory ailment characterized by the destruction of myelin and axons to varying degrees. The trajectory of the condition is uncertain and exhibits an initial presentation of neurological deficits that are reversible, but eventually transition to a progressive state [26, 29]. While a cure for MS has yet to be discovered, there exist approved therapies aimed at mitigating the symptoms and advancement of the disease. A diverse range of anti-neoplastic medications have been repositioned for treating multiple sclerosis and its associated symptoms. The aforementioned scenario pertains to the synthetic compounds, specifically mitoxantrone, which is classified as an anthracenedione. This particular compound has been recognized as a broad-spectrum antitumor agent that is utilized in the treatment of various types of cancer such as breast, prostate cancer, acute leukemia, and lymphoma. Mitoxantrone was approved to treat multiple sclerosis; it has immunosuppressive characteristics. This is linked to the unpredictable reactions of the central nervous system T and B cells to antigens, myelin damage caused by macrophages, and axonal injuries. The compound Mitoxantrone has the ability to impede the stimulation of T cells, thereby halting the replication of both T and B cells. Additionally, it can reduce the production of antibodies and deactivate macrophages. The compound Mitoxantrone exhibited a notable level of tolerability, as reported [93]. Cyclophosphamide, an alkylating agent, is utilized in the management of diverse types of solid neoplasms. It has been sanctioned for the therapy of leukemia, lymphoma, breast carcinoma, and other malignancies. This pertains to nitrogen mustards and their affinity for DNA binding, thereby impeding mitosis and cellular replication, with a preference for cells that divide at a rapid rate. The utilization of cyclophosphamide in the context of multiple sclerosis is attributed to its capacity to exert both immunosuppressive and immunomodulatory effects. The substance in question exerts its effects on T and B cells, resulting in the inhibition of both cell-mediated and humoral immunity. Studies have demonstrated that the administration of cyclophosphamide can lead to a reduction in the production of the pro-inflammatory T helper 1 cytokine interferon-$\gamma$ and interleukin-12, while simultaneously promoting the secretion of anti-inflammatory cytokines in both the brain and bloodstream. Additionally, it modifies T lymphocytes to exhibit a phenotype that is less inflammatory. Cyclophosphamide has the ability to penetrate the blood-brain barrier and exhibit favorable bioavailability in the central nervous system. It can effectively exert its immunomodulatory and immunosuppressive effects, thereby stabilizing the condition and preventing its advancement [94]. The antimetabolite Cladribine is utilized in the treatment of hairy cell leukemia and other hematologic malignancies. The aforementioned substance is an analog of deoxyadenosine that necessitates intracellular phosphorylation into a triphosphate for its activation, ultimately resulting in cellular death. The drug in question underwent repurposing for the purpose of treating MS; however, it was ultimately rejected in the year 2013. In 2017, the European Medicines Agency granted authorization for the marketing of cladribine as a therapeutic intervention for the aforementioned illness. The mechanism of action is associated with the reduction in the levels of circulating B and T lymphocytes. Other proposed mechanisms include the activation of interferon-$\alpha$-producing myeloid dendritic cells and the disruption of the synaptic effects of interleukin-1$\beta$. These findings suggest that cladribine may possess neuroprotective qualities [95, 96]. Amiloride, a diuretic medication, is utilized in the management of hypertension and edema resulting from liver diseases or heart failure. The neuroprotective properties of the substance in question have been investigated in the context of MS. The compound known as Amiloride has the ability to inhibit the activity of the acid-sensing ion channel 1 (ASIC1) found in neurons. This particular ion channel is known to be present in high levels in both axons and oligodendrocytes located in lesions associated with MS. As a result, Amiloride is believed to have a beneficial impact on the protection of both neurons and myelin. In addition, the advantageous aspect of amiloride's protective effect lies in its downstream impact on inflammation, rendering it effective even during the initial stages of inflammation [97]. Ibudilast has been authorized for use in certain nations as a therapeutic intervention for bronchial asthma and cerebrovascular ailments. The mechanism of action involves the inhibition of phosphodiesterases, which are recognized for their anti-inflammatory properties. Additionally, this inhibition can impede the synthesis of leukotriene and nitric oxide mechanisms that are associated with MS. Ibudilast has been observed to exhibit inhibitory effects on the release of the tumor necrosis factor from both microglia and astrocytes in the brain, thereby reducing neuronal degeneration. Moreover, it has the potential to safeguard astrocytes against apoptosis and impede oligodendrocyte apoptosis and demyelination, thereby demonstrating its efficacy in the treatment of MS. Research has indicated that at a high dosage, it is both safe and well-tolerated, and has the potential to decrease the rate of brain atrophy [98].

## 5.5 Drug Repurposing for Amyotrophic Lateral Sclerosis (ALS)

It is a neurodegenerative disorder characterized by the abnormal degeneration of lower and upper motor neurons, which are responsible for regulating voluntary muscle movements. Muscle atrophy is a condition characterized by a gradual weakening and reduction in the size of muscles. Additional indications include muscular rigidity and spasms, as well as challenges with respiration, ingestion, and verbal communication. In the majority of cases the pathology of ALS is not known, with approximately 10% having a genetic basis [30]. While certain drugs are presently under investigation for their potential application in the management of ALS, only two drugs, namely riluzole and edaravone, are presently accessible for the purpose of impeding the advancement of the ailment. However, it is important to note that these drugs are incapable of reversing the symptoms once they have become apparent [99, 100]. Masitinib is depicted as a tyrosine kinase inhibitor that is utilized in veterinary medicine for the treatment of canine cancer. The application of tyrosine kinase inhibitors in the context of ALS is based on the hypothesis that the anomalous glial cells that undergo proliferation in this disease could exhibit sensitivity to such inhibitors. The efficacy of mastinib in inhibiting glial cell activation and improving survival has been demonstrated in a relevant rat model [101]. The neuroprotective properties of Ibudilast is under clinical trial to investigate as a potential therapeutic agent for ALS. The antiretroviral medication, Triumeq, which is commonly utilized as an anti-HIV therapy, was examined as a potential treatment for ALS. This was based on the observation that ALS patients exhibit serum concentrations of reverse transcriptase that are similar to those of individuals infected with HIV, and that the expression of a human endogenous retrovirus was detected in the brains of individuals affected by ALS. Considering this, the utilization of anti-HIV medications may prove to be beneficial in the treatment of ALS. Triumeq is a medication that comprises dolutegravir, an integrase inhibitor, abacavir, and lamivudine, both of which are antiretroviral agents. According to a study, Triumeq has demonstrated safety and tolerability in patients with ALS. The drug Retigabine has been granted approval for the treatment of epilepsy. Its mechanism of action involves binding to the voltage-gated potassium channels and subsequently augmenting the M-current, resulting in membrane hyperpolarization. Retigabine exhibits the ability to extend the survival of motor neurons and reduce their excitability. This characteristic is deemed beneficial in the management of ALS, as it is postulated that hyper-excitability of neurons is a hallmark feature of this condition, culminating in their demise. The medication in question is presently undergoing clinical trials for its potential use in addressing ALS [102]. The drugs presented in this chapter are summarized in Table 1.

**Table 1** Summary of drugs studied for repurposing and repositioning

| Sl. no. | Drug name | Drug structure | Mechanism | Clinically approved for | Repurposedfor |
|---|---|---|---|---|---|
| 1. | Captopril | 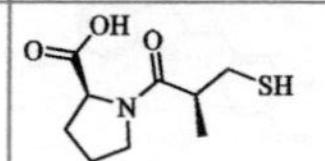 | Captopril reduces hippocampus release and inhibits ACE activity<br>Angiotensin II increases $\beta$ and $\gamma$ secretase activity, which may accelerate the accumulation of A$\beta$ by decreasing the formation of reactive oxygen species (ROS) | Antihypertensive | AD |
| 2. | Valproic acid | 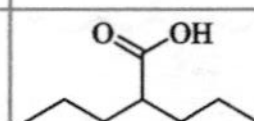 | S9 site of GSK-3$\beta$ is phosphorylated by VPA, and then inhibits the activity of GSK-3$\beta$ | Epilepsy and bipolar disorder | AD |
| 3. | Melatonin | 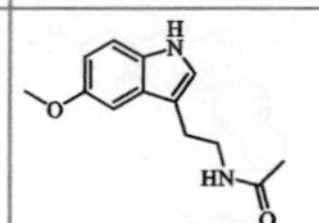 | Melatonin reduces the formation of A$\beta$ aggregates through receptor-dependent and independent pathway | Improve sleep and reduce free radical | AD |
| 4. | Imatinib | 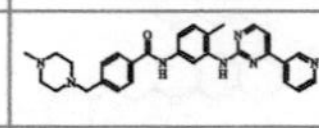 | By blocking the interaction between GSAP and $\gamma$-secretase substrate, reduces the level of A$\beta$ | Cancer | AD |
| 5. | Plumbagin | 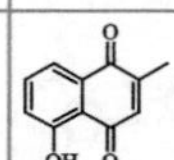 | The temperature-induced amyloid aggregation is reduces | Multiple health benefits | AD |
| 6. | Zileuton | 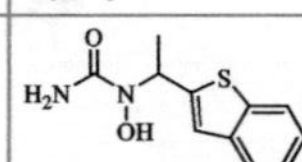 | Reduced the A $\beta$ deposition and decreased the tau phosphorylation | Asthma | AD |
| 7. | Doxycycline | 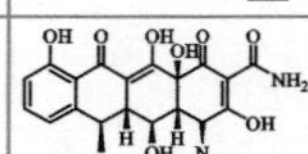 | Interacts with the tau's microtubule-binding domain and prevents tau aggregation | Antibiotic | AD |
| 8. | Clioquinol | 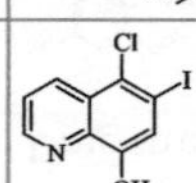 | Prevent Cu and Zn-mediated A$\beta$ deposition | Antibiotic | AD |

(continued)

**Table 1** (continued)

| Sl. no. | Drug name | Drug structure | Mechanism | Clinically approved for | Repurposedfor |
|---|---|---|---|---|---|
| 9. | Resveratrol | 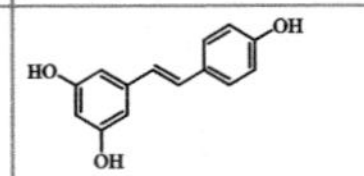 | Resveratrol increases the level of intracellular calcium which activates calcium/calmodulin $\beta$-dependent protein kinase kinase-$\beta$ (CaMKK $\beta$) and it is promoting the initiation of A$\beta$ clearance by phosphorylating AMPK | Cancer | AD |
| 10. | Nilotinib | 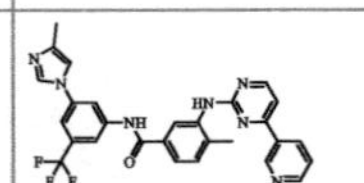 | Nilotinib activates parkin which is responsible for autophagic clearance of A$\beta$ | Adult chronic myelogenous leukemia | AD |
| 11. | Bosutinib | 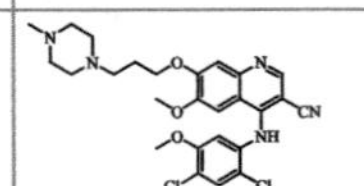 | Bosutinib activates parkin which is responsible for autophagic clearance of A$\beta$ | Adult chronic myelogenous leukemia | AD |
| 12. | Berberine | 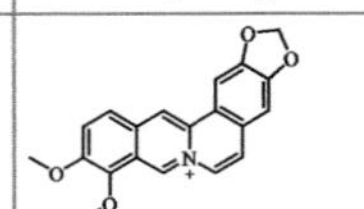 | Increases the activity of autophagy through the class III PI3K/beclin-1 pathway, it promotes the autophagic clearance of A$\beta$ | Dementia | AD |
| 13. | 1,3 bis(2-chloroethyl)-1-nitrosourea (BCNU) | 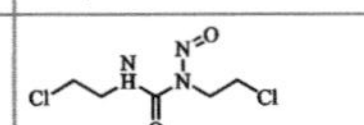 | BCNU improves the secretion of APP$\alpha$ and prevents the aggregation of A$\beta$ | Cancer | AD |
| 14. | Retinoic acid | 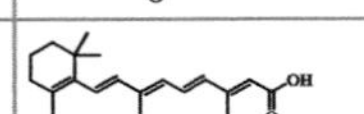 | Binds on ADAM10 to activate the non-amyloidogenic pathway | Psoriasis | AD |
| 15. | Sulfuretin | 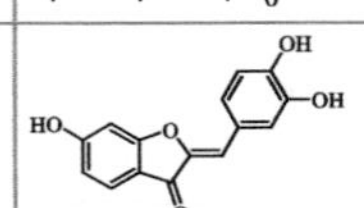 | Activates PPAR-$\alpha$ signaling pathway and increases the transcription of ADAM10 | Inflammation | AD |
| 16. | Disulfiram | 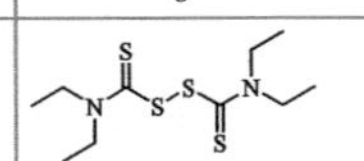 | Disulfiram prevents A$\beta$ aggregation and enhances the production of ADAM10 and APPs$\alpha$ | Alcohol deterrent | AD |

| | | | | | |
|---|---|---|---|---|---|
| 17. | Valsartan | 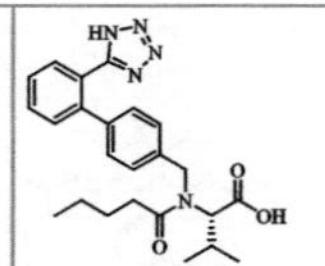 | Shows *in vitro* A$\beta$-oligomerization inhibiting activity | Antihypertensive | AD |
| 18. | Losartan | 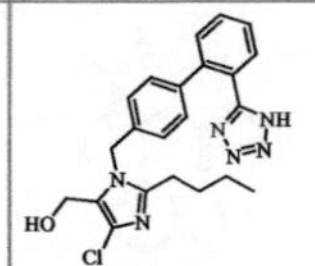 | Shows *in vitro* A$\beta$-oligomerization inhibiting activity and also has amyloidogenic mechanism | Antihypertensive | AD |
| 19. | Telmisartan | 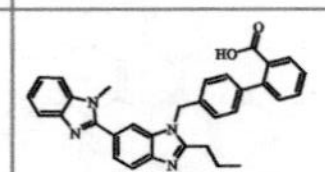 | Shows *in vitro* A$\beta$-oligomerization inhibiting activity | Antihypertensive | AD |
| 20. | Fluoxetine | 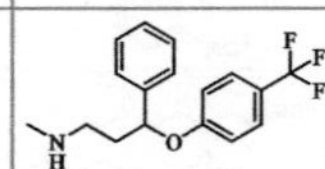 | Fluoxetine increases the level of $\alpha$-secretase and reduces A$\beta$ | Depression | AD |
| 21. | Bexarotene | 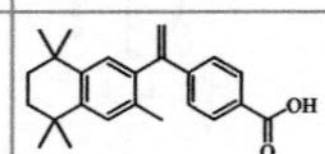 | Selectively suppresses the primary nucleation through this to prevent A$\beta$ deposition | Cancer | AD |
| 22. | Cilostazol | 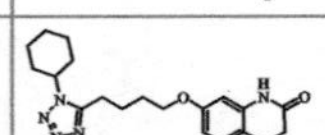 | Cilostazol is a selective inhibitor of phosphodiesterase type 3 | Selective phosphodiesterase 3 inhibitor | AD |
| 23. | Donepezil | 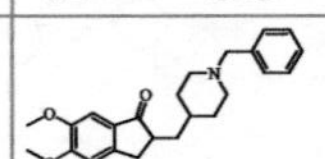 | Inhibits the acetylcholinesterase enzyme | Acetylcholinesterase inhibitor | AD |
| 24. | Rivastigmine | 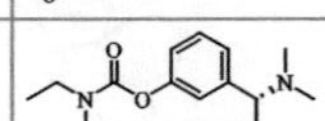 | Inhibits the acetylcholinesterase enzyme | Acetylcholinesterase inhibitor | AD and PD |

(continued)

**Table 1** (continued)

| Sl. no. | Drug name | Drug structure | Mechanism | Clinically approved for | Repurposedfor |
|---|---|---|---|---|---|
| 25. | Galantamine | 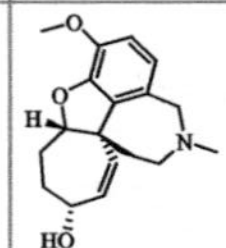 | Inhibits the acetylcholinesterase enzyme | Acetylcholinesterase inhibitor | AD |
| 26. | Dimebon | 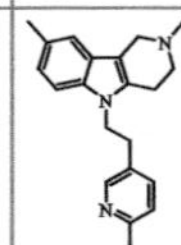 | Inhibition of acetylcholinesterase and butyrylcholines eras | Antihistamine drug | AD |
| 27. | Amlodipine | 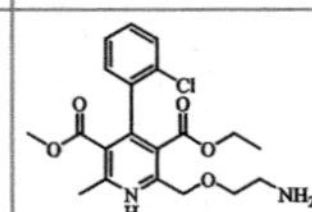 | Calcium channel blocking activity | Antihypertensive agent | AD |
| 28. | Nilvadipine | 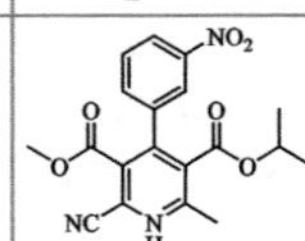 | Calcium channel blocker | Antihypertensive agent | AD |
| 29. | Isradipine | 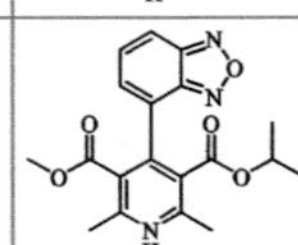 | Calcium channel blocker | Antihypertensive agent, heart attack, and stroke | AD |
| 30. | Nimodipine | 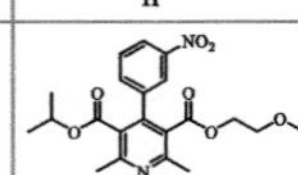 | Calcium channel blocker | Antihypertensive agent, heart attack, and stroke | AD |

| | | | | | |
|---|---|---|---|---|---|
| 31. | Amantidine | 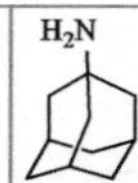 | Amantadine disrupts the transmembrane domain of the viral M2 protein | Influenza | PD |
| 32. | Zonisamide | 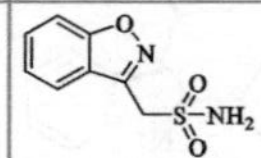 | Obstruction of sodium and calcium channels, regulation of the GABA-A receptor, suppression of carbonic anhydrase, and hindrance of glutamate release | Antiepileptic | PD |
| 33. | Selegiline | 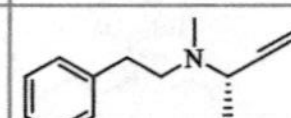 | Monoamine oxidase inhibitor and monoamine oxidase-B inhibitor | Depressive disorder | PD |
| 34. | Methylphe-nidate | 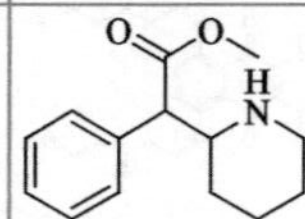 | Stimulant of the central nervous system | Attention-deficit hyperactivity disorder. | PD |
| 35. | Lycorine | 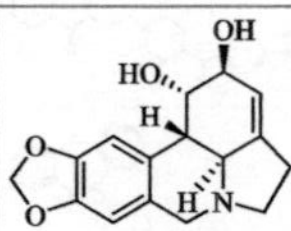 | Activate cAMP-dependent protein kinase (PKA) and enhance the degradation of $\alpha$-synuclein | Anti-amoeba therapy | PD |
| 36. | Ambroxol | 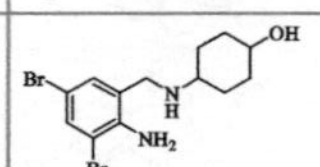 | The potential to bolster lysosomal activity by stimulating the CLEAR network, ultimately leading to the elimination of $\alpha$-synuclein | Asthma | PD |
| 37. | Tetrabenaz-ine | 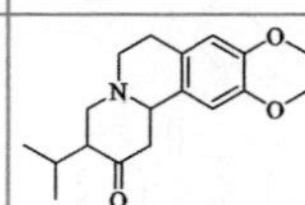 | Acts as mild D2 receptor blocker in postsynaptic neurons and also a reversible inhibitor of monoamine uptake in presynaptic neurons | Antipsychotic | HD |
| 38. | Triapride | 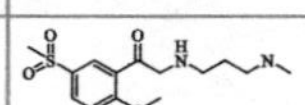 | D2 receptor antagonist | Antipsychotic | HD |

(continued)

**Table 1** (continued)

| Sl. no. | Drug name | Drug structure | Mechanism | Clinically approved for | Repurposedfor |
|---|---|---|---|---|---|
| 39. | Clozapine | 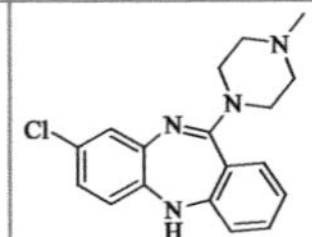 | D1 and D4 receptor blocker, while demonstrating minimal antagonistic effects on the D2 dopaminergic receptors | Neuroleptic, schizophrenia | HD |
| 40. | Olanzapine | 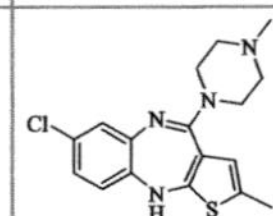 | Exhibits a notable inclination toward serotonergic receptors | Antipsychotic | HD |
| 41. | Quetiapine | 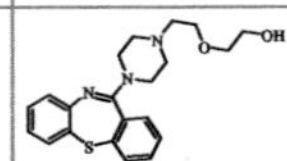 | Exhibits a notable affinity toward both serotonin and dopamine receptors | Antipsychotic | HD |
| 42. | Risperidone | 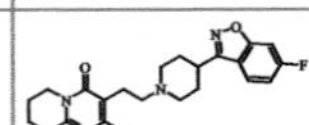 | Antagonist to D2 receptors and an agonist to serotonin. | Antipsychotic, schizophrenia, and bipolar disorder | HD |
| 43. | Mitoxantrone | 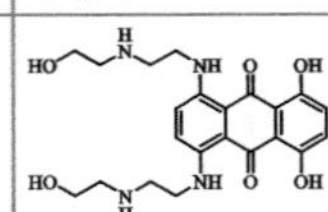 | Activation of T cells, thereby halting the proliferation of both B and T cells. Additionally, it can reduce the production of antibodies and deactivate macrophages | Antitumor, prostate cancer, acute leukemia, lymphoma, breast cancer | MS |
| 44. | Cyclophosphamide | 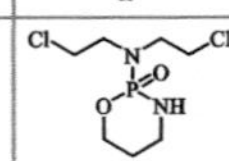 | Impeding mitosis and cellular replication | Alkylating agent, neoplasms, leukemia, lymphomas, breast carcinoma, and other malignancies. | MS |
| 45. | Ibudilast | 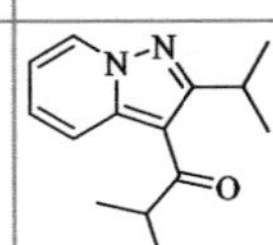 | Inhibition of phosphodiesterases, which are recognized for their anti-inflammatory properties | Bronchial asthma and cerebrovascular ailments | MS |

| | | | | | |
|---|---|---|---|---|---|
| 46. | Memantine | 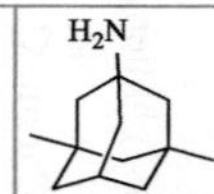 | Glutamate-mediated excitotoxicity is reduced by inhibition of N-methyl-D-aspartate (NMDA) receptors | Alzheimer's disease | HD |
| 47. | Amiloride | 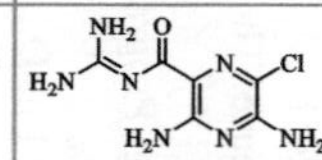 | Potassium channel agonist | Diuretic, hypertension and edema | MS |
| 48. | Mastinib | 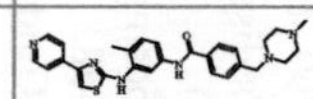 | Tyrosine kinase inhibitor | Canine cancer | ALS |
| 49. | Abacavir | 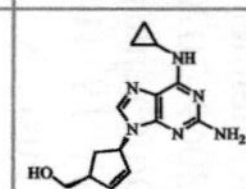 | Reverse transcriptase inhibitor | Antiretroviral | ALS |
| 50. | Dolutegravir | 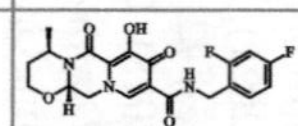 | Integrase inhibitor | Anti-HIV | ALS |
| 51. | Retigabine | 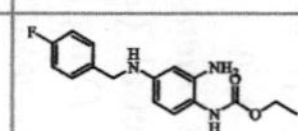 | Binding to the voltage-gated potassium channels and subsequently augmenting the M-current, resulting in membrane hyperpolarization. | Epilepsy | ALS |
| 52. | Lamivudine | 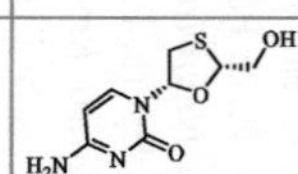 | Reverse transcriptase inhibitor | Antiretroviral, anti-HIV | ALS |

## 6 Conclusion

Neurodegenerative diseases have very complex pathophysiology and the incident rate is increasing day by day and now it is leading cause of death; it was so complex to develop a new drug to treat neurodegenerative diseases, it was a time consuming and also very expensive process, the drug repurposing and repositioning approach is simple, cost-effective, and potential way to find out treatment for various diseases. The main advantage of drug repurposing and repurposing is known about pharmacokinetic and pharmacodynamic profile of drug; it is helpful for development of drug for the treatment of various diseases. Drug repurposing and repositioning is also a way to develop multi-target drugs that can interact with various targets at a same time and are helpful to treat complex diseases such as neurodegenerative diseases. *In vitro, in vivo* screening, chemical optimization, toxicity studies, formulation development, and bulk manufacturing process are bypassed by the drug repurposing and repositioning approach. The repurposing of drugs on neurodegenerative diseases is a rewarding job for the global healthcare system and a step forward for people to get safer and cheaper medicines.

## References


1. Fu, H., Hardy, J., & Duff, K. E. (2018). Selective vulnerability in neurodegenerative diseases. *Nature Neuroscience, 21*, 1350–1358.
2. Scheltens, P., De Strooper, B., Kivipelto, M., Holstege, H., Chételat, G., Teunissen, C. E., Cummings, J., & van der Flier, W. M. (2021). Alzheimer's disease. *The Lancet, 397*, 1577–1590.
3. Kalia, L. V., & Lang, A. E. (2015). Parkinson's disease. *The Lancet, 386*, 896–912.
4. Lin, M. T., & Beal, M. F. (2006). Mitochondrial dysfunction and oxidative stress in neurodegenerative diseases. *Nature, 443*, 787–795.
5. Estes, M. L., & McAllister, A. K. (2016). Maternal immune activation: Implications for neuropsychiatric disorders. *Science (1979), 353*, 772–777.
6. Jamebozorgi, K., Taghizadeh, E., Rostami, D., Pormasoumi, H., Barreto, G. E., Hayat, S. M. G., & Sahebkar, A. (2019). Cellular and molecular aspects of Parkinson treatment: Future therapeutic perspectives. *Molecular Neurobiology, 56*, 4799–4811.
7. Tagliavini, F., Tiraboschi, P., & Federico, A. (2021). Alzheimer's disease: The controversial approval of Aducanumab. *Neurological Sciences, 42*, 3069–3070.
8. Pushpakom, S., Iorio, F., Eyers, P. A., et al. (2018). Drug repurposing: Progress, challenges and recommendations. *Nature Reviews. Drug Discovery, 18*, 41–58.
9. Agnati, L. F., Benfenati, F., Solfrini, V., Biagini, G., Fuxe, K., Guidolin, D., Carani, C., & Zini, I. (1992). Brain aging and neuronal plasticity. *Annals of the New York Academy of Sciences*, 180–186.
10. Gitler, A. D., Dhillon, P., & Shorter, J. (2017). neurodegenerative disease: Models, mechanisms, and a new hope. *DMM Disease Models and Mechanisms, 10*, 499–502.
11. Wang, J., Gu, B. J., Masters, C. L., & Wang, Y. J. (2017). A systemic view of Alzheimer disease – Insights from amyloid-β metabolism beyond the brain. *Nature Reviews Neurology, 13*, 612–623.
12. Agnati, J. F., & Zoli, M. (1992). Neuronal plasticity and ageing processes in the frame of the 'Red Queen Theory'. *Acta Physiologica Scandinavica, 145*, 301–309.

13. Breijyeh, Z., & Karaman, R. (2020). Comprehensive review on Alzheimer's disease: causes and treatment. *Molecules, 25*(24), 5789.
14. Kumar, A., Singh, A., & Ekavali. (2015). A review on Alzheimer's disease pathophysiology and its management: An update. *Pharmacological Reports, 67*, 195–203.
15. Chui, D.-H., Tabira, T., Izumi, S., Koya, G., & Ogata, J. (1994). Short communication decreased f3-amyloid and increased abnormal tau deposition in the brain of aged patients with leprosy, *1*, 45.
16. Scheltens, P., Blennow, K., Breteler, M. M. B., de Strooper, B., Frisoni, G. B., Salloway, S., & Van der Flier, W. M. (2016). Alzheimer's disease. *The Lancet, 388*, 505–517.
17. Antony, P. M. A., Diederich, N. J., Krüger, R., & Balling, R. (2013). The hallmarks of Parkinson's disease. *FEBS Journal, 280*, 5981–5993.
18. Poewe, W., Seppi, K., Tanner, C. M., Halliday, G. M., Brundin, P., Volkmann, J., Schrag, A. E., & Lang, A. E. (2017). Parkinson disease. *Nature Reviews Disease Primers, 3*, 1–21.
19. Simon, D. K., Tanner, C. M., & Brundin, P. (2020). Parkinson disease epidemiology, pathology, genetics, and pathophysiology. *Clinics in Geriatric Medicine, 36*, 1–12.
20. Bates, G. P., Dorsey, R., Gusella, J. F., et al. (2015). Huntington disease. *Nature Reviews. Disease Primers, 1*, 15005.
21. Roos, R. A. C. (2010). Huntington's disease: A clinical review. *Orphanet Journal of Rare Diseases, 5*, 40.
22. Caron, N. S., Dorsey, E. R., & Hayden, M. R. (2018). Therapeutic approaches to huntington disease: From the bench to the clinic. *Nature Reviews. Drug Discovery, 17*, 729–750.
23. DiFiglia, M., Sapp, E., Chase, K. O., Davies, S. W., Bates, G. P., Vonsattel, J. P., & Aronin, N. (1997). Aggregation of huntingtin in neuronal intranuclear inclusions and dystrophic neurites in brain. *Science, 277*, 1990–1993.
24. Gárdián, G., Vécsei, L., et al. (2004). Huntington's disease: Pathomechanism and therapeutic perspectives. *Journal of Neural Transmission (Vienna)*, 1485–1494.
25. Kumar, A., Kumar, V., Singh, K., Kumar, S., Kim, Y. S., Lee, Y. M., & Kim, J. J. (2020). Therapeutic advances for huntington's disease. *Brain Sciences, 10*(1), 43.
26. Goldenberg, M. M. (2012). Disease overview. *Multiple Sclerosis Review, 37*(3), 175–184.
27. Dawood Tafti, A., Ehsan, M., & Xixis Affiliations KL. (2022). *Multiple sclerosis continuing education activity*: 1–17.
28. Huang, W. J., Chen, W. W., & Zhang, X. (2017). Multiple sclerosis: Pathology, diagnosis and treatments (review). *Experimental and Therapeutic Medicine, 13*, 3163–3166.
29. Trapp, B. D., & Nave, K. A. (2008). Multiple sclerosis: An immune or neurodegenerative disorder? *Annual Review of Neuroscience, 31*, 247–269.
30. Ewis, L., Owland, P. R., & Hneider, E. A. S. (2001). *Medical Progress, 344*, 1688–1700.
31. Longinetti, E., & Fang, F. (2019). Epidemiology of amyotrophic lateral sclerosis: An update of recent literature. *Current Opinion in Neurology, 32*, 771–776.
32. Ashburn, T. T., & Thor, K. B. (2004). Drug repositioning: Identifying and developing new uses for existing drugs. *Nature Reviews. Drug Discovery, 3*, 673–683.
33. Doan, T. L., Pollastri, M., Walters, M. A., & Georg, G. I. (2011). The future of drug repositioning. Old drugs, new opportunities. In *Annual reports in medicinal chemistry* (pp. 385–401). Academic.
34. Middeldorp, C. M., & Wray, N. R. (2018). The value of polygenic analyses in psychiatry. *World Psychiatry, 17*, 26–28.
35. Richardson, P., Hideshima, T., & Anderson, K. (2001). THALIDOMIDE: Emerging role in cancer medicine, *53*, 629–657.
36. Baker, N. C., Ekins, S., Williams, A. J., & Tropsha, A. (2018). A bibliometric review of drug repurposing. *Drug Discovery Today, 23*, 661–672.
37. Sleire, L., Førde-Tislevoll, H. E., Netland, I. A., Leiss, L., Skeie, B. S., & Enger, P. Ø. (2017). Drug repurposing in cancer. *Pharmacological Research, 124*, 74–91.
38. Gammon, K. (2014). Brain windfall. *Nature, 515*, 299.
39. Heinrich, M. (2010). Galanthamine from Galanthus and other Amaryllidaceae – Chemistry and biology based on traditional use. *The Alkaloids. Chemistry and Biology, 68*, 157–165.

40. Mucke, H. A. (2015). The case of galantamine: Repurposing and late blooming of a cholinergic drug. *Future Science OA, 1*(41), 73.
41. Monacelli, F., Cea, M., Borghi, R., Odetti, P., & Nencioni, A. (2017). Do cancer drugs counteract neurodegeneration? Repurposing for Alzheimer's disease. *Journal of Alzheimer's Disease, 55*, 1295–1306.
42. Blakeley, J., & Grossman, S. A. (2012). Chemotherapy with cytotoxic and cytostatic agents in brain cancer. In *Handbook of clinical neurology* (pp. 229–254). Elsevier B.V.
43. Hayes, C. D., Dey, D., Pablo Palavicini, J., Wang, H., Patkar, K. A., Minond, D., Nefzi, A., & Lakshmana, M. K. (2013). Striking reduction of amyloid plaque burden in an Alzheimer's mouse model after chronic administration of carmustine, *11*, 81.
44. Tousi, B. (2015). The emerging role of bexarotene in the treatment of Alzheimer's disease: Current evidence. *Neuropsychiatric Disease and Treatment, 11*, 311–315.
45. Fukasawa, H., Nakagomi, M., Yamagata, N., Katsuki, H., Kawahara, K., Kitaoka, K., Miki, T., & Shudo, K. (2012). Tamibarotene: A candidate retinoid drug for Alzheimer's disease, *35*(8), 1206–1212.
46. Netzer, W. J., Dou, F., Cai, D., Veach, D., Jean, S., Li, Y., Bornmann, W. G., Clarkson, B., Xu, H., & Greengard, P. (2003). Gleevec inhibits-amyloid production but not Notch cleavage, *100*(21), 12444–12449.
47. Brunden, K. R., Yao, Y., Potuzak, J. S., Ferrer, N. I., Ballatore, C., James, M. J., Hogan, A. M. L., Trojanowski, J. Q., Smith, A. B., & Lee, V. M. Y. (2011). The characterization of microtubule-stabilizing drugs as possible therapeutic agents for Alzheimer's disease and related tauopathies. *Pharmacological Research, 63*, 341–351.
48. Zhang, B., Maiti, A., Shively, S., et al. (2004). Microtubule-binding drugs offset tau sequestration by stabilizing microtubules and reversing fast axonal transport deficits in a tauopathy model, *102*(1), 227–231.
49. Appleby, B. S., Nacopoulos, D., Milano, N., Zhong, K., & Cummings, J. L. (2013). A review: Treatment of Alzheimer's disease discovered in repurposed agents. *Dementia and Geriatric Cognitive Disorders, 35*, 1–22.
50. Ryu, J. K., & McLarnon, J. G. (2008). Thalidomide inhibition of perturbed vasculature and glial-derived tumor necrosis factor-α in an animal model of inflamed Alzheimer's disease brain. *Neurobiology of Disease, 29*, 254–266.
51. Diomede, L., Cassata, G., Fiordaliso, F., Salio, M., Ami, D., Natalello, A., Doglia, S. M., De Luigi, A., & Salmona, M. (2010). Tetracycline and its analogues protect Caenorhabditis elegans from β amyloid-induced toxicity by targeting oligomers. *Neurobiology of Disease, 40*, 424–431.
52. Costa, R., Speretta, E., Crowther, D. C., & Cardoso, I. (2011). Testing the therapeutic potential of doxycycline in a Drosophila melanogaster model of Alzheimer disease. *Journal of Biological Chemistry, 286*, 41647–41655.
53. Loeb, M. B., Molloy, Ã. D. W., Smieja, M., et al. (2013). A randomized, controlled trial of doxycycline and rifampin for patients with Alzheimer's disease, *28*(5), 463–470.
54. Tomiyama, T., Shoji, A., Kataoka, K.-I., Suwa, Y., Asano, S., Kaneko, H., & Endo, N. (1996). *Journal of Biological Chemistry, 271*(12), 6839–6844.
55. Endoh, M., Kunishita, T., Tabira, T., et al. (1999). *Journal of Neurological Sciences, 165*, 28–30.
56. Wozniak, M. A., & Itzhaki, R. F. (2010). *Therapeutic Advances in Neurological Disorders, 3*, 141–152.
57. Smith, N. W., Annunziata, O., & Dzyuba, S. V. (2009). Amphotericin B interactions with soluble oligomers of amyloid Aβ1-42 peptide. *Bioorganic & Medicinal Chemistry, 17*, 2366–2370.
58. Grossi, C., Francese, S., Casini, A., Rosi, M. C., Luccarini, I., Fiorentini, A., Gabbiani, C., Messori, L., Moneti, G., & Casamenti, F. (2009). Clioquinol decreases amyloid-β burden and reduces working memory impairment in a transgenic mouse model of alzheimer's disease. *Journal of Alzheimer's Disease, 17*, 423–440.

59. Smith, A. M., Gibbons, H. M., & Dragunow, M. (2010). Valproic acid enhances microglial phagocytosis of amyloid-β1-42. *Neuroscience, 169*, 505–515.
60. Qing, H., He, G., Ly, P. T. T., et al. (2008). Valproic acid inhibits aβ production, neuritic plaque formation, and behavioral deficits in Alzheimer's disease mouse models. *Journal of Experimental Medicine, 205*, 2781–2789.
61. Danielyan, L., Klein, R., Hanson, L. R., Buadze, M., Schwab, M., Gleiter, C. H., & Frey Ii, W. H. (2014). Protective effects of intranasal losartan in the APP/PS1 transgenic mouse model of Alzheimer disease. *Cell Transplantation, 23*, 123–139.
62. Wang, J., Ho, L., Chen, L., et al. (2007). Valsartan lowers brain β-amyloid protein levels and improves spatial learning in a mouse model of Alzheimer disease. *Journal of Clinical Investigation, 117*, 3393–3402.
63. Bachmeier, C., Beaulieu-Abdelahad, D., Mullan, M., & Paris, D. (2011). Selective dihydropyiridine compounds facilitate the clearance of β-amyloid across the blood-brain barrier. *European Journal of Pharmacology, 659*, 124–129.
64. Hanyu, H., Hirao, K., Shimizu, S., Iwamoto, T., Koizumi, K., & Abe, K. (2007). Favourable effects of nilvadipine on cognitive function and regional cerebral blood flow on SPECT in hypertensive patients with mild cognitive impairment. Lippincott Williams & Wilkins. *Nuclear Medicine Communications, 28*, 281–287.
65. Carro, E., & Torres-Aleman, I. (2004). The role of insulin and insulin-like growth factor I in the molecular and cellular mechanisms underlying the pathology of Alzheimer's disease. *European Journal of Pharmacology, 490*, 127–133.
66. Perry, T., Lahiri, D. K., Sambamurti, K., Chen, D., Mattson, M. P., Egan, J. M., & Greig, N. H. (2003). Glucagon-Like Peptide-1 decreases endogenous amyloid-peptide (A) levels and protects hippocampal neurons from death induced by a and iron. *Journal of Neuroscience Research, 72*, 603–612.
67. Di Meco, A., Lauretti, E., Vagnozzi, A. N., & Praticò, D. (2014). Zileuton restores memory impairments and reverses amyloid and tau pathology in aged Alzheimer's disease mice. *Neurobiology of Aging, 35*, 2458–2464.
68. Halliday, M., Radford, H., Zents, K. A. M., et al. (2017). Drugs targeting eIF2a-P-mediated translational repression prevent neurodegeneration in mice. *Journal of Neurology, 140*(6), 1768–1783.
69. Zhang, J., Guo, J., Zhao, X., Chen, Z., Wang, G., Liu, A., Wang, Q., Zhou, W., Xu, Y., & Wang, C. (2013). Phosphodiesterase-5 inhibitor sildenafil prevents neuroinflammation, lowers beta-amyloid levels and improves cognitive performance in APP/PS1 transgenic mice. *Behavioural Brain Research, 250*, 230–237.
70. García-Barroso, C., Ricobaraza, A., Pascual-Lucas, M., et al. (2013). Tadalafil crosses the blood-brain barrier and reverses cognitive dysfunction in a mouse model of AD. *Neuropharmacology, 64*, 114–123.
71. Li, Y., Zhang, J., Wan, J., Liu, A., & Sun, J. (2020). Melatonin regulates Aβ production/clearance balance and Aβ neurotoxicity: A potential therapeutic molecule for Alzheimer's disease. *Biomedicine and Pharmacotherapy, 132*, 110887.
72. Chen, J., Luo, B., Zhong, B. R., et al. (2021). Sulfuretin exerts diversified functions in the processing of amyloid precursor protein. *Genes & Diseases, 8*, 867–881.
73. Pagan, F., Hebron, M., Valadez, E. H., et al. (2016). Nilotinib effects in Parkinson's disease and dementia with lewy bodies. *Journal of Parkinson's Disease, 6*, 503–517.
74. Hebron, M. L., Lonskaya, I., & Moussa, C. E. H. (2013). Nilotinib reverses loss of dopamine neurons and improvesmotorbehavior via autophagic degradation of α-synuclein in parkinson's disease models. *Human Molecular Genetics, 22*, 3315–3328.
75. González-Lizárraga, F., Socías, S. B., Ávila, C. L., et al. (2017). Repurposing doxycycline for synucleinopathies: Remodelling of α-synuclein oligomers towards non-toxic parallel beta-sheet structured species. *Scientific Reports, 7*, 41.
76. Bermejo, P. E., & Anciones, B. (2009). A review of the use of zonisamide in Parkinson-s disease. *Therapeutic Advances in Neurological Disorders, 2*, 313–317.

77. Biagini, G., et al. L-deprenyl increase GFAP immunoreactivity selectivity in activated astrocytes in rat brain. *Neuropharmacology and Neurotoxicology, 4*, 955–958.
78. Devos, D., Moreau, C., Delval, A., Dujardin, K., Defebvre, L., & Bordet, R. (2013). Methylphenidate: A treatment for Parkinson's disease? *CNS Drugs, 27*, 1–14.
79. Athauda, D., Maclagan, K., Skene, S. S., et al. (2017). Exenatide once weekly versus placebo in Parkinson's disease: A randomised, double-blind, placebo-controlled trial. *The Lancet, 390*, 1664–1675.
80. Bomba, M., et al. (2013). Exenatide promotes cognitive enhancement and positive brain metabolic changes in PS1-KI mice but has no effects in 3xTg-AD animals. *Cell Death & Disease, 4*, 612.
81. Mittal, et al. (2017). Adrenoreceptor is a regulator of the α-synuclein gene driving risk of Parkinson's disease. *Science, 357*, 891–898.
82. Schneider, S. A., & Alcalay, R. N. (2020). Precision medicine for Parkinson disease: Ambroxol for glucocerebrosidase-associated Parkinson's disease first trial completed. *Movement Disorders, 35*(7), 1134–1135.
83. Zhu, Q., Xu, Z. X., Yue, C. J., Ning, Y. N., Chen, Y., Zan, C. C., Qiong, T. J., Xing, S. H., & Hong, L. J. (2021). Lycorine, a natural alkaloid, promotes the degradation of alpha-synuclein via PKA-mediated UPS activation in transgenic Parkinson's disease models. *Phytomedicine, 78*, 153578.
84. Perni, M., Galvagnion, C., Maltsev, A., et al. (2017). A natural product inhibits the initiation of α-synuclein aggregation & suppresses its toxicity. *Proceedings of the National Academy of Sciences of the United States of America, 114*, 1009–1017.
85. Diana, P., et al. (2007). Tetrabenazine in the treatment of Huntington's disease. *Neuropsychiatric Disease and Treatment, 3*(5), 545–551.
86. Rood, C. R. A., Buruma, J. S., Brwn, G. W., & Van Der Vejde, E. A. (1982). Tiapride in the treatment of Huntington's chorea. *Acta Neurologica Scandinavica, 65*, 45–50.
87. Wildridge, B., Rozewicz, S., Mohamed, A., James, J., & Connolly, G. (2022). Use of clozapine for psychosis and chorea in Huntington's disease systematic narrative review. *Progress in Neurology and Psychiatry, 26*, 38–43.
88. Coppen, E. M., & Roos, R. A. C. (2017). Current pharmacological approaches to reduce chorea in Huntington's disease. *Drugs, 77*, 29–46.
89. Palacacu, D., et al. (2002). Olanzapine in Huntington's disease, *105*, 441–444.
90. Duff, K., Beglinger, L. J., O'Rourke, M. E., Nopoulos, P., Paulson, H. L., & Paulsen, J. S. (2008). Risperidone and the treatment of psychiatric, motor, and cognitive symptoms in Huntington's disease. *Annals of Clinical Psychiatry, 20*, 1–3.
91. Alpay, M., & Koroshetz, W. J. (2006). Quetiapine in the treatment of behavioral disturbances in patients with Huntington's disease. *Psychosomatics, 47*, 70–72.
92. Anitha, M., Nandhu, M. S., Anju, T. R., Jes, P., & Paulose, C. S. (2011). Targeting glutamate mediated excitotoxicity in huntington's disease: Neural progenitors and partial glutamate antagonist – Memantine. *Medical Hypotheses, 76*, 138–140.
93. Hrynchak, I., Sousa, E., Pinto, M., & Costa, V. M. (2017). The importance of drug metabolites synthesis: The case-study of cardiotoxic anticancer drugs. *Drug Metabolism Reviews, 49*, 158–196.
94. Awad, A., & Stue, O. (2009). Cyclophosphamide in multiple sclerosis: Scientific rationale, history and novel treatment paradigms. *Therapeutic Advances in Neurological Disorders, 2*, 357–368.
95. Leist, T. P., & Weissert, R. (2011). Cladribine: Mode of action and implications for treatment of multiple sclerosis. *Clinical Neuropharmacology, 34*, 28–35.
96. Holmøy, T., Torkildsen, Ø., & Myhr, K. M. (2017). An update on cladribine for relapsing-remitting multiple sclerosis. *Expert Opinion on Pharmacotherapy, 18*, 1627–1635.
97. Arun, T., Tomassini, V., Sbardella, E., et al. (2013). Targeting ASIC1 in primary progressive multiple sclerosis: Evidence of neuroprotection with amiloride. *Brain, 136*, 106–115.

98. Barkhof, F., Hulst, H. E., Drulovic, J., Uitdehaag, B. M. J., Matsuda, K., & Landin, R. (2010). Ibudilast in relapsing-remitting multiple sclerosis A neuroprotectant. *Neurology, 74*, 1033–1040.
99. Zoccolella, S., Beghi, E., Palagano, G., et al. (2007). Riluzole and amyotrophic lateral sclerosis survival: A population-based study in southern Italy. *European Journal of Neurology, 14*, 262–268.
100. Sawada, H. (2017). Clinical efficacy of edaravone for the treatment of amyotrophic lateral sclerosis. *Expert Opinion on Pharmacotherapy, 18*, 735–738.
101. Trias, E., Ibarburu, S., Barreto-Núñez, R., et al. (2016). Post-paralysis tyrosine kinase inhibition with masitinib abrogates neuroinflammation and slows disease progression in inherited amyotrophic lateral sclerosis. *Journal of Neuroinflammation, 13*, 177.
102. Martinez, A., Del Valle Palomo Ruiz, M., Perez, D. I., & Gil, C. (2017). Drugs in clinical development for the treatment of amyotrophic lateral sclerosis. *Expert Opinion on Investigational Drugs, 26*, 403–414.

# Recent Therapies Including Gene Therapy and Immunotherapy in Overcoming the Gaps in the Treatment of Neurodegeneration

Sakshi Goswami, Yeshvandra Verma, and S. V. S. Rana

**Abstract**

Neurodegenerative disorders cause problems in the central nervous system resulting in progressive degeneration of neural tissues. Neurodegeneration causes progressive loss of neurons' structural and functional integrity. It is estimated that neurodegenerative disorders have an impact on more than 50 million people worldwide. Recently, researchers from all around the world have been working on the treatment of neurodegenerative disorders. Conventional therapies are either failed or inadequate to cure neurodegenerative disorders. Moreover, current treatments for neurodegenerative disorders are symptomatic without affecting the aetiology of diseases. Advancements in recent therapies support a potential role to overcome the gaps in the treatment of neurodegeneration. In this article, we aim to describe the gene therapy and immunotherapy being developed/developing to cure neurodegenerative disorders.

**Keywords**

Brain · Immunotherapy · Neurodegeneration · Nervous system, Neurodegenerative disorders · Immunization

## 1 Introduction

The brain is a very intricate organ with countless pathways that enable us to perform the astounding feats that we do. We rely on the efficient functioning of the brain's billions of cells every second of every day. Neurons are some of the most significant brain cells. Cells that make up the brain and communicate with one another allow

S. Goswami · Y. Verma (✉) · S. V. S. Rana
Department of Toxicology, Chaudhary Charan Singh University, Meerut, India



R. Shukla et al. (eds.), *Neuroendocrine-Immune Axis*,
https://doi.org/10.1007/978-981-95-5088-3_14

us to move, think, talk, and perform all other human functions. All of the functions of the brain are carried out by neurons interacting with one another. Because of the intricacy of the brain, diseases can result from even the smallest cell-to-cell miscommunications.

The term "neurodegeneration" or, more specifically, "neurodegenerative disorders" (NDDs) refers to a group of complicated and significant medical problems that primarily damage the neurons in the human brain. This term doesn't just refer to a single type of condition. Instead, it serves as a general phrase for a number of different conditions.

Such ailments cause problems of the central nervous system (CNS), which in turn cause the progressive degeneration of neural tissues, including the death of neurons, over time. The neurodegenerative illnesses and the associated neurodegenerative diseases do not have natural treatments on their own, since the neurons cannot self-regenerate after neurodegenerative cell death or serious damage to the brain tissue. Neurodegenerative diseases are a great example of the debilitating consequences of improper brain cell communication. These conditions can affect a person's ability to move, speak, breathe (called ataxias), their mental functioning (called dementias), and much more [1]. Because of the ageing of the world population, NDDs, which are incurable and debilitating illnesses, are becoming more common. Neurodegeneration causes a progressive loss of neurons' structural and functional integrity. This loss results in a variety of clinical and pathological manifestations as well as a decline in the functional architecture [2]. Progressive neuronal cell death frequently results in a variety of neurodegenerative disorders (NDDs), which can be distinguished based on the cardinal syndrome and disease protein (Fig. 1), including Parkinson's disease (PD), Huntington's disease (HD), Alzheimer's disease (AD), amyotrophic lateral sclerosis (ALS), brain injury (BT), prion disease (PrD), progressive supranuclear palsy (PSP), and spinocerebellar ataxias (SCA). These disorders typically take a long time to develop, and the consequences and symptoms typically manifest later in life. Although rare, degenerative brain illnesses are widespread enough that everyone is aware of them. They are thought to have an impact on more than 50 million people worldwide.

Most of these illnesses have a significant ageing component and are far more likely to affect persons over the age of 65 (although other illnesses, including Huntington's disease and ALS, sometimes manifest earlier). The World Health Organization (WHO) predicts that in the next 30 years the number of persons over 65 will at least double. That implies that the population's prevalence of neurological diseases will increase at a similar rate. Over 36 million people worldwide are diagnosed with AD or PD, the two most common neurodegenerative disorders. Over a million Americans, including those over 60, suffer from idiopathic PD, whereas an estimated 5.7 million Americans of all ages are thought to have Alzheimer's dementia [3]. Dementia—which currently affects 44 million people worldwide and is projected to reach 135 million by the year 2050—is a major factor in the infirmity of the aged [4]. In Europe, the prevalence of NDDs that progress to dementia is 1.6% and 1% for men and women in the 65–69 age groups, respectively, and rises to 11% and 12.6% in the 85–89 age group [5]. Unless prompt worldwide action is taken, the

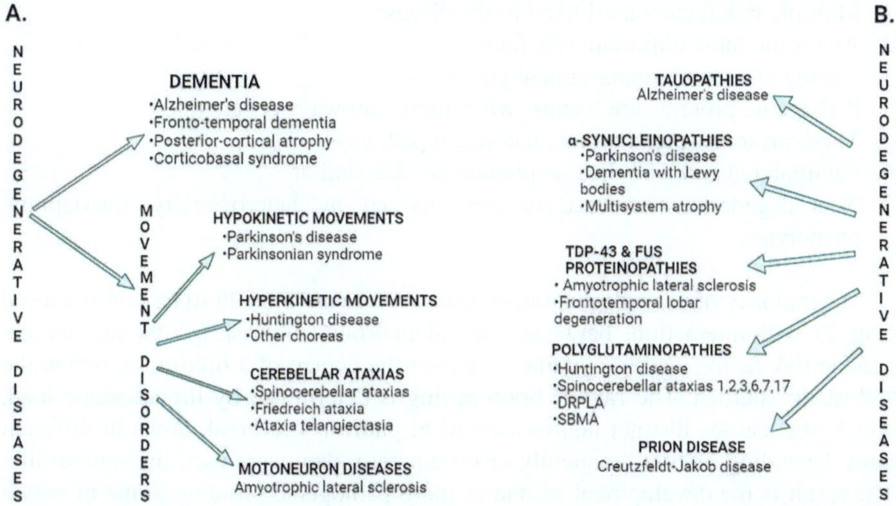


**Fig. 1** Classification of neurodegenerative disorders according to (**a**) cardinal syndrome and (**b**) disease protein

neurological load is anticipated to become an even more acute and unmanageable hazard to public health, according to a WHO report [6].

The NDDs have drawn a lot of attention among the diseases because of their complicated causes and symptoms, lack of appropriate, effective treatment(s), lack of reports of irreversibility, and negative effects on social and financial aspects. Today, numerous cutting-edge methods can identify the onset and progression of many NDD problems [7]. According to their unique cell death pathways, these diseases' symptoms and exacerbations range greatly, and having unique cell death mechanisms results in distinct therapeutic approaches.

## 2 Causes of Neurodegenerative Diseases

Research shows that multiple factors are probably responsible for causing neurodegenerative diseases. The most common neurodegenerative disorders are caused by protein abnormalities and include amyloidoses (build-up of amyloid in organs), tauopathies (deposition of tau protein in the brain), and synucleinopathies (accumulation of insoluble α-synuclein in neurons and glial cells). Aging is a major risk factor for Alzheimer's disease, macular degeneration, glaucoma, and other neurodegenerative diseases. Neuronal degeneration is influenced by a number of factors, including human longevity.

The pathophysiology of neurodegenerative illness and the variety of its manifestations are both explained by a hypothesis. The following seven axioms serve as its foundation: age is the most significant risk factor for neurodegenerative disease, according to the following facts:

1. Multiple risk factors are linked to the disease
2. Age is the most important risk factor
3. Ageing affects neuroanatomical pathways
4. Pathogenic proteins are formed when these pathways degenerate
5. These proteins spread along anatomical pathways
6. Familial and sporadic disease phenotypes are similar
7. Neurodegenerative disease is characterized by heterogeneity, overlapping phenotypes

The majority of neurodegenerative disease cases are thought to be multifactorial (Fig. 2), with interactions between external environmental risk factors and internal genetic risk factors acting cumulatively over the course of a lifetime to define the "allostatic" burden. The rate of brain ageing is determined by the allostatic load, which also causes distinct neuro-anatomical pathways to break down at different rates depending on how frequently or infrequently they were used throughout life. The result is the development of one or more pathogenic proteins, some of which may display "prion-like" behaviour and spread throughout the brain from their initial sites of production along neuro-anatomical pathways to harm related brain regions. The clinical and pathological diversity of disease phenotypes is ultimately caused by variations in the pathogenic proteins produced and their anatomical distribution.

With ageing, cells that remove cellular waste become less effective. Immune cells in the eyes and brain may misinterpret debris accumulation in and around neurons as an infection or other threat and overreact by mounting a harmful immune

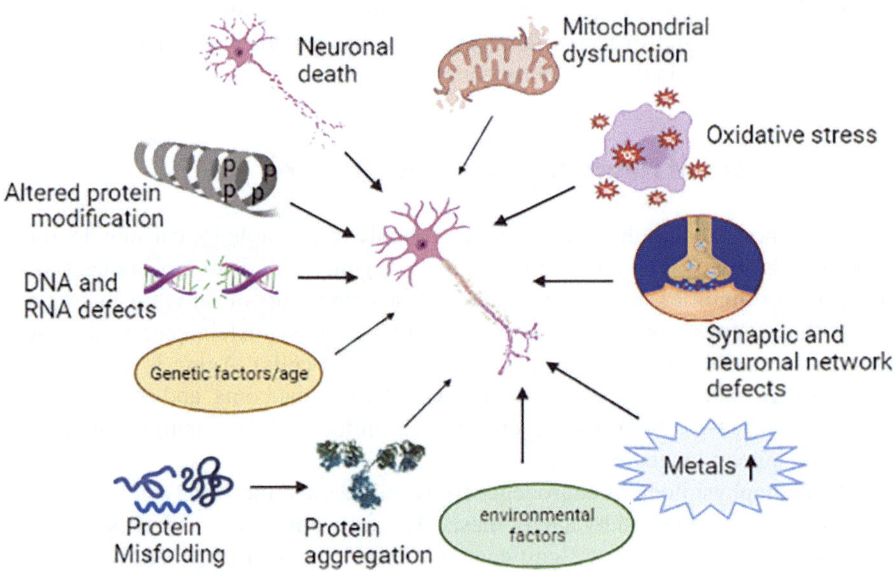


**Fig. 2** Different factors associated with Neurodegenerative diseases

response. The collection of fatty debris known as drusen may be the cause of the chronic sterile inflammation that experts have connected to macular degeneration.

Some other risk factors known to cause NDDs are genetics, medical history, habits, routine, and choices. The aetiology of neurodegeneration includes genetic variables, which are significant. Inherited genetic alterations are the root cause of some NDDs. These conditions are hereditary; parents pass a defective gene to their offspring. Examples include Huntington's disease, unusual forms of Alzheimer's disease, and motor neuron disease. Significant medical advancements have resulted from the discovery of disease genes and risk loci. In order to create novel disease-modifying medicines, translational research ultimately relies on genetics [8]. Approximately 5–10% of all ALS cases are familial (also known as hereditary or genetic). More than a dozen gene variants have been linked to familial ALS. A deficiency in the C9ORF72 gene, which produces a protein that is located in motor neurons and nerve cells in the brain, accounts for about 25–40% of all familial cases (and a tiny percentage of sporadic cases). Some carriers of this gene also experience frontotemporal degeneration (FTD), a form of dementia brought on by the atrophy of the temporal and frontal lobes of the brain (NINDS, National Institute of Neurological Disorders and Stroke). Also, RNA toxicity, abnormal protein aggregation (such as tau in AD and traumatic brain injuries), misfolded proteins (typical of PolyQ diseases), protein aggregation (such as synuclein in PD or TDP-43 in ALS), or translational products from repeat expansion within genes are some of the common underlying causes of these conditions [9]. Environmental elements also have a role in NDDs. Also, these diseases can arise for a variety of reasons, including the environment, pollution, chemical and toxin exposure, specific infections, UV radiation, alcohol, drugs, and even where you live may all be factors. These compounds cause oxidative stress in cells, which may lead to protein misfolding and inflammation, both of which are associated with neurodegenerative disease. For instance, there is evidence connecting Parkinson's disease with chronic exposure to poisons, chemicals, and pesticides [10]. Also, dementia-like conditions have been linked to low vitamin D levels, which are more common the farther away from the equator you live [11].

Neurodegeneration risk is heightened by unhealthy lifestyle choices and general well-being. Age-related macular degeneration has been linked to smoking and high-glycaemic diets (those heavy in refined sugars and carbohydrates). High blood pressure increases the risk of glaucoma. A higher risk of Alzheimer's neurodegeneration has also been associated with sleep difficulties, depression, post-traumatic stress disorder, and other ailments [12]. Alzheimer's disease is a complex ailment that is brought on by environmental and genetic variables interacting to speed up the natural ageing process. Reactive proteins like A and tau are produced as a result, which causes the degradation of blood vessels and neurons. Both early-onset familial AD (EO-FAD) and late-onset sporadic AD (LO-SAD) are caused by gene alterations, which also affect how age-related neuronal degeneration develops [13].

## 3 Traditional Therapies for Neurodegenerative Diseases

To prevent the occurrence of neurodegenerative illness, it may be important to minimize the variables that increase allostatic stress throughout the course of a lifetime and increase cognitive and physical exercise [5]. Different novel approaches in the treatment of neurodegenerative diseases are shown in Fig. 3.

Protein aggregation, inflammation, and oxidative stress in the central nervous system (CNS) are hallmarks of neurodegenerative disorders. Neurodegenerative illnesses are associated with a number of molecular processes, including aberrant ubiquitination, oxidative stress, and neurotransmitter depletion or inadequate synthesis. Damage to the blood-brain barrier (BBB) in the CNS also contributes to a number of illnesses that affect the CNS. Although synthetic medications are used to treat chronic conditions including autism, Parkinson's disease, Alzheimer's disease, and others, they do have certain negative effects [14]. Another brain illness connected to CNS inflammation, learning, and social impairments is autism spectrum disorder (ASD). Children with ASD display a variety of pathological conditions, such as allergy symptoms, elevated levels of anti-brain protein autoantibodies, high levels of anxiety, elevated levels of oxidative stress, and elevated levels of food intolerance. The flavonoid luteolin inhibits human mast cells from releasing tumour necrosis factor (TNF), which has helped in the treatment of children with ASD [14].

Due to the progressive death of neuronal cells, neurodegenerative disorders like prion disease, Alzheimer's disease, and Parkinson's disease are linked to high rates of morbidity and mortality. Although there is no known cure for these illnesses,

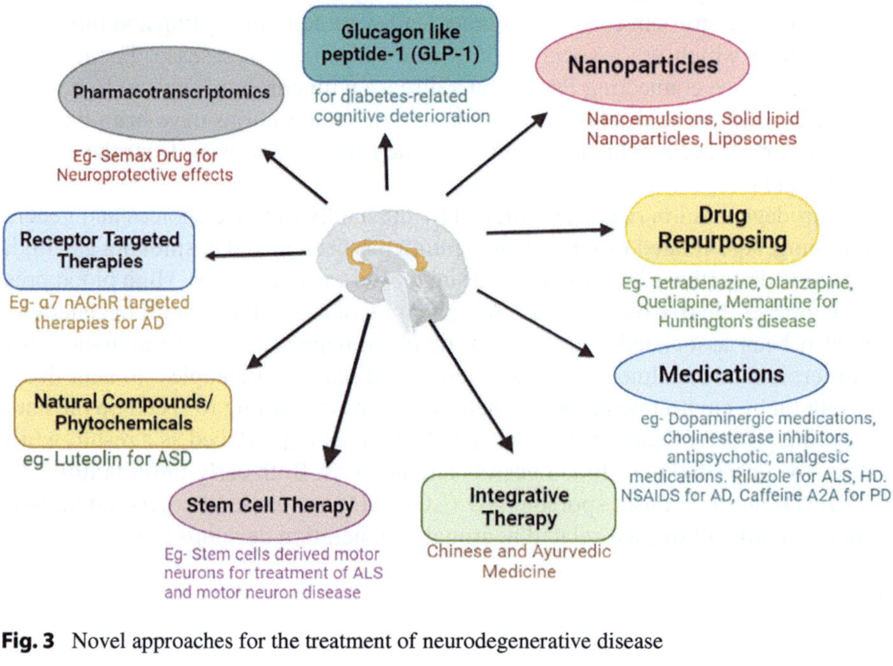


**Fig. 3** Novel approaches for the treatment of neurodegenerative disease

there are symptomatic treatments available, including dopaminergic medications for movement disorders [15], cholinesterase inhibitors for cognitive issues [16], antipsychotic medications for behavioural and psychological dementia symptoms [17], analgesic medications for pain [18], anti-inflammatories for infections [19], and even the use of deep brain stimulation to stop tremor and refractory movement disorders [20]. Scientists also worked to produce medicines to slow down the development of diseases such as riluzole for ALS, cerebellar ataxia, and Huntington's disease [21], NSAIDs (non-steroidal anti-inflammatory drugs) for Alzheimer's disease [22], and caffein A2A receptor antagonists for the neuroprotection of Parkinson's disease [23]. Edaravone, a free radical scavenger, was approved for ALS, but its effectiveness is yet unknown, and at most the medication has a mild impact on the development of illness [24]. These, however, did not work out very well because there were still too many issues to handle these diseases' severe and escalating symptoms.

For the major issues associated with many degenerative diseases, integrative therapy [25], which includes traditional Chinese medicine and Ayurvedic medicine, has also shown good outcomes. By taking into account all associated symptoms and addressing non-motor disorders, sleep disorders, autonomic symptoms, neuropsychiatric symptoms, gastrointestinal symptoms, sensory symptoms, and exhaustion, these treatments seek to improve the quality of activity of daily living (ADL) for patients [26]. To manage the growing and dangerous symptoms of these diseases, there are still too many issues. In order to safeguard patients' health, western medicine and integrative medicine are being used more and more in the treatment of certain diseases. Recent research have demonstrated that complementary and alternative medicine can improve patients' quality of life and increase their capacity for reserve movement [27]. However, these are not effective treatments for degenerative diseases in modern society.

Currently, experts and researchers from all around the world are putting in a lot of effort to find a cure for neurodegenerative diseases. Although there is currently no treatment, researchers have developed a few concepts that show great promise. The replacement of dead neurons in the brain with new ones is one of the most intriguing concepts being researched by scientists [28]. Stem cells, which are immature cells that may develop into any type of cell in the body, including neurons, can be used to replace dying neurons. Since stem cells are undeveloped cells that can develop into any type of cell in the body, they have the potential to repair damaged or dead brain neurons. This concept has even been tested on lab mice [28]. But because humans and other animals differ in important ways, scientists believe that these kinds of research should not be carried out on people who have neurodegenerative disorders.

Numerous gene mutations that cause familial types of serious neurodegenerative illnesses have been discovered in the last 20 years, which has spurred fresh research into the biology of neurodegeneration and strategies for neuronal protection. Numerous potential pathways of neurodegeneration have been proposed based on the toxic and transgenic disease models. The majority of them share similar traits and mechanisms, such as the buildup of intrinsically disordered proteins in the form of aggregates, the involvement of mitochondrial dysfunction and oxidative stress, and neuroinflammation, which was taken into consideration for the development of therapeutic strategies intended to stop or at least slow disease progression, as opposed to therapy used to treat symptoms of the disease [29].

Another alternative to conventional therapies taken for the treatment of neurodegenerative illnesses is drug repurposing. Because of its proven safety and tolerability, which necessitates smaller sample sizes and lowers clinical development expenses, it is cost-effective. Testing for new targets is made simple by methods like computational chemistry and high-throughput screening. The arsenal against neurodegenerative illnesses still grows in the future, despite the fact that no treatment can completely reverse neurodegeneration pathways. Several drugs are currently undergoing phase II and III clinical trials. These medications ought to act as a guide for the molecular alterations and ligand-based creation of novel molecules. Unfortunately, a medication that may completely reverse the causes of neurodegeneration has not yet been discovered [30].

However, the fundamental hurdle in creating a successful treatment plan is the permeability of the treatment components at the blood–brain barrier. Several approaches have been explored to overcome this obstacle. For example, nanoparticles improve drug transport to the BBB due to their size. They have been widely utilized in neurology, including as diagnostic nanoprobes and in nanomedicine. Neurodegenerative illnesses have been treated with nanomaterials in a variety of forms, including nanoparticles, nanoemulsions, solid lipid nanoparticles (SLN), and liposomes [31].

Chronic hyperglycemia causes mitochondrial failure, neuroinflammation, neurotransmitter change, and vascular illness. Diabetes mellitus is a key risk factor for cognitive dysfunction. Glucagon-like peptide-1 (GLP-1), a gut hormone, is a potential link between metabolic and brain impairment. Studies have demonstrated the impact of GLP-1 on a variety of neural processes, including energy balance, blood pressure regulation, neurogenesis, neurodegeneration, and retinal repair. In animal models, GLP-1 receptor agonists (GLP-1RAs) have demonstrated positive effects on brain ischemia, neurodegeneration, learning, and memory. GLP-1RAs seem to be a viable treatment for diabetes-related cognitive deterioration, while more clinical research is required [32].

Among the most potential pharmacological targets is nuclear factor erythroid 2 p45 related factor 2 (Nrf2). The genes that encode antioxidant enzymes and proteins involved in detoxification, cellular organelle repair, inflammation, and mitochondrial bioenergetics are the targets of this transcriptional target, which coordinates networks of cellular defence systems. Activating Nrf2 and simulating mitochondrial function play protective roles in the pathogenesis of Parkinson's disease, according to Claudia Bento-Pereira and Albena Dinkova-Kostova [33]. They also discuss how the creation of small molecules that act as Nrf2 activators can become one of the most promising therapeutic targets for mitochondrial dysfunction in PD, a treatment option for the currently incurable neurodegenerative diseases.

The hallmarks of the neurodegenerative process are mitochondrial malfunction and oxidative stress. One of the largest sources of reactive oxygen species and one of the most susceptible organelles to oxidative stress is the mitochondria. Lipid peroxidation, one of the symptoms of oxidative stress, is a characteristic of neurological disorders like ALS, Parkinson's disease, and Alzheimer's disease. In their examination of the relationship between mitochondria and lipid peroxidation in the mechanism causing neuronal death [34], also include potential therapeutic approaches to stop it.

The mitochondria are an organelle with numerous tasks, including important ones in energy metabolism and participation in the process of cell death. Numerous neurodegenerative diseases, including Alzheimer's, have mitochondrial aetiology, which has been demonstrated. Shevtsova et al. [35] concentrate on the possible applications of different mitochondria-targeting medicines in Alzheimer's disease. The authors talk about how preclinical and clinical trials are implementing mitochondrially targeted techniques.

The use of pharmacological drugs that operate on various targets is collectively referred to as polypharmacology. In order to create medications for difficult neurodegenerative diseases like Alzheimer's [36], draw readers' attention to polypharmacological techniques. The development of successful multitarget drug discovery for Alzheimer's disease is reviewed with examples drawn from the most current literature. These prospects range from medication combos to multitarget-directed ligands (both codrugs and hybrids).

Pharmacotranscriptomics, one of the most cutting-edge approaches to drug research, strives to achieve more accurate medication dosing based on interindividual transcriptome differences. Reviews of research on the effects of peptides with neuroprotective qualities on gene transcription in nerve cells are provided by Dergunova et al. [37].

One of the characteristics of neurodegeneration linked to Alzheimer's is the loss of cholinergic neurons. Alpha7 nicotinic acetylcholine receptor ($\alpha$7 nAChR), a receptor that is involved in the cholinergic nervous system of the brain, has thus long been a target for pharmaceutical companies. Despite encouraging preclinical outcomes, the medications have not been effective in human trials. The omega-3 fatty acids eicosapentaenoic acid (EPA) and docosahexaenoic acid (DHA) have also been linked to a reduced incidence of Alzheimer's disease, according to epidemiological research. Preclinical animal research with promising results demonstrated that long-term, high-dose DHA supplementation decreased $\beta$-amyloid levels and decreased neuronal death. But in clinical trials, people who had previously received a diagnosis did not gain from taking additional DHA, and their signs and symptoms worsened [38].

Unfortunately, there is no effective treatment for neurodegenerative illnesses till present day. The symptoms of neurodegenerative illnesses can be treated with medication, both physically and mentally. Usually, a combination of medicine and psychotherapy is used to treat them. Depending on the type of neurodegenerative disease one has been diagnosed with a specific treatment regimen and dosage may be required. Different neurodegenerative disease forms require different treatments. However, the goal of treatment is usually to reduce the condition's symptoms.

### 3.1 Gene Therapy for Neurodegenerative Diseases

Gene therapy is a cutting-edge and innovative area of study that has the potential to treat diseases like neurodegenerative disorders that are now incurable. Current treatments for neurodegenerative diseases target symptom relief rather than the underlying pathology. Pharmacological therapies become less effective as neurodegeneration worsens. Therapeutics can be delivered directly into the brain, bypassing the blood-brain barrier, and specifically target disease processes based on their functional and anatomical characteristics. This approach requires chronic re-administration of the therapeutic agent, while single-gene therapy infusions have durability of a decade or more [3]. The current state of gene therapy for motor neuron illnesses is discussed by Zakharova [39] along with the potential for the creation of novel treatment approaches including viral vector therapy and RNA-modifying techniques. Gene therapy can be used to correct defective genes by introducing a functional copy of the gene, by introducing a disease-modifying gene, by incorporating RNA interference (RNAi) to silence a mutant allele, or by employing gene-editing technology [40, 41].

Neurodegenerative conditions like Parkinson's disease, amyotrophic lateral sclerosis, spinal muscular atrophy, and Alzheimer's disease may be effectively treated by gene therapy. Clinical experiments have so far been unable to demonstrate any improvement in results beyond the placebo effect. Three key areas are being targeted for improvement: vector design and the discovery of new vector serotypes, the manner of gene therapy delivery, and the discovery of new therapeutic targets. To increase efficacy, these developments are being put to the test both alone and collectively. With these advancements, gene treatment for certain conditions might eventually be effective [42]. A summary of the different routes of delivery for gene therapy for neurodegenerative diseases is shown in Fig. 4.

For neurodegenerative illnesses, gene therapy offers therapeutic advantages such as pathogenic mechanism rectification, neuroprotection, neurorestoration, and symptom management. Effectiveness is influenced by disease aetiology, the specificity of gene expression, and the ability to complete transduction without any leakage into surrounding areas or perivascular spaces.

Vectors used in gene therapy can be either viral or non-viral. Non-viral vectors may be produced at a relatively low cost, have no size restriction for the therapeutic gene, and often provide little danger of immunogenicity [43]. However, the majority of gene treatments today rely on viral vectors, since substantial therapeutic dosages are necessary when employing non-viral technologies, and the resulting gene expression is often temporary. Some viral vectors that have been tested in clinics include vaccinia, measles, vesicular stomatitis virus (VSV), polio, reovirus, adenovirus, lentivirus, γ-retrovirus, herpes simplex virus (HSV), and adeno-associated virus (AAV) [44].

Adeno-associated viruses (AAVs), which are most frequently utilized in clinical trials for CNS illnesses, are modified viral vectors that are used to transfer genes to the brain. Lentiviral vectors are an exception, as they may carry a bigger DNA payload and have only been employed in one clinical trial thus far [45]. Developing a

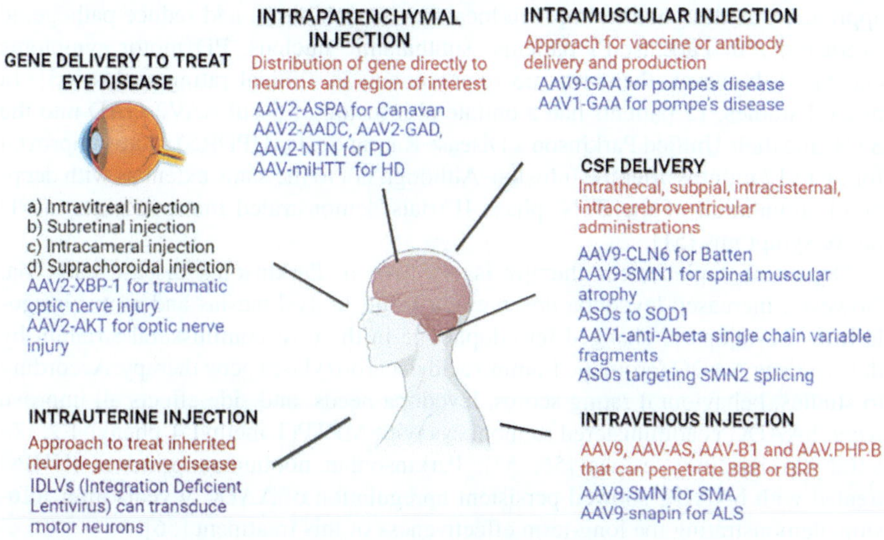


**Fig. 4** Delivery routes of gene therapy for neurodegenerative disorders

predictable and repeatable gene therapy paradigm requires an understanding of how a particular serotype distributes gene products from the targeted region for vector delivery [46, 47]. The delivery path must be taken into account, particularly for the central nervous system. AAV2 has been the gold standard for neurosurgical gene therapy and has been utilized in numerous phase 1 [48, 49] and phase 2 [50, 51] trials due to its relative phenotypic selectivity for neurons and clinical safety profile. The advantage of intravenous injection for remote administration is that it is non-invasive. However, the blood-brain barrier is a substantial barrier that prevents most vectors from entering the central nervous system. Therefore, it was important to learn that AAV9 can cross the BBB [52]. For neurodegenerative illnesses, there have been numerous clinical trials using gene therapy. Canavan disease, which is brought on by a mutation in the aspartoacylase (ASPA) gene, was the subject of one of the initial experiments utilizing AAV-mediated gene therapy. The patients received an injection of AAV2-ASPA into the brain and underwent monitoring for up to 10 years following surgery. A follow-up revealed some improvement in seizure frequency along with a delayed rate of brain shrinkage [53]. Recently, patient recruitment for the first trial using AAV9 to deliver the gigaxonin gene intrathecally to treat gigantic axonal neuropathy (GAN) has started.

Gene therapy has been thoroughly investigated as a potential PD treatment. Several gene therapy studies met the Phase I safety requirements and offered some evidence of efficacy [54]. However, the bulk of them have fallen short of demonstrating benefit beyond the placebo effect when they were moved to controlled, blinded Phase II studies. Delivery of AAV2-glutamic acid decarboxylase (GAD) by direct injection into the subthalamic nucleus (STN) was the sole experiment to demonstrate effectiveness in a monitored, blinded Phase II study. The AAV2-GAD approach was designed to improve local GABA inhibition and reduce pathogenic overactivity in Parkinson's patients' subthalamic nucleus. PD motor symptoms decreased, however, there was no improvement in clinical rating scales [55]. In phase I studies, 12 patients had a unilateral infusion of 50 μl AAV2-GAD into the STN, and their Unified Parkinson's Disease Rating Scale (UPDRS) scores improved for up to 12 months after the infusion. Although not to the same extent as with deep-brain stimulation of the STN, phase II trials demonstrated improvements in PD motor symptoms [51].

Dopamine replacement therapy is effective in Parkinson's disease early on. However, increased levodopa doses might result in dyskinesias and motor irregularities. Levodopa is changed into dopamine in the post-commissural striatum by the pro-drug AADC (aromatic l-amino acid decarboxylase) gene therapy. According to studies, behavioural rating scores, levodopa needs, and side effects all improve when hAADC is administered to monkeys with MPTP(1-methyl-4-phenyl-1,2,3,6-tetrahydropyridine) lesions [56, 57]. Parkinsonian nonhuman primates (NHPs) treated with hAADC showed persistent upregulation of AADC 9 years after infusion, demonstrating the long-term effectiveness of this treatment [56].

Neurturin (NTN), a structural and functional homolog of GDNF (glial-derived neurotrophic factor), has been studied in parkinsonian non-human primates as a potential neuroprotectant. When AAV2-NTN was infused into the human putamen, the dopaminergic innervation of the nigrostriatal region was retained, and function gradually recovered. No major side effects were observed during a 12-month double-blind, sham surgery-controlled phase II trial, however, UPDRS motor score did significantly improve [58].

The gold standard for precise vector delivery is interventional MRI-guided convection-enhanced delivery (iMRI-CED), which could hasten preclinical therapies for Parkinson's, Huntington's, and Alzheimer's illnesses. The development of convection-enhanced delivery in the 1990s allowed for efficient, direct, and controlled distributions of viral vector particles throughout the brain [3]. iMRI-CED is being used in a new phase I experiment that aims to deliver AAV2-hAADC to the putamen. With the aid of iMRI-CED, it is possible to track and dynamically modify vector infusions in real time, and accurate calculations of the volume of vector coverage infused into the target brain region are also made possible. Studies in phase 2 are still ongoing.

New treatments are being developed in the domain of proof-of-principle testing. An AAV5 vector encoding tyrosine hydroxylase (TH) and GTP cyclohydrolase 1 (GCH1) has been investigated in rat and nonhuman monkey models of PD as a dopamine production inhibitor [59]. The rat model showed encouraging findings, however, non-human primate research did not demonstrate efficacy. This was explained by TH not being expressed in the caudate putamen [60]. Two groups have evaluated the effectiveness of cerebral dopamine neurotrophic factor (CDNF) in rat models of Parkinson's disease [61, 62].

Controlling transgene expression might provide a more subtle method of gene therapy. This is pertinent when transgenic expression is ongoing and could result in undesirable side effects. It has been discovered that glial-derived neurotrophic factor works well in animal models of Parkinson's disease. Consequences of constitutive GDNF expression, however, include abnormal axonal sprouting, downregulation of TH, and accelerated dopamine turnover [63, 64]. Therefore, GDNF delivery and expression control systems are required for PD therapy [65]. The destabilizing domain (DD) of E. coli dihydrofolate reductase has been coupled to a lentiviral vector producing GDNF. The medication trimethoprim can pass across the BBB and binds to the DD domain. This controls the expression of GDNF and, thus, the stability of the DD domain [66]. It protected neurons and enhanced motor performance when evaluated in a rat model of PD [65].

For continuous dopamine production, TH, AADC, and GTP cyclohydrolase-1 three key dopamine biosynthetic enzymes were expressed in a clinical trial using a lentivirus (ProSavin) delivered intrastriatally [45, 67]. Motor function showed a considerable improvement, and a long-term analysis is still being conducted.

Mucopolysaccharidosis (MPS), which is a rare metabolic disease, is brought on by the defective breakdown of mucopolysaccharides, which causes a buildup of glycosaminoglycans in the lysosome. Adults may experience speech difficulties, ataxia, weakness, and dyskinesia, whereas children often exhibit neurological damage, delayed growth, and early mortality [68]. For the treatment of MPS I (Hurler infantile syndrome), MPS II (Hunter syndrome), MPS IIIA, and MPS IIIB (Sanfilippo syndromes A and B), gene therapy clinical studies are presently being conducted. In order to introduce a functional copy of alpha-L-iduronidase (IDUA) or iduronate 2-sulfatase (IDS), respectively, into the albumin locus of patient hepatocytes, the MPS I and MPS II studies rely on intravenous AAV delivery of ZFN therapeutic agents. MPS IIIA aims to deliver a functional copy of human N-sulfoglucosaminesulfohydrolase (h.SGSH) into young children's brains. Although the therapeutic benefit of h.SGSH administration is modest, it is safe [69]. MPS IIIB aims to restore h.SGSH expression. Intracerebral and intravenous administration has shown promising results in preclinical investigations, but their therapeutic effects on patients have not yet been shown [70].

A GAA gene mutation that results in lysosomal glycogen buildup is the cause of Pompe disease, a neuromuscular condition [71]. Dysfunction of the heart, lungs, and skeletal muscles is caused by the accumulation of glycogen in several tissues. Preclinical findings demonstrate glycogen restoration and enhanced lifespan post-GAA delivery via AAV, which is presently being investigated in three clinical trials [72].

A phase 1 clinical trial used AAV2 for gene delivery of NGF (nerve growth factor) in mild to moderate Alzheimer's patients. A follow-up multicentre phase 2 study enrolled 49 participants, who received intracerebral injections without CED or iMRI guidance. The procedure was safe and well-tolerated for 24 months, but no significant difference was observed on the Alzheimer's Disease Assessment Scale–cognitive subscale [49]. The majority of gene therapy studies concentrate on AAV-NGF intracerebral delivery. NGF is comparable to GDNF and NTN in that it has the potential to assist neurons in a neurotrophic manner. According to studies, NGF can stop adult cholinergic neurons in the forebrain from degenerating following damage [73]. The effectiveness of these therapies is still unclear, though. Intracisternal administration of AAVrh.10hAPOE2, which has been proven to raise the likelihood of developing sporadic AD, is the subject of new research. Although APOE2 homozygotes are resistant to the condition, APOE2 delivery may be helpful for APOE4 homozygous individuals [74].

Batten disease, which affects infants, kids, and adults, is an uncommon neurological disorder. It results in early mortality, retinal degeneration, loss of cognitive and motor function, and epilepsy. The diseases have been associated with more than 430 loss-of-function mutations and 12 CLN genes [75]. Clinical studies using AAV mostly use AAV-CLN2 and AAV-CLN3, some of which have minimal therapeutic value. A follow-up trial employing a different vector (AAVRh.10CUhCLN2) is continuing; however, aberrant T2 hyperintensities were seen in individuals who had received the initial dosage. A study on CLN3 and CLN6 is ongoing. According to preliminary statistics, in the first 2 years following dosage, there was no additional advancement [76].

Spinal muscular atrophy (SMA) is caused by the loss of function of the gene that codes for the survival motor neuron 1 (SMN1), which affects babies and often results in mortality before the age of 20 months. AAV9-SMN delivery, which is being used by AveXis in clinical studies, has demonstrated considerable advantages in speaking, sitting, rolling over, and controlling one's head when sitting. Increased serum aminotransferase following intravenous injections of AAV9-SMN can be treated with oral prednisolone. AAV9-SMN intravenous administration is being looked into by AveXis for paediatric SMA type 1 patients in order to improve CNS transduction. Results for additional varieties of SMA are still forthcoming. In 2019, the FDA approved AAV9-SMS for the treatment of children with SMA type 1 [77].

### 3.2 RNAi Therapy in Neurodegenerative Diseases

Gene therapy is an appealing alternative for treating Huntington's disease, given that it enables precise targeting of damaged nuclei to lower mHtt levels. Huntington's disease is a genetic disorder inherited in an autosomal dominant manner, resulting from an abnormally high number of CAG repeats located on chromosome 4, leading to the production of a polyglutaminated form of the huntingtin protein (mHtt) [78]. Patients frequently present with motor, cognitive, and psychological problems around the age of 40. Deficits in attention and mental flexibility are examples of cognitive alterations, whereas sadness, apathy, and impulsivity are examples of psychiatric symptoms [79]. RNA interference is presently the main technique used in preclinical gene therapy to reduce pathogenic levels of mHtt. These methods lessen unintended negative effects while allowing for precise targeting of mHtt mRNA. Furthermore, research on nonhuman primates have shown that up to 45% decreases in normal Htt may be successfully tolerated [80].

AMT-130, the first AAV-miRNA-based gene therapy for Huntington's disease has been approved by the FDA by UniQure. AAV5-miHTT, the miRNA that serves as the foundation for AMT-130, targets both the wild-type and mutant Huntingtin alleles [81–83]. After a single infusion into the brain, a transgenic HD (tgHD) minipig model's brain had a significant reduction in human mutant huntingtin mRNA and protein [82]. In 26 individuals with early evident Huntington's disease, the Phase I/II clinical study of AMT-130 will examine the safety, tolerability, and effectiveness indications.

Voyager Therapeutics is pursuing a similar strategy, and early findings with its lead candidate VY-HTT01 in preclinical animals are encouraging. Many other neurodegenerative diseases, such as spinocerebellar ataxia, amyotrophic lateral sclerosis, frontotemporal dementia, synucleopathies, such as Parkinson's disease, tauopathies, and AD, are thought to benefit from miRNA-mediated silencing [84, 85]. Given that miRNAs can potentially cause toxicity associated with off-target effects related to the passenger strand, researchers should concentrate on choosing scaffolds with low passenger strand activity and utilizing bioinformatic techniques to estimate binding probabilities. More miRNA therapies are anticipated to enter the clinic as several of these obstacles are overcome.

### 3.3 Common Alternative Silencing Strategies for Neurodegenerative Diseases

Since the 1970s, antisense oligonucleotides (ASOs) have been employed in preclinical research, with several programs progressing to the clinical development phase. In 2012, Ionis Pharmaceuticals conducted a phase I study on ALS that utilized intrathecal infusion of SOD1-targeting ASOs [86]. Regrettably, this study did not succeed in demonstrating a reduction in SOD1 protein levels, which was attributed to the insufficient concentration of ASO in the targeted area. Nevertheless, the treatment was confirmed to be safe, and a follow-up phase I/II trial is currently underway.

A phase I clinical study (BIIB078) for C9orf72-ALS patients using ASOs was launched by Ionis Pharmaceuticals in 2019. The most frequent genetic cause of frontotemporal dementia and amyotrophic lateral sclerosis is the enlarged GGGGCC hexanucleotide repeat in the C9orf72 gene. The experiment tries to decrease C9orf72 transcripts that include repeats. With preclinical trials demonstrating its effectiveness in C9orf72-related amyotrophic lateral sclerosis and spinocerebellar ataxia, Ionis has ASOs for Huntington's disease, myotonic dystrophy type 1 (DM1), and spinal muscular atrophy.

The fact that endo- and exo-nucleases break down ASOs and necessitate re-administration is a significant barrier to their use in clinical settings [87]. This is particularly troublesome in CNS illnesses that need invasive treatment techniques. However, chemical changes may be added to ASOs to alter their pharmacological profiles, and work is now being done to optimize ASOs' delivery to, for instance, penetrate the BBB and increase target engagement [88].

### 3.4 Gene Editing (CRISPR)

Gene editing is a relatively new method for removing, adding, or changing DNA in a sequence-specific way. The most recent nuclease system, CRISPR/Cas9, is based on a mechanism that occurs naturally in the bacterial Streptococcus pyogenes' adaptive immune system. The CRISPR/Cas9 system has been employed to create numerous animal models, and an increasingly large number of preclinical investigations illustrate the promise of this method when integrated with gene therapy for the treatment of neurodegenerative illnesses.

For example, in AD, the impaired amyloid precursor protein (APP) gene was successfully modified in human fibroblasts through the application of a CRISPR/Cas9 system, leading to a reduction in amyloid beta concentrations, which is a crucial element of the plaques found in the brains of individuals with AD [89].

CRISPR/Cas9 was used to permanently reduce mutant Huntingtin and its aggregates in the striatum of HD140Q-knock-in mice [90]. In patient-derived iPSCs, the CAG repeat in the ATXN3 gene was effectively deleted using CRISPR/Cas9 technology to treat spinocerebellar ataxia [91].

Since transgenic products and delivery vectors might be hazardous, safety and tolerability in patients are given priority throughout drug development. AAV injection in the human CNS is risk-free, although transgenic products and immune responses might cause adverse effects, as in the case of RNAi-based gene therapy. Safety and tolerability can be improved by predicting unfavourable side effects utilizing resources like iPSC technology, animal models, and bioinformatics.

### 3.5 Immunotherapies for Neurodegenerative Disease

Utilizing the immune system to fight against misfolded proteins, immunotherapy offers the potential to control the course of illness. Neurodegenerative disorders often cause chronic inflammation, leading to collateral damage and permanent brain atrophy. This damage is caused by cytotoxic factors released by astrocytes, microglia, and other immune cells leading to neurodegeneration [92]. As the blood-brain barrier's gatekeepers, astrocytes are linked to the pathological breakdown of the BBB that takes place throughout progressive neurodegeneration. It is believed that the BBB disruption makes it easier for immune cells and blood components to enter, which might aggravate CNS neuroinflammation and neurodegeneration [93]. According to several researches, astrocytic atrophy and degeneration play a significant role in the early stages of neurodegenerative disorders, which may have effects on the loss of neurons and the course of the illness [94].

The brain is now understood to be an immune-privileged location, with endogenous CNS cell types harbouring specialized pattern recognition receptors (PRRs) that cause an innate immune response when certain host-derived chemicals known as danger/damage-associated molecular patterns (DAMPs) are present [95]. These DAMPs have the ability to cause neuroinflammation, which transforms immunocompetent cells from their advantageous functions to a persistently reactive state, accelerating neurodegeneration [96]. With an emphasis on avoiding neuronal death or offering trophic support to populations that are degenerating, this viewpoint has altered how we study and treat neurodegenerative illnesses. Another finding commonly seen is exaggerated pro-inflammatory cytokine and chemokine production in the CNS, including overexpression of IL-1β and TNF, which is another characteristic of neurodegenerative disorders [97, 98]. Another key component of the innate immune system that can have a harmful role in progressive neurodegenerative disorders is the complement system [99].

The majority of research publications outlining immunotherapies for neurodegenerative illnesses have traditionally concentrated on autoimmune neurodegenerative diseases like multiple sclerosis [100]. The effectiveness of immunotherapeutic treatments for aging-related and other non-autoimmune neurodegenerative illnesses was poorly understood until recently. Immunotherapy development is being evaluated for passive and active vaccines, including monoclonal antibodies and adaptive antigens.

## 3.6 Neuroinflammation—Targeting Therapeutic Approaches in the Treatment of Neurodegenerative Diseases

### 3.6.1 TNF α

Targeting the inflammatory mediators responsible for neuronal malfunction and death is a key therapeutic technique. For neurodegenerative diseases such as Alzheimer's and Parkinson's, inhibition of soluble TNF-α via small-molecule inhibitors or viral overexpression has proved beneficial [101, 102]. Etanercept, a TNF inhibitor, did not, however, enhance cognition or behaviour in AD dementia patients [103].

### 3.6.2 NLRP3 and Interleukins

Chronic IL-1 production in the central nervous system might cause neuronal vulnerability and result in midbrain dopamine (DA) neuron degradation in the substantia nigra. In transgenic AD models, inhibition of NLRP3 (Nod-like receptor protein-3) driven IL-1 secretion is protective [104]. It has been discovered that endogenous ketone bodies can suppress the NLRP3 inflammasome, indicating that CNS-permeable substances like β-hydroxybutyrate may be useful in treating neuroinflammatory illnesses [105]. In experimental AD models, neutralizing antibodies like IL-12 and IL-23 have been linked to therapeutic advantages [106]. Psoriasis and other neurological disorders may benefit from the use of IL-12 neutralizing medications that have been licenced by the US FDA [107]. In models of AD, ALS, and HD, inhibition of complement receptor C5a reduces disease pathogenesis [108]. Recent studies examine the anti-inflammatory and immunomodulatory functions of molecules such as glucagon-like peptide-1, peroxisome proliferator-activated receptor gamma (PPAR-γ), and granulocyte-macrophage colony-stimulating factor (GM-CSF).

### 3.6.3 GM-CSF

GM-CSF, a cytokine and immunomodulatory growth factor, exhibits deregulation in neurodegenerative disorders. It regulates inflammation and microglial homeostasis, as well as the induction of regulatory T ($T_{reg}$) cells and microglial proliferation [109–111]. With regard to AD, GM-CSF has demonstrated neuroprotective potential by reducing neuroinflammation and cognitive deterioration. In clinical studies, Sargamostim, a synthetic recombinant version of GM-CSF, significantly improved groups' cognitive and memory functions. Sargamostim is well-tolerated by patients and safe [112], and it has shown promise in reducing PD-related motor symptoms and protecting nigrostriatal dopaminergic neurons through $T_{reg}$ induction [113].

According to preclinical research, GM-CSF possesses strong anti-inflammatory properties and increases the longevity of transgenic mice. It is neuroprotective [114, 115]. However, due to ambiguous illness processes, clinical trials have shown contradictory outcomes. A powerful anti-inflammatory medication, GM-CSF has been shown to be effective in treating ALS by upregulating cytokines like IL-10 and downregulating pro-inflammatory cytokines including TNF, monocyte chemoattractant protein-1, interferon, IL-7, and IL-17 [116]. Additionally, CSF could be involved in reducing motor symptoms [116]. However, other researchers claim that clinical results connected to ALS have not improved [117].

### 3.6.4 GLP-1 and PPAR-γ

It has been discovered that GLP-1 and PPAR-γ agonists have neuroprotective and anti-inflammatory properties, reducing the symptoms of AD and PD and restoring Akt and m-TOR signalling [32, 118, 119]. However, recent findings from AD and PD clinical trials using GLP-1 agonist medications have indicated very modest, if any, improvements in disease pathogenesis [120, 121]. In pre-clinical studies using AD animal models, PPAR-γ agonists have demonstrated the ability to decrease neuroinflammation, A-42 burden, phosphorylated tau, and synaptophysin [121]. Furthermore, two promising PPAR-γ agonists are currently under phase II clinical trials, one of which, T3D-959, enhanced cognition and insulin metabolism in AD patients [122], whereas the other one, Pioglitazone, decreased neuroinflammation and microglial proliferation in PD models [123, 124].

### 3.6.5 Targeting PRRs-DAMPs Interaction

As DAMPs and misfolded proteins interact with PRRs to cause neuroinflammation, PRRs like RAGE, Mac1, and TLRs are promising therapeutic targets. Reactive gliosis and persistent neuroinflammation might be reduced by targeting these PRRs. RAGE inhibitors like PF-04494700 had positive preclinical findings but results from clinical trials were inconclusive [125], whereas NLRP3 inhibitors like MCC950 showed promising results in transgenic AD models [104].

### 3.6.6 Vaccine Therapy

Vaccine therapy focuses on passive and active immunization. Active immunization uses specific antigens to induce adaptive immune responses and antibody production, whereas the use of passive immunization through monoclonal antibodies provides a decrease in the target molecules while ensuring effective dosage management; however, this method is costly, necessitates regular administration, and is associated with adverse side effects [126]. A summary of all the active and passive immunization therapies for different neurodegenerative diseases is given in Figs. 5 and 6. The current state of vaccination treatment for neurodegenerative illnesses is outlined below.

### 3.7 Immunization for Synucleinopathies

**Passive immunization** The purpose of therapies for synucleinopathies is to lessen α-synuclein buildup and cell-to-cell transfer. These therapies can be paired with medications that reduce neuroinflammation, expecting synergistic results.

(a) *Trials in Transgenic mice*—Monoclonal antibodies are used in passive immunization against synucleinopathies to target certain areas of α-syn. The antibody that targets the α-syn N-terminal has the strongest efficacy in decreasing α-syn-induced nigral cell death and activated microglia in the substantia nigra [127]. The intracellular oligomerization of α-syn in vitro and spinal cord α-syn protofibril levels in vivo have been reduced by the development of a novel monoclonal antibody, mAb47, directed against protofibril structures of α-syn [128]. A537 transgenic PD mouse model was immunized using a novel passive immunization method called NAbs-syn, which was created from intravenous immune

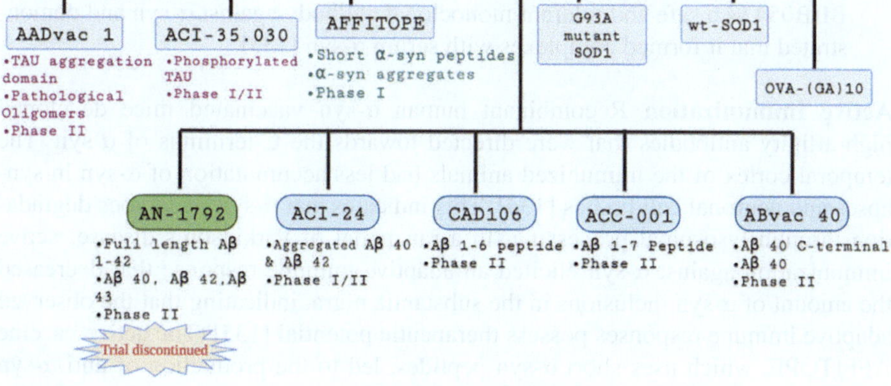


**Fig. 5** Active immunization therapies in clinical trials for different diseases

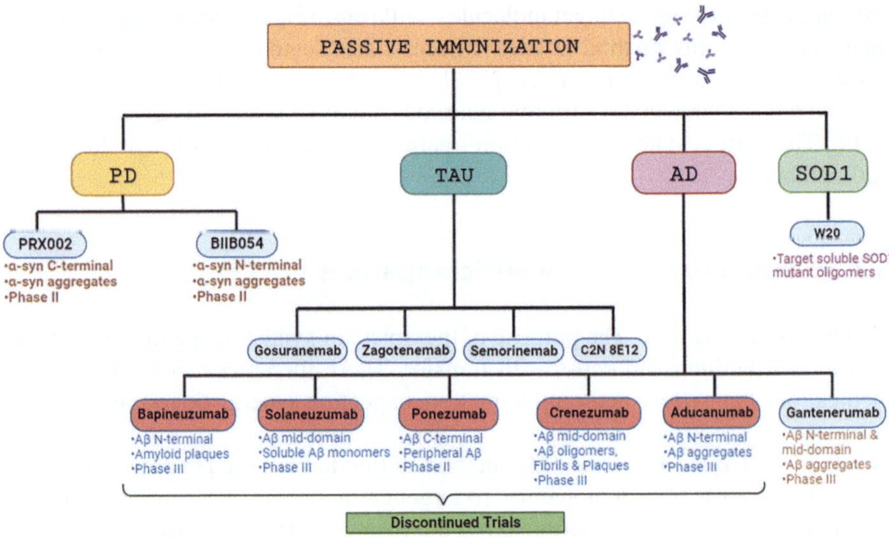


**Fig. 6** Passive immunization therapies in clinical trials for different diseases

globulin (IVIG) of healthy persons. PD-related cognitive, motor, and neuroinflammation impairments are lessened by NAbs-syn, which also decreases soluble α-syn, oligomers, and intracellular phosphorylated α-syn deposits [129].

(b) *Clinical Trials*—Phase II clinical research is being conducted on the C terminus targeting monoclonal antibody prasinezumab (PRX002). Free serum α-syn is markedly decreased by PRX002 but not free CSFα-syn [130]. According to preclinical data, internalization of the antibody-α-syn aggregation complex occurs via Fc-γ-receptors on microglial cells, which lowers the concentration of α-syn. PRX002 is well-tolerated, secure, and does not result in host immunogenicity [130, 131].

BIIB054 is a monoclonal antibody that selectively targets α-syn aggregates and is now in phase II clinical testing [132]. The current work established BIIB054 as a safe and tolerant monoclonal antibody against α-syn and demonstrated that it formed complexes with serum α-syn [133].

**Active Immunization** Recombinant human α-syn vaccinated mice developed high-affinity antibodies that were directed towards the C-terminus of α-syn. The temporal cortex of the immunized animals had less accumulation of α-syn in synapses and neuronal cell bodies [134]. This indicates that α-syn undergoes degradation through lysosomal processing. In a rat model of Parkinson's disease, active immunization against α-syn elicited an adaptive immune response that decreased the amount of α-syn inclusions in the substantia nigra, indicating that the observed adaptive immune responses possess therapeutic potential [135]. The active vaccine AFFITOPE, which uses short α-syn peptides, led to the production of anti-α-syn antibodies that entered the CSF and plasma, targeted α-syn aggregates, and inhibited neuronal degeneration [136, 137]. Clinical studies for AFFITOPE are now in phase I and progressing well [137].

## 3.8 Immunization for Amyloidoses

**Passive Immunization** The most advanced immunotherapy currently in development is passive immunization for AD.

(a) *Trials in Transgenic Mice*—In transgenic mice, the anti-Aβ antibody that targets the Aβ 31–35 sequence produced favourable immunological outcomes and decreased disease pathology. It was a viable option for passive immunization therapy for AD since it reduced Aβ-42 toxicity, stopped Aβ-42-induced cell death, restored hippocampus synaptic plasticity, and saved spatial learning and memory [138].

(b) *Clinical Trials*—Bapineuzumab, a monoclonal antibody targeting Aβ N-terminus [139], was initially studied for AD treatment. However, studies found severe adverse events and no cognitive decline enhancement, making it not recommended for AD treatment. Clinical trials have since been discontinued [140].

Solanezumab, a monoclonal antibody targeting Aβ peptide mid domain, and selectively binds to monomeric, soluble Aβ failed to reduce Aβ burden and improve cognitive decline in mild AD patients, leading to its termination due to less promising results [141].

Gantenerumab is a monoclonal antibody targeting Aβ in the brain and showing high affinity for aggregated amyloid β species. Clinical trials indicate that it lowers Aβ plaque levels in the brain to below the threshold considered healthy [142] and improves the mental status, particularly in early stages of Alzheimer's disease [142, 143].

Crenezumab is an IgG4 monoclonal antibody targeting Aβ's mid-domain, limiting aggregation and disaggregation. Despite tolerability and Aβ plaque reduction [144], its efficacy remains questionable [145]. A phase III investigation for familial AD patients [146] concluded that achieving the primary endpoints was improbable and therefore clinical trials have been terminated.

Aducanumab, a monoclonal antibody, is now undergoing phase III testing. It binds to the N-terminal of both soluble and insoluble Aβ species [147] and causes few side effects from the medication. It has been demonstrated to penetrate specific areas of the brain, lessen Aβ plaques, and halt disease development [148]. Trials have been terminated due to futility of analysis.

A monoclonal antibody called ponezumab that targets soluble amyloid beta may be able to lower levels of CNS amyloid beta species by removing free blood Aβ [149]. With relatively few side effects, Landen et al. evaluated the tolerance and efficacy of ponezumanb. It had little impact on illness severity or cognitive function, though [150]. Notably, ponezumab increased the amount of free Aβ species in the circulation without entering the CSF. On the basis of the aforementioned, ponezumab clinical trials have been discontinued.

**Active Immunization** AN-1792, the first human Alzheimer's disease vaccine, cleared Aβ40, Aβ42, and Aβ43 plaques, but caused severe side effects [151], leading to its termination.

ACI-24 is a liposome-based vaccine targeting aggregated Aβ peptides. Preclinical studies show high anti-Aβ antibody titres in AD transgenic mice [152], and promising phase I results [153]. ACI-24 is presently undergoing phase II clinical research to assess its immunogenicity, safety, and tolerability.

CAD106 is an anti-Aβ vaccine derived from multiple copies of the Aβ1–6 peptides which is found well-tolerated and safe in a phase II study. It effectively targets and clears Aβ species without causing meningoencephalitis, autoimmune disease, or CNS inflammation [154].

ACC-001, a phase II vaccine with N-terminal Aβ1–7 peptide, is well-tolerated by patients and induces a healthy immune response mediated by anti-Aβ IgG [155].

ABvac40 is a vaccine formulated from Aβ40 C-terminus fragments, which generates specific anti-Aβ40 antibodies in patients with mild to moderate AD. Preliminary assessments of tolerability and safety yielded positive outcomes, showing no signs of vasogenic oedema or microhaemorrhage. Given the encouraging results from phase I, a phase II clinical trial is set to be conducted [156].

## 3.9 Immunization for Tauopathies

### Passive Immunization

(a) *Trials in Transgenic Mice*—To enhance the therapeutic potential of CBTAU-22.1, which is a naturally occurring tau-specific human antibody, Van Ameijde et al. improved its affinity through random mutagenesis, creating dmCBTAU-22.1, which lessened tau aggregation, seeding, and spread in vitro and in vivo in P301L transgenic mice [157].

In triple transgenic (3xTG) mice, the mouse monoclonal antibody 43D, which targets the N-terminal projection of tau, dramatically decreased total tau, p-tau, p-tau seeding and propagation, and attenuated amyloid beta [158]. Antibody 77E9 also attenuated disease pathology, although the combination use of 77E9 and 43D was less successful than using either antibody alone. In 3xTG mice, both antibodies reduced short-term and spatial memory deficits [159]. In passive immunization, tau seeding and propagation were decreased by a third antibody, Antibody D, which targets the core region of tau protein [160].

Monoclonal antibodies that target phosphorylated tau are a promising therapy for Alzheimer's disease, although they are only used on test animals at this time. RZ3, CP13, and PG5 are the three anti-phosphorylated tau antibodies that D'Abramo investigated for using as antibody treatments for AD in JNLP3 transgenic mice. In transgenic mice, CP13, which targets phosphorylated tau, effectively decreased insoluble or soluble tau species [161].

In the A53T transgenic mice model of PD, Gerson et al. discovered that tau oligomer-specific monoclonal antibodies (TOMA) greatly decreased toxic tau oligomers, fibril-like α-synuclein oligomers, and Lewy Body structures. Mice immunized with TOMA were shielded from PD-related motor and cognitive impairments [162]. Additionally, it had encouraging outcomes in a number of AD mice models, lowering memory impairments and cognitive deterioration while also delaying the onset of the illness [163].

(b) *Clinical Trials*—Monoclonal antibodies targeting tau protein are transitioning from animal testing to clinical trials. From these, C2N 8E12, gosuranemab, zagotenemab, and semorinemab show promising preliminary safe and tolerable results. Additional studies are required to assess their effectiveness and safety [164].

**Active Immunization**

(a) *Trials in Transgenic Mice*—Phosphorylated tau peptide, Tau379-408, was tested as an immunogen in 3xTG mice, generating anti-tau antibodies and reducing pathological tau aggregates, clearance of soluble and insoluble tau species, and reduction of Aβ deposition [165]. The study suggests potential use in active immunization against AD pathology.

(b) *Clinical Trials*—AADvac1 is a new tau active immunization vaccine that recognizes pathological tau oligomers. It reduces neurofibrillary pathology in transgenic rats' brains [100]. AADvac1 triggered anti-tau antibody production, reducing cognitive decline in patients and causing no adverse effects. A phase II clinical trial is underway to assess its efficacy [166].

The liposomal-based vaccination ACI-35, which targets phosphorylated tau species, has favourable immunogenic activity in test animals without causing any negative neurological consequences [167]. ACI-35.030, a vaccine that is 5 times more immunogenic than ACI-35, will be used in lieu of ACI-35 in the phase I/II safety and tolerability clinical study [168] since results from the phase I trial showed that the ACI-35 therapy is safe and tolerated but was not effectively immunogenic.

## 3.10 Immunization for Superoxide Dismutase 1

**Passive Immunization** The passive immunization of SOD1-G93A mice using the anti-SOD1 antibody (W20), which specifically targets harmful soluble SOD1 mutant oligomers, offers protection to motor neurons against apoptosis, prolongs survival, diminishes the presence of neurotoxic SOD1 oligomer aggregates, and suppresses gliosis and neuroinflammation in transgenic mice, thereby enhancing motor neuron survival and motor performance [169].

**Active Immunization** 20% of familial ALS cases are caused by mutated SOD1 [170], which may have an impact on cellular metabolism [171], oxidative stress pathways [172], and metal dyshomeostasis [173]. The efficacy of a vaccination that targets extracellular SOD1 in protecting transgenic mice from ALS was examined by Takeuchi et al. [174]. Both the mutant SOD1 (G93A mutant SOD1) and wild-type SOD1 (wt-SOD1) vaccines alleviate symptoms and lessen the pathogenicity of ALS. Additionally, the wt-SOD1 vaccination increased lifespan, postponed the beginning of illness, and both the wt-SOD1 and mutant SOD1 vaccines minimized anterior horn motor neuron loss [174]. In G37R SOD1 and G93A SOD1 transgenic mice, a vaccination based on mutant SOD1 prolonged life and delayed the onset of illness without overexpressing the mutant SOD1 phenotype [175].

## 3.11 Immunization for C9orf72

**Passive Immunization** Mutations in the C9orf72 gene, specifically an expansion of a hexanucleotide repeat (GGGGCC), have been associated with amyotrophic lateral sclerosis and frontotemporal dementia which results in the production of dipeptide repeat protein aggregates (DPR) like poly GA, poly GP, poly GR, poly PA, and poly PR [176]. Poly GA is an abundant and cytotoxic DPR in brain and spinal cord neuronal inclusions in ALS patients [177]. It can be transferred across neurons, serves as a nucleation seed, and improves G4C2 repeat expression. Studies have shown that poly GA can aid in the spread of disease pathology in affected individuals [177]. Poly GA has been investigated by Zhou et al. as a potential immunotherapy target for ALS.

Anti-GA RAN antibodies have been investigated as potential immunotherapies for ALS and FTD. These antibodies have successfully restored proteasome function, decreased pathogenic poly-GA aggregates, and reduced neuroinflammation and neurodegeneration in the anterior and posterior horns of the lumbar spine [178]. Therefore, clinical trials should be conducted to further evaluate anti-GA RAN antibodies as potential treatments for ALS and FTD [178].

**Active Immunization** OVA-(GA)10, a poly-GA and ovalbumin combination, may be used as an active immunization agent, according to a recent study. In transgenic mice, OVA-(GA)10 induced a potent immune response, elevated anti-GA antibody titres, and decreased motor impairments and neurodegeneration [179]. Additionally, it reduced neuroinflammation, normalized macrophages and microglia, and reversed TDP-43's spinal cord cytoplasmic mislocalization. These results support the potential use of OVA-(GA)10 as a candidate vaccination to lessen pathology and disease severity in ALS and FTD [179].

### 3.12 Immunization for Mutant Huntingtin Protein

**Active Immunization** There are still conflicting reports on active and passive immunization against mutant Huntingtin proteins. According to preliminary findings, a research done on HDR6/2 mice using a candidate plasmid vaccination showed that it had no effect on the quantity of m-Htt aggregates [180]. Another combination of peptide, protein, and DNA plasmid vaccines produced substantial antibody production following immunization [181]; however, this study did not examine the effectiveness of the vaccine in lowering m-Htt aggregates [181].

## 4 Conclusion

New research backs the use of immunotherapeutic drugs in the treatment of neurodegenerative diseases. To stop neuroinflammation's evolution and stop the onset of illness, it is essential to comprehend the pathophysiology of the condition and its function in neurodegeneration. The treatment of neurodegenerative diseases may be transformed by advanced immunotherapeutic techniques, such as passive and active medicines that target distinctive disease biomarkers. The use of vaccination therapy is an intriguing strategy since it allows for the customization of stronger vaccinations that can develop targeted immune responses against protein accumulation. The type of immunogenic particle, method of administration, and antigenic presentation of the immunogen by the vaccine should be the main topics of future research. Another crucial factor for improving efficacy is the way the vaccination is administered. Passive immunization using monoclonal antibodies has received attention recently and may be a safer method than active immunization. Understanding the stages of neuroinflammation provides information on possible therapies for neurodegeneration. The innate immune system can be modified, but there are also cutting-edge techniques that might be crucial treatment choices, such as microglial control by the gut microbiome and dysregulation in sleep patterns. Neuroinflammatory responses may be controlled by systemic immune responses and lymphatic arteries in the brain that are related to deep cervical lymph nodes. Early intervention is essential in the fight against these disorders. More efficient treatments for these now incurable neurodegenerative illnesses will be created as our understanding of the specific underlying immunological systems increases. Although there has been improvement in our understanding of innate immune responses in the central nervous system throughout healthy ageing and neurodegenerative disorders, there are still some information gaps.

Treatment modalities aimed at gene replacement or augmentation in neurodegenerative brain disorders are continually advancing as progress is achieved in comprehending essential elements of the underlying molecular pathophysiology. The neurosurgical profession is now prepared to maximize the potential translation of these tactics for clinical success thanks to the development of iMRI-CED, which enables reliable therapeutic vector administration with real-time verification.

Another important factor is safety. Considering the irreversible nature of modern gene treatments, it is vital to assess both short- and long-term damage. Prediction of on- and off-target effects, immunogenicity, dosage determination, timing of therapy, and the capacity to modify gene expression are other key factors that must be taken into consideration during preclinical investigations. The improvement of preclinical models will be made easier with a better understanding of the mechanisms that cause neurodegenerative disorders.

The present progress in biochemical, pathological, and pharmacological research may guarantee a better future for neuronal health globally, but to prevent the formation of such NDDs, a healthy lifestyle must be adopted from an early age. Improving therapeutic treatment choices is essential since the WHO predicts that the number of individuals with dementia will quadruple in the next decades, reaching 152 million in 2050.

## References

1. Bak, T. H., & Chandran, S. (2012). What wires together dies together: Verbs, actions and neurodegeneration in motor neuron disease. *Cortex, 48*, 936–944.
2. Przedborski, S., Vila, M., & Jackson-Lewis, V. (2003). Neurodegeneration: What is it and where are we? *The Journal of Clinical Investigation, 111*, 3–10.
3. Sudhakar, V., & Richardson, R. M. (2019). Gene therapy for neurodegenerative diseases. *Neurotherapeutics, 16*, 166–175.
4. Prince, M., Albanese, E., Guerchet, M., & Prina, M. (2014). *World Alzheimer report 2014: Dementia and risk reduction: An analysis of protective and modifiable factors*. Alzheimer's Disease International.
5. Armstrong, R. (2020). What causes neurodegenerative disease? *Folia Neuropathologica, 58*(2), 93–112.
6. WHO. (2020). *Chapter 4*. Available from: https://www.who.int/mental_health/neurology/chapter_4_neuro_disorders_public_h_challenges.pdf?ua=1
7. Poddar, M. K., Chakraborty, A., & Banerjee, S. (2021). Neurodegeneration: Diagnosis, prevention, and therapy. In *Oxidoreductase*. https://doi.org/10.5772/intechopen.94950
8. Pihlstrøm, L., Wiethoff, S., & Houlden, H. (2017). Genetics of neurodegenerative diseases: An overview. *Handbook of Clinical Neurology, 145*, 309–323.
9. Hussain, R., Zubair, H., Pursell, S., & Shahab, M. (2018). Neurodegenerative diseases: Regenerative mechanisms and novel therapeutic approaches. *Brain Sciences, 8*, 177. https://doi.org/10.3390/brainsci8090177
10. Hatcher, J. M., Pennell, K. D., & Miller, G. W. (2008). Parkinson's disease and pesticides: A toxicological perspective. *Trends in Pharmacological Sciences, 29*(6), 322–329.
11. Dickens, A. P., Lang, I. A., Langa, K. M., Kos, K., & Llewellyn, D. J. (2011). Vitamin D, cognitive dysfunction and dementia in older adults. *CNS Drugs, 25*(8), 629–639.
12. Brenowitz, W. D., & Yaffe, K. (2022). Observational studies in Alzheimer disease: Bridging preclinical studies and clinical trials. *Nature Reviews Neurology, 18*(12), 747–757.
13. Tanzi, R. E. (2012). The genetics of Alzheimer disease. *Cold Spring Harbor Perspectives in Medicine, 2*(10), a006296.
14. Rasool, M., et al. (2014). Recent updates in the treatment of neurodegenerative disorders using natural compounds. *Evidence-Based Complementary and Alternative Medicine*, 97973014. https://doi.org/10.1155/2014/979730
15. Mizuno, Y. (2014). Recent research progress in and future perspective on treatment of Parkinson's disease. *Integrative Medicine International, 1*, 67–79.

16. Crane, P. K., & Doody, R. S. (2009). Donepezil treatment of patients with MCI: A 48-week randomized, placebo-controlled trial. *Neurology, 73*, 1514–1515.
17. Desai, A. K., & Grossberg, G. T. (2005). Diagnosis and treatment of Alzheimer's disease. *Neurology, 64*(suppl 3), S34–S39.
18. Chaudhuri, K. R., & Schapira, A. H. V. (2009). Non-motor symptoms of Parkinson's disease: Dopaminergic pathophysiology and treatment. *Lancet Neurology, 8*, 464–474.
19. Tizabi, Y., Hurley, L. L., Qualls, Z., & Akinfiresoye, L. (2014). Relevance of the anti-inflammatory properties of curcumin in neurodegenerative diseases and depression. *Molecules, 19*, 20864–20879.
20. Okun, M. S. (2014). Deep-brain stimulation – Entering the era of human neural-network modulation. *The New England Journal of Medicine, 371*, 1369–1373.
21. Armstrong, M. J., & Miyasaki, J. M. (2012). American Academy of Neurology: Evidence-based guideline: Pharmacologic treatment of chorea in Huntington disease: Report of the guideline development subcommittee of the American Academy of Neurology. *Neurology, 79*, 597–603.
22. Stewart, W. F., Kawas, C., Corrada, M., & Metter, E. J. (1997). Risk of Alzheimer's disease and duration of NSAID use. *Neurology, 48*, 626–632.
23. Schwarzschild, M. A., Xu, K., Oztas, E., & Petzer, J. P. (2003). Neuroprotection by caffeine and more specific A2A receptor antagonists in animal models of Parkinson's disease. *Neurology, 61*(suppl 6), S55–S61.
24. Cruz, M. P. (2018). Edaravone (Radicava): A novel neuroprotective agent for the treatment of amyotrophic lateral sclerosis. *Pharmacy and Therapeutics, 43*, 25–28.
25. Pan, W., & Zhou, H. (2014). Inclusion of integrative medicine in clinical practice. *Integrative Medicine International, 1*, 1–4.
26. Pan, W., & Zhou, H. (2013). Integrative medicine: A paradigm shift in clinical practice. *International Journal of Integrative Medicine, 1*, 21.
27. Chen, X., & Pan, W. (2015). The treatment strategies for neurodegenerative diseases by integrative medicine. *Integrative Medicine International, 1*(4), 223–225.
28. Jaiswal, M. (2017). Therapeutic opportunities and challenges of induced pluripotent stem cells-derived motor neurons for treatment of amyotrophic lateral sclerosis and motor neuron disease. *Neural Regeneration Research, 12*, 723–736.
29. Abramov, A. Y., Berezhnov, A. V., Fedotova, E. I., Zinchenko, V. P., & Dolgacheva, L. P. (2017). Interaction of misfolded proteins and mitochondria in neurodegenerative disorders. *Biochemical Society Transactions, 45*(4), 1025–1033.
30. Durães, F., Pinto, M., & Sousa, E. (2018). Old drugs as new treatments for neurodegenerative diseases. *Pharmaceuticals (Basel), 11*(2), 44.
31. Poovaiah, N., Davoudi, Z., Peng, H., Schlichtmann, B., Mallapragada, S., Narasimhan, B., & Wang, Q. (2018). Treatment of neurodegenerative disorders through the blood–brain barrier using nanocarriers. *Nanoscale, 10*(36), 16962.
32. Grieco, M., et al. (2019). Glucagon-like Peptide-1: A focus on neurodegenerative diseases. *Frontiers in Neuroscience, 13*, 1112.
33. Bento-Pereira, C., & Dinkova-Kostova, A. T. (2020). Activation of transcription factor Nrf2 to counteract mitochondrial dysfunction in Parkinson's disease. *Medical Research Reviews*. https://doi.org/10.1002/med.21714
34. Angelova, P. R., Esteras, N., & Abramov, A. Y. (2020). Mitochondria and lipid peroxidation in the mechanism of neurodegeneration: Finding ways for prevention. *Medical Research Reviews*. https://doi.org/10.1002/med.21712
35. Shevtsova, E. F., Maltsev, A. A., Vinogradova, D. V., Shevtsov, P. N., & Bachurin, S. O. (2020). Mitochondria as a promising target for developing novel agents for treating Alzheimer's disease. *Medical Research Reviews*. https://doi.org/10.1002/med.21715
36. Albertini, C., Salerno, A., de Sena Murteira Pinheiro, P., & Bolognesi, M. L. (2020). From combinations to multitarget-directed ligands: A continuum in Alzheimer's disease polypharmacology. *Medical Research Reviews*. https://doi.org/10.1002/med.21699

37. Dergunova, L. V., Filippenkov, I. B., Limborska, S. A., & Myasoedov, N. F. (2020). Pharmacotranscriptomics of peptide drugs with neuroprotective properties. *Medical Research Reviews*. https://doi.org/10.1002/med.21704
38. Overcoming gaps in the treatment of neurodegenerative disease. (2020). *EBioMedicine, 60*, https://doi.org/10.1016/j.ebiom.2020.103088
39. Zakharova, M. (2020). Modern approaches in gene therapy of motor neuron diseases. *Medical Research Reviews*. https://doi.org/10.1002/med.21705
40. Dow, L. E., Fisher, J., O'Rourke, K. P., Muley, A., Kastenhuber, E. R., Livshits, G., et al. (2015). Inducible in vivo genome editing with CRISPR-Cas9. *Nature Biotechnology, 33*, 390–394.
41. Saraiva, J., Nobre, R. J., & Pereira de Almeida, L. (2016). Gene therapy for the CNS using AAVs: The impact of systemic delivery by AAV9. *Journal of Controlled Release, 241*, 94–109.
42. O'Connor, D. M., & Boulis, N. M. (2015). Gene therapy for neurodegenerative diseases. *Trends in Molecular Medicine, 21*(8), 504–512.
43. Nayerossadat, N., Ali, P., & Maedeh, T. (2012). Viral and nonviral delivery systems for gene delivery. *Advanced Biomedical Research, 1*, 27. https://doi.org/10.4103/2277-9175.98152
44. Lundstrom, K. (2018). Viral vectors in gene therapy. *Diseases, 6*, 42. https://doi.org/10.3390/diseases6020042
45. Palfi, S., Gurruchaga, J. M., Ralph, G. S., et al. (2014). Long-term safety and tolerability of ProSavin, a lentiviral vector-based gene therapy for Parkinson's disease: A dose escalation, open-label, phase 1/2 trial. *Lancet, 383*, 1138–1146.
46. Samaranch, L., Blits, B., San Sebastian, W., Hadaczek, P., Bringas, J., Sudhakar, V., Macayan, M., Pivirotto, P. J., Petry, H., & Bankiewicz, K. S. (2017). MR-guided parenchymal delivery of adeno-associated viral vector serotype 5 in non-human primate brain. *Gene Therapy, 24*(4), 253–261.
47. Green, F., Samaranch, L., Zhang, H. S., Manning-Bog, A., Meyer, K., Forsayeth, J., & Bankiewicz, K. S. (2016). Axonal transport of AAV9 in nonhuman primate brain. *Gene Therapy, 23*(6), 520–526.
48. Muramatsu, S., Fujimoto, K., Kato, S., et al. (2010). A phase I study of aromatic L-amino acid decarboxylase gene therapy for Parkinson's disease. *Molecular Therapy, 18*, 1731–1735.
49. Rafii, M. S., Tuszynski, M. H., Thomas, R. G., et al. (2018). Adeno-associated viral vector (serotype 2)-nerve growth factor for patients with Alzheimer disease: A randomized clinical trial. *JAMA Neurology, 75*, 834–841.
50. Marks, W. J., Bartus, R. T., Siffert, J., et al. (2010). Gene delivery of AAV2-neurturin for Parkinson's disease: A double-blind, randomised, controlled trial. *The Lancet Neurology, 9*, 1164–1172.
51. LeWitt, P. A., Rezai, A. R., Leehey, M. A., et al. (2011). AAV2-GAD gene therapy for advanced Parkinson's disease: A double-blind, sham-surgery controlled, randomised trial. *Lancet Neurology, 10*, 309–319.
52. Duque, S., et al. (2009). Intravenous administration of selfcomplementary AAV9 enables transgene delivery to adult motor neurons. *Molecular Therapy, 17*, 1187–1196.
53. Leone, P., et al. (2012). Long-term follow-up after gene therapy for canavan disease. *Science Translational Medicine, 4*, 165ra163.
54. Bartus, R. T., et al. (2014). Parkinson's disease gene therapy: Success by design meets failure by efficacy. *Molecular Therapy, 22*, 487–497.
55. Levy, R., Lang, A. E., Dostrovsky, J. O., et al. (2001). Lidocaine and muscimol microinjections in subthalamic nucleus reverse Parkinsonian symptoms. *Brain, 124*, 2105–2118.
56. Hadaczek, P., Eberling, J. L., Pivirotto, P., Bringas, J., Forsayeth, J., & Bankiewicz, K. S. (2010). Eight years of clinical improvement in MPTP-lesioned primates after gene therapy with AAV2-hAADC. *Molecular Therapy, 18*, 1458–1461.
57. San Sebastian, W., Richardson, R. M., Kells, A. P., et al. (2012). Safety and tolerability of magnetic resonance imaging-guided convection-enhanced delivery of AAV2-hAADC with a novel delivery platform in nonhuman primate striatum. *Human Gene Therapy, 23*, 210–217.

58. Kordower, J. H., Herzog, C. D., Dass, B., et al. (2006). Delivery of neurturin by AAV2 (CERE-120)-mediated gene transfer provides structural and functional neuroprotection and neurorestoration in MPTP-treated monkeys. *Annals of Neurology, 60*, 706–715.
59. Cederfjall, E., et al. (2012). Design of a single AAV vector for coexpression of TH and GCH1 to establish continuous DOPA synthesis in a rat model of Parkinson's disease. *Molecular Therapy, 20*, 1315–1326.
60. Cederfjall, E., et al. (2013). Continuous DOPA synthesis from a single AAV: Dosing and efficacy in models of Parkinson's disease. *Scientific Reports, 3*, 2157.
61. Ren, X., et al. (2013). AAV2-mediated striatum delivery of human CDNF prevents the deterioration of midbrain dopamine neurons in a 6-hydroxydopamine induced parkinsonian rat model. *Experimental Neurology, 248*, 148–156.
62. Back, S., et al. (2013). Gene therapy with AAV2-CDNF provides functional benefits in a rat model of Parkinson's disease. *Brain and Behavior, 3*, 75–88.
63. Hovland, D. N., et al. (2007). Six-month continuous intraputamenal infusion toxicity study of recombinant methionyl human glial cell linederived neurotrophic factor (r-metHuGDNF) in rhesus monkeys. *Toxicologic Pathology, 35*, 676–692.
64. Manfredsson, F. P., et al. (2009). Tight long-term dynamic doxycycline responsive nigrostriatal GDNF using a single rAAV vector. *Molecular Therapy, 17*, 1857–1867.
65. Quintino, L., et al. (2013). Functional neuroprotection and efficient regulation of GDNF using destabilizing domains in a rodent model of Parkinson's disease. *Molecular Therapy, 21*, 2169–2180.
66. Iwamoto, M., et al. (2010). A general chemical method to regulate protein stability in the mammalian central nervous system. *Chemistry & Biology, 17*, 981–988.
67. Piguet, F., Alves, S., & Cartier, N. (2017). Clinical gene therapy for neurodegenerative diseases: Past, present, and future. *Human Gene Therapy, 28*, 988–1003.
68. Platt, F. M. (2018). Emptying the stores: Lysosomal diseases and therapeutic strategies. *Nature Reviews Drug Discovery, 17*, 133–150.
69. Vincent, F., Adamsbaum, C., Hocquemiller, M., Crystal, R. G., Zérah, M., Fraldi, A., et al. (2014). Intracerebral administration of adeno-associated viral vector serotype rh.*10* carrying human SGSH and SUMF1 cDNAs in children with Mucopolysaccharidosis Type IIIA disease: Results of a phase I/II trial. *Human Gene Therapy, 25*, 506–516.
70. Fu, H., Dirosario, J., Killedar, S., Zaraspe, K., & McCarty, D. M. (2011). Correction of neurological disease of mucopolysaccharidosis IIIB in adult mice by rAAV9 trans-blood-brain barrier gene delivery. *Molecular Therapy, 19*, 1025–1033.
71. Corti, M., Cleaver, B., Clément, N., Conlon, T. J., Faris, K. J., Wang, G., et al. (2015). Evaluation of readministration of a recombinant adeno-associated virus vector expressing acid alpha-glucosidase in pompe disease: Preclinical to clinical planning. *Human Gene Therapy Clinical Development, 26*, 185–193.
72. Puzzo, F., Colella, P., Biferi, M. G., Bali, D., Paulk, N. K., Vidal, P., et al. (2017). Rescue of Pompe disease in mice by AAV-mediated liver delivery of secretable acid α-glucosidase. *Science Translational Medicine, 9*, eaam6375. https://doi.org/10.1126/scitranslmed.aam6375
73. Hefti, F. (2018). Nerve growth factor promotes survival of septal cholinergic neurons after fimbrial transections. *The Journal of Neuroscience, 6*, 2155–2162.
74. Rosenberg, J. B., Kaplitt, M. G., De, B. P., Chen, A., Flagiello, T., Salami, C., et al. (2018). AAVrh.*10*-mediated APOE2 central nervous system gene therapy for APOE4-associated Alzheimer's disease. *Human Gene Therapy Clinical Development, 29*, 24–47.
75. Mole, S. E., & Cotman, S. L. (2015). Genetics of the neuronal ceroid lipofuscinoses (Batten disease). *Biochimica et Biophysica Acta – Molecular Basis of Disease, 1852*(10 Pt B), 2237–2241.
76. Johnson, T. B., Cain, J. T., White, K. A., Ramirez-Montealegre, D., Pearce, D. A., & Weimer, J. M. (2019). Therapeutic landscape for Batten disease: Current treatments and future prospects. *Nature Reviews. Neurology, 15*(3), 161–178.
77. Al-Zaidy, S. A., & Mendell, J. R. (2019). From clinical trials to clinical practice: Practical considerations for gene replacement therapy in SMA type 1. *Pediatric Neurology, 100*, 3–11.

78. MacDonald, M. E., Ambrose, C. M., Duyao, M. P., et al. (1993). A novel gene containing a trinucleotide repeat that is expanded and unstable on Huntington's disease chromosomes. *Cell, 72*, 971–983.
79. Saudou, F., & Humbert, S. (2016). The biology of Huntingtin. *Neuron, 89*, 910–926.
80. Grondin, R., Kaytor, M. D., Ai, Y., et al. (2012). Six-month partial suppression of Huntingtin is well tolerated in the adult rhesus striatum. *Brain, 135*, 1197–1209.
81. Miniarikova, J., Evers, M. M., & Konstantinova, P. (2018). Translation of MicroRNA-based huntingtin-lowering therapies from preclinical studies to the clinic. *Molecular Therapy, 26*, 947–962.
82. Evers, M. M., Miniarikova, J., Juhas, S., et al. (2018). AAV5-miHTT gene therapy demonstrates broad distribution and strong human mutant huntingtin lowering in a Huntington's disease Minipig model. *Molecular Therapy, 26*, 2163–2177.
83. Keskin, S., Brouwers, C., Sogorb-Gonzalez, M., Martier, R., Depla, J., Vallès, A., et al. (2019). AAV5-miHTT lowers huntingtin mRNA and protein without off-target effects in patient-derived neuronal cultures and astrocytes. *Molecular Therapy – Methods & Clinical Development, 15*, 275–284.
84. Liu, Z. H., Li, S. L., Liang, Z. B., Zhao, Y., Zhang, Y. L., Yang, Y. Q., et al. (2013). Targeting β-secretase with RNAi in neural stem cells for Alzheimer's disease therapy. *Neural Regeneration Research, 8*, 3095–3106.
85. Nóbrega, C., Nascimento-Ferreira, I., Onofre, I., Albuquerque, D., Hirai, H., Déglon, N., et al. (2013). Silencing mutant ataxin-3 rescues motor deficits and neuropathology in Machado-Joseph disease transgenic mice. *PLoS One, 8*, e52396. https://doi.org/10.1371/journal.pone.0052396
86. Miller, T. M., Pestronk, A., David, W., Rothstein, J., Simpson, E., Appel, S. H., et al. (2013). An antisense oligonucleotide against SOD1 delivered intrathecally for patients with SOD1 familial amyotrophic lateral sclerosis: A phase 1, randomised, first-in-man study. *Lancet Neurology, 12*, 435–442.
87. Evers, M. M., Toonen, L. J. A., & van Roon-Mom, W. M. C. (2015). Antisense oligonucleotides in therapy for neurodegenerative disorders. *Advanced Drug Delivery Reviews, 87*, 90–103.
88. Juliano, R. L. (2016). The delivery of therapeutic oligonucleotides. *Nucleic Acids Research, 44*, 6518–6548.
89. Gyorgy, B., Ingelsson, M., Loov, C., Takeda, S., Lannfelt, L., Hyman, B. T., et al. (2016). CRISPR-Cas9 mediated gene editing in a monogenic form of Alzheimer's disease. *Molecular Therapy, 24*, S226–S227.
90. Yang, S., Chang, R., Yang, H., Zhao, T., Hong, Y., Kong, H. E., et al. (2017). CRISPR/Cas9-mediated gene editing ameliorates neurotoxicity in mouse model of Huntington's disease. *The Journal of Clinical Investigation, 127*, 2719–2724.
91. Ouyang, S., Xie, Y., Xiong, Z., Yang, Y., Xian, Y., Ou, Z., et al. (2018). CRISPR/Cas9-targeted deletion of polyglutamine in spinocerebellar ataxia type 3-derived induced pluripotent stem cells. *Stem Cells and Development, 27*, 756–770.
92. Block, M. L., Zecca, L., & Hong, J. S. (2007). Microglia-mediated neurotoxicity: Uncovering the molecular mechanisms. *Nature Reviews Neuroscience, 8*, 57–69.
93. Lokovic, B. V. (2008). The blood-brain barrier in health and chronic neurodegenerative disorders. *Neuron, 57*, 178–201.
94. Maragakis, N. J., & Rothstein, J. D. (2006). Mechanisms of disease: Astrocytes in neurodegenerative disease. *Nature Reviews Neurology, 2*, 679.
95. Maslanik, T., Mahaffey, L., Tannura, K., Beninson, L., Greenwood, B. N., & Fleshner, M. (2013). The inflammasome and danger associated molecular patterns (DAMPs) are implicated in cytokine and chemokine responses following stressor exposure. *Brain, Behavior, and Immunity, 28*, 54–62.
96. Heneka, M. T., Kummer, M. P., & Latz, E. (2014). Innate immune activation in neurodegenerative disease. *Nature Reviews Immunology, 14*, 463–477.

97. Gordon, R., Anantharam, V., Kanthasamy, A. G., & Kanthasamy, A. (2012). Proteolytic activation of proapoptotic kinase protein kinase Cδ by tumor necrosis factor α death receptor signaling in dopaminergic neurons during neuroinflammation. *Journal of Neuroinflammation, 9*, 82. https://doi.org/10.1186/1742-2094-9-82
98. Godoy, M. C., Tarelli, R., Ferrari, C. C., Sarchi, M. I., & Pitossi, F. J. (2008). Central and systemic IL-1 exacerbates neurodegeneration and motor symptoms in a model of Parkinson's disease. *Brain, 131*, 1880–1894.
99. Mantovani, S., Gordon, R., Macmaw, J., Pfluger, C., Henderson, R., Noakes, P., et al. (2014). Elevation of the terminal complement activation products C5a and C5b-9 in ALS patient blood. *Journal of Neuroimmunology, 276*, 213–218.
100. Liang, Z., Zhao, Y., Ruan, L., Zhu, L., Jin, K., Zhuge, Q., et al. (2017). Impact of aging immune system on neurodegeneration and potential immunotherapies. *Progress in Neurobiology, 157*(Suppl. C), 2–28.
101. Barnum, C. J., Chen, X., Chung, J., Chang, J., Williams, M., Grigoryan, N., et al. (2014). Peripheral administration of the selective inhibitor of soluble tumor necrosis factor (TNF) XPro® 1595 attenuates nigral cell loss and glial activation in 6-OHDA hemiparkinsonian rats. *Journal of Parkinson's Disease, 4*, 349–360.
102. Detrait, E. R., Danis, B., Lamberty, Y., & Foerch, P. (2014). Peripheral administration of an anti-TNF-alpha receptor fusion protein counteracts the amyloid induced elevation of hippocampal TNF-alpha levels and memory deficits in mice. *Neurochemistry International, 72*, 10–13.
103. Butchart, J., Brook, L., Hopkins, V., Teeling, J., Puntener, U., Culliford, D., et al. (2015). Etanercept in Alzheimer disease: A randomized, placebo-controlled, double-blind, phase 2 trial. *Neurology, 84*, 2161–2168.
104. Heneka, M. T., Kummer, M. P., Stutz, A., Delekate, A., Schwartz, S., Vieira-Saecker, A., et al. (2013). NLRP3 is activated in Alzheimer' s disease and contributes to pathology in APP/PS1 mice. *Nature, 493*, 674–678.
105. Youm, Y. H., Nguyen, K. Y., Grant, R. W., Goldberg, E. L., Bodogai, M., Kim, D., et al. (2015). The ketone metabolite [beta]-hydroxybutyrate blocks NLRP3 inflammasome-mediated inflammatory disease. *Nature Medicine, 21*, 263–269.
106. Vom Berg, J., Prokop, S., Miller, K. R., Obst, J., Kälin, R. E., Lopategui-Cabezas, I., et al. (2012). Inhibition of IL-12/IL-23 signaling reduces Alzheimer's disease-like pathology and cognitive decline. *Nature Medicine, 18*, 1812–1819.
107. Papatriantafyllou, M. (2013). Immunological bullets against Alzheimer's disease. *Nature Reviews Drug Discovery, 12*, 24.
108. Woodruff, T. M., Nandakumar, K. S., & Tedesco, F. (2011). Inhibiting the C5–C5a receptor axis. *Molecular Immunology, 48*, 1631–1642.
109. Bhattacharya, P., Budnick, I., Singh, M., Thiruppathi, M., Alharshawi, K., Elshabrawy, H., et al. (2015). Dual role of GM-CSF as a pro-inflammatory and a regulatory cytokine: Implications for immune therapy. *Journal of Interferon & Cytokine Research, 35*, 585–599.
110. Chitu, V., Biundo, F., Shlager, G. G. L., Park, E. S., Wang, P., Gulinello, M. E., et al. (2020). Microglial homeostasis requires balanced CSF-1/CSF-2 receptor signaling. *Cell Reports, 30*, 3004–19.e05.
111. Dikmen, H. O., Hemmerich, M., Lewen, A., Hollnagel, J. O., Chausse, B., & Kann, O. (2020). GM-CSF induces noninflammatory proliferation of microglia and disturbs electrical neuronal network rhythms in situ. *Journal of Neuroinflammation, 17*, 235.
112. Potter, H., Woodcock, J. H., Boyd, T. D., Coughlan, C. M., O'Shaughnessy, J. R., Borges, M. T., et al. (2021). Safety and efficacy of sargramostim (GM-CSF) in the treatment of Alzheimer's disease. *Alzheimer's & Dementia, 7*, e12158. https://doi.org/10.1002/trc2.12158
113. Gendelman, H. E., Zhang, Y., Santamaria, P., Olson, K. E., Schutt, C. R., Bhatti, D., et al. (2017). Evaluation of the safety and immunomodulatory effects of sargramostim in a randomized, double-blind phase 1 clinical Parkinson's disease trial. *NPJ Parkinson's Disease, 3*, 10. https://doi.org/10.1038/s41531-017-0013-5

114. Pollari, E., Savchenko, E., Jaronen, M., Kanninen, K., Malm, T., Wojciechowski, S., et al. (2011). Granulocyte colony stimulating factor attenuates inflammation in a mouse model of amyotrophic lateral sclerosis. *Journal of Neuroinflammation, 8*, 74. https://doi.org/10.1186/1742-2094-8-74
115. Yamasaki, R., Tanaka, M., Fukunaga, M., Tateishi, T., Kikuchi, H., Motomura, K., et al. (2010). Restoration of microglial function by granulocyte-colony stimulating factor in ALS model mice. *Journal of Neuroimmunology, 229*, 51–62. https://doi.org/10.1016/j.jneuroim.2010.07.002
116. Johannesen, S., Budeus, B., Peters, S., Iberl, S., Meyer, A. L., Kammermaier, T., et al. (2018). Biomarker supervised G-CSF (filgrastim) response in ALS patients. *Frontiers in Neurology, 9*, 971. https://doi.org/10.3389/fneur.2018.00971
117. Amirzagar, N., Nafissi, S., Tafakhori, A., Modabbernia, A., Amirzargar, A., Ghaffarpour, M., et al. (2015). Granulocyte colony-stimulating factor for amyotrophic lateral sclerosis: A randomized, double-blind, placebo-controlled study of Iranian patients. *Journal of Clinical Neurology, 11*, 164–171.
118. Guo, J., Yang, G., He, Y., Xu, H., Fan, H., An, J., et al. (2021). Involvement of α7nAChR in the protective effects of genistein against β-amyloid-induced oxidative stress in neurons via a PI3K/Akt/Nrf2 pathway-related mechanism. *Cellular and Molecular Neurobiology, 41*, 377–393.
119. Paladugu, L., Gharaibeh, A., Kolli, N., Learman, C., Hall, T. C., Li, L., et al. (2021). Liraglutide has anti-inflammatory and anti-amyloid properties in streptozotocin-induced and 5xFAD mouse models of Alzheimer's disease. *International Journal of Molecular Sciences, 22*, 860.
120. Mulvaney, C. A., Duarte, G. S., Handley, J., Evans, D. J., Menon, S., Wyse, R., et al. (2020). GLP-1 receptor agonists for Parkinson's disease. *Cochrane Database of Systematic Reviews, 7*, Cd012990. https://doi.org/10.1002/14651858.CD012990.pub2
121. Watson, K. T., Wroolie, T. E., Tong, G., Foland-Ross, L. C., Frangou, S., Singh, M., et al. (2019). Neural correlates of liraglutide effects in persons at risk for Alzheimer's disease. *Behavioural Brain Research, 356*, 271–278.
122. Chamberlain, S., Gabriel, H., Strittmatter, W., & Didsbury, J. (2020). An exploratory phase iia study of the PPAR delta/gamma agonist T3D-959 assessing metabolic and cognitive function in subjects with mild to moderate Alzheimer's disease. *Journal of Alzheimer's Disease, 73*, 1085–1103.
123. Machado, M. M. F., Bassani, T. B., Cóppola-Segovia, V., Moura, E. L. R., Zanata, S. M., Andreatini, R., et al. (2019). PPAR-γ agonist pioglitazone reduces microglial proliferation and NF-κB activation in the substantia nigra in the 6-hydroxydopamine model of Parkinson's disease. *Pharmacological Reports, 71*, 556–564.
124. Pinto, M., Nissanka, N., Peralta, S., Brambilla, R., Diaz, F., & Moraes, C. T. (2016). Pioglitazone ameliorates the phenotype of a novel Parkinson's disease mouse model by reducing neuroinflammation. *Molecular Neurodegeneration, 11*, 25. https://doi.org/10.1186/s13024-016-0090-7
125. Sabbagh, M. N., Agro, A., Bell, J., Aisen, P. S., Schweizer, E., & Galasko, D. (2011). PF-04494700, an oral inhibitor of receptor for advanced glycation end products (RAGE), in Alzheimer's disease. *Alzheimer Disease & Associated Disorders, 25*, 206.
126. Bachmann, M. F., & Whitehead, P. (2013). Active immunotherapy for chronic diseases. *Vaccine, 31*, 1777–1784.
127. Shahaduzzaman, M., Nash, K., Hudson, C., Sharif, M., Grimmig, B., Lin, X., et al. (2015). Anti-human α-synuclein N-terminal peptide antibody protects against dopaminergic cell death and ameliorates behavioral deficits in an AAV-α-synuclein rat model of Parkinson's disease. *PLoS One, 10*, e0116841. https://doi.org/10.1371/journal.pone.0116841
128. Gustafsson, G., Eriksson, F., Möller, C., da Fonseca, T. L., Outeiro, T. F., Lannfelt, L., et al. (2017). Cellular uptake of α-synuclein oligomer-selective antibodies is enhanced by the extracellular presence of α-synuclein and mediated via Fcγ receptors. *Cellular and Molecular Neurobiology, 37*, 121–131.

129. Huang, Y. R., Xie, X. X., Ji, M., Yu, X., Zhu, J., Zhang, L. X., et al. (2019). Naturally occurring autoantibodies against α-synuclein rescues memory and motor deficits and attenuates α-synuclein pathology in mouse model of Parkinson's disease. *Neurobiology of Disease, 124*, 202–217.
130. Jankovic, J., Goodman, I., Safirstein, B., Marmon, T. K., Schenk, D. B., Koller, M., et al. (2018). Safety and tolerability of multiple ascending doses of PRX002/RG7935, an anti–α-synuclein monoclonal antibody, in patients with Parkinson disease: A randomized clinical trial. *JAMA Neurology, 75*, 1206–1214.
131. Schenk, D. B., Koller, M., Ness, D. K., Griffith, S. G., Grundman, M., Zago, W., et al. (2017). First-in-human assessment of PRX002, an anti–α-synuclein monoclonal antibody, in healthy volunteers. *Movement Disorders, 32*, 211–218.
132. Weihofen, A., Liu, Y., Arndt, J. W., Huy, C., Quan, C., Smith, B. A., et al. (2019). Development of an aggregate-selective, human-derived α-synuclein antibody BIIB054 that ameliorates disease phenotypes in Parkinson's disease models. *Neurobiology of Disease, 124*, 276–288.
133. Brys, M., Fanning, L., Hung, S., Ellenbogen, A., Penner, N., Yang, M., et al. (2019). Randomized phase I clinical trial of anti–α-synuclein antibody BIIB054. *Movement Disorders, 34*, 1154–1163.
134. Valera, E., & Masliah, E. (2013). Immunotherapy for neurodegenerative diseases: Focus on α-synucleinopathies. *Pharmacology & Therapeutics, 138*, 311–322.
135. Sanchez-Guajardo, V., Annibali, A., Jensen, P. H., & Romero-Ramos, M. (2013). α-Synuclein vaccination prevents the accumulation of parkinson disease-like pathologic inclusions in striatum in association with regulatory T cell recruitment in a rat model. *Journal of Neuropathology and Experimental Neurology, 72*, 624–645.
136. Mandler, M., Valera, E., Rockenstein, E., Mante, M., Weninger, H., Patrick, C., Adame, A., Schmidhuber, S., Santic, R., Schneeberger, A., Schmidt, W., Mattner, F., & Masliah, E. (2015). Active immunization against alpha-synuclein ameliorates the degenerative pathology and prevents demyelination in a model of multiple system atrophy. *Molecular Neurodegeneration, 19*(10), 10.
137. McFarthing, K., & Simuni, T. (2019). Clinical trial highlights: Targetting alpha-synuclein. *Journal of Parkinson's Disease, 9*(1), 5–16.
138. Cheng, L., Zhang, J., Li, X. Y., Yuan, L., Pan, Y. F., Chen, X. R., et al. (2017). A novel antibody targeting sequence 31–35 in amyloid β protein attenuates Alzheimer's disease-related neuronal damage. *Hippocampus, 27*, 122–133.
139. Rinne, J. O., Brooks, D. J., Rossor, M. N., Fox, N. C., Bullock, R., et al. (2010). 11C-PiB PET assessment of change in fibrillar amyloid-beta load in patients with Alzheimer's disease treated with bapineuzumab: A phase 2, double-blind, placebo-controlled, ascending-dose study. *Lancet Neurology, 9*(4), 363–372.
140. Lu, M., & Brashear, H. R. (2019). Pharmacokinetics, pharmacodynamics, and safety of subcutaneous bapineuzumab: A single-ascending-dose study in patients with mild to moderate Alzheimer disease. *Clinical Pharmacology in Drug Development, 8*(3), 326–335.
141. Doggrell, S. A. (2018). Grasping at straws: The failure of solanezumab to modify mild Alzheimer's disease. *Expert Opinion on Biological Therapy, 18*, 1189–1192.
142. Klein, G., Delmar, P., Voyle, N., Rehal, S., Hofmann, C., et al. (2019). Gantenerumab reduces amyloid-β plaques in patients with prodromal to moderate Alzheimer's disease: A PET substudy interim analysis. *Alzheimer's Research & Therapy, 11*(1), 101.
143. Ostrowitzki, S., Lasser, R. A., Dorflinger, E., Scheltens, P., Barkhof, F., Nikolcheva, T., et al. (2017). A phase III randomized trial of gantenerumab in prodromal Alzheimer's disease. *Alzheimer's Research & Therapy, 9*, 95. https://doi.org/10.1186/s13195-017-0318-y
144. Yang, T., Dang, Y., Ostaszewski, B., Mengel, D., Steffen, V., Rabe, C., et al. (2019). Target engagement in an Alzheimer trial: Crenezumab lowers amyloid β oligomers in cerebrospinal fluid. *Annals of Neurology, 86*, 215–224.
145. Cummings, J. L., Cohen, S., van Dyck, C. H., Brody, M., Curtis, C., Cho, W., et al. (2018). ABBY: A phase 2 randomized trial of crenezumab in mild to moderate Alzheimer disease. *Neurology, 90*, e1889–e1897. https://doi.org/10.1212/WNL.0000000000005550

146. Tariot, P. N., Lopera, F., Langbaum, J. B., Thomas, R. G., Hendrix, S., Schneider, L. S., et al. (2018). The Alzheimer's prevention initiative autosomal-dominant Alzheimer's disease trial: A study of crenezumab versus placebo in preclinical PSEN1 E280A mutation carriers to evaluate efficacy and safety in the treatment of autosomal-dominant Alzheimer's disease, including a placebo-treated noncarrier cohort. *Alzheimer's & Dementia, 4*, 150–160.
147. Arndt, J. W., Qian, F., Smith, B. A., Quan, C., Kilambi, K. P., Bush, M. W., et al. (2018). Structural and kinetic basis for the selectivity of aducanumab for aggregated forms of amyloid-β. *Scientific Reports, 8*, 6412.
148. Budd, S. H., O'Gorman, J., Chiao, P., Bussiere, T., Tian, Y., Zhu, Y., et al. (2017). Clinical development of aducanumab, an anti-Aβ human monoclonal antibody being investigated for the treatment of early Alzheimer's disease. *Journal of Prevention of Alzheimer's Disease, 4*, 255–263.
149. Liu, Y. H., Wang, Y. R., Xiang, Y., Zhou, H. D., Giunta, B., Mañucat-Tan, N. B., et al. (2015). Clearance of amyloid-beta in Alzheimer's disease: Shifting the action site from center to periphery. *Molecular Neurobiology, 51*, 1–7.
150. Landen, J. W., Andreasen, N., Cronenberger, C. L., Schwartz, P. F., Börjesson-Hanson, A., Östlund, H., et al. (2017). Ponezumab in mild-to-moderate Alzheimer's disease: Randomized phase II PET-PIB study. *Alzheimer's & Dementia, 3*, 393–401.
151. Jäkel, L., Boche, D., Nicoll, J. A., & Verbeek, M. M. (2019). Aβ43 in human Alzheimer's disease: Effects of active Aβ42 immunization. *Acta Neuropathologica Communications, 7*, 141. https://doi.org/10.1186/s40478-019-0791-6
152. Muhs, A., Hickman, D. T., Pihlgren, M., Chuard, N., Giriens, V., Meerschman, C., et al. (2007). Liposomal vaccines with conformation-specific amyloid peptide antigens define immune response and efficacy in APP transgenic mice. *Proceedings of the National Academy of Sciences of the United States of America, 104*, 9810–9815.
153. AC Immune. (2019). *ACI 24 anti Abeta vaccine for prevention and therapy in Alzheimer's disease and Down syndrome*. Retrieved from https://ir.acimmune.com/static-files/276dac58-d882-48a6-8cb4-d47ccad53b69
154. Vandenberghe, R., Riviere, M. E., Caputo, A., Sovago, J., Maguire, R. P., Farlow, M., et al. (2017). Active Aβ immunotherapy CAD106 in Alzheimer's disease: A phase 2b study. *Alzheimer's & Dementia, 3*, 10–22.
155. Hull, M., Sadowsky, C., Arai, H., Le Prince, L., Holstein, A., Booth, K., et al. (2017). Long-term extensions of randomized vaccination trials of ACC-001 and QS-21 in mild to moderate Alzheimer's disease. *Current Alzheimer Research, 14*, 696–708.
156. Lacosta, A. M., Pascual-Lucas, M., Pesini, P., Casabona, D., Pérez-Grijalba, V., Marcos-Campos, I., et al. (2018). Safety, tolerability and immunogenicity of an active anti-Aβ 40 vaccine (ABvac40) in patients with Alzheimer's disease: A randomised, double-blind, placebo-controlled, phase I trial. *Alzheimer's Research & Therapy, 10*, 12. https://doi.org/10.1186/s13195-018-0340-8
157. van Ameijde, J., Crespo, R., Janson, R., Juraszek, J., Siregar, B., Verveen, H., et al. (2018). Enhancement of therapeutic potential of a naturally occurring human antibody targeting a phosphorylated Ser 422 containing epitope on pathological tau. *Acta Neuropathologica Communications, 6*, 59. https://doi.org/10.1186/s40478-018-0562-9
158. Dai, C. l., Hu, W., Tung, Y. C., Liu, F., Gong, C. X., et al. (2018). Tau passive immunization blocks seeding and spread of Alzheimer hyperphosphorylated Tau-induced pathology in 3 × Tg-AD mice. *Alzheimer's Research & Therapy, 10*, 13. https://doi.org/10.1186/s13195-018-0341-7
159. Dai, C. L., Tung, Y. C., Liu, F., Gong, C. X., & Iqbal, K. (2017). Tau passive immunization inhibits not only tau but also Aβ pathology. *Alzheimer's Research & Therapy, 9*, 1. https://doi.org/10.1186/s13195-016-0227-5
160. Albert, M., Mairet-Coello, G., Danis, C., Lieger, S., Caillierez, R., Carrier, S., et al. (2019). Prevention of tau seeding and propagation by immunotherapy with a central tau epitope antibody. *Brain, 142*, 1736–1750.

161. d'Abramo, C., Acker, C. M., Jimenez, H., & Davies, P. (2015). Passive immunization in JNPL3 transgenic mice using an array of phospho-tau specific antibodies. *PLoS One, 10*, e0135774. https://doi.org/10.1371/journal.pone.0135774
162. Gerson, J. E., Farmer, K. M., Henson, N., Castillo-Carranza, D. L., Murillo, M. C., Sengupta, U., et al. (2018). Tau oligomers mediate α-synuclein toxicity and can be targeted by immunotherapy. *Molecular Neurodegeneration, 13*, 13. https://doi.org/10.1186/s13024-018-0245-9
163. Castillo-Carranza, D. L., Guerrero-Munoz, M. J., Sengupta, U., Hernandez, C., Barrett, A. D., Dineley, K., et al. (2015). Tau immunotherapy modulates both pathological tau and upstream amyloid pathology in an Alzheimer's disease mouse model. *The Journal of Neuroscience, 35*, 4857–4868.
164. Qureshi, I. A., Tirucherai, G., Ahlijanian, M. K., Kolaitis, G., Bechtold, C., & Grundman, M. (2018). A randomized, single ascending dose study of intravenous BIIB092 in healthy participants. *Alzheimer's & Dementia, 4*, 746–755.
165. Rajamohamedsait, H., Rasool, S., Rajamohamedsait, W., Lin, Y., & Sigurdsson, E. M. (2017). Prophylactic active tau immunization leads to sustained reduction in both tau and amyloid-β pathologies in 3xTg mice. *Scientific Reports, 7*, 17034. https://doi.org/10.1038/s41598-017-17313-1
166. Novak, P., Schmidt, R., Kontsekova, E., Kovacech, B., Smolek, T., Katina, S., et al. (2018). Fundamant: An interventional 72-week phase 1 follow-up study of AADvac1, an active immunotherapy against tau protein pathology in Alzheimer's disease. *Alzheimer's Research & Therapy, 10*, 108. https://doi.org/10.1186/s13195-018-0436-1
167. Theunis, C., Crespo-Biel, N., Gafner, V., Pihlgren, M., López-Deber, M. P., Reis, P., et al. (2013). Efficacy and safety of a liposome-based vaccine against protein Tau, assessed in tau. P301L mice that model tauopathy. *PLoS One, 8*, e72301. https://doi.org/10.1371/journal.pone.0072301
168. Ayalon, G., Lee, S. H., Adolfsson, O., Foo-Atkins, C., Atwal, J. K., Blendstrup, M., et al. (2021). Antibody semorinemab reduces tau pathology in a transgenic mouse model and engages tau in patients with Alzheimer's disease. *Science Translational Medicine, 13*, eabb2639. https://doi.org/10.1126/scitranslmed.abb2639
169. Dong, Q. X., Zhu, J., Liu, S. Y., Yu, X. L., & Liu, R. T. (2018). An oligomer-specific antibody improved motor function and attenuated neuropathology in the SOD1-G93A transgenic mouse model of ALS. *International Immunopharmacology, 65*, 413–421.
170. Rosen, D. R., Siddique, T., Patterson, D., Figlewicz, D. A., Sapp, P., Hentati, A., et al. (1993). Mutations in Cu/Zn superoxide dismutase gene are associated with familial amyotrophic lateral sclerosis. *Nature, 362*, 59–62.
171. Kim, J. M., Billington, E., Reyes, A., Notarianni, T., Sage, J., Agbas, E., et al. (2019). Impaired Cu–Zn superoxide dismutase (SOD1) and calcineurin (Cn) interaction in ALS: A presumed consequence for TDP-43 and zinc aggregation in Tg SOD1 G93A rodent spinal cord tissue. *Neurochemical Research, 44*, 228–233.
172. Crow, J. P., Sampson, J. B., Zhuang, Y., Thompson, J. A., & Beckman, J. S. (1997). Decreased zinc affinity of amyotrophic lateral sclerosis-associated superoxide dismutase mutants leads to enhanced catalysis of tyrosine nitration by peroxynitrite. *Journal of Neurochemistry, 69*, 1936–1944.
173. Kaneko, M., Noguchi, T., Ikegami, S., Sakurai, T., Kakita, A., Toyoshima, Y., et al. (2015). Zinc transporters ZnT3 and ZnT6 are downregulated in the spinal cords of patients with sporadic amyotrophic lateral sclerosis. *Journal of Neuroscience Research, 93*, 370–379.
174. Takeuchi, S., Fujiwara, N., Ido, A., Oono, M., Takeuchi, Y., Tateno, M., et al. (2010). Induction of protective immunity by vaccination with wild-type apo superoxide dismutase 1 in mutant SOD1 transgenic mice. *Journal of Neuropathology and Experimental Neurology, 69*, 1044–1056.
175. Urushitani, M., Ezzi, S. A., & Julien, J. P. (2007). Therapeutic effects of immunization with mutant superoxide dismutase in mice models of amyotrophic lateral sclerosis. *Proceedings of the National Academy of Sciences of the United States of America, 104*, 2495–2500.

176. Mackenzie, I. R., Frick, P., Grässer, F. A., Gendron, T. F., Petrucelli, L., Cashman, N. R., et al. (2015). Quantitative analysis and clinico-pathological correlations of different dipeptide repeat protein pathologies in C9ORF72 mutation carriers. *Acta Neuropathologica, 130*, 845–861. https://doi.org/10.1007/s00401-015-1476-2
177. Zhang, Y. J., Jansen-West, K., Xu, Y. F., Gendron, T. F., Bieniek, K. F., Lin, W. L., et al. (2014). Aggregation-prone c9FTD/ALS poly (GA) RAN-translated proteins cause neurotoxicity by inducing ER stress. *Acta Neuropathologica, 128*, 505–524.
178. Nguyen, L., Montrasio, F., Pattamatta, A., Tusi, S. K., Bardhi, O., Meyer, K. D., et al. (2020). Antibody therapy targeting RAN proteins rescues C9 ALS/FTD phenotypes in C9orf72 mouse model. *Neuron, 105*, 645–62.e11.
179. Zhou, Q., Mareljic, N., Michaelsen, M., Parhizkar, S., Heindl, S., Nuscher, B., et al. (2020). Active poly-GA vaccination prevents microglia activation and motor deficits in a C9orf72 mouse model. *EMBO Molecular Medicine, 12*, e10919.
180. Miller, T. W., Shirley, T. L., Wolfgang, W. J., Kang, X., & Messer, A. (2003). DNA vaccination against mutant huntingtin ameliorates the HDR6/2 diabetic phenotype. *Molecular Therapy, 7*, 572–579.
181. Ramsingh, A. I., Manley, K., Rong, Y., Reilly, A., & Messer, A. (2015). Transcriptional dysregulation of inflammatory/immune pathways after active vaccination against Huntington′ s disease. *Human Molecular Genetics, 24*, 6186–6197.

MIX
Papier aus verantwortungsvollen Quellen
Paper from responsible sources
FSC® C105338

If you have any concerns about our products,
you can contact us on
**ProductSafety@springernature.com**

In case Publisher is established outside the EU,
the EU authorized representative is:
**Springer Nature Customer Service Center GmbH**
**Europaplatz 3, 69115 Heidelberg, Germany**

Printed by Libri Plureos GmbH
in Hamburg, Germany